Children's Proportions for Artists

Studies Are Presented for Figurative Artists About the Yearly Growth and Development of Children of Each Gender From Infancy to Young Adulthood. Illustrations Are Included With Charts of About 130 Representative Measurements and Proportions.

by

Eugene Fairbanks, BA, MD

Fairbanks Art and Books
Bellingham, WA 98226

Children's Proportions for Artists
Eugene F. Fairbanks, BA, MD

Library of Congress Control Number: 2011914449.
ISBN: 0-9725841-2-9

Typography, design and prepress: Kathleen Weisel (weiselcreative.com)

Printed in the U.S.A.

Fairbanks Art and Books
Bellingham, Washington 98226
www.fairbanksartbooks.com

This book is intended to be a companion text to a previously published book, ***Human Proportions for Artists*** that was based on average and nearly ideal measurements, of about fifty individuals that the author's father, Avard T. Fairbanks, sculptor, tabulated and used to illustrate for his doctoral thesis in 1933 at the University of Michigan. A few of his illustrations are included in this book. Illustrations of children's faces and figures were created by Andrea G., who has illustrated many children's story books. She accepted this challenge with enthusiasm utilizing photographs and personal observation, and tabulated measurements. Other line drawings are created by the author.

Inside front cover: **The Riddle of the Sphynx:** "What creature walks on four in the morning, at noon on two, and at evening on three?" It is answered and illustrated by: "Man, who goes about as a baby on hands and knees, later on his feet, and in old age with a staff." The illustration is by Dugald Stermer. It is reprinted by permission of the artist.

Contents

Measurement comparisons, illustrations, and charts

Appendix

Stages of Development

List of Illustrations

Designations of Proportions

Illustrations of Proportions

Conversion Tables

Recognition, Analysis, and Synthesis

When meeting a child or youth, by calling on past experiences, people automatically estimate the age of a younger person. This estimate considers several factors, including Height, facial features, social behavior, and the tone of the voice. Age is often estimated even during telephone conversations by tone, vocabulary, and discussion. This is the **Recognition** phase of an estimated age and it is sufficient for most people. An artist that is interested in using a young person as a model needs additional evaluation to understand why this child or youth appears to be of that particular age. This phase is the **Analysis** of the many features that the young person presents. The face of a young person, while resembling an adult face has many subtle differences that are consistent with his or her age. A skillful artist needs to avoid drawing or modeling an adult head on a small body. Another habit to avoid is using a scaled down adult proportions of seven heads to a smaller height. This would make a child appear to be a little adult.

After the analytic phase, there is a **Synthesis** phase for the artist that is planning to create a fine composition. He or she needs to accurately recreate the features of the face and body for a fine representational masterpiece that also displays action and emotion. This phase of **Synthesis** is often achieved by careful observation of details. The study of children's proportions, by applying knowledge and the science of anthropometry, is intended to facilitate appreciation of the varied patterns of growth and development from infancy to near maturity. Hopefully, it may improve the general quality of art in an enlightened society.

INTRODUCTION

Since most subjects for artists are adults, the proportions of children are underemphasized and are often an overlooked aspect for observation. When the classic Greek artists attempted to portray a child, they would model or sketch a little adult that was about one-fourth of an adult height. They used an adult proportions ratio, with seven little adult heads for the total height. Their creations do not look like a baby. Occasionally when viewing illustrations of children by later artists, it appears that an adult head was portrayed on a small body. This seems equally unnatural. The present study intends to provide more information about children's growth and proportions for artists to facilitate more natural and accurate portrayal. Gradually changing characteristics of infants, children, and youth of both sexes are illustrated, designed for recognition to assist in realistic and fine artistic endeavor.

Anthropometry, human measurement, is a major component of Physical Anthropology. There have been numerous studies measuring individuals of various ages and ethnic groups for many different purposes. There is a great variety of measurements that can be considered. Other studies have generally been for various specific purposes. The recognition of optimum growth and development is important in health care. Physicians are urged to record height, weight, head and chest circumference. Deviations from normal may be an early warning of abnormalities that need additional attention and care. Other significant studies are directed to body size and configurations that affect industries for clothing, shoes, athletic equipment and furniture. The University of Michigan sponsored Highway Safety Research Institute conducted and recorded a very comprehensive study for Product Safety Design. The comprehensive study directed special attention to the automobile industry for improved vehicle design and increased safety. It served as an excellent information resource. Studies of head and face development have interested dentists, orthodontists, maxillofacial, and plastic surgeons. Careful investigations of facial proportions have been pursued by these specialties to improve long term outcome. Some studies compare regional or ethnic variations of features. Others, primarily anthropologists, have had an academic interest in advancing the general scientific knowledge of human growth and structure.

The purpose of this study of infants, children, youth, and young adults is for the benefit of artists. Many of the other studies stress anatomical points that are bony prominences. These are important, but artists also need surface measurements of soft tissues. This is especially important in facial features. The author recorded many measurements of individuals in both sexes, within two months of their birthdays. Those selected had approximate average height and weight at ages in yearly increments up to eighteen years of age. Few recorded facial measurements, particularly of the soft features are available in other studies. Considerable other information is utilized from various available studies both to verify averages of recordings and to add some features not otherwise included. One hundred twenty features are listed. Drawings are included to illustrate the listed proportions. The proportion designations are listed, both in anatomic and common terms for better general understanding. Significant features and prominences of the human skeleton are also illustrated in the appendix, titled Anthropometry, and properly labeled to enhance comprehension.

Life size measurements of various ages are recorded in both metric and English units. Although inch and foot measurements are derived from the human body, metric measurements are in a decimal system and are more easily reduced to fractional proportions. For convenience, a conversion scale is included in the study. In a previous publication, *Human Proportions for Artists*, using information gathered by my father, Dr. Avard T. Fairbanks, sculptor, there was a spread from heroic size at three-halves down to one twelfth adult life size. For brevity, the fractions used here are from life size down to one-fourth life size. Presently, an artist can enter the life size proportions into a computer and select whatever fractional proportions, that are desired, such as one-fifth, one-seventh, or one-tenth life size.

This study of children's proportions is intended to assist art students and serious amateurs to acquaint them with many anatomical features. The Growth, development and early behavior are important considerations. Rather than offering stereotyped images, artists are advised and encouraged to use live models. This information may serve as to assist initial sketching. After the live model sessions are complete, the information may serve as a re-check on some details. Students may also become more aware of characteristic features. An astute observer will recognize many variations from average measurements. There is hardly anyone that has average features. There is considerable variation in sizes and features at every age and with ethnic groups. Special individual characteristics are important to include in portrayals. Photographic studies have been made by other persons for averaging a face by using one-fiftieth exposures of fifty individuals of the same age and sex on the same photographic print. The result is neither handsome nor ugly, but merely plain and uninteresting. Let this study serve as a baseline from which more accurate and expressive portrayals of children can be created.

The individuals, available for measuring by the author, were primarily descendants of Northwestern Europeans. Other ethnic groups have similar growth patterns, but may have some special and interesting characteristics. Anyone motivated to pursue a similar study of another ethnic group is both welcome and encouraged to use this or a similar format with the author's compliments and permission.

ACKNOWLEDGEMENTS

The tables of proportions are primarily a product of an extensive search of publications. The most extensive source was *The University of Michigan sponsored Highway Safety Research Institute comprehensive study for Product Safety Design.* The first study, reported May 1975, recorded forty–one measurements of 4027 infants and children ages two weeks to thirteen-years- of-age. A Second study was published in May 1977. This expanded information included eighty-seven measurements on 4127 individuals up to eighteen years of age.

Wilton Marion Krogman published a monograph: *Growth of the Head, Face, Trunk and Limbs of White and Negro Children of Elementary and High School Age in Philadelphia.* Over 4,000 individuals were recorded in tables of sixty-one measurements in 1970. Katherine Simmons published a monograph in 1941, *Study of Child Physiological Growth and Development,* for the The Brush Foundation. Over 5,000 male and 4,000 female individuals, ages three-months to seventeen years of age, were included .in a study of twenty-five measurements.

The author's tabulations and these three studies had minor variations in some measurements. These were consolidated by factoring ratios, using the tables of stature by the National Center for Health Statistics and calculating slight adjustments. A few appropriate measurements in several other publications were included. The author's own tabulation of childrens' measurements included family members and patients in a Family Medical Practice over a period of many years. These served for additional measurements, especially of the face, for this publication, that the other studies did not include.

DESIGNATIONS OF MEASUREMENTS FOR TABLES OF CHILDREN'S PROPORTIONS

In order to compile information about human proportions, Avard Fairbanks made a comprehensive study in 1933 of about 25 each of young adults, males and females, predominantly descendants of northwestern Europeans. He selected those with what he considered an ideal stature and calculated average measurements of them. His results were not a general population average, but were about the 75th percentile. His data was the basis for a book, *Human Proportions for Artists.* The purpose of this present study is to expand the scope of human proportions for the benefit of artists. It correlates with other anthropometric studies and presents the proportions of children from birth to age 18 as a reference source for artists. The author measured many children and compared information with other large studies. These were then correlated with the average heights recorded by the National Center for Health Statistics, United States Health Resources Administration, Department of Health, Education, and Welfare.

This study is not intended to replace a live model, but rather as a guide and another check for reasonable accuracy and quality in art. It can be helpful to graphic artists for enlarging or diminishing sizes. For sculpture, these tables may help in the construction of an armature and in the roughed-in modeling for a figurine, a statuette, or a full size statue. Public monuments are usually placed on pedestals and the figures are above eye level. If a monument that is erected out of doors has a life size figure, the honored person looks too small. Consequently, sculptors prefer heroic statues of about one and one-half life size. Colossal sizes are multiples of life size. These tables can be readily multiplied for intended enlargements

The tables of proportions presented in this book need both brevity and an adequate definition. For rapid comprehension, brief descriptions are on the same line as the human measurements and the fractional data. Since some of the terminology is anatomic, the extended explanation of each proportional measurement is listed here. In the event of question or misunderstanding, a code precedes the proportion so that it can be used as a quick reference to the explanation. Life size inch and metric measure are listed, but since reductions of inches into halves and quarters are cumbersome, smaller fractions are listed in metric measurements which use a convenient decimal system. A conversion chart of inches and metric system is included for convenience.

As an example, the heading and a line is copied from a page and presented here:

Life size			EIGHTEEN-YEAR-OLD GIRL	Fractions of life size				
				3/4	2/3	1/2	1/3	1/4
Inches	cm	Code		cm	cm	cm	cm	cm
			VERTICAL MEASUREMENTS					
64.6	**164.0**	1 V	Stature, Height, Vertex to Sole	123.0	109.3	82.0	54.7	41.0

The **designation of a proportions measurement,** considered to be an approximate average measurement, is written in the **center column** of each page, with abbreviated names for prompt and easy reference. The terminology, briefly listed, is a combination of common, anthropometric (human physical measurement), and anatomic names in an attempt to carefully define measurements. The more adequate descriptions allow easy comprehension for a novice. A study of these lists may familiarize the reader with technical terms used in scientific studies.

Life-size measurements, are in the first and second column, in inches and centimeters respectively, and are printed in bold type on each page. The third column lists a **reference code.** The five other columns, on the right side of the pages, are in **fractions of life size** in centimeters. The **code** is a combination of numbers and letters that serves as an index or a guide to designations or names of the itemized proportional measurements. The **numbers** indicate the serial category of measurement. The first **capital letter** designates the **body portion** being considered. The **number** refers to the **list in a series of measurements.** The **following capital letter** indicates the **direction of measurement,** as listed.

- H. Head and neck measurements.
- B. Body or torso measurements.
- U. Upper extremity: Measurements of arms and hands.
- L. Lower extremity: Measurements of legs and feet.

Numbers indicate reference in the order of listing for various groupings of the features of the head and neck, body or torso, upper extremity, and lower extremity. After the numbers, there are letters designating the direction of measurement.

- V. Vertical measurements, heights: Assuming erect posture, and longitudinal measurements of the extremities.
- T. Transverse measurements: Usually breadth, horizontal measurements on a frontal, or side to side, plane.
- D. Depth: Measurements in the anterior-posterior direction on a sagittal, or center line plane.

These serve as a code for referring to a more complete description of the location of measurements listed in the center column. The format of three columns, separated by the designations, followed by the next four rows, is intended to facilitate use and render easy comprehension. Rows are separated by a space after about every five lines, because the mind has less chance for confusion with a small series of numbers.

For ease of comprehension, common names are often used in conjunction with scientific terms of anatomy that more accurately describe points of reference. Anatomical illustrations are included with labels designating points of reference and with anthropometric designations. Underlined words are copied from the center column. Vertical measurements will be termed heights. Transverse measurements on a frontal plane will be termed breadth or occasionally width. Anterior-posterior, front to back (ventral to dorsal) measurements, will be termed depth, except when applied to flexed extremities. Length and width are very good terms, but may be construed differently by various people and therefore have usually been avoided in this study.

Although this study is based on science and is intended to increase the application of science to art, it is not a strictly scientific endeavor. An accurate scientific study would measure features on a skull and a skeleton. Specific points could be determined. Measurements that are made on living individuals may not be entirely accurate, because they depend on points located on soft tissues and skin. There will be variations by different persons recording measurements.

The primary emphasis of this study is to familiarize students, amateurs, and professional artists with human features and let them appreciate the many subtle changes that progress during growth and maturation during infancy, childhood, and youth.

Illustration of Stature

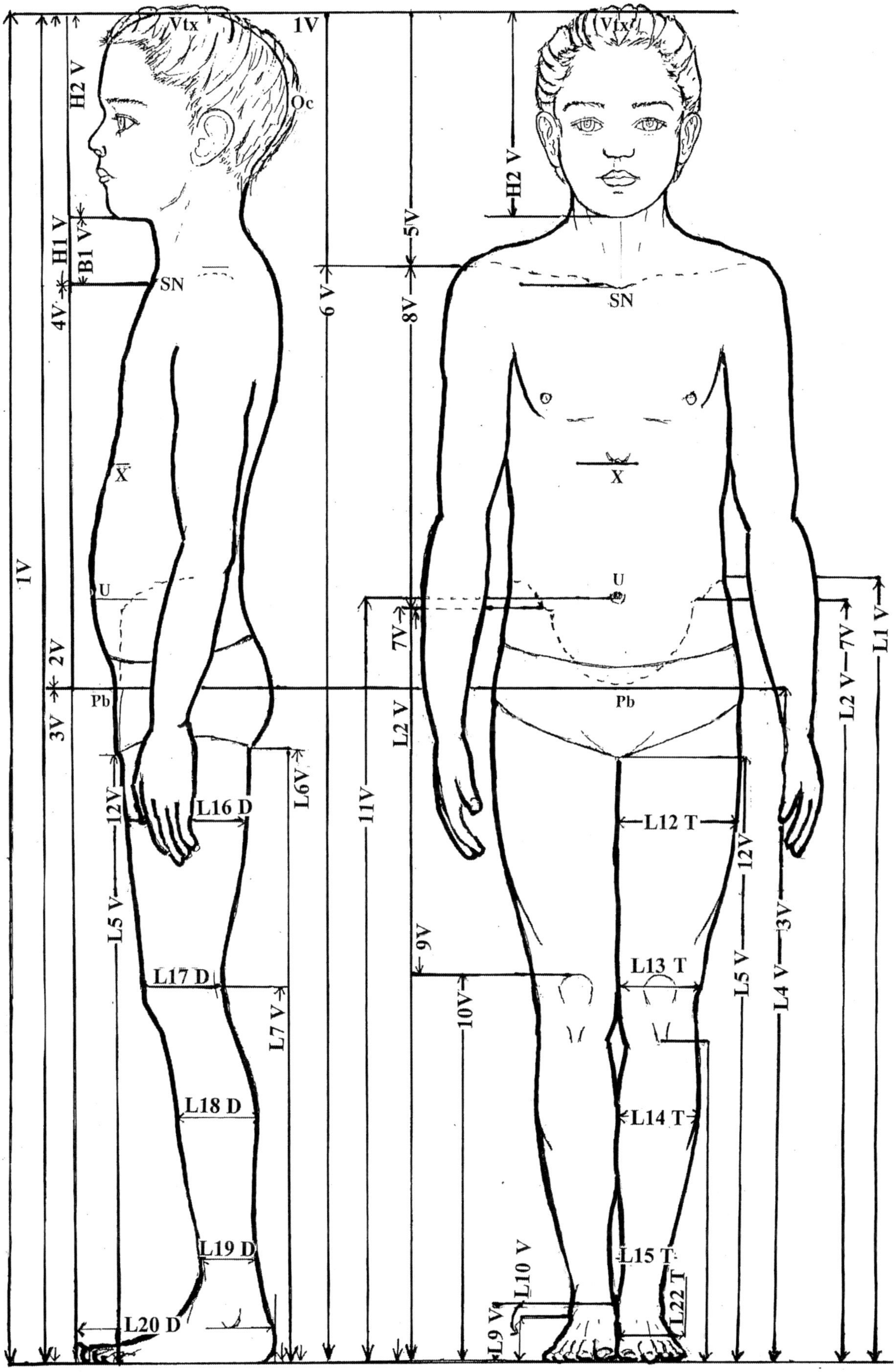

GENERAL VERTICAL BODY MEASUREMENTS

1V Stature or height, vertex or crown of the head to the sole of a foot._
2V Vertex to the symphysis pubis, the anterior pelvis at the base of the genitals; it is nearly one-half total height in an adult.
*3V Symphysis pubis of the pelvis to the sole.
4V Sternal notch or the pit of the neck to the sole of the foot.
5V Vertex to the acromion process, the point of the shoulder.

6V Acromion process at the point of a shoulder to the sole of a foot. Note that it is nearly the same measure as the sternal notch to the sole of the foot.
*7V Anterior superior spine of ilium of the pelvis to the sole of the foot.
8V The acromion process to the anterior superior spine of the pelvis
9V The anterior superior spine of the pelvis to the knee at the patella, knee cap.
10V Knee at upper border of the knee cap, or patella, to the sole of the foot.

11V The umbilicus, or belly button, to the sole of the foot.
*12V The crotch to the sole of the foot.
13V Kneeling height, vertex to knee when kneeling.
14V Sitting height, vertex to buttocks when sitting.

15V Buttocks, while sitting with horizontal straight leg to sole of foot.
16V Buttocks, while sitting with Thigh horizontal to flexed knee.
17V Flexed knee to sole of the foot with a bent knee while sitting.

** These features are also listed in Lower Extremity Measurements.*

Boy, Seated and Kneeling

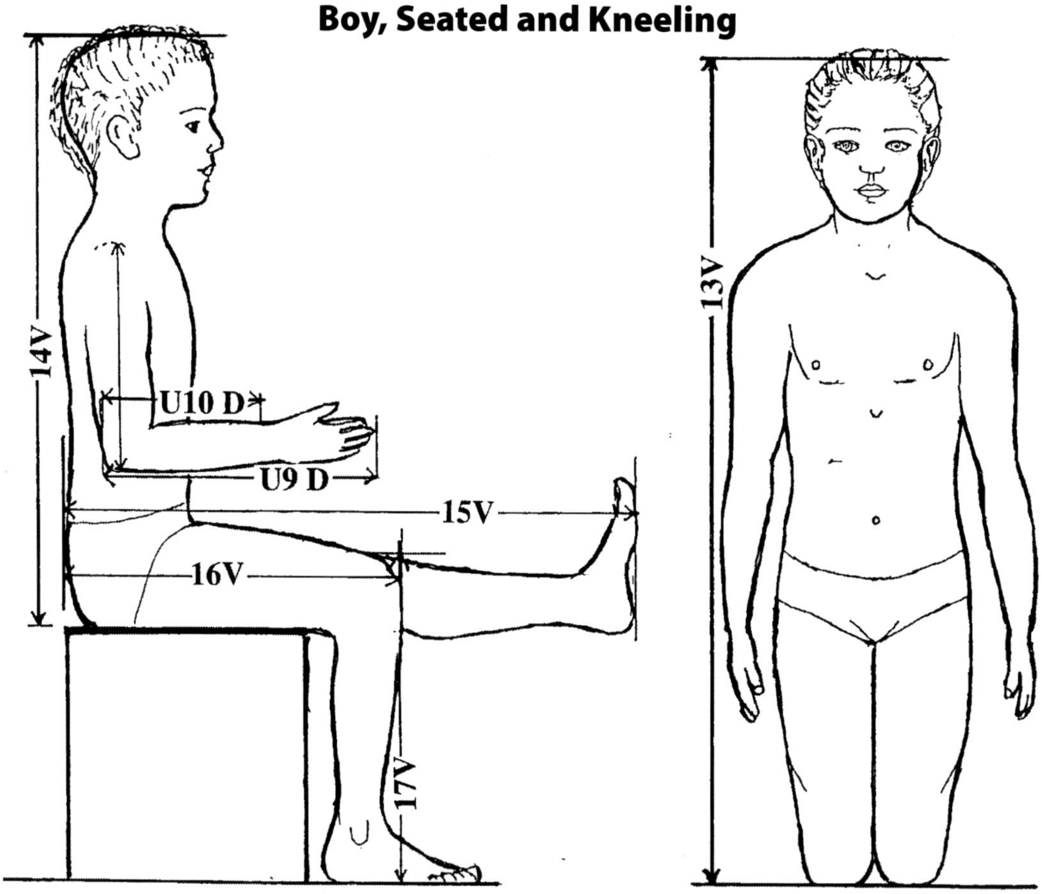

VERTICAL HEAD AND NECK MEASUREMENTS

H1 V Head and neck; vertex of head to the sternal notch, or pit of the neck.

H2 V Vertex of the head to the chin, head height.

H3 V Hairline to chin, face height; from, trichion of forehead to menton of chin.

H4 V Vertex of the head to the level of the lateral corner of an eye; the vertex to the lateral canthus of an eye, that is one-half of head height in an adult.

H5 V Vertex of the head to tragion, the tragus of an ear is a small flap just anterior to the ear canal. The tragion at the upper border of the tragus.

H6 V Hairline to the upper border of an eye brow; trichion to clabella, about one-third of the face height in an adult.

H7 V Brow to chin, the brow, glabella to the chin, menton, the mask of the face.

H8 V Brow to subnasal point below the nose; glabella to subnasal point, at the juncture with the upper lip, which is one third of face height in an adult.

H9 V Sellion (or nasion), the saddle of the nose to infranasal point; at the upper lip, the length of the nose. The sellion and the nasion are very close together and in this study the sellion will be used as the reference point.

H10 V Sellion, (Latin for saddle) saddle of nose to the chin, or menton, the lower face height.

H11 V Subnasal point to the lower border of the chin; subnasal point to menton of the chin.

H12 V Infranasal point to the lip or mouth aperture; infranasal to stomion.

H13 V Lip aperture of the mouth to sublabial crease, an arc below the lower lip and above the chin. This semicircular crease has an upward apex. The lips and this crease are soft tissues and are variable.

H14 V Sublabial crease to the chin, menton, this is soft tissue subject to variation.

TRANSVERSE HEAD MEASUREMENTS

H15 T Bi-parietal, maximum transverse head breadth between the parietal prominences above and posterior to the ear; The broadest diameter of the skull.

H16 T Bi-zygomatic width of cheek bones at zygoma, anterior to an ear.

H17 T Bi-fronto temporal breadth of forehead at the junctions of frontal and temporal bones, just a little above and posterior to the lateral brow near the hairline of the sideburns.

H18 T Inter-brow ridge, the breadth between the lateral brow ridges as they turn down at the orbits or eye sockets; inter-supercilliary arch.

Head, Anterior View

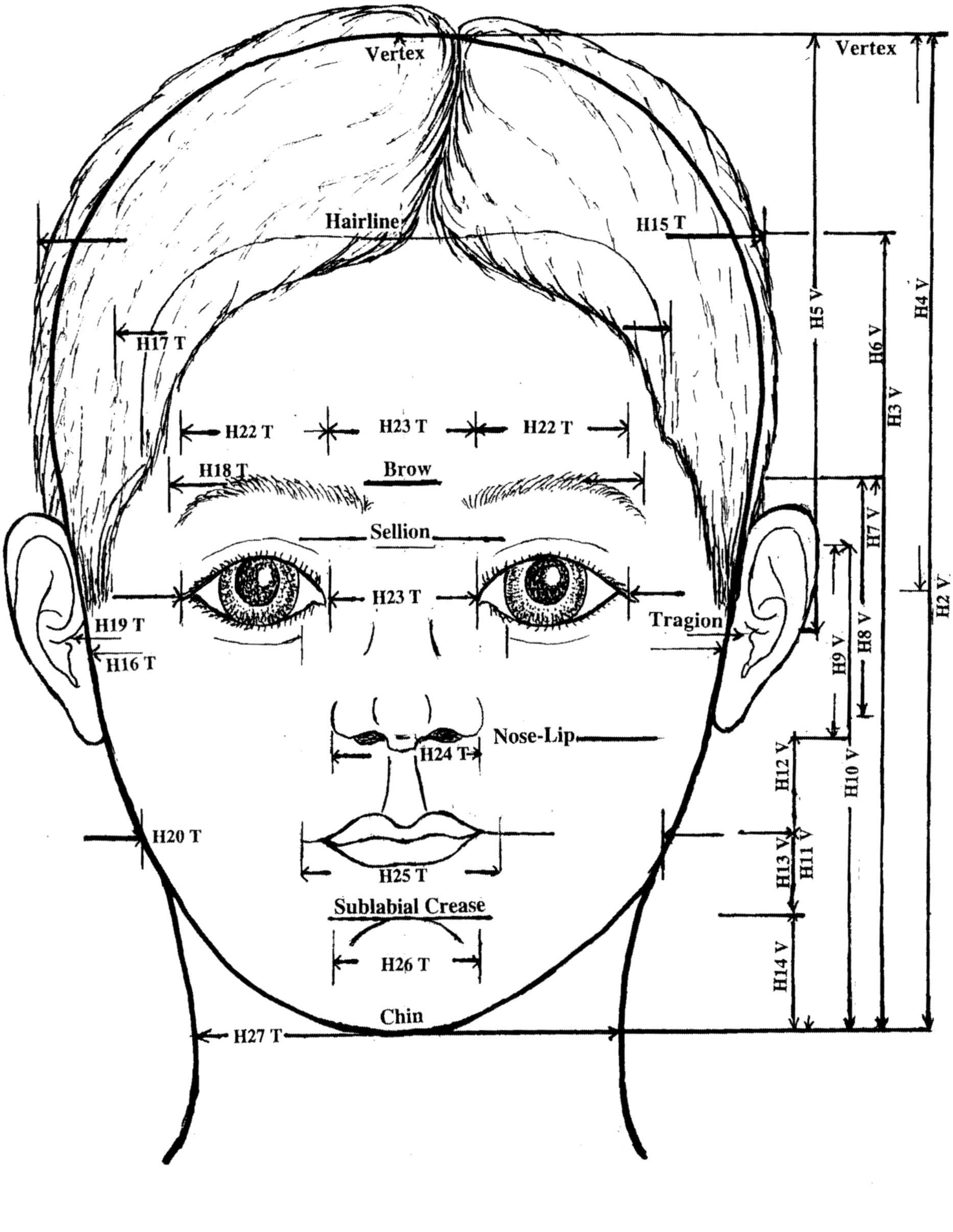

Head, Lateral View

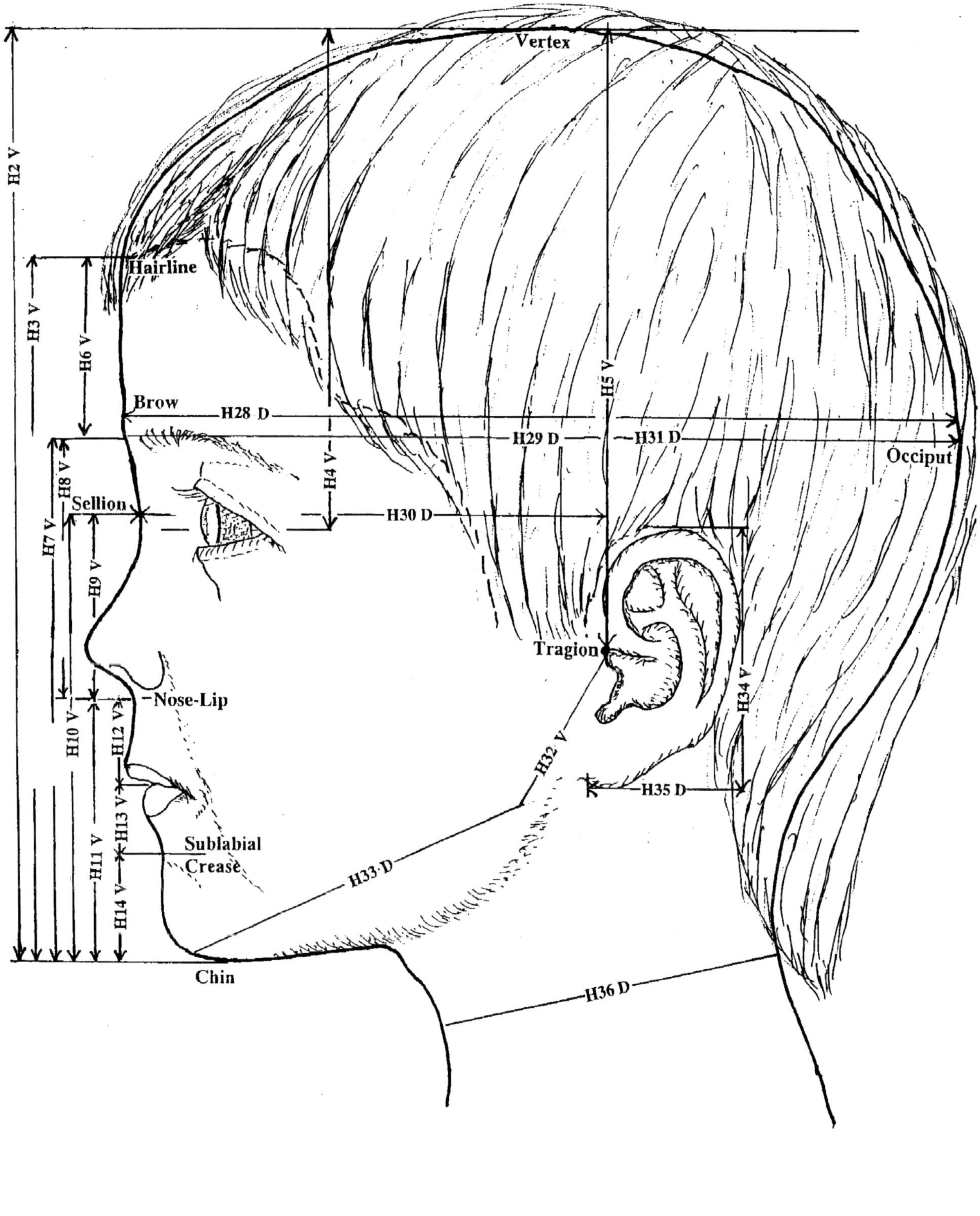

TRANSVERSE HEAD MEASUREMENTS

H19 T Inter-tragion, breadth between the upper borders of the tragus, the ear flaps of both ears just anterior to ear canals.

H20 T Inter-angle of mandibles breadth between Angles of the mandibles, jaw bone, below the ears; inter-gonion, this is the widest aspect of the Jaw.

H21 T Inter-lateral corners of eyes, the breadth between the lateral centhus of both eyes.

H22 T Width of an eye, the breadth of palpebral fissure, opening of the eyelids of one eye, measured between the medial and lateral canthus or corners of an eye.

H23 T Inter-inner corners of eyes, the width between inner canthus or corners of both eyelids.

H24 T Breadth of nose, a measure between the flare of nostrils or wings of the nose, ala nasae.

H25 T Breadth of mouth between corners of mouth, the lip Aperture.

H26 T Breadth of the chin, at borders of the arc of the sublabial crease, where the angle of the jaw is truncated, or intersected .

H27 T Breadth of neck at level of the thyroid cartilage of the larynx, (the voice box or adam's apple).

ANTERIOR POSTERIOR HEAD MEASUREMENTS

H28 D Brow to occiput, the maximum anterior posterior head measurement; the glabella, at the brow to the back of the head.

H29 D Brow to the tragus of an ear, the flap anterior to the ear canal; the anterior cranium measurement.

H30 D Sellion, bridge of nose to tragus of ear; or nasion to tragus.

H31 D Tragus of an ear to the occiput, back of the head, the posterior cranium.

H32 V Tragus of an ear to the angle of a mandible; tagion to gonion.

H33 D Angle of a mandible to the anterior chin; gonion to menton.

H34 V Length of ear, vertical measurement of helix at top of the ear to the lobe. Since the Lobe is soft tissue there is considerable variation.

H35 D Width of ear, anterior-posterior measurement of the anterior helix, outer ear border near the tragus to the posterior border of the helix of an ear.

H36 D Anterior-posterior of neck a measure at a level of the thyroid cartilage of the larynx.

VERTICAL BODY MEASUREMENTS

B1 V Chin to the sternal notch, with the face looking straight ahead, the chin to the pit of the neck at the top of the breast bone or sternum in the midline of the chest. This notch is nearly the same level (1/4 inch lower in adults) as the acromion process of the shoulder. Since the head is very mobile, this is not a reliable measure, but it must be considered.

B2 V Sternal notch to axilla, the pit of the neck to level of the upper, anterior crease of an armpit.

B3 V Sternal notch to the level of the nipples. Breasts being soft tissues are therefore variable in measurements. These are approximate measurements.

B4 V Sternal notch to the lower tip of the breast bone or xyphoid process. There are considerable variations in size and length of the xyphoid process.

B5 V Sternal notch to umbilicus or belly button. Some variations are recognized.

B6 V Xyphoid process, tip of the sternum breast bone to the umbilicus .

B7 V Sternal notch to the level of the crest of the ilium of the pelvis at the hip.

B8 V Sternal notch to the anterior superior spine of the ilium of the pelvis, a prominence anterior to the crest of the ilium of the pelvis.

B9 V Sternal notch to the symphysis pubis in the midline of the pelvis, above the base of the genitals.

B10 V Umbilicus to the superior border of the symphysis pubis of the pelvis.

TRANSVERSE BODY MEASUREMENTS

B11 T Shoulder breadth; the maximum measurements between the deltoid muscles below the point of the shoulder. The muscles are soft tissue and measurements are variable.

B12 T Clavicle to the acromion process, the collar bone at the sternal notch to point of shoulder.

B13 T Bi-acromion process breadth, width between points of both shoulders.

B14 T Inter-anterior axillary folds, distance between armpits, seen anteriorly.

B15 T Inter-nipples distance between the nipples; this soft tissue is variable.

B16 T Breadth of thorax, or chest at the mid thorax at about the eighth rib and below the axilla or armpit.

B17 T Inter-costal rib margin breadth of the chest at the lower rib margins.

B18 T Inter-crest of ilium upper lateral breadth of the pelvis.

B19 T Inter-anterior superior spines of the ilium, a measurement between the anterior prominences of the Pelvis.

B20 T Inter-greater trochanters of the femurs, the widest breadth of hips, these points are on the level considered to be the pivots of the hip joint.

Torso Measurements

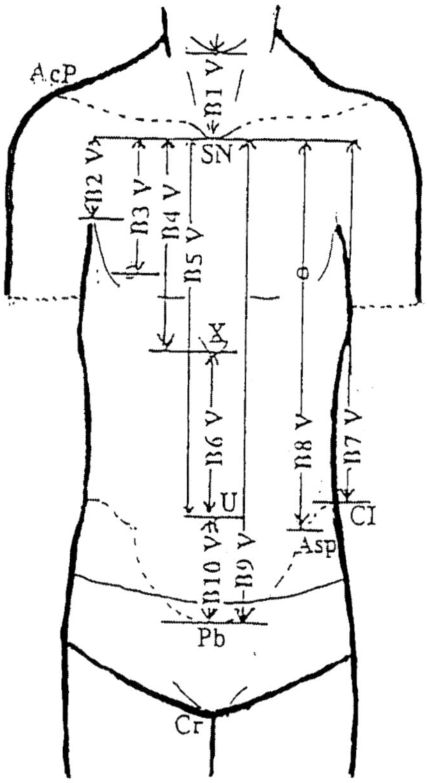

VERTICAL DIMENSIONS

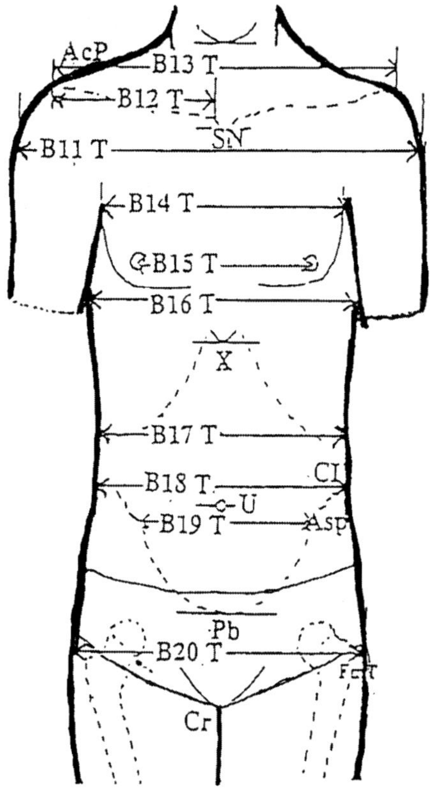

TRANSVERSE DIMENSIONS

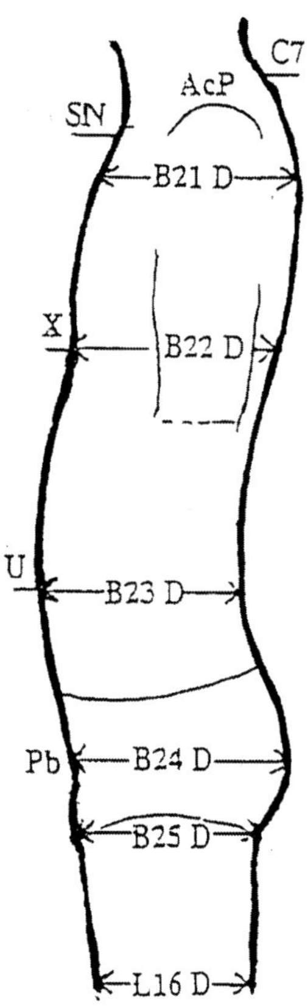

ANTERIOR-POSTERIOR DIMENSIONS

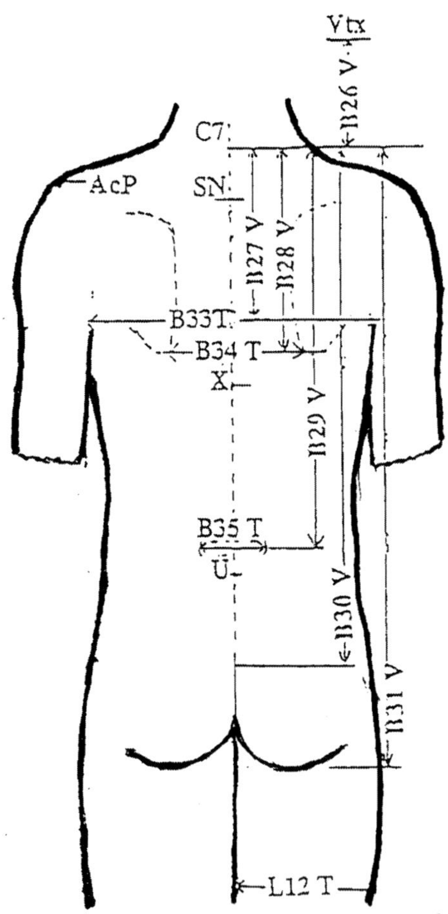

DORSAL DIMENSIONS

ANTERIOR-POSTERIOR BODY MEASUREMENTS

B21 D Sternal manubrium joint, at level of second rib, to dorsal spines of thorax.

B22 D Xyphoid process to the posterior scapula, shoulder blade; xyphoid-dorsal.

B23 D Umbilicus to dorsum, the belly button to the low back. This involves soft tissue, and measurements are variable.

B24 D Pubis to buttock, the symphysis pubis of pelvis to the posterior buttock. This involves soft tissue, and measurements are variable.

B25 D Anterior thigh to the gluteal furrow, a crease just below the buttock.

VERTICAL BACK MEASUREMENTS

B26V Vertex of the head to the seventh cervical vertebra spinous process. This is a prominence at the base of the neck posteriorly at the upper thorax.

The seventh cervical vertebra spinous process to other features:

B27 V Seventh cervical vertebra spinous process to the posterior axillary folds.

B28 V Seventh servical spinous process to the to the inferior tip of scapulae, or shoulder blades.

B29 V Seventh cervical spinous process to posterior superior spines of the ilium, recognized as a dimple on both sides of the posterior pelvis.

B30 V Seventh cervical spinous process to the upper buttocks crease in the midline, essentially near inferior tip of the coccyx or tail bone.

B31 V Seventh cervical spinous process to the gluteal furrow, an inverted arc-shaped crease below the buttocks. It is soft tissue and has variations.

B32 V Seventh cervical spinous process to the popliteal crease behind the knee. The creases in soft tissue are variable.

TRANSVERSE BACK MEASUREMENTS

B33 T Inter-posterior axillary folds, breadth between the armpits folds in the back, which is wider and lower than between anterior armpit folds..

B34 T Inter scapulae, breadth between the median borders of both scapulae, shoulder blades.

B35 T Inter posterior superior spines of the ilium of the pelvis, recognized as the distance between dimples over the posterior pelvis.

UPPER EXTREMITIES MEASUREMENTS

U1 T Span of outstretched arms and hands to the fingertips.
U2 T Sternal notch to the acromion process of the scapula.
U3 V The acromion process to the fingertip; the point of the shoulder to the fingertip of a dependant arm.
U4 V Vertical reach to the fingertip; while standing one arm of an individual is raised and measured from the level of the sternal notch to the maximum upward reach.
U5 T Lateral reach to fngertip; while standing erect, an individual is measured from one shoulder to the opposite horizontal arm to the fingertip.

U6 T Forward reach to fingertip; while standing erect, an individual is measured from the back, dorsum, to the extended fingertip of a forward reaching horizontal arm.
U7 V The shoulder to flexed elbow from acromion process, point of the shoulder to the olecranon process or point of the elbow.
U8 V The acromion process at the shoulder to lateral epicondyle of the humerus at the elbow. This is essentially at the pivot of the elbow joint.
U9 V Flexed, elbow to fingertip, the measurement from the olecranon process of ulna, point of the elbow, to the tip of the third finger.
U10 V Flexed elbow at the pivot, or lateral epicondyle of humerus, to wrist at its pivot, the styloid process of the radius at the wrist on the side of the thumb.

HAND MEASUREMENTS

U11 V Wrist at the pivot, radial styloid process, to the fingertip of the third finger.
U12 V Wrist at pivot to the tip of the thumb.
U13 V Length of palm from the wrist skin crease to the palmar crease of the third finger.
U14 V Length of third finger, from the palmar finger crease to the fingertip.

U15 V Thumb crotch to the third fingertip.
U16 V Thumb crotch to the tip of the thumb.
U17 D Anterior-posterior depth of an upper arm at mid-biceps muscle.
U18 D Anterior-posterior of a forearm at mid forearm.

U19 D Depth of wrist, volar (palmar)to dorsal (back of hand and wrist).
U20 T Breadth of wrist between the spines of radius and ulna bones.
U21 T Breadth of hand including thumb, either outstretched or clenched.
U22 T Breadth of the palm of the hand.

Since there are considerable variations of measurement at any age, particularly of stature, the use of a live model is recommended. If the stature of the average table does not correlate with the age of the model, it may be practical to use measurements of an older or younger age, as it appears appropriate. These adjustments may be more significant for the longer body measurements in the arms and legs.

Upper Extremity Measurements

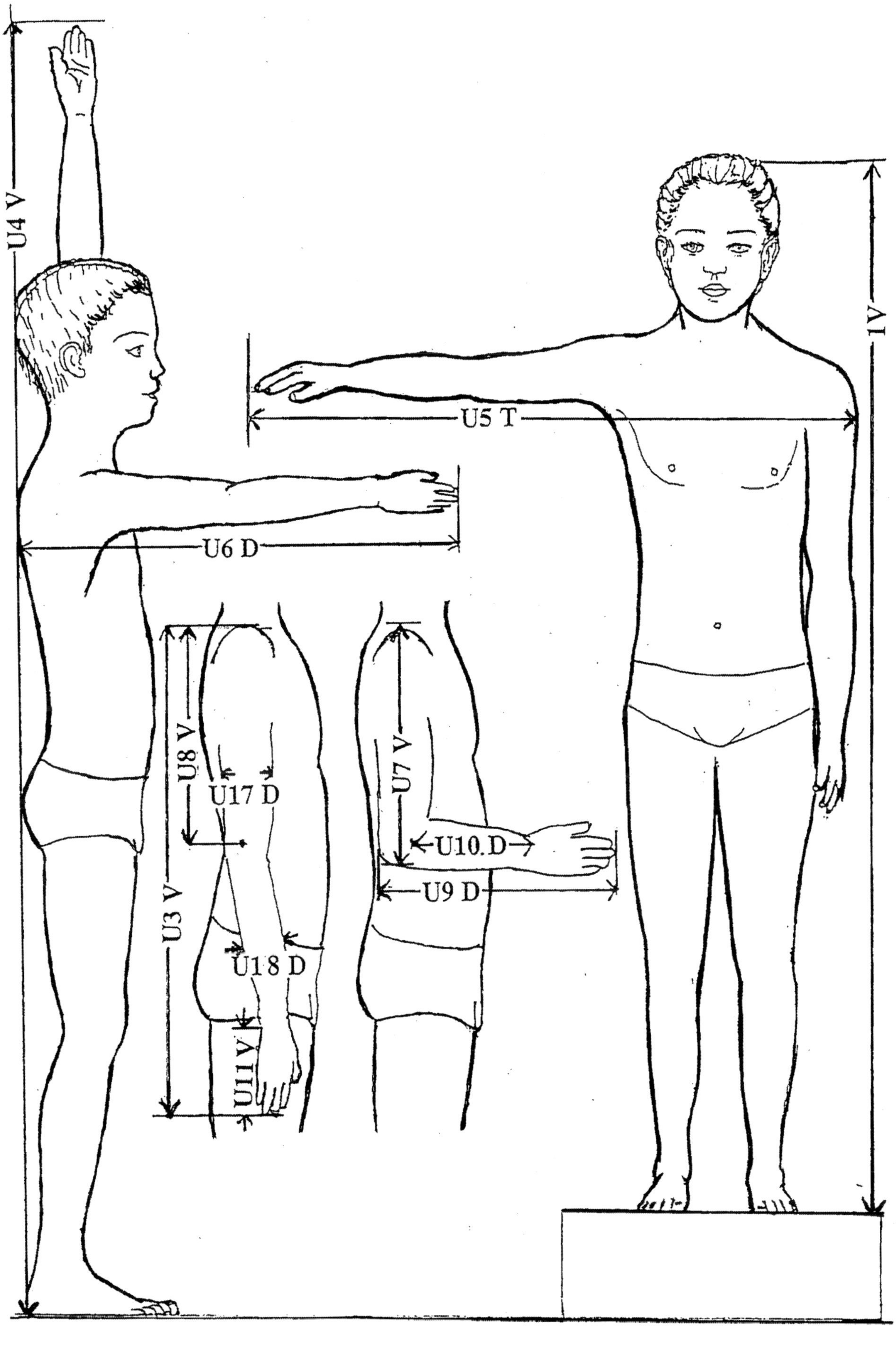

Hand and Foot Measurements

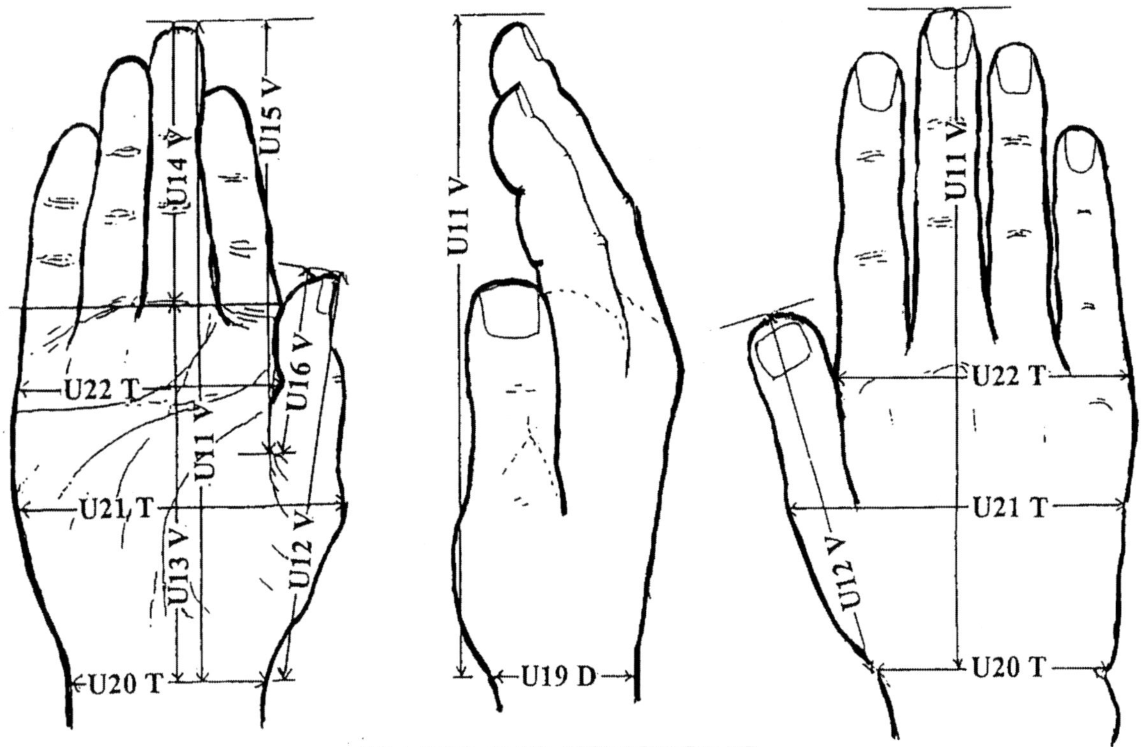

HAND DIMENSIONS

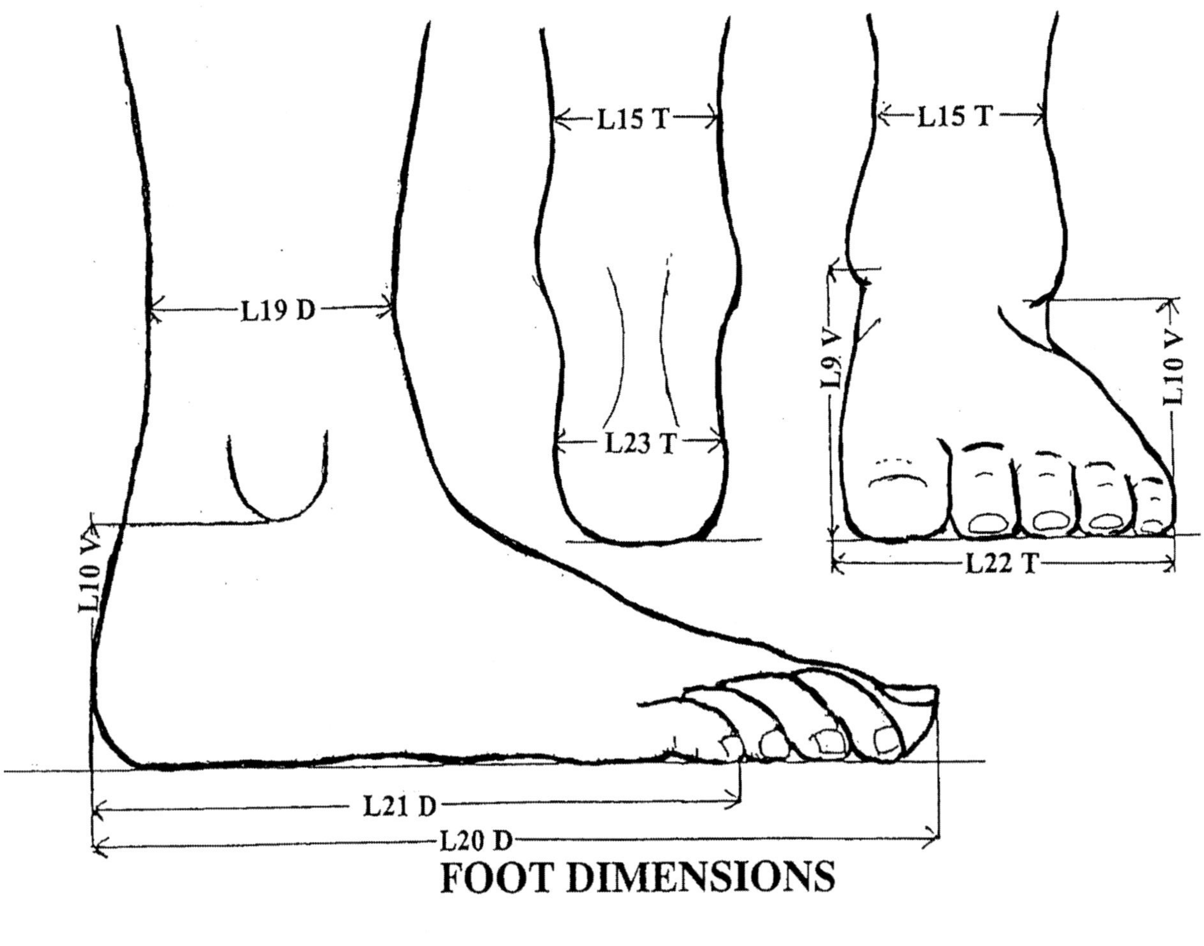

FOOT DIMENSIONS

LOWER EXTREMITY MEASUREMENTS

L1 V	Crest of the ilium of the pelvis to the sole of a foot.
*L2 V	Anterior superior spine of the ilium to the sole of a foot.
L3 V	Greater trochanter of the femur to the sole of a foot. Note; this is nearly the level of the hip pivot
*L4 V	Symphysis pubis of the pelvis, at the base of the genitals in the midline to the sole of a foot
*L5 V	Crotch to sole of a foot. Since this is composed of soft tissue, it is variable.
L6 V	Gluteal furrow, a curved crease below a buttock to the sole of a foot.
L7 V	Posterior knee crease at the popliteal fossa behind a knee, to the sole of a foot. This measurement is variable.
L8 V	Tibia at knee, shin bone, below the knee cap or patella, to the sole of the foot; tibial height.
L9 V	Medial malleolus of an ankle, a medial prominence at the point of the tibia at the ankle to the sole of a foot.
L10 V	Lateral malleolus, a lateral ankle prominence, the end of a fibula at an ankle to the sole of a foot. This is lower than the medial malleolus.
L11.T	Breadth of the hips while sitting.
L12 T	Breadth of a mid thigh.
L13 T	Breadth of a knee.
L14 T	Breadth of a calf at the gastrocnemius muscle.
L15 T	Breadth of an ankle above the level of the malleoli.
L16 D	Mid thigh, anterior-posterior.
L17 D	Knee at patella, anterior-posterior.
L18 D	Mid calf anterior-posterior, at the gastrocnemius muscle.
L19 D	Ankle at its least diameter, anterior-posterior.
L20 D	Length of foot; heel to the tip of a first or second toe.
L21 D	Measurement from the heel to the base of the fifth toe.
L22 T	Breadth of a foot, relaxed but horizontal.
L23 T	Width of a heel.

* *These measurement are also listed in Vertical Measurements.*

The measurements listed are based on averages of several studies in different parts of the nation. Some variations of a few millimeters occur at each point of reference. If calculations of addition or subtraction are needed, the inaccuracies are compounded. Another significant factor is the difference of maturation of individuals in any age of childhood and youth.

The use of a live model then becomes important to an artist. A more authentic illustrated reproduction can be achieved. The expressions and actions characterized by the intended age can be recognized and incorporated into the artistic creation. There also may be ethnic characteristics that need to be considered that are best evaluated by the artist.

Lower Extremity Measurements

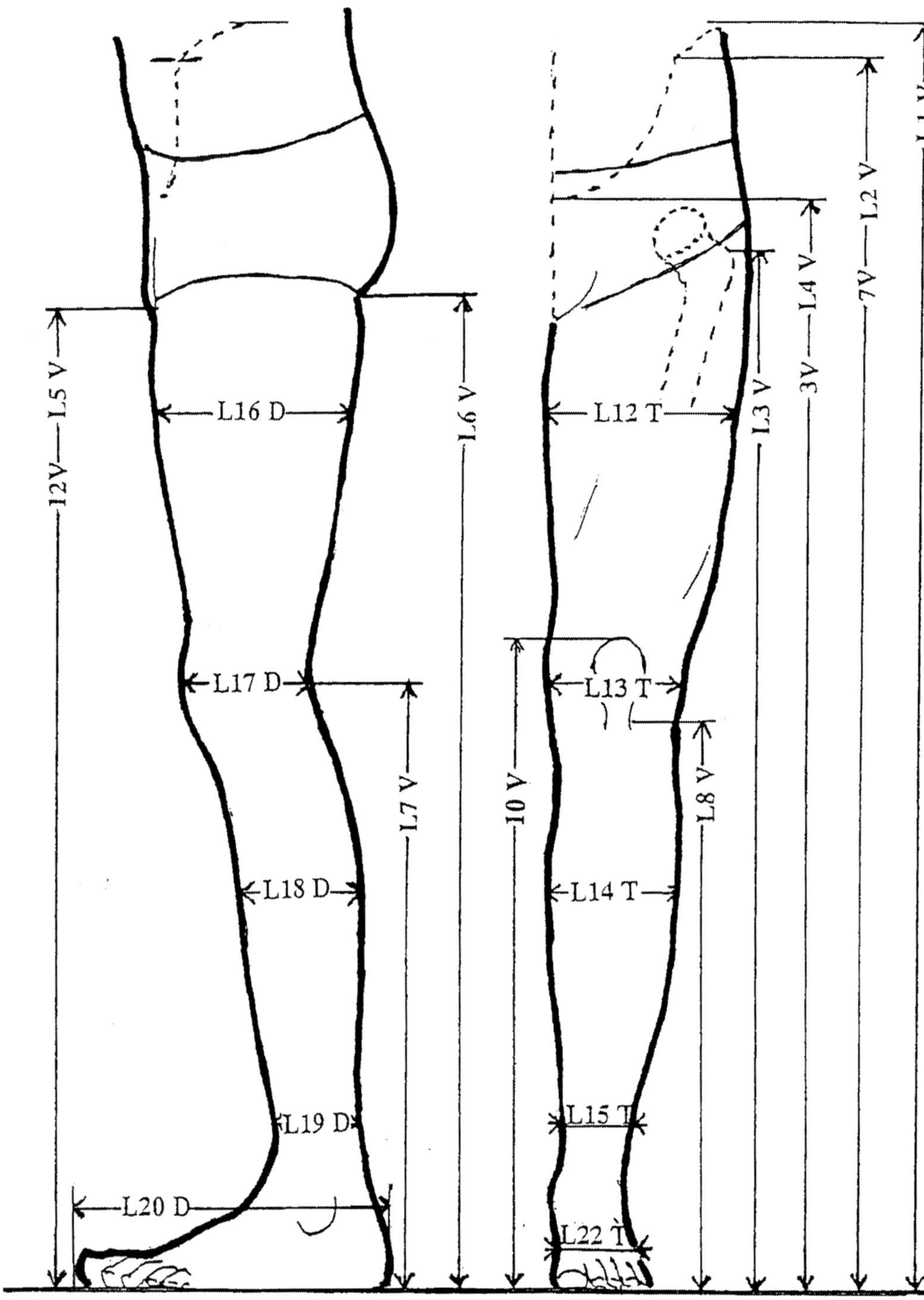

LEG DIMENSIONS

Studies About the Growth and Development of Children from Infancy to Young Adult

Including Illustrations and Charts of Representative Measurements and Proportions for Figurative Artists

Development of the Face in Embryo

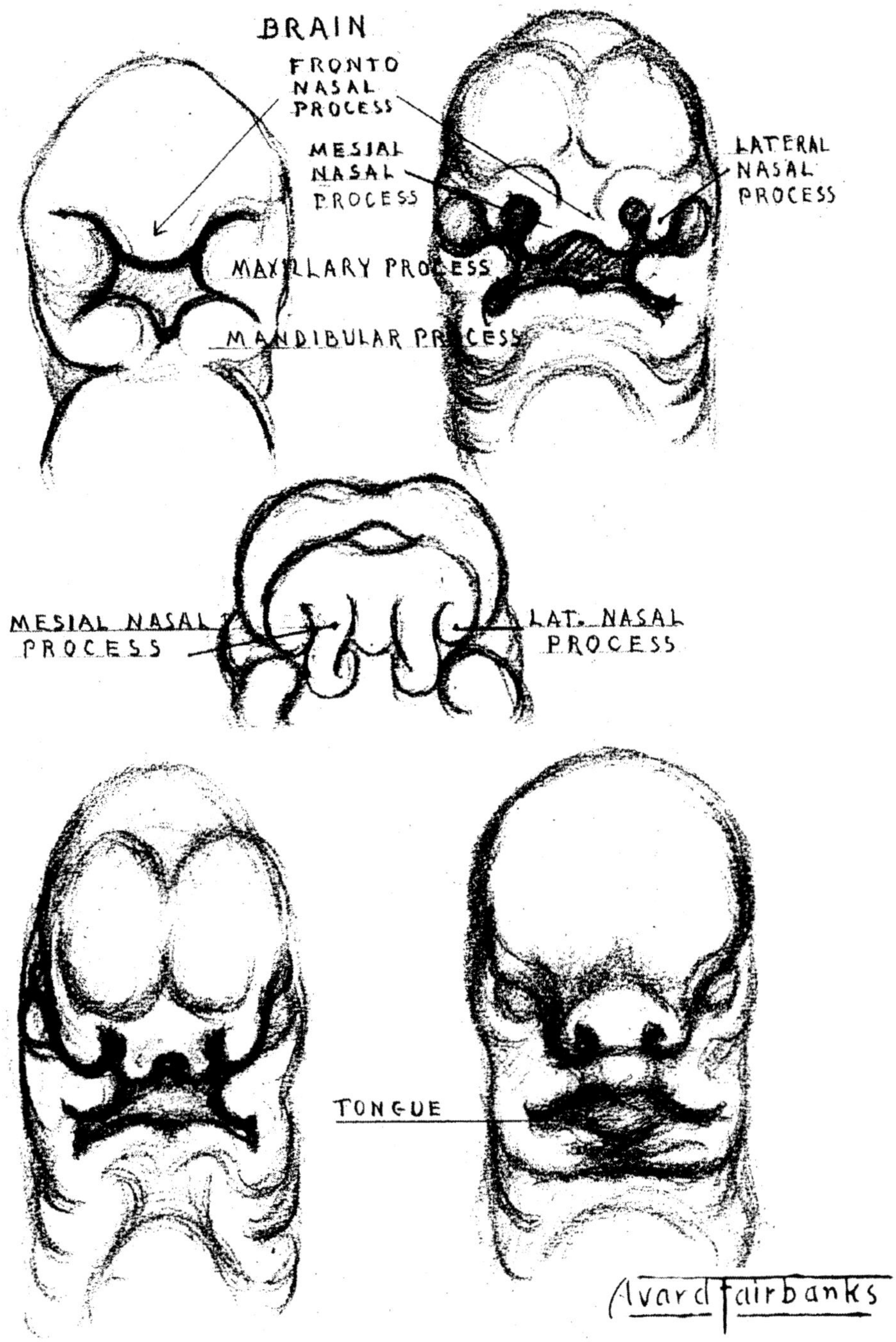

The features of the face may be better understood by a review of the embryological developments as illustrated. In the fourth week, when the embryo is 3.5 cm in length, there are five tiny prominent folds or processes in the developing face: the upper midline nasal process approaching the paired maxillary, and mandibular processes.

At five weeks, the embryo is 8 mm long and more form is evident. The nasal process is central and has paired lateral nasal processes. The paired maxillary processes have grown. The paired mandibular processes have fused. A view from below demonstrates added details.

At six weeks, the embryo is 12 mm long. Additional growth and development is evident. Primordial eyes are apparent. The maxillary processes approach the midline nasal process.

At seven weeks, The embryo is 19 mm long. The maxillary processes have joined the central nasal process. The tongue and lips have begun to develop. The eyes have also grown.

A Young Infant's Face

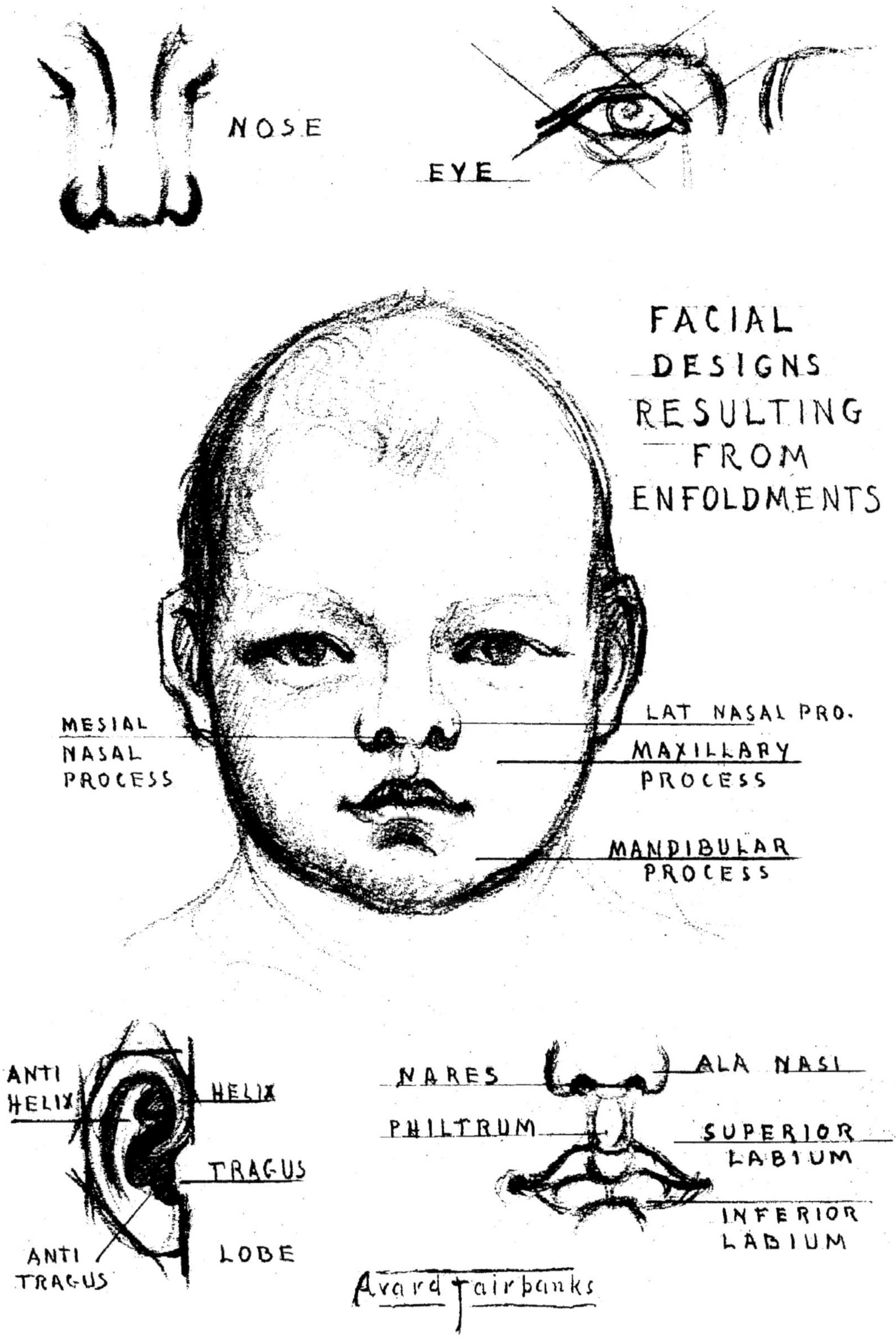

By three months gestation, the elements of the face of an embryo have fused and a tiny face can be recognized. The cranium is much larger than the face. Both are growing rapidly. At birth the features of a human face can be well appreciated. An illustration by Avard Fairbanks, of a young infant, is included that correlates the embryonic components of its face.

NEWBORN INFANTS

On the day of birth, the head of newborn infants demonstrate molding to conform to the birth canal. Some do not look pretty. Three days later they have improved facial and skull contours, and are more attractive for artists and photographers. While developing in the uterus, infants were confined to a restricted space. During early months, when at rest, they tend to retain their fetal position with arms and legs flexed. The plump, stubby fists are usually clenched. The measurement from wrist crease to knuckles is about one half the width of the fist. Crown to rump measurements are more significant than total height because a baby usually has hips and knees flexed. When an infant is active and lying face down, alternating movements of extremities, like crawling or swimming, are normal.

Young infants nurse often and sleep most of the remaining time. When hungry, they awaken and cry in a monotone. In the first month they smile in response to their mother's attention. Later, they smile spontaneously. In order to satisfy the public, newborn infants are illustrated as beautiful, innocent and happy. A gentle smile is expected. Spontaneous smiles develop in early weeks, and responsive smiles a little later.

In movie scenarios, the post-delivery scene with a beaming young mother in a hospital bed, is invariably depicted holding a three or four-month old happy, smiling, and cooing infant. This may be good drama, but it is poor science. Magazine and catalogue advertising usually depicts a fat baby with a round face. Artists may exercise some license and portray a more publicly acceptable appearing infant.

A baby needs emotional support and security. While awake it expects attention from parents and observers. Pleasant sounds render happy responses. Therefore, very young infants are best portrayed demonstrating a mother's attention, loving care, and support.

At birth, male infants, on the average, are slightly longer and slightly heavier than are female infants. Buttocks and shoulders may be more rounded in female infants. Growth and development of both sexes are similar up to the age of five or six years and in this study, male and female information is combined up to the age of four years.

The portrayal of newborn infants may best consider the face proportions with chin-to-brow as one-half and brow-to-vertex as one-half of the chin-to-vertex measure. A convenient division of the face, or brow-to-chin, is in one-fourths: Brow to eyes, eyes to nose, nose to lower lip, and lip to chin. This relationship is satisfactory for several years. Soon the upper point of reference needs to move down from the brow to the selion or saddle of the nose.

A comparison of newborn proportions with adult proportions is significant for artists. An infant's height is one fourth of an adult height. During growth and development, not all features increase by similar multiples. The head increases in height by approximately two times, the trunk increases by three times, the arms increase by four times, and the legs increase by five times.[1]

During early months, infants grow rapidly. The brain grows more rapidly resulting in an increased cranial size. Progressive growth and weight gain is relatively rapid. The extremities grow in both length and diameter.

1. Breckenridge, M.E., Murphy, M.N., p 252.

At birth babies vary in length and weight depending on maternal nutrition and the length of gestation. The usual full-term birth weight is expected to be between about 6 pounds (2.7 kilograms) and 9 pounds (4 kilograms). The expected full-term length is between 19 inches (48 cm) and 21 inches (53 cm).

There is considerable similarity of male and female, full term, newborn infants. For a practical list of body proportions, both sexes will be combined in the tables of measurements. The ranges in percentiles of height and weight are listed for comparison. Twenty-fifth is short; fiftieth is medium; and to seventy-fifth is tall.

Percentiles	**Twenty-fifth**		**Fiftieth**		**Seventy-fifth**	
Boys:	**Short**		**Medium**		**Tall**	
Crown to Sole	49 cm,	19 1/4 in.	50.5 cm,	20 in.	51.8 cm,	20 1/2 in.
Weight	3 kg,	6 1/2 lb.	3.27 kg,	7 1/4 lb.	3.64 kg,	8 lb.
Girls:						
Crown to Sole	48.2 cm,	19 in.	49.9 cm,	19 3/4 in.	51 cm,	20 in.
Weight	2.93 kg,	6 1/2 lb.	3.23 kg,	7 lb.	3.52 kg,	7 3/4 lb.[2]

Throughout this study, ages of newborn to eighteen years old, short individuals at twenty-fifth-percentile are about three percent shorter than those in fiftieth percentile. Tall individuals, at seventy-fifth percentile, are about three percent taller than those of medium height at fiftieth percentile.

NEWBORN INFANT
BODY MEASUREMENTS COMPARED TO HEIGHT

	Feature Measure in cm	Feature Compared to Stature	Approx. Ratio of Feature	Compared to Median Adult
Stature: crown to sole	51			28%
Head Height: crown to chin	13	25%	1/4	62%
Sitting Height: crown to rump	35	69%	3/4-	40%
Trunk Measure: sternal notch to symphysis pubis	19	37%	1/3+	40%
Leg Length: greater trochanter of femur to sole	22	31%	1/3-	18%
Thigh Length: greater trochanter to tibia at knee	11	16%	1/6	18%
Tibia at Knee to Sole:	11	21%	1/5	24%
Arm Length: acromion process at shoulder to dependant fingertip:	21	41%	2/5	28%
Hand Length: usually clenched:	6.5	13%	1/8+	34%
Foot Length:	8	16%	1/6	30%

2. Nelson, W.E., Vaughan, V.C., Behrman, R.E., p. 38.

Newborn Infant, anterior and lateral head

all measurements in centimeters

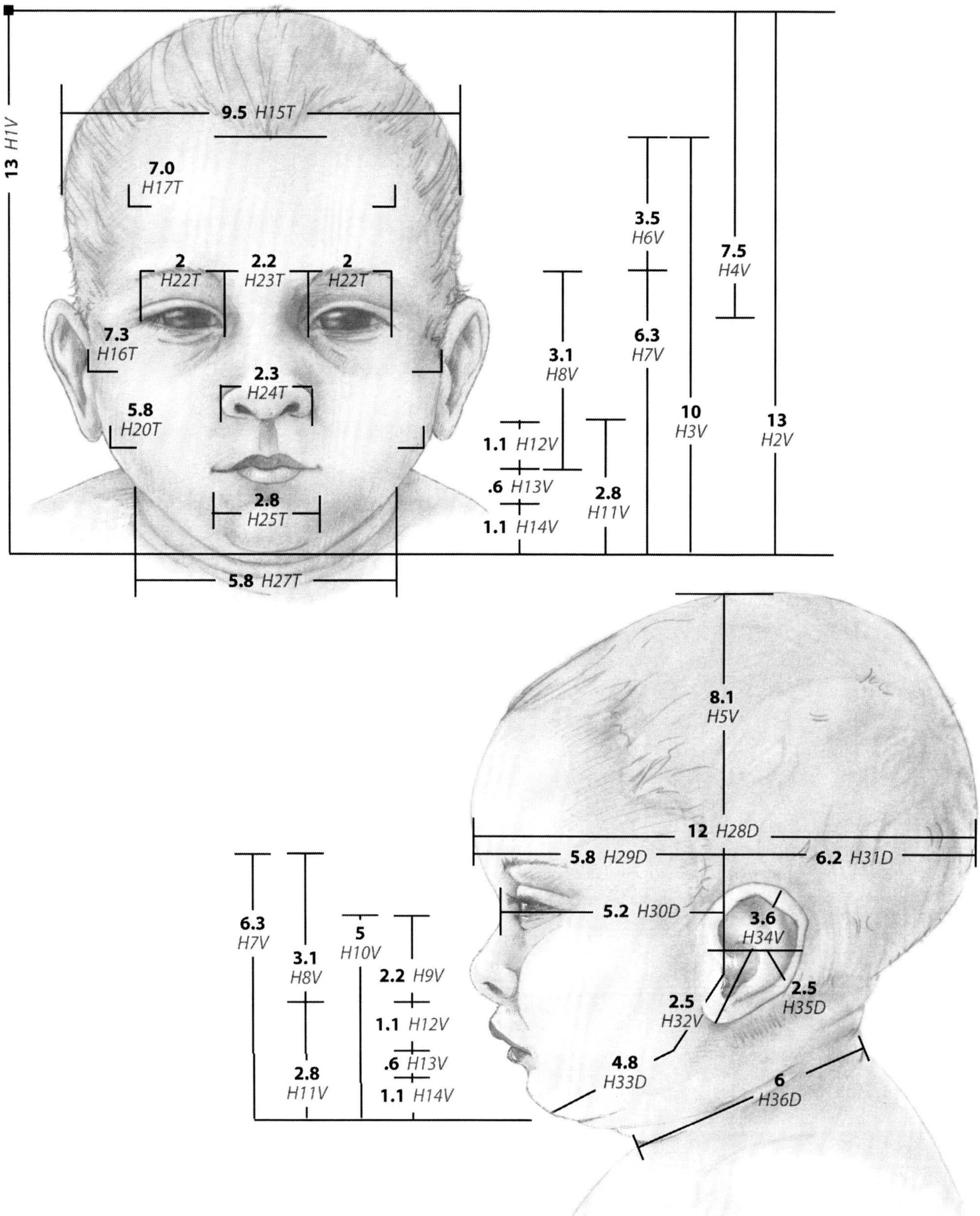

Newborn Infant, anterior and lateral figure

all measurements in centimeters

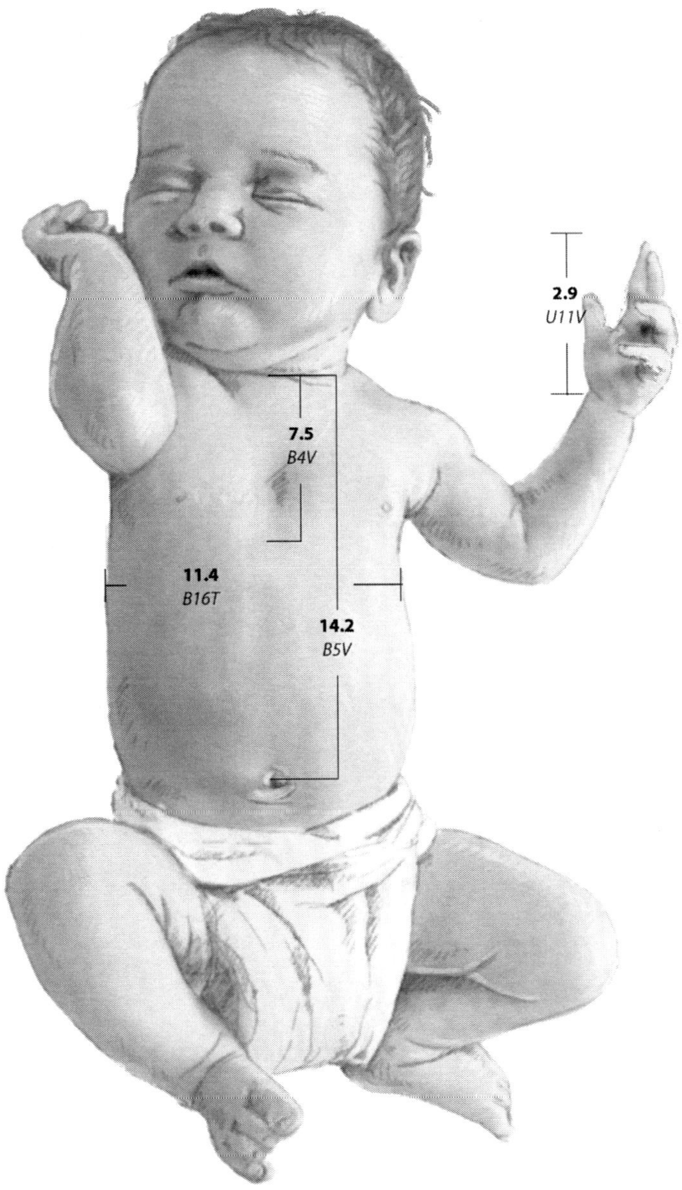

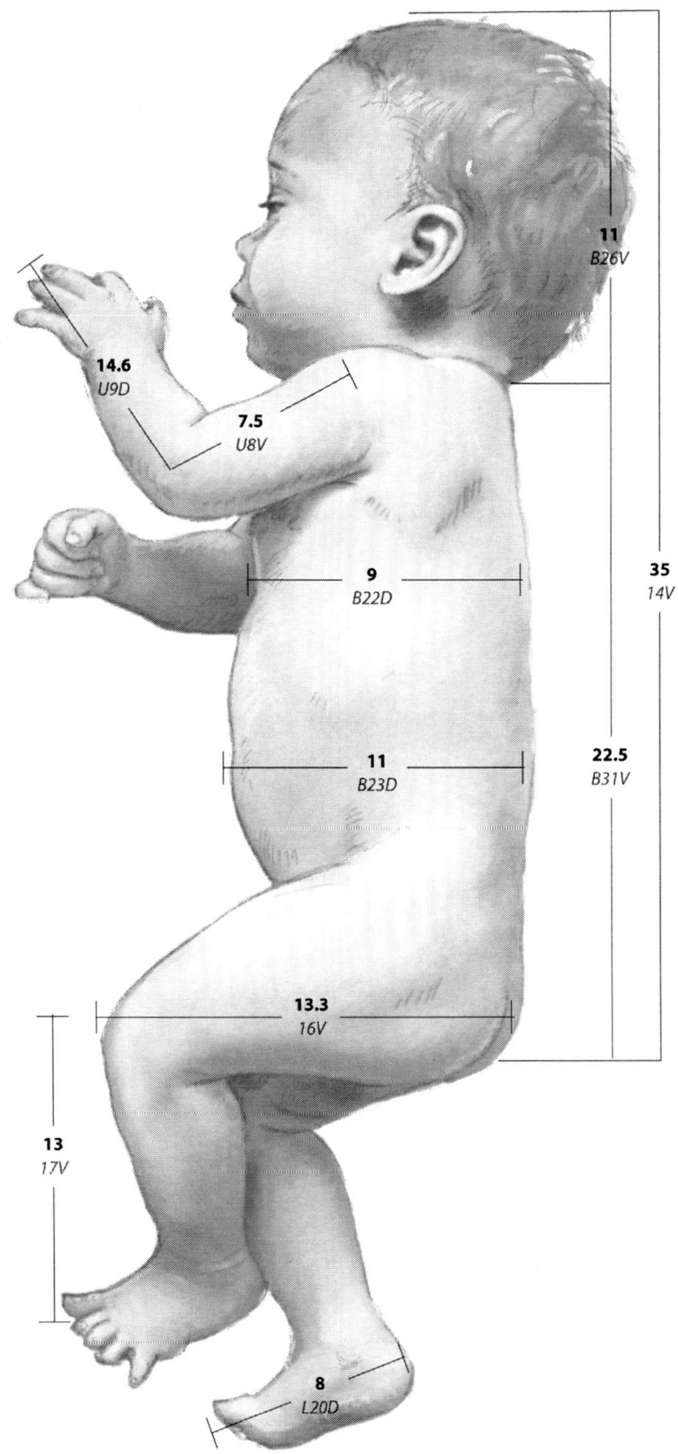

Life size			NEWBORN INFANT	Fractions of life size				
				3/4	2/3	1/2	1/3	1/4
Inches	cm	Code		cm	cm	cm	cm	cm
			VERTICAL MEASUREMENTS					
20.1	**51.0**	1 V	Stature, Height, Vertex to Sole	38.3	34.0	25.5	17.0	12.8
12.6	**32.0**	2 V	Vertex to Symphysis Pubis	24.0	21.3	16.0	10.7	8.0
7.5	**19.0**	* 3 V	Symphysis Pubis of Pelvis-Sole	14.3	12.7	9.5	6.3	4.8
15.0	**38.0**	4 V	Sternal Notch, Neck Pit, to Sole	28.5	25.3	19.0	12.7	9.5
5.1	**13.0**	5 V	Vertex to Acromion Process	9.8	8.7	6.5	4.3	3.3
15.4	**39.0**	6 V	Acromion Process to Sole	29.3	26.0	19.5	13.0	9.8
9.1	**23.0**	* 7 V	Ant.Sup.Spine of Ilium to Sole	17.3	15.3	11.5	7.7	5.8
5.9	**15.0**	8 V	Acrom. Proc. to Ant. Sup. Spine	11.3	10.0	7.5	5.0	3.8
4.7	**12.0**	9 V	Ant.Sup.Spine of Ilium to Knee	9.0	8.0	6.0	4.0	3.0
4.7	**12.0**	10 V	Knee at Patella to Sole	9.0	8.0	6.0	4.0	3.0
9.4	**24.0**	11 V	Umbilicus to Sole	18.0	16.0	12.0	8.0	6.0
6.0	**15.2**	* 12 V	Crotch to Sole	11.4	10.1	7.6	5.1	3.8
13.8	**35.0**	14 V	Sitting Height - Crown Rump	26.3	23.3	17.5	11.7	8.8
10.2	**26.0**	15 V	Buttock, Straight Leg to Sole	19.5	17.3	13.0	8.7	6.5
5.2	**13.3**	16 V	Buttock to Flexed Knee	10.0	8.9	6.7	4.4	3.3
5.1	**13.0**	17 V	Flexed Knee to Sole	9.8	8.7	6.5	4.3	3.3
		** These are also listed with the Lower Extremity Measurements.*						
			VERTICAL HEAD MEASUREMENTS					
5.1	**13.0**	H1 V	Vertex to Sternal Notch	9.8	8.7	6.5	4.3	3.3
5.1	**13.0**	H2 V	Vertex to Chin, Head Height	9.8	8.7	6.5	4.3	3.3
3.9	**10.0**	H3 V	Hairline to Chin, Face Height	7.5	6.7	5.0	3.3	2.5
3.1	**7.8**	H4 V	Vertex to Lateral Corner of Eye	5.9	5.2	3.9	2.6	2.0
3.2	**8.1**	H5 V	Vertex to Tragion of Ear	6.1	5.4	4.1	2.7	2.0
1.4	**3.5**	H6 V	Hairline to Brow, Glabella	2.6	2.3	1.8	1.2	0.9
2.5	**6.3**	H7 V	Brow, Glabella, to Chin, Menton	4.7	4.2	3.2	2.1	1.6
1.2	**3.1**	H8 V	Brow to Subnasal Point	2.3	2.1	1.6	1.0	0.8
0.9	**2.2**	H9 V	Sellion to Subnasal, Nose Length	1.7	1.5	1.1	0.7	0.6
2.0	**5.0**	H10 V	Sellion, Nose Saddle to Chin	3.8	3.3	2.5	1.7	1.3
1.1	**2.8**	H11 V	Subnasal Point to Chin	2.1	1.9	1.4	0.9	0.7
0.4	**1.1**	H12 V	Subnasal to Lip Aperature	0.8	0.7	0.6	0.4	0.3
0.2	**0.6**	H13 V	Lip Aperat. to Sublabial Crease	0.5	0.4	0.3	0.2	0.2
0.4	**1.1**	H14 V	Sublabial Crease to Chin	0.8	0.7	0.6	0.4	0.3
			TRANSVERSE HEAD MEASUREMENTS					
3.7	**9.5**	H15 T	Bi-Parietal, Head Breadth	7.1	6.3	4.8	3.2	2.4
2.8	**7.0**	H17 T	Bi-Fronto Temporal, Forehead	5.3	4.7	3.5	2.3	1.8
2.7	**6.8**	H18 T	Inter-Brow Ridge	5.1	4.5	3.4	2.3	1.7
3.4	**8.6**	H19 T	Inter-Tragion of Ears	6.5	5.7	4.3	2.9	2.2

Life size			NEWBORN INFANT	Fractions of life size				
				3/4	2/3	1/2	1/3	1/4
Inches	cm	Code		cm	cm	cm	cm	cm
			TRANSVERSE HEAD MEASUREMENTS					
2.3	**5.8**	H20 T	Inter-Angle of Mandible	4.4	3.9	2.9	1.9	1.5
2.5	**6.3**	H21 T	Inter-Lateral Corners of Eyes	4.7	4.2	3.2	2.1	1.6
0.8	**2.0**	H22 T	Width of Eye	1.5	1.3	1.0	0.7	0.5
0.9	**2.3**	H23 T	Inter-Inner Corner of Eyes	1.7	1.5	1.2	0.8	0.6
0.9	**2.3**	H24 T	Breadth of Nose	1.7	1.5	1.2	0.8	0.6
1.1	**2.8**	H25 T	Breadth of Mouth	2.1	1.9	1.4	0.9	0.7
0.8	**2.0**	H26 T	Breadth of Chin	1.5	1.3	1.0	0.7	0.5
2.3	**5.8**	H27 T	Breadth of Neck	4.4	3.9	2.9	1.9	1.5
			ANTERIOR-POSTERIOR HEAD MEASUREMENTS					
4.7	**12.0**	H28 D	Brow to Occiput, Posterior Head	9.0	8.0	6.0	4.0	3.0
2.3	**5.8**	H29 D	Brow to Tragus of Ear	4.4	3.9	2.9	1.9	1.5
2.0	**5.2**	H30 D	Sellion, Nose Saddle, to Tragus	3.9	3.5	2.6	1.7	1.3
2.4	**6.2**	H31 D	Tragus to Occiput, Back of Head	4.7	4.1	3.1	2.1	1.6
1.0	**2.5**	H32 V	Tragus to Angle Mandible	1.9	1.7	1.3	0.8	0.6
1.9	**4.8**	H33 D	Angle of Mandible to Chin	3.6	3.2	2.4	1.6	1.2
1.4	**3.6**	H34 V	Length of Ear, Vertical	2.7	2.4	1.8	1.2	0.9
1.0	**2.5**	H35 D	Width of Ear, Anterior-Posterior	1.9	1.7	1.3	0.8	0.6
2.4	**6.0**	H36 D	Anterior to Posterior of Neck	4.5	4.0	3.0	2.0	1.5
			VERTICAL BODY MEASUREMENTS					
		B1 V	Chin to Sternal Notch, Neck Pit					
0.8	**2.0**	B2 V	Sternal Notch to Axilla, Arm Pit	1.5	1.3	1.0	0.7	0.5
1.6	**4.0**	B3 V	Sternal Notch to Nipples	3.0	2.7	2.0	1.3	1.0
3.0	**7.5**	B4 V	Sternal Notch to Xyphoid Proc.	5.6	5.0	3.8	2.5	1.9
5.6	**14.2**	B5 V	Sternal Notch to Umbilicus	10.7	9.5	7.1	4.7	3.6
2.6	**6.7**	B6 V	Xyphoid Process to Umbilicus	5.0	4.5	3.4	2.2	1.7
6.1	**15.4**	B7 V	Sternal Notch to Crest Ilium	11.6	10.3	7.7	5.1	3.9
6.6	**16.8**	B8 V	Sternal Notch to Ant.S.Sp.Ilium	12.6	11.2	8.4	5.6	4.2
7.5	**19.0**	B9 V	Sternal Notch to Pubis	14.3	12.7	9.5	6.3	4.8
2.0	**5.0**	B10 V	Umbilicus to Pubis of Pelvis	3.8	3.3	2.5	1.7	1.3
			TRANSVERSE BODY MEASUREMENTS					
5.5	**14.0**	B11 T	Shoulder Breadth	10.5	9.3	7.0	4.7	3.5
2.2	**5.7**	B12 T	Clavicle to Acromion Process	4.3	3.8	2.9	1.9	1.4
4.5	**11.5**	B13 T	Bi-acromial Process Breadth	8.6	7.7	5.8	3.8	2.9
4.0	**10.2**	B14 T	Inter-Anterior Axillary Fold	7.7	6.8	5.1	3.4	2.6
3.0	**7.7**	B15 T	Inter-Nipples	5.8	5.1	3.9	2.6	1.9

Life size			NEWBORN INFANT	Fractions of life size				
				3/4	2/3	1/2	1/3	1/4
Inches	cm	Code		cm	cm	cm	cm	cm
			TRANSVERSE BODY MEASUREMENTS					
4.5	**11.4**	B16 T	Breadth of Thorax	8.6	7.6	5.7	3.8	2.9
4.2	**10.6**	B17 T	Inter-Costal. Rib Margin Breadth	8.0	7.1	5.3	3.5	2.7
3.0	**7.5**	B18 T	Inter-Crest of Ilium Breadth	5.6	5.0	3.8	2.5	1.9
2.9	**7.3**	B19 T	Inter-Anterior Superior Spine	5.5	4.9	3.7	2.4	1.8
4.1	**10.3**	B20 T	Inter-Greater Trochanter Femur	7.7	6.9	5.2	3.4	2.6
			ANTERIOR-POSTERIOR BODY MEASUREMENTS					
		B21 D	Manubrium Dorsal Spine					
3.5	**9.0**	B22 D	Xyphoid Process to Dorsum	6.8	6.0	4.5	3.0	2.3
4.3	**11.0**	B23 D	Umbilicus to Dorsum	8.3	7.3	5.5	3.7	2.8
3.1	**8.0**	B24 D	Pubis of Pelvis to Buttock	6.0	5.3	4.0	2.7	2.0
		B25 D	Thigh to Gluteal Furrow					
			VERTICAL BACK MEASUREMENTS					
4.3	**11.0**	B26 V	Vertex to 7th Cervical Spine	8.3	7.3	5.5	3.7	2.8
			7th Cervical Vertebra Spine to:					
1.4	**3.5**	B27 V	To Posterior Axillary Fold, Armpit	2.6	2.3	1.8	1.2	0.9
2.1	**5.4**	B28 V	To Inferior Tip of Scapula	4.1	3.6	2.7	1.8	1.4
6.3	**16.0**	B29 V	To Posterior Superior Spine Ilium	12.0	10.7	8.0	5.3	4.0
7.7	**19.5**	B30 V	To Midline Upper Buttock Crease	14.6	13.0	9.8	6.5	4.9
8.9	**22.5**	B31 V	To Gluteal Furrow	16.9	15.0	11.3	7.5	5.6
			TRANSVERSE BACK MEASUREMENTS					
4.3	**10.9**	B33 T	Inter-Posterior Axillary Folds	8.2	7.3	5.5	3.6	2.7
1.9	**4.9**	B34 T	Inter-Scapulae Medial Borders	3.7	3.3	2.5	1.6	1.2
1.1	**2.9**	B35 T	Inter-Post. Sup. Spines Sacrum	2.2	1.9	1.5	1.0	0.7
			UPPER EXTREMITY MEASUREMENTS					
19.7	**50.0**	U1 V	Span of Arms	37.5	33.3	25.0	16.7	12.5
2.4	**6.0**	U2 V	Sternum to Acromion Process	4.5	4.0	3.0	2.0	1.5
8.3	**21.0**	U3 V	Acromion Process to Fingertip	15.8	14.0	10.5	7.0	5.3
3.5	**9.0**	U7 V	Shoulder to Flexed Elbow	6.8	6.0	4.5	3.0	2.3
3.0	**7.5**	U8 V	Acromion to Lat. Condyle Elbow	5.6	5.0	3.8	2.5	1.9
5.7	**14.6**	U9 D	Flexed Elbow to Fingertip	11.0	9.7	7.3	4.9	3.7
2.9	**7.3**	U10 D	Flexed Elbow Pivot to Wrist Pivot	5.5	4.9	3.7	2.4	1.8

Life size			NEWBORN INFANT	Fractions of life size				
				3/4	2/3	1/2	1/3	1/4
Inches	cm	Code		cm	cm	cm	cm	cm
			UPPER EXTREMITY MEASUREMENTS					
2.6	**6.5**	U11 V	Wrist to Fingertip	4.9	4.3	3.3	2.2	1.6
1.1	**2.9**	U12 V	Wrist to Thumb Tip	2.2	1.9	1.5	1.0	0.7
1.5	**3.7**	U13 V	Length of Palm	2.8	2.5	1.9	1.2	0.9
1.3	**3.2**	U14 V	Length 3rd Finger	2.4	2.1	1.6	1.1	0.8
1.7	**4.3**	U15 V	Thumb Crotch to 3rd Finger	3.2	2.9	2.2	1.4	1.1
1.2	**3.0**	U16 V	Thumb Crotch to Thumb Tip	2.3	2.0	1.5	1.0	0.8
1.4	**3.5**	U17 D	Anterior-Post. Arm at Biceps	2.6	2.3	1.8	1.2	0.9
1.2	**3.0**	U18 D	Anterior-Posterior at Forearm	2.3	2.0	1.5	1.0	0.8
0.8	**2.1**	U19 D	Depth of Wrist	1.6	1.4	1.1	0.7	0.5
1.1	**2.7**	U20 T	Breadth of Wrist	2.0	1.8	1.4	0.9	0.7
1.7	**4.2**	U21 T	Breadth of Hand	3.2	2.8	2.1	1.4	1.1
1.3	**3.3**	U22 T	Breadth of Palm	2.5	2.2	1.7	1.1	0.8
			LOWER EXTREMITY MEASUREMENTS					
9.1	**23.0**	L1 V	Crest of Ilium of Pelvis to Sole	17.3	15.3	11.5	7.7	5.8
8.7	**22.0**	* L2 V	Ant.Sup.Spine of Ilium to Sole	16.5	14.7	11.0	7.3	5.5
7.9	**20.0**	L3 V	Greater Trochanter Femur-Sole	15.0	13.3	10.0	6.7	5.0
7.5	**19.0**	* L4 V	Symphysis Pubis of Pelvis-Sole	14.3	12.7	9.5	6.3	4.8
6.0	**15.2**	* L5 V	Crotch to Sole	11.4	10.1	7.6	5.1	3.8
5.5	**14.0**	L6 V	Gluteal Furrow to Sole	10.5	9.3	7.0	4.7	3.5
3.5	**9.0**	L7 V	Posterior Knee Crease to Sole	6.8	6.0	4.5	3.0	2.3
4.3	**11.0**	L8 V	Tibia at Knee to Sole	8.3	7.3	5.5	3.7	2.8
0.9	**2.3**	L9 V	Med. Malleolus of Ankle to Sole	1.7	1.5	1.2	0.8	0.6
0.7	**1.9**	L10 V	Lat. Malleolus of Ankle to Sole	1.4	1.3	1.0	0.6	0.5
4.3	**11.0**	L11 T	Breadth of Hips	8.3	7.3	5.5	3.7	2.8
2.0	**5.0**	L12 T	Breadth Mid Thigh	3.8	3.3	2.5	1.7	1.3
1.6	**4.0**	L13 T	Breadth of Knee	3.0	2.7	2.0	1.3	1.0
1.3	**3.2**	L14 T	Breadth of Calf	2.4	2.1	1.6	1.1	0.8
0.9	**2.3**	L15 T	Breadth Ankle	1.7	1.5	1.2	0.8	0.6
2.0	**5.0**	L16 D	Mid Thigh, Anterior-Posterior	3.8	3.3	2.5	1.7	1.3
1.4	**3.5**	L17 D	Knee, Anterior-Posterior	2.6	2.3	1.8	1.2	0.9
1.3	**3.2**	L18 D	Mid Calf, Anterior-Posterior	2.4	2.1	1.6	1.1	0.8
1.0	**2.5**	L19 D	Ankle, Anterior-Posterior	1.9	1.7	1.3	0.8	0.6
3.1	**8.0**	L20 D	Length of Foot	6.0	5.3	4.0	2.7	2.0
1.3	**3.3**	L22 T	Breadth Foot	2.5	2.2	1.7	1.1	0.8
0.8	**2.0**	L23 T	Width of Heel	1.5	1.3	1.0	0.7	0.5
		** These are also listed in Vertical Measurements.*						

**Growth of the Skull,
Superimposed Face Profiles of Infants to One Year Old**

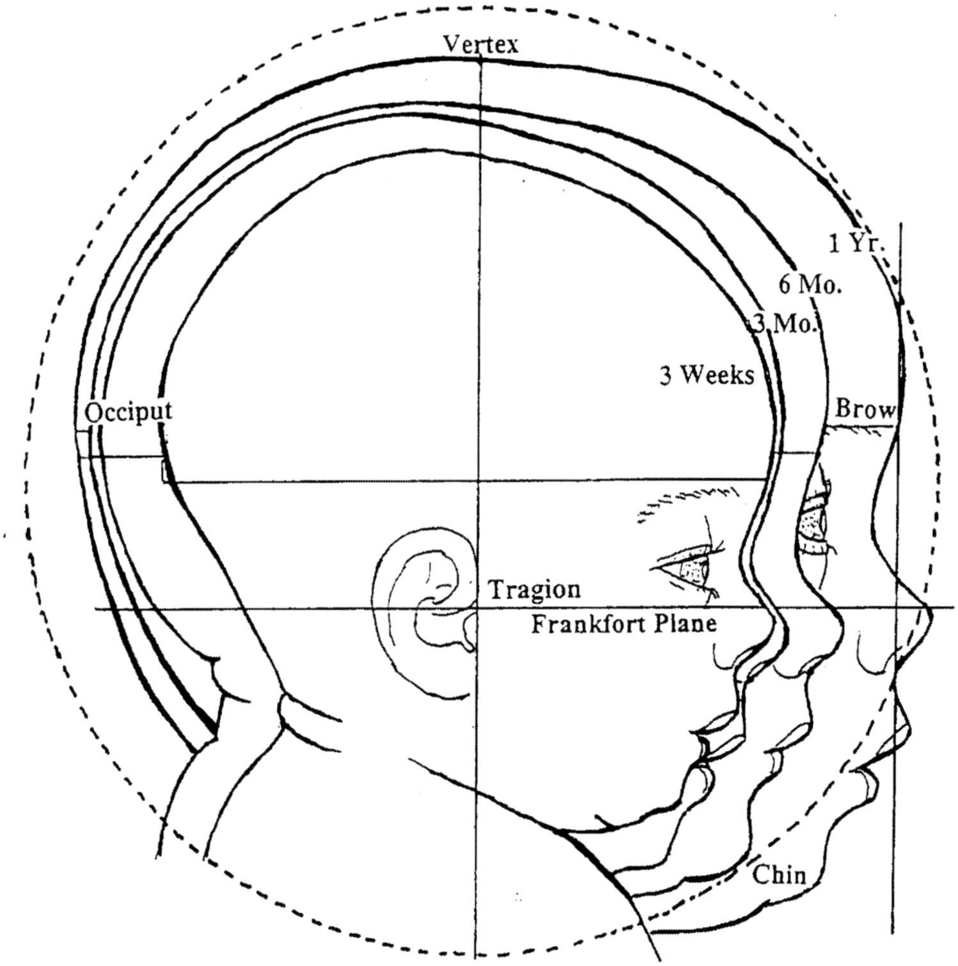

THE GROWTH OF THE SKULL

At birth the head is one-fourth of the total body height. Only one-half of the newborn's head height—chin to vertex—is face height. The head needs to be considered in two parts, the neurocranium and the viscerocranium. The upper half is the cranial vault, also called the neurocranium. At birth, the brain is sixty-three percent of adult size. The neurocranium is also eight times the volume of the facial structures. During early life, the brain grows rapidly so that at the end of the first year volume is eighty-two percent, by three years it is eighty-nine percent, and by the fifth year, over ninety percent of growth occurred. By ten years, ninety-five percent, and by fifteen years ninety-eight percent of brain growth is accomplished.[1]

Since the main function of the face in infancy is for eating and breathing, anatomically it is called the viscerocranium. The mask of the face, the viscerocranium, is from the brow to the chin, and will be referred to as the face. Both the cranium and the face grow rapidly in infancy, but the cranium grows more rapidly up to five years of age.

1. Faulkner, F., p. 525-532.

Infant Face Proportion Divisions

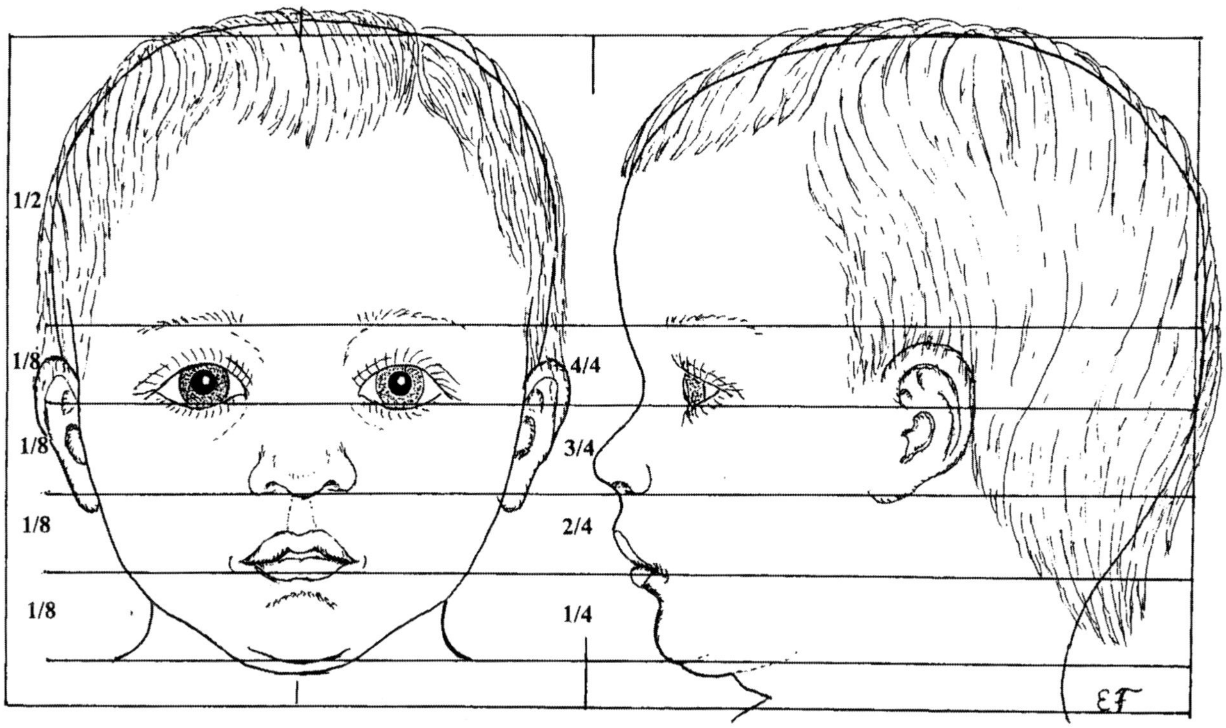

INFANT FACE PROPORTION DIVISIONS

Leonardo da Vinci's recommendation of thirds for adult face proportions is not appropriate for infants and children. The young infant's head proportions are one half from chin to brow and one half from brow to vertex. Another popular system of fractions for the face has then been recommended for artists. The brow to chin is considered in four divisions. The chin to the vermillion border of the lower lip is one-fourth; the lower lip to the subnasal point a second fourth; the nose-lip to the level of the eye corners is a third fourth; and the eye corners to the brow is the last fourth.[1] As the face grows, the upper point needs to be gradually shifted from the brow down to the sellion or saddle of the nose. Both of these points are covered with soft tissue, but the brow is more prominent. The chin and the subnasal point are reliable and easily determined. These relationships gradually change with age, and are appropriate until early teenage. They are convenient ratios to consider when observing a child that is a live model.

Throughout development of the face there is an interesting relationship: the chin to the mouth is two-thirds of the measurement of the chin to the subnasal point. This is approximately the Golden Ratio of 1 to 0.62. It is often recognized as a manifestation of beauty in nature and is a factor in proportional growth.

1. Loomis, Andrew. *Drawing the Head.* Laguna Hills, CA: Walter Foster Publishing Inc. 2001, p. 17.

THE GROWTH OF THE CRANIUM

Consider the head in two parts, the neurocranium or cranium above the eyebrows, and the viscerocranium or face. At birth, the brain volume is about 350 cc or twenty-five percent of adult volume, and the cranium is eight times the volume of the facial structures. During early life, the brain grows rapidly so that at the end of the first year volume is 750 cc; at the second year it is 900 cc, and by the fifth year it is about 1300 cc. The brain growth is over ninety-percent of adult size.[1] By ten years, brain growth is ninety five percent, and by fifteen years ninety-eight percent of adult size.[2] The cranium, including the forehead, grows to accommodate the growth of the brain. The mask of the face, viscerocranium, is from the brow to the chin, and is referred to as the face. The main function of the face in infancy is for eating and breathing.

Since the cranium and face grow at different rates, the logical division is at the base of the cranium. This designation is the Frankfort plane. The brow is a more practical division for artists who consider surface features, but being aware of this division, an artist may better appreciate the changes during growth and development.

The breadth of an infant's cranium also grows rapidly in early months. At birth, the breadth is about 10 cm, or about four inches. Most of the growth is by five years. Infants' heads appear to have relatively more breadth often resembling a square with rounded corners. During growth, the ratio of breadth to height is progressively diminished from about three-fourths to two-thirds. The adult ratio is nearly sixty-two percent, approximating the Golden Mean of 0.618…, the ratio of dynamic symmetry.

Comparisons of Average Head Breadth to Head Height, During Growth

Age	Head Breadth cm	Head Height cm	Percent of Breadth to Height	Ratio	Child's Head Breadth % of Adult's Head
Newborn	10	13	77	3/4+	66
Six Months	12	15.8	76	3/4	77
One Year	12.5	16.5	76	3/4	81
Two Years	13	17.5	74	3/4	84
Three Years	13.5	18	75	3/4	87
Five Years	14	18.5	76	3/4	90
Ten Years	14.3	20	72	3/4-	92
Fifteen Years	15	22	68	2/3+	97
Eighteen Years	15.5	22.5	69	2/3+	100
Adult, 25+ Years	15.5	23	67	2/3	

Male and female heads show similar development until puberty. At age sixteen, the typical female skull retains more of the child's features. It is smaller, lighter, and smoother with smaller jaws, teeth, mastoid processes, and air sinuses. The brows and arches above the eye sockets are also less prominent.[3] From age eighteen to twenty-five, the male face develops contours that are more angular.

1. Breckenridge, M. ,Murphy, M. p. 271.
2. Faulkner, F., p. 527.
3. Lockhart, R.D., Hamilton, G.F., Fyfe, F.W., p. 60.

THREE-MONTH-OLD INFANTS

There is considerable similarity between male and female three-month-old infants. The ranges of twenty-fifth to fiftieth and to seventy-fifth percentiles are listed here for comparison:

Percentile	Twenty-fifth		Fiftieth		Seventy-fifth	
Boys:	**Short**		**Medium**		**Tall**	
Crown to Sole	59 cm,	23 1/2 in.	61 cm,	24 1/4 in.	63 cm,	24 3/4 in.
Weight	5.3 kg,	11 3/4 lb.	6 kg,	13 1/4 lb.	5.5 kg,	14 1/2 lb.
Girls:						
Crown to Sole	58 cm,	22 1/4 in.	59.5 cm,	23 1/2 in.	61 cm,	24 in.
Weight	4.9 kg,	10 1/4 lb.	5.4 kg,	12 lb.	5.9 kg,	13 lb.[1]

Infants nurse frequently and grow rapidly. There is an increase in total length by ten cm, or twenty percent to a comparison of thirty-five percent of adult stature. After arousing from sleep, infants become active, and extremity movements are like crawling. Responsive smiling to an adoring mother's voice develops in one or two months. Spontaneous smiling begins in two to five months. In ten to twelve weeks, babies play pat-a-cake. At three months, coordination has improved and babies are able to roll over and lift their head. Sitting requires some support. Some may start crawling. For a short period after awakening, babies act cute, smile and coo. When hunger develops, they become more active and begin to cry. This is indicates the time to nurse.

Body Measurements Compared To Height

	Feature Measure in cm	Feature Compared to Stature	Approx. Ratio of Feature	Compared to Median Adult
Stature: Crown to sole	61			33%
Head Height: Crown to chin	15	25%	1/4	71%
Sitting Height: Crown to rump	42	69%	3/4–	47%
Trunk Measure: Sternal notch to symphysis pubis	22	36%	1/3+	44%
Leg Length: Greater trochanter to heel	23	38%	1/3+	26%
Thigh Length: Greater trochanter to tibia at knee	11	17%	1/6	24%
Tibia at Knee to Sole:	12	20%	1/5	27%
Arm Length: Shoulder to dependent fingertip	23	38%	1/3+	31%
Hand Length:	7	11%	1/9	38%
Foot Length:	9	15%	1/6–	35%

The head circumference nearly equals chest circumference. The chest is barrel shaped until after age five. The pelvis is relatively small at birth, but grows rapidly in the first three months. When a baby moves its arms and kicks its legs, it stretches, lengthens, and strengthens muscles in preparation for crawling.[2]

1. Nelson, W.E., p. 38.
2. Breckenridge, M.E., Murphy, M.N., p. 272.

Three-month-old infant, anterior and lateral head

all measurements in centimeters

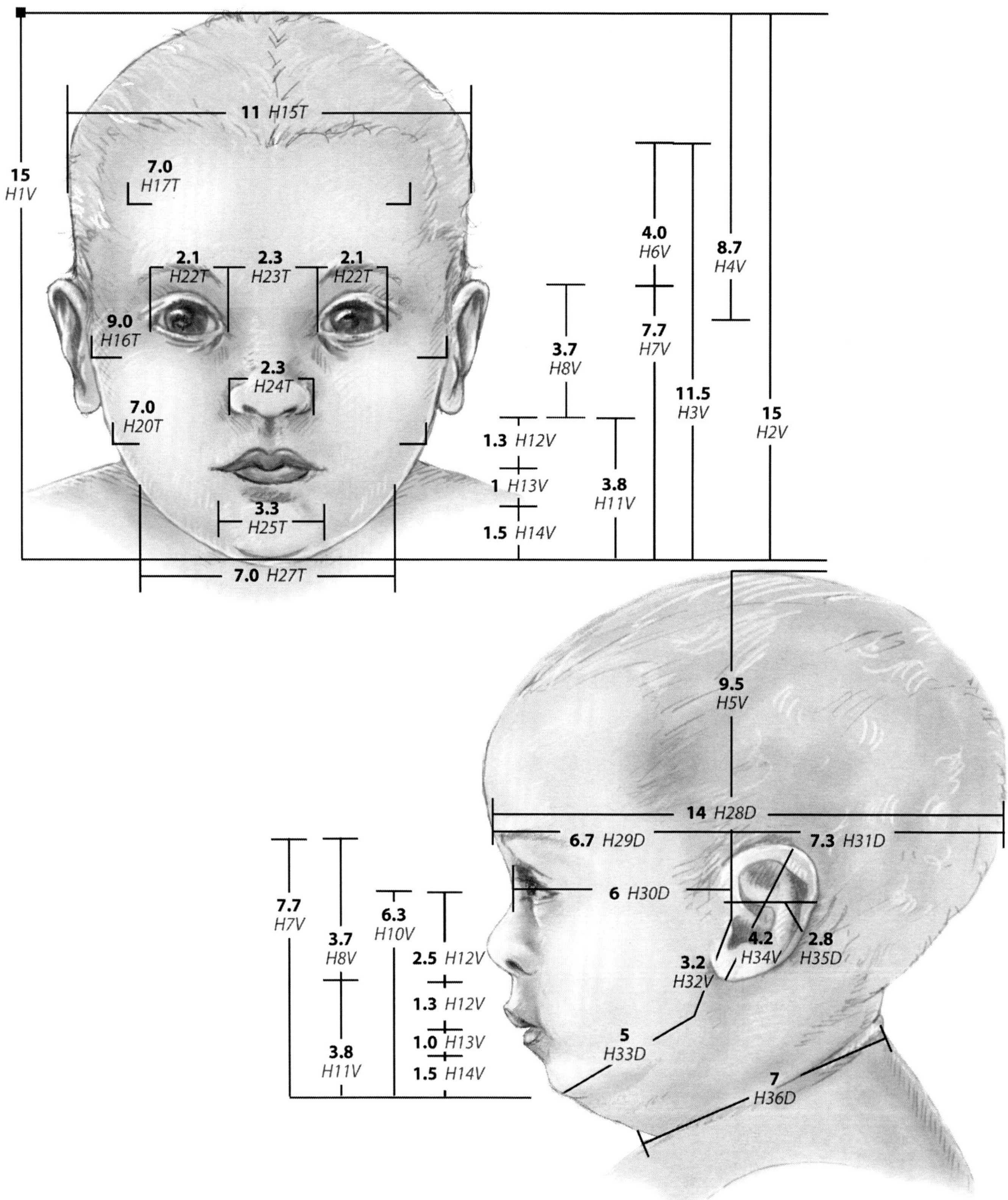

Three-month-old infant, anterior and lateral figure

all measurements in centimeters

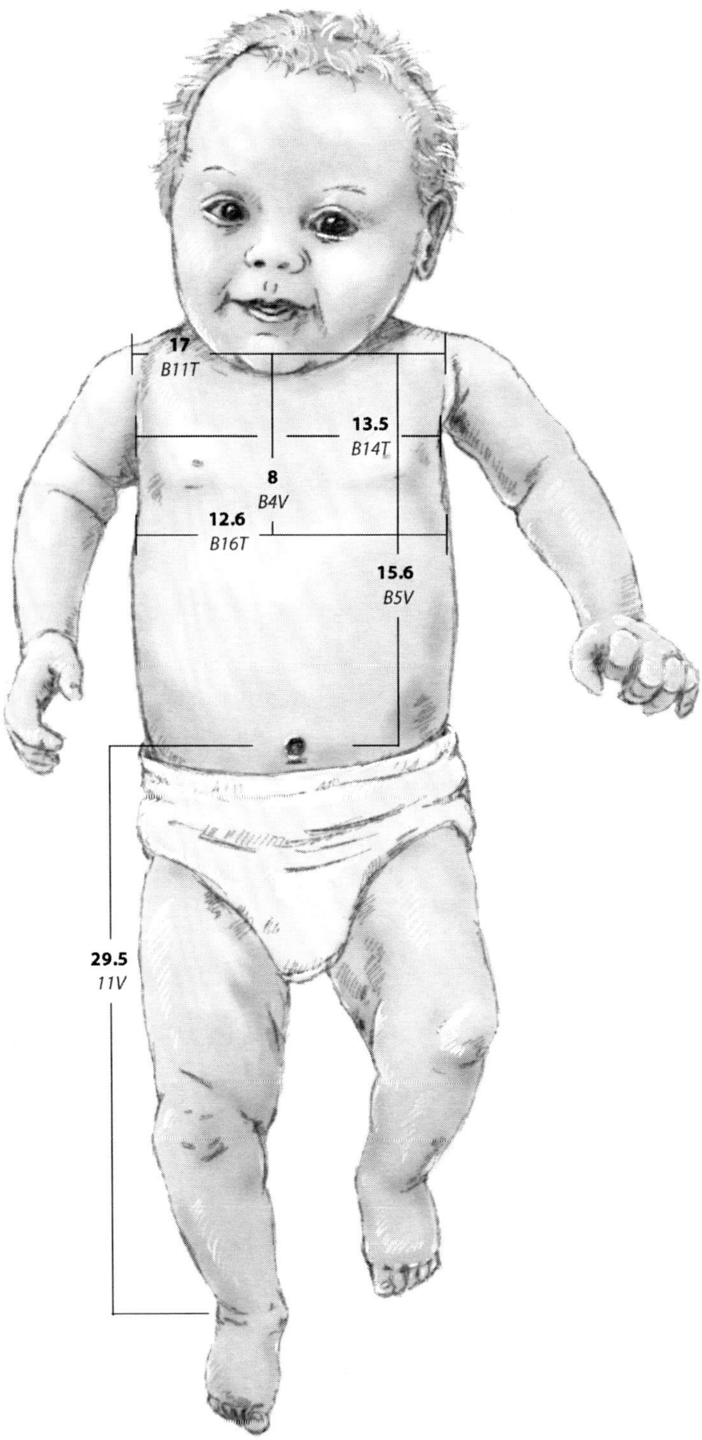

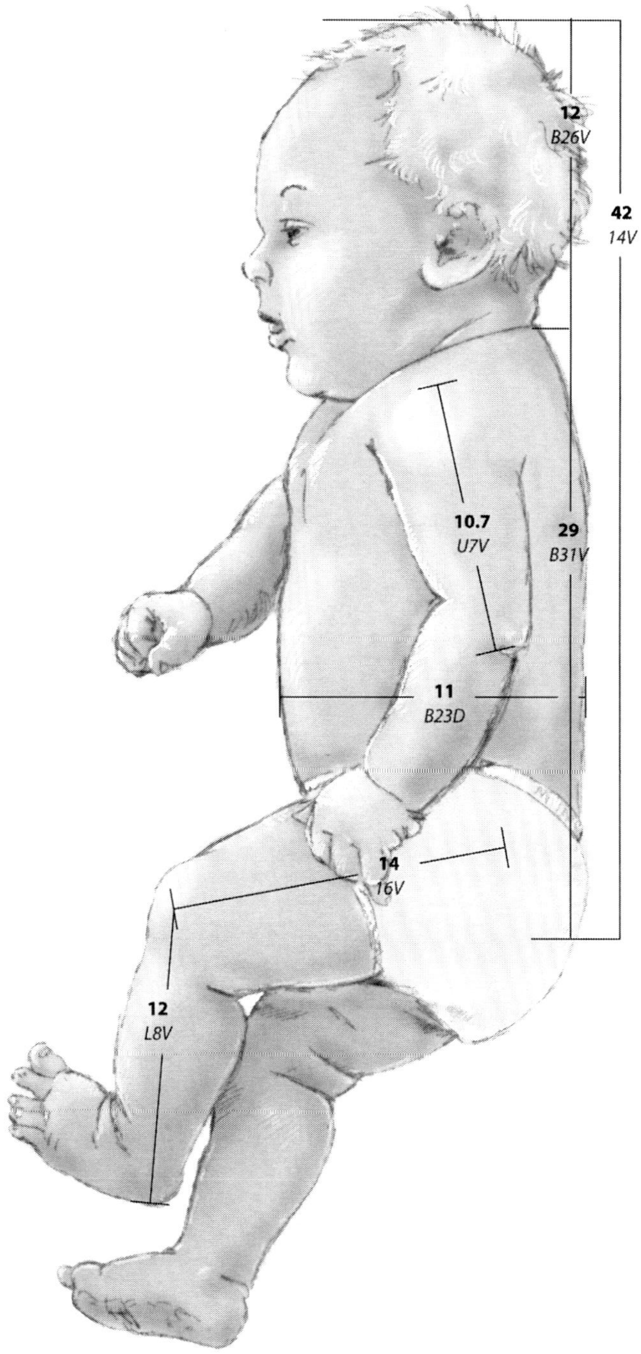

Life size			THREE-MONTH-OLD INFANT	Fractions of life size				
				3/4	2/3	1/2	1/3	1/4
Inches	cm	Code		cm	cm	cm	cm	cm
			VERTICAL MEASUREMENTS					
24.0	**61.0**	1 V	Stature, Height, Vertex to Sole	45.8	40.7	30.5	20.3	15.3
14.8	**37.5**	2 V	Vertex to Symphysis Pubis	28.1	25.0	18.8	12.5	9.4
9.3	**23.5**	* 3 V	Symphysis Pubis of Pelvis-Sole	17.6	15.7	11.8	7.8	5.9
17.7	**45.0**	4 V	Sternal Notch, Neck Pit, to Sole	33.8	30.0	22.5	15.0	11.3
5.9	**15.0**	5 V	Vertex to Acromion Process	11.3	10.0	7.5	5.0	3.8
17.9	**45.5**	6 V	Acromion Process to Sole	34.1	30.3	22.8	15.2	11.4
10.2	**26.0**	* 7 V	Ant.Sup.Spine of Ilium to Sole	19.5	17.3	13.0	8.7	6.5
7.7	**19.5**	8 V	Acrom. proc. to Ant. Sup. Spine	14.6	13.0	9.8	6.5	4.9
5.1	**13.0**	9 V	Ant.Sup.Spine of Ilium to Knee	9.8	8.7	6.5	4.3	3.3
5.1	**13.0**	10 V	Knee at Patella to Sole	9.8	8.7	6.5	4.3	3.3
11.6	**29.5**	11 V	Umbilicus to Sole	22.1	19.7	14.8	9.8	7.4
7.9	**20.0**	* 12 V	Crotch to Sole	15.0	13.3	10.0	6.7	5.0
16.5	**42.0**	14 V	Sitting Height - Crown Rump	31.5	28.0	21.0	14.0	10.5
10.2	**26.0**	15 V	Buttock, Straight Leg to Sole	19.5	17.3	13.0	8.7	6.5
5.5	**14.0**	16 V	Buttock to Flexed Knee	10.5	9.3	7.0	4.7	3.5
5.5	**14.0**	17 V	Flexed Knee to Sole	10.5	9.3	7.0	4.7	3.5
		** These are also listed in Lower Extremity Measurements.*						
			VERTICAL HEAD MEASUREMENTS					
5.9	**15.0**	H1 V	Vertex to Sternal Notch	11.3	10.0	7.5	5.0	3.8
5.9	**15.0**	H2 V	Vertex to Chin, Head Height	11.3	10.0	7.5	5.0	3.8
4.5	**11.5**	H3 V	Hairline to Chin, Face Height	8.6	7.7	5.8	3.8	2.9
3.4	**8.7**	H4 V	Vertex to Lateral Corner of Eye	6.5	5.8	4.4	2.9	2.2
3.7	**9.5**	H5 V	Vertex to Tragion of Ear	7.1	6.3	4.8	3.2	2.4
1.6	**4.0**	H6 V	Hairline to Brow, Glabella	3.0	2.7	2.0	1.3	1.0
3.0	**7.7**	H7 V	Brow, Glabella, to Chin, Menton	5.8	5.1	3.9	2.6	1.9
1.5	**3.7**	H8 V	Brow to Subnasal Point	2.8	2.5	1.9	1.2	0.9
1.0	**2.5**	H9 V	Sellion to Subnasal, Nose Length	1.9	1.7	1.3	0.8	0.6
2.5	**6.3**	H10 V	Sellion, Nose Saddle, to Chin	4.7	4.2	3.2	2.1	1.6
1.5	**3.8**	H11 V	Subnasal Point to Chin	2.9	2.5	1.9	1.3	1.0
0.5	**1.2**	H12 V	Subnasal to Lip Aperature	1.0	0.9	0.7	0.4	0.3
0.4	**1.0**	H13 V	Lip Aperat. to Sublabial Crease	0.8	0.7	0.5	0.3	0.3
0.6	**1.5**	H14 V	Sublabial Crease to Chin	1.1	1.0	0.8	0.5	0.4
			TRANSVERSE HEAD MEASUREMENTS					
4.3	**11.0**	H15 T	Bi-Parietal, Head Breadth	8.3	7.3	5.5	3.7	2.8
3.5	**9.0**	H16 T	Bi-Zygomatic, Cheekbones	6.8	6.0	4.5	3.0	2.3
2.8	**7.0**	H17 T	Bi-Fronto Temporal, Forehead	5.3	4.7	3.5	2.3	1.8
3.0	**7.5**	H18 T	Inter-Brow Ridge	5.6	5.0	3.8	2.5	1.9

Life size			THREE-MONTH-OLD INFANT	Fractions of life size				
				3/4	2/3	1/2	1/3	1/4
Inches	cm	Code		cm	cm	cm	cm	cm
			TRANSVERSE HEAD MEASUREMENTS					
3.9	**10.0**	H19 T	Inter-Tragion of Ears	7.5	6.7	5.0	3.3	2.5
2.8	**7.0**	H20 T	Inter-Angle of Mandible	5.3	4.7	3.5	2.3	1.8
2.6	**6.7**	H21 T	Inter-Lateral Corners of Eyes	5.0	4.5	3.4	2.2	1.7
0.8	**2.1**	H22 T	Width of Eye	1.7	1.4	1.1	0.7	0.5
0.9	**2.3**	H23 T	Inter-Inner Corner of Eyes	1.7	1.5	1.2	0.8	0.6
0.9	**2.3**	H24 T	Breadth of Nose	1.7	1.5	1.2	0.8	0.6
1.3	**3.3**	H25 T	Breadth of Mouth	2.5	2.2	1.7	1.1	0.8
0.9	**2.3**	H26 T	Breadth of Chin	1.7	1.5	1.2	0.8	0.6
2.3	**5.8**	H27 T	Breadth of Neck	5.3	3.9	2.9	1.9	1.5
			ANTERIOR-POSTERIOR HEAD MEASUREMENTS					
5.5	**14.0**	H28 D	Brow to Occiput, Posterior Head	10.5	9.3	7.0	4.7	3.5
2.6	**6.7**	H29 D	Brow to Tragus of Ear	5.0	4.5	3.4	2.2	1.7
2.4	**6.0**	H30 D	Sellion, Nose Saddle, to Tragus	4.5	4.0	3.0	2.0	1.5
2.9	**7.3**	H31 D	Tragus to Occiput, Back of Head	5.5	4.9	3.7	2.4	1.8
1.3	**3.2**	H32 V	Tragus to Angle Mandible	2.4	2.1	1.6	1.1	0.8
2.0	**5.0**	H33 D	Angle of Mandible to Chin	3.8	3.3	2.5	1.7	1.3
1.7	**4.2**	H34 V	Length of Ear, Vertical	3.2	2.8	2.1	1.4	1.1
1.1	**2.8**	H35 D	Width of Ear, Anterior-Posterior	2.1	1.9	1.4	0.9	0.7
2.8	**7.0**	H36 D	Anterior to Posterior of Neck	5.3	4.7	3.5	2.3	1.8
			VERTICAL BODY MEASUREMENTS					
0.4	**1.0**	B1 V	Chin to Sternal Notch, Neck Pit	0.8	0.7	0.5	0.3	0.3
1.0	**2.5**	B2 V	Sternal Notch to Axilla, Arm Pit	1.9	1.7	1.3	0.8	0.6
2.0	**5.0**	B3 V	Sternal Notch to Nipples	3.8	3.3	2.5	1.7	1.3
3.1	**8.0**	B4 V	Sternal Notch to Xyphoid Proc.	6.0	5.3	4.0	2.7	2.0
6.1	**15.5**	B5 V	Sternal Notch to Umbilicus	11.6	10.3	7.8	5.2	3.9
3.0	**7.5**	B6 V	Xyphoid Process to Umbilicus	5.6	5.0	3.8	2.5	1.9
7.1	**18.0**	B7 V	Sternal Notch to Crest Ilium	13.5	12.0	9.0	6.0	4.5
7.7	**19.5**	B8 V	Sternal Notch to Ant.S.Sp.Ilium	14.6	13.0	9.8	6.5	4.9
8.7	**22.0**	B9 V	Sternal Notch to Pubis	16.5	14.7	11.0	7.3	5.5
2.6	**6.5**	B10 V	Umbilicus to Pubis of Pelvis	4.9	4.3	3.3	2.2	1.6
			TRANSVERSE BODY MEASUREMENTS					
6.7	**17.0**	B11 T	Shoulder Breadth	12.8	11.3	8.5	5.7	4.3
2.4	**6.0**	B12 T	Clavicle to Acromion Process	4.5	4.0	3.0	2.0	1.5
5.7	**14.5**	B13 T	Bi-acromial Process Breadth	10.9	9.7	7.3	4.8	3.6
5.3	**13.5**	B14 T	Inter-Anterior Axillary Fold	10.1	9.0	6.8	4.5	3.4
3.5	**9.0**	B15 T	Inter-Nipples	6.8	6.0	4.5	3.0	2.3

Life size			THREE-MONTH-OLD INFANT	Fractions of life size				
				3/4	2/3	1/2	1/3	1/4
Inches	cm	Code		cm	cm	cm	cm	cm
			TRANSVERSE BODY MEASUREMENTS					
5.0	**12.6**	B16 T	Breadth of Thorax	9.5	8.4	6.3	4.2	3.2
5.1	**13.0**	B17 T	Inter-Costal. Rib Margin Breadth	9.8	8.7	6.5	4.3	3.3
4.3	**11.0**	B18 T	Inter-Crest of Ilium Breadth	8.3	7.3	5.5	3.7	2.8
3.3	**8.5**	B19 T	Inter-Anterior Superior Spine	6.4	5.7	4.3	2.8	2.1
4.3	**11.0**	B20 T	Inter-Greater Trochanter Femur	8.3	7.3	5.5	3.7	2.8
			ANTERIOR-POSTERIOR BODY MEASUREMENTS					
3.5	**9.0**	B21 D	Manubrium Dorsal Spine	6.8	6.0	4.5	3.0	2.3
3.9	**10.0**	B22 D	Xyphoid Process to Dorsum	7.5	6.7	5.0	3.3	2.5
4.3	**11.0**	B23 D	Umbilicus to Dorsum	8.3	7.3	5.5	3.7	2.8
3.1	**8.0**	B24 D	Pubis of Pelvis to Buttock	6.0	5.3	4.0	2.7	2.0
2.8	**7.0**	B25 D	Thigh to Gluteal Furrow	5.3	4.7	3.5	2.3	1.8
			VERTICAL BACK MEASUREMENTS					
4.7	**12.0**	B26 V	Vertex to 7th Cervical Spine	9.0	8.0	6.0	4.0	3.0
			7th Cervical Vertebra Spine to:					
2.4	**6.0**	B27 V	To Posterior Axillary Fold, Armpit	4.5	4.0	3.0	2.0	1.5
2.4	**6.0**	B28 V	To Inferior Tip of Scapula	4.5	4.0	3.0	2.0	1.5
6.7	**17.0**	B29 V	To Posterior Superior Spine Ilium	12.8	11.3	8.5	5.7	4.3
7.9	**20.0**	B30 V	To Midline Upper Buttock Crease	15.0	13.3	10.0	6.7	5.0
11.4	**29.0**	B31 V	To Gluteal Furrow	21.8	19.3	14.5	9.7	7.3
			TRANSVERSE BACK MEASUREMENTS					
5.7	**14.6**	B33 T	Inter-Posterior Axillary Folds	11.0	9.7	7.3	4.9	3.7
2.4	**6.0**	B34 T	Inter-Scapulae Medial Borders	4.5	4.0	3.0	2.0	1.5
1.6	**4.0**	B35 T	Inter-Post. Sup. Spines Sacrum	3.0	2.7	2.0	1.3	1.0
			UPPER EXTREMITY MEASUREMENTS					
21.7	**55.0**	U1 V	Span of Arms	41.3	36.7	27.5	18.3	13.8
2.6	**6.5**	U2 V	Sternum to Acromion Process	4.9	4.3	3.3	2.2	1.6
9.1	**23.0**	U3 V	Acromion Process to Fingertip	17.3	15.3	11.5	7.7	5.8
4.2	**10.7**	U7 V	Shoulder to Flexed Elbow	8.0	7.1	5.4	3.6	2.7
3.5	**9.0**	U8 V	Acromion to Lat. Condyle Elbow	6.8	6.0	4.5	3.0	2.3
6.3	**16.0**	U9 D	Flexed Elbow to Fingertip	12.0	10.7	8.0	5.3	4.0
3.5	**8.8**	U10 D	Flexed Elbow Pivot to Wrist Pivot	10.5	5.9	4.4	2.9	2.2

Life size			THREE-MONTH-OLD INFANT	Fractions of life size				
				3/4	2/3	1/2	1/3	1/4
Inches	cm	Code		cm	cm	cm	cm	cm
			UPPER EXTREMITY MEASUREMENTS					
2.8	**7.0**	U11 V	Wrist to Fingertip	5.3	4.7	3.5	2.3	1.8
2.2	**5.5**	U12 V	Wrist to Thumb Tip	4.1	3.7	2.8	1.8	1.4
1.5	**3.8**	U13 V	Length of Palm	2.9	2.5	1.9	1.3	1.0
1.3	**3.3**	U14 V	Length 3rd Finger	2.4	2.2	1.7	1.1	0.8
1.7	**4.3**	U15 V	Thumb Crotch to 3rd Finger	3.2	2.9	2.2	1.4	1.1
1.2	**3.0**	U16 V	Thumb Crotch to Thumb Tip	2.3	2.0	1.5	1.0	0.8
1.6	**4.0**	U17 D	Anterior-Post. Arm at Biceps	3.0	2.7	2.0	1.3	1.0
1.5	**3.8**	U18 D	Anterior-Posterior at Forearm	2.9	2.5	1.9	1.3	1.0
1.1	**2.7**	U19 D	Depth of Wrist	2.0	1.8	1.4	0.9	0.7
1.2	**3.0**	U20 T	Breadth of Wrist	2.3	2.0	1.5	1.0	0.8
1.8	**4.5**	U21 T	Breadth of Hand	3.4	3.0	2.3	1.5	1.1
1.5	**3.7**	U22 T	Breadth of Palm	2.8	2.5	1.9	1.2	0.9
			LOWER EXTREMITY MEASUREMENTS					
11.0	**28.0**	L1 V	Crest of Ilium of Pelvis to Sole	21.0	18.7	14.0	9.3	7.0
10.5	**26.6**	* L2 V	Ant.Sup.Spine of Ilium to Sole	20.0	17.7	13.3	8.9	6.7
9.1	**23.0**	L3 V	Greater Trochanter Femur-Sole	17.3	15.3	11.5	7.7	5.8
9.3	**23.5**	* L4 V	Symphysis Pubis of Pelvis-Sole	17.6	15.7	11.8	7.8	5.9
7.7	**19.5**	* L5 V	Crotch to Sole	14.6	13.0	9.8	6.5	4.9
7.5	**19.0**	L6 V	Gluteal Furrow to Sole	14.3	12.7	9.5	6.3	4.8
4.7	**12.0**	L7 V	Posterior Knee Crease to Sole	9.0	8.0	6.0	4.0	3.0
4.7	**12.0**	L8 V	Tibia at Knee to Sole	9.0	8.0	6.0	4.0	3.0
1.0	**2.5**	L9 V	Med. Malleolus of Ankle to Sole	1.9	1.7	1.3	0.8	0.6
0.8	**2.0**	L10 V	Lat. Malleolus of Ankle to Sole	1.5	1.3	1.0	0.7	0.5
5.1	**13.0**	L11 T	Breadth of Hips	9.8	8.7	6.5	4.3	3.3
2.4	**6.0**	L12 T	Breadth Mid Thigh	4.5	4.0	3.0	2.0	1.5
2.0	**5.0**	L13 T	Breadth of Knee	3.8	3.3	2.5	1.7	1.3
1.7	**4.3**	L14 T	Breadth of Calf	3.2	2.9	2.2	1.4	1.1
1.2	**3.0**	L15 T	Breadth Ankle	2.3	2.0	1.5	1.0	0.8
2.4	**6.0**	L16 D	Mid Thigh, Anterior-Posterior	4.5	4.0	3.0	2.0	1.5
1.6	**4.0**	L17 D	Knee, Anterior- Posterior	3.8	2.7	2	1.3	1
1.7	**4.3**	L18 D	Mid Calf, Anterior-Posterior	3.2	2.9	2.2	1.4	1.1
1.4	**3.5**	L19 D	Ankle, Anterior-Posterior	2.6	2.3	1.8	1.2	0.9
3.5	**9.0**	L20 D	Length of Foot	6.8	6.0	4.5	3.0	2.3
1.5	**3.9**	L22 T	Breadth Foot	2.9	2.6	2.0	1.3	1.0
0.9	**2.3**	L23 T	Width of Heel	1.7	1.5	1.2	0.8	0.6
		These are also listed in Vertical Measurements.						

Infant Eye and Nose Details

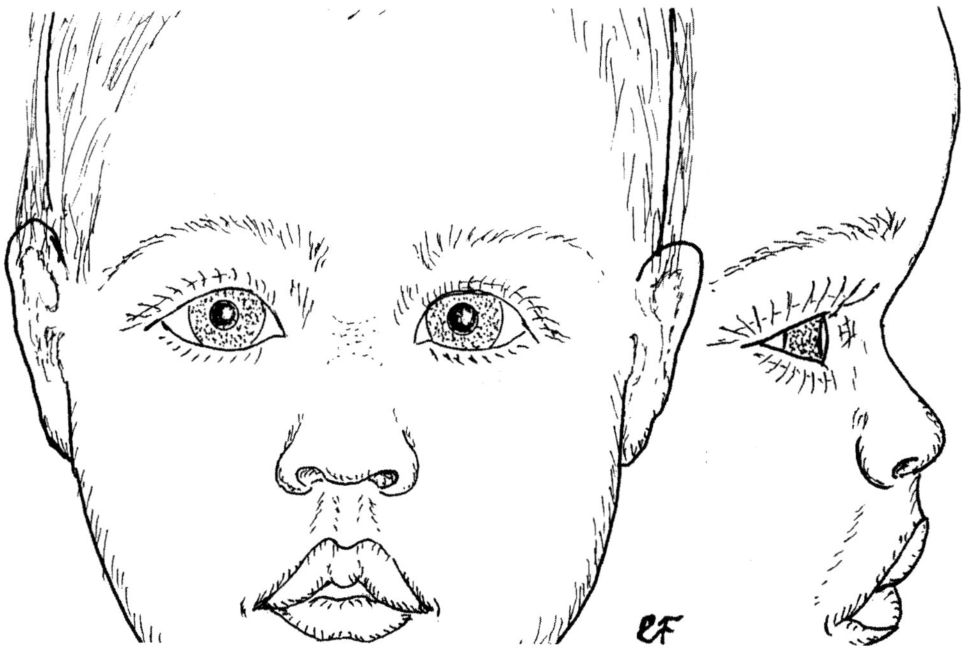

INFANT EYE DETAILS

Infants and young children appear to have larger eyes that are farther apart, because they are relatively larger than adults eyes. The openings of the eyelids are small so that only a small part of the sclera, or white of the eye, is exposed. Newborn infant's eyeballs are 16.5 mm anterior-posterior measurement, or sixty-seven percent the size of adult eyes. Growth of the eyes is most rapid during the first eighteen months with an average increase to 20 mm. By age seven years, eyes have grown to nearly adult size. They then more slowly approach adult size of 24.5 mm by age 15 to 18 years.[1]

The portrayal of infants and children needs to indicate the relatively large and expressive eyes. At first, the color of the iris is dark blue, obscuring the circle of the pupil. The permanent color of the iris develops gradually over several months.

AN INFANT'S NOSE

In infancy, to age six months, the length of the nose, sellion to subnasal point, is about one-fourth the distance of the brow to the chin. The contour is concave, the tip is slightly bulbous, and the ala are relatively wide. They have cute little pug noses. The upper lip is angled forward and the lip to the nose angle is about one hundred thirty-five to one hundred forty degrees. This makes the nostrils, on frontal view, appear fully exposed and directed forward.

By six months, the contour continues concave but the sellion to subnasal point has increased and the lip to nose angle has decreased. After one year of age, the concave anterior border is less concave and the bulbous tip has remained the same size, so the expected form of the nose is progressing.

In children, the nose components grow with the needs of respiration. Most children continue to have a gentle concave nose for several years, gradually progressing to a straight leading border. In infants the ala of the nose are less developed than in older children.

1. Faulkner, F., p. 535.

SIX-MONTH-OLD INFANTS

Percentile	Twenty-fifth		Fiftieth		Seventy-fifth	
Boys:	**Short**		**Medium**		**Tall**	
Crown to Sole	66 cm,	26 in.	69 cm,	27 1/2 in.	70 cm,	27 1/2 in.
Weight	7.2 kg,	15 1/4 lb.	7.9 kg,	17 1/4 lb.	8.5 kg,	18 3/4 lb.
Girls:						
Crown to Sole	64 cm,	25 1/4 in.	66 cm,	26 in.	68 cm,	26 1/4 in.
Weight	6.6 kg,	14 1/2 lb.	7.2 kg,	16 lb.	7.8 kg,	17 1/4 lb.[1]

Both sexes are combined in the practical list of body proportions in measurement tables. Male infants, on the average, are slightly longer and slightly heavier than female infants. At six months, infants may spend more time at plays and increase vocalization. They may sit, and have improved balance but are better held or propped up with a pillow. Between six and ten months they begin to respond and enjoy playing peek-a-boo. By eight months they will assume a sitting position spontaneously. They have learned to crawl very well, and like to bang on toys to make noise. They often stretch their legs, kick, and may attempt to pull themselves to a standing position while holding on to a support. In another month, they may attempt walking while supported by furniture. The brain has been growing and is about seventy six-percent of adult size. The neurocranium has increased; the skull is wider. The face has also grown wider and is more round.[2]

Body Measurements Compared To Height

	Feature Measure in cm	Feature Compared to Stature	Approx. Ratio of Feature	Compared to Median Adult
Stature: Crown to sole	67			36%
Head Height: Crown to chin	15	22%	1/5	70%
Sitting Height: Crown to rump	42	63%	2/3	45%
Trunk Measure: Sternal notch to symphysis pubis	24	36%	1/3+	46%
Leg Length: Greater trochanter to sole	25	37%	1/3+	28%
Thigh Length: Greater trochanter to tibia at knee	11	16%	1/6	25%
Tibia at Knee to Sole:	14	21%	1/5	31%
Arm Length: Shoulder to dependent fingertip	26	39%	2/5	36%
Hand length:	8	12%	1/8	44%
Foot Length:	9.5	14%	1/7	40%

1. Nelson, W.E. 38.
2. Falkner, F. Tanner, J. M 532.

Six-month-old infant, anterior and lateral head

all measurements in centimeters

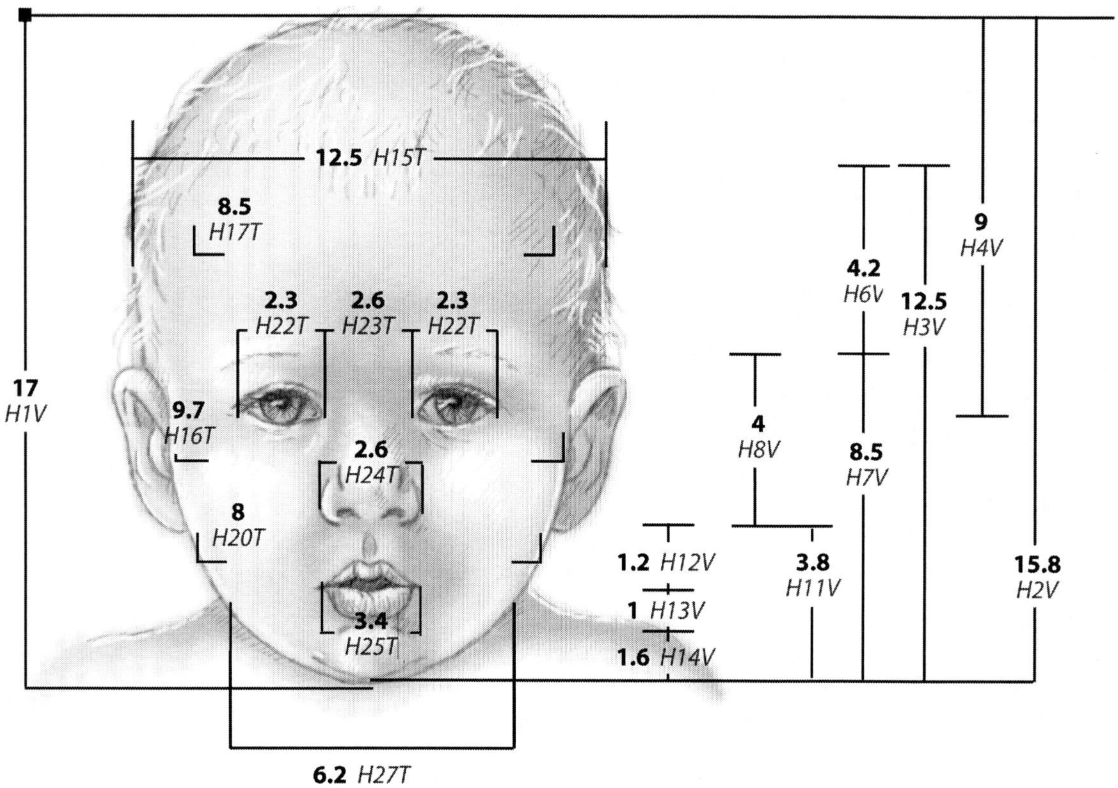

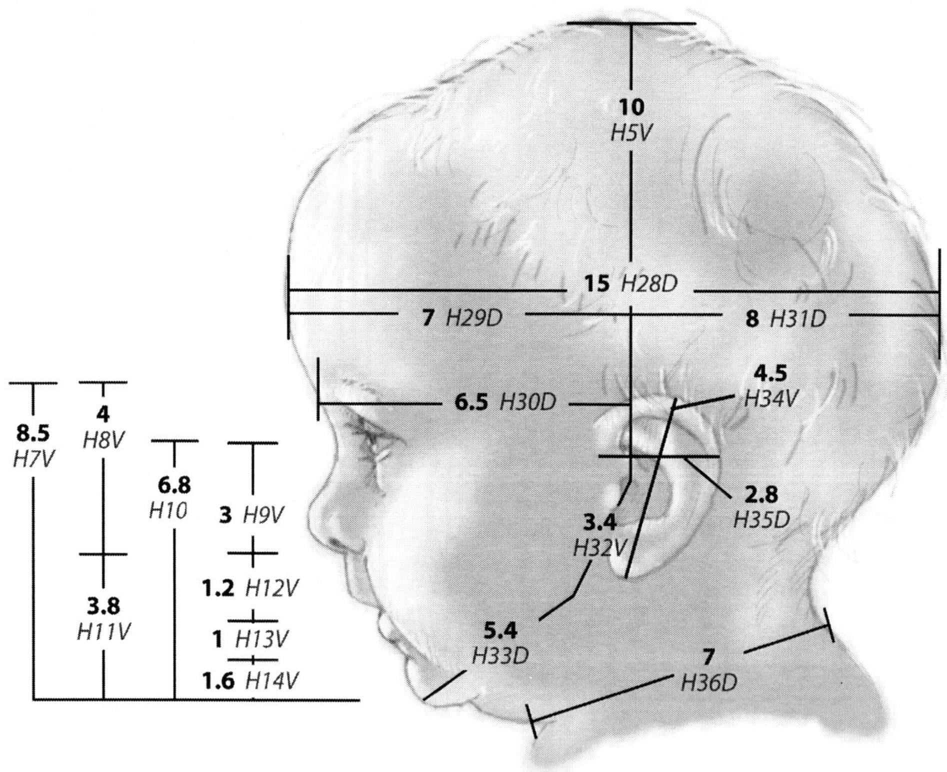

Six-month-old infant, anterior and lateral figure

all measurements in centimeters

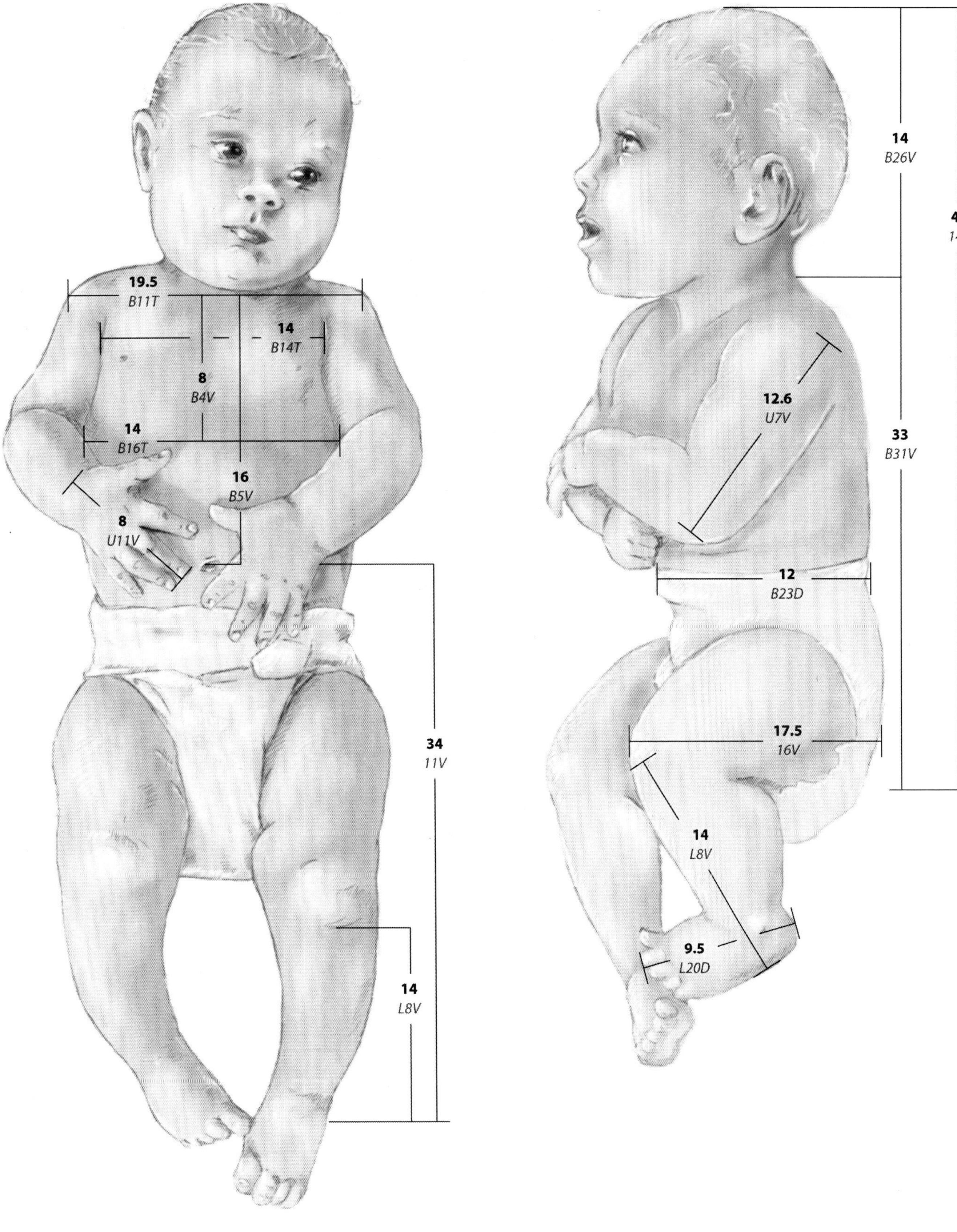

Life size			SIX-MONTH-OLD INFANT	Fractions of life size				
				3/4	2/3	1/2	1/3	1/4
Inches	cm	Code		cm	cm	cm	cm	cm
			VERTICAL MEASUREMENTS					
26.4	67.0	1 V	Stature, Height, Vertex to Sole	50.3	44.7	33.5	22.3	16.8
16.1	41.0	2 V	Vertex to Symphysis Pubis	30.8	27.3	20.5	13.7	10.3
10.2	26.0	* 3 V	Symphysis Pubis of Pelvis-Sole	19.5	17.3	13.0	8.7	6.5
19.7	50.0	4 V	Sternal Notch, Neck Pit, to Sole	37.5	33.3	25.0	16.7	12.5
6.3	16.0	5 V	Vertex to Acromion Process	12.0	10.7	8.0	5.3	4.0
19.7	50.0	6 V	Acromion Process to Sole	37.5	33.3	25.0	16.7	12.5
11.8	30.0	* 7 V	Ant.Sup.Spine of Ilium to Sole	22.5	20.0	15.0	10.0	7.5
7.9	20.0	8 V	Acrom. proc. to Ant. Sup. Spine	15.0	13.3	10.0	6.7	5.0
5.9	15.0	9 V	Ant.Sup.Spine of Ilium to Knee	11.3	10.0	7.5	5.0	3.8
6.1	15.5	10 V	Knee at Patella to Sole	11.6	10.3	7.8	5.2	3.9
13.4	34.0	11 V	Umbilicus to Sole	25.5	22.7	17.0	11.3	8.5
8.7	22.0	* 12 V	Crotch to Sole	16.5	14.7	11.0	7.3	5.5
		13 V	Kneeling Height (not yet kneeling)					
16.5	42.0	14 V	Sitting Height - Crown Rump	31.5	28.0	21.0	14.0	10.5
11.4	29.0	15 V	Buttock, Straight Leg to Sole	21.8	19.3	14.5	9.7	7.3
6.9	17.5	16 V	Buttock to Flexed Knee	13.1	11.7	8.8	5.8	4.4
6.9	17.5	17 V	Flexed Knee to Sole	13.1	11.7	8.8	5.8	4.4
		* *These are also listed in Lower Extremity Measurements.*						
			VERTICAL HEAD MEASUREMENTS					
6.7	17.0	H1 V	Vertex to Sternal Notch	12.8	11.3	8.5	5.7	4.3
6.2	15.8	H2 V	Vertex to Chin, Head Height	11.9	10.5	7.9	5.3	4.0
4.9	12.5	H3 V	Hairline to Chin, Face Height	9.4	8.3	6.3	4.2	3.1
3.5	9.0	H4 V	Vertex to Lateral Corner of Eye	6.8	6.0	4.5	3.0	2.3
3.9	10.0	H5 V	Vertex to Tragion of Ear	7.5	6.7	5.0	3.3	2.5
1.7	4.2	H6 V	Hairline to Brow, Glabella	3.2	2.8	2.1	1.4	1.1
3.3	8.5	H7 V	Brow, Glabella, to Chin, Menton	6.4	5.7	4.3	2.8	2.1
1.6	4.0	H8 V	Brow to Subnasal Point	3.0	2.7	2.0	1.3	1.0
1.2	3.0	H9 V	Sellion to Subnasal, Nose Length	2.3	2.0	1.5	1.0	0.8
2.7	6.8	H10 V	Sellion, Nose Saddle to Chin	5.1	4.5	3.4	2.3	1.7
1.5	3.8	H11 V	Subnasal Point to Chin	2.9	2.5	1.9	1.3	1.0
0.5	1.2	H12 V	Subnasal to Lip Aperature	0.9	0.8	0.6	0.4	0.3
0.4	1.0	H13 V	Lip Aperat. to Sublabial Crease	0.8	0.7	0.5	0.3	0.3
0.6	1.6	H14 V	Sublabial Crease to Chin	1.2	1.1	0.8	0.5	0.4
			TRANSVERSE HEAD MEASUREMENTS					
4.9	12.5	H15 T	Bi-Parietal, Head Breadth	9.4	8.3	6.3	4.2	3.1
3.8	9.7	H16 T	Bi-Zygomatic, Cheekbones	7.3	6.5	4.9	3.2	2.4
3.3	8.5	H17 T	Bi-Fronto Temporal, Forehead	6.4	5.7	4.3	2.8	2.1

Life size			SIX-MONTH-OLD INFANT	Fractions of life size				
				3/4	2/3	1/2	1/3	1/4
Inches	cm	Code		cm	cm	cm	cm	cm
			TRANSVERSE HEAD MEASUREMENTS					
3.1	8.0	H18 T	Inter-Brow Ridge	6.0	5.3	4.0	2.7	2.0
4.1	10.5	H19 T	Inter-Tragion of Ears	7.9	7.0	5.3	3.5	2.6
3.1	8.0	H20 T	Inter-Angle of Mandible	6.0	5.3	4.0	2.7	2.0
2.8	7.2	H21 T	Inter-Lateral Corners of Eyes	5.4	4.8	3.6	2.4	1.8
0.9	2.3	H22 T	Width of Eye	1.7	1.5	1.2	0.8	0.6
1.0	2.6	H23 T	Inter-Inner Corner of Eyes	2.0	1.7	1.3	0.9	0.7
1.0	2.6	H24 T	Breadth of Nose	2.0	1.7	1.3	0.9	0.7
1.3	3.4	H25 T	Breadth of Mouth	2.6	2.3	1.7	1.1	0.9
0.9	2.4	H26 T	Breadth of Chin	1.8	1.6	1.2	0.8	0.6
2.4	6.2	H27 T	Breadth of Neck	4.7	4.1	3.1	2.1	1.6
			ANTERIOR-POSTERIOR HEAD MEASUREMENTS					
5.9	15.0	H28 D	Brow to Occiput, Posterior Head	11.3	10.0	7.5	5.0	3.8
2.8	7.0	H29 D	Brow to Tragus of Ear	5.3	4.7	3.5	2.3	1.8
2.6	6.5	H30 D	Sellion, Nose Saddle to Tragus	4.9	4.3	3.3	2.2	1.6
3.1	8.0	H31 D	Tragus to Occiput, Back of Head	6.0	5.3	4.0	2.7	2.0
1.3	3.4	H32 V	Tragus to Angle Mandible	2.6	2.3	1.7	1.1	0.9
2.1	5.4	H33 D	Angle of Mandible to Chin	4.1	3.6	2.7	1.8	1.4
1.8	4.5	H34 V	Length of Ear, Vertical	3.4	3.0	2.3	1.5	1.1
1.1	2.8	H35 D	Width of Ear, Anterior-Posterior	2.1	1.9	1.4	0.9	0.7
2.8	7.0	H36 D	Anterior to Posterior of Neck	5.3	4.7	3.5	2.3	1.8
			VERTICAL BODY MEASUREMENTS					
0.8	2.0	B1 V	Chin to Sternal Notch, Neck Pit	1.5	1.3	1.0	0.7	0.5
1.0	2.5	B2 V	Sternal Notch to Axilla, Arm Pit	1.9	1.7	1.3	0.8	0.6
2.0	5.0	B3 V	Sternal Notch to Nipples	3.8	3.3	2.5	1.7	1.3
3.1	8.0	B4 V	Sternal Notch to Xyphoid Proc.	6.0	5.3	4.0	2.7	2.0
6.3	16.0	B5 V	Sternal Notch to Umbilicus	12.0	10.7	8.0	5.3	4.0
3.1	8.0	B6 V	Xyphoid Process to Umbilicus	6.0	5.3	4.0	2.7	2.0
7.9	20.0	B7 V	Sternal Notch to Crest Ilium	15.0	13.3	10.0	6.7	5.0
8.3	21.0	B8 V	Sternal Notch to Ant.S.Sp.Ilium	15.8	14.0	10.5	7.0	5.3
9.4	24.0	B9 V	Sternal Notch to Pubis	18.0	16.0	12.0	8.0	6.0
3.1	8.0	B10 V	Umbilicus to Pubis of Pelvis	6.0	5.3	4.0	2.7	2.0
			TRANSVERSE BODY MEASUREMENTS					
7.7	19.5	B11 T	Shoulder Breadth	14.6	13.0	9.8	6.5	4.9
2.8	7.0	B12 T	Clavicle to Acromion Process	5.3	4.7	3.5	2.3	1.8
6.1	15.5	B13 T	Bi-acromial Process Breadth	11.6	10.3	7.8	5.2	3.9
5.5	14.0	B14 T	Inter-Anterior Axillary Fold	10.5	9.3	7.0	4.7	3.5
3.7	9.5	B15 T	Inter-Nipples	7.1	6.3	4.8	3.2	2.4

Life size			SIX-MONTH-OLD INFANT	Fractions of life size				
				3/4	2/3	1/2	1/3	1/4
Inches	cm	Code		cm	cm	cm	cm	cm
			TRANSVERSE BODY MEASUREMENTS					
5.5	**14.0**	B16 T	Breadth of Thorax	10.5	9.3	7.0	4.7	3.5
5.9	**15.0**	B17 T	Inter-Costal. Rib Margin Breadth	11.3	10.0	7.5	5.0	3.8
5.5	**14.0**	B18 T	Inter-Crest of Ilium Breadth	10.5	9.3	7.0	4.7	3.5
3.9	**10.0**	B19 T	Inter-Anterior Superior Spine	7.5	6.7	5.0	3.3	2.5
5.9	**15.0**	B20 T	Inter-Greater Trochanter Femur	11.3	10.0	7.5	5.0	3.8
			ANTERIOR-POSTERIOR BODY MEASUREMENTS					
4.3	**11.0**	B21 D	Manubrium Dorsal Spine	8.3	7.3	5.5	3.7	2.8
3.9	**10.0**	B22 D	Xyphoid Process to Dorsum	7.5	6.7	5.0	3.3	2.5
4.7	**12.0**	B23 D	Umbilicus to Dorsum	9.0	8.0	6.0	4.0	3.0
3.7	**9.5**	B24 D	Pubis of Pelvis to Buttock	7.1	6.3	4.8	3.2	2.4
2.8	**7.0**	B25 D	Thigh to Gluteal Furrow	5.3	4.7	3.5	2.3	1.8
			VERTICAL BACK MEASUREMENTS					
5.5	**14.0**	B26 V	Vertex to 7th Cervical Spine	10.5	9.3	7.0	4.7	3.5
			7th Cervical Vertebra Spine to:					
2.4	**6.0**	B27 V	To Posterior Axillary Fold, Armpit	4.5	4.0	3.0	2.0	1.5
2.8	**7.0**	B28 V	To Inferior Tip of Scapula	5.3	4.7	3.5	2.3	1.8
7.1	**18.0**	B29 V	To Posterior Superior Spine Ilium	13.5	12.0	9.0	6.0	4.5
8.7	**22.0**	B30 V	To Midline Upper Buttock Crease	16.5	14.7	11.0	7.3	5.5
13.0	**33.0**	B31 V	To Gluteal Furrow	24.8	22.0	16.5	11.0	8.3
			TRANSVERSE BACK MEASUREMENTS					
6.1	**15.4**	B33 T	Inter-Posterior Axillary Folds	11.6	10.3	7.7	5.1	3.9
2.8	**7.0**	B34 T	Inter-Scapulae Medial Borders	5.3	4.7	3.5	2.3	1.8
1.6	**4.0**	B35 T	Inter-Post. Sup. Spines Sacrum	3.0	2.7	2.0	1.3	1.0
			UPPER EXTREMITY MEASUREMENTS					
25.6	**65.0**	U1 V	Span of Arms	48.8	43.3	32.5	21.7	16.3
3.0	**7.5**	U2 V	Sternum to Acromion Process	5.6	5.0	3.8	2.5	1.9
10.2	**26.0**	U3 V	Acromion Process to Fingertip	19.5	17.3	13.0	8.7	6.5
5.0	**12.0**	U7 V	Shoulder to Flexed Elbow	9.5	8.4	6.3	4.2	3.2
3.7	**9.5**	U8 V	Acromion to Lat. Condyle Elbow	7.1	6.3	4.8	3.2	2.4
7.1	**18.0**	U9 D	Flexed Elbow to Fingertip	13.5	12	9	6	4.5
3.7	**9.5**	U10 D	Flexed Elbow Pivot to Wrist Pivot	7.1	6.3	4.8	3.2	2.4

Life size			SIX-MONTH-OLD INFANT	Fractions of life size				
				3/4	2/3	1/2	1/3	1/4
Inches	cm	Code		cm	cm	cm	cm	cm
			UPPER EXTREMITY MEASUREMENTS					
3.1	**8.0**	U11 V	Wrist to Fingertip	6.0	5.3	4.0	2.7	2.0
2.2	**5.5**	U12 V	Wrist to Thumb Tip	4.1	3.7	2.8	1.8	1.4
1.8	**4.5**	U13 V	Length of Palm	3.4	3.0	2.3	1.5	1.1
1.4	**3.6**	U14 V	Length 3rd Finger	2.7	2.4	1.8	1.2	0.9
1.8	**4.6**	U15 V	Thumb Crotch to 3rd Finger	3.5	3.1	2.3	1.5	1.2
1.4	**3.5**	U16 V	Thumb Crotch to Thumb Tip	2.6	2.3	1.8	1.2	0.9
1.6	**4.0**	U17 D	Anterior-Post. Arm at Biceps	3.0	2.7	2.0	1.3	1.0
1.7	**4.3**	U18 D	Anterior-Posterior at Forearm	3.2	2.9	2.2	1.4	1.1
1.0	**2.5**	U19 D	Depth of Wrist	1.9	1.7	1.3	0.8	0.6
1.4	**3.5**	U20 T	Breadth of Wrist	2.6	2.3	1.8	1.2	0.9
2.2	**5.5**	U21 T	Breadth of Hand	4.1	3.7	2.8	1.8	1.4
1.7	**4.2**	U22 T	Breadth of Palm	3.2	2.8	2.1	1.4	1.1
			LOWER EXTREMITY MEASUREMENTS					
12.6	**32.0**	L1 V	Crest of Ilium of Pelvis to Sole	24.0	21.3	16.0	10.7	8.0
11.8	**30.0**	* L2 V	Ant.Sup.Spine of Ilium to Sole	22.5	20.0	15.0	10.0	7.5
9.8	**25.0**	L3 V	Greater Trochanter Femur-Sole	18.8	16.7	12.5	8.3	6.3
10.2	**26.0**	* L4 V	Symphysis Pubis of Pelvis-Sole	19.5	17.3	13.0	8.7	6.5
8.7	**22.0**	* L5 V	Crotch to Sole	16.5	14.7	11.0	7.3	5.5
7.9	**20.0**	L6 V	Gluteal Furrow to Sole	15.0	13.3	10.0	6.7	5.0
5.1	**13.0**	L7 V	Posterior Knee Crease to Sole	9.8	8.7	6.5	4.3	3.3
5.5	**14.0**	L8 V	Tibia at Knee to Sole	10.5	9.3	7.0	4.7	3.5
1.2	**3.0**	L9 V	Med. Malleolus of Ankle to Sole	2.3	2.0	1.5	1.0	0.8
1.0	**2.5**	L10 V	Lat. Malleolus of Ankle to Sole	1.9	1.7	1.3	0.8	0.6
5.9	**15.0**	L11 T	Breadth of Hips, Sitting	11.3	10.0	7.5	5.0	3.8
2.6	**6.5**	L12 T	Breadth Mid Thigh	4.9	4.3	3.3	2.2	1.6
2.2	**5.5**	L13 T	Breadth of Knee	4.1	3.7	2.8	1.8	1.4
2.0	**5.0**	L14 T	Breadth of Calf	3.8	3.3	2.5	1.7	1.3
1.6	**4.0**	L15 D	Breadth Ankle	3.0	2.7	2.0	1.3	1.0
2.4	**6.0**	L16 D	Mid Thigh, Anterior-Posterior	4.5	4.0	3.0	2.0	1.5
1.8	**4.5**	L17 D	Knee, Anterior-Posterior	3.4	3.0	2.3	1.5	1.1
1.8	**4.5**	L18 D	Mid Calf, Anterior-Posterior	3.4	3.0	2.3	1.5	1.1
1.6	**4.0**	L19 D	Ankle, Anterior-Posterior	3.0	2.7	2.0	1.3	1.0
3.7	**9.5**	L20 D	Length of Foot	7.1	6.3	4.8	3.2	2.4
2.9	**7.3**	L21 D	Heel to Fifth Toe	5.5	4.9	3.7	2.4	1.8
1.7	**4.2**	L22 T	Breadth Foot	3.2	2.8	2.1	1.4	1.1
0.9	**2.3**	L23 T	Width of Heel	1.7	1.5	1.2	0.8	0.6

** These are also listed in Vertical Measurements.*

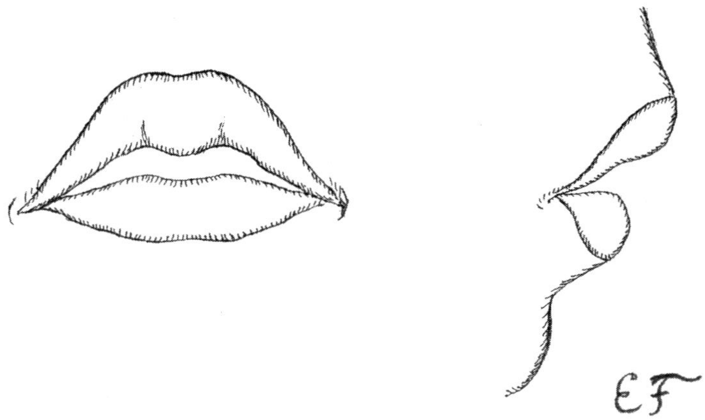

Infant Lip Detail

The well developed lips, in infancy, fulfill the needs of nursing. This is more evident on profile. The upper lip is forward of the lower lip, and both lips have a rounded anterior surface when viewed from the side. The mucous membrane of the lips is the same tissue as in the mouth; and it is rolled out forming the red part of the lips. The chin lacks prominence until dentition increases. During the first few months, an infant's mouth is slightly open, most of the time, and the corners of the lips are lower than the center of the mouth. Often the lower lip appears to be thicker. The form of the lips is important to recognize for individuality.

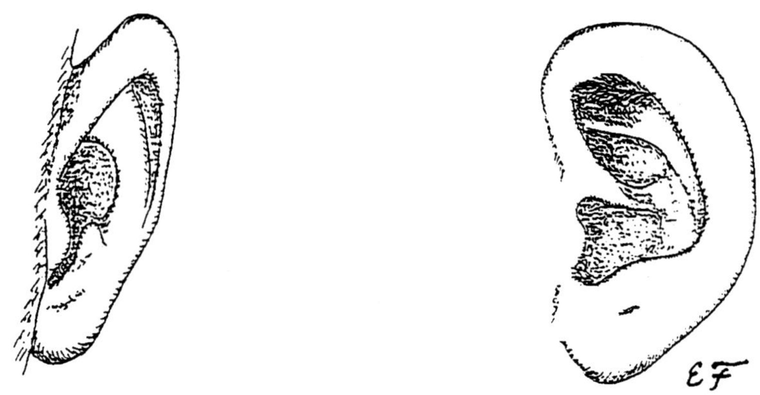

Infant Ear Detail

The ear of an infant at birth is well developed, appearing similar but shorter than an adult ear. The form is distinct, since the pinna of the ear or external ear is supported by cartilage that grows slowly. The less developed lobule. It is composed of fatty and fibrous tissue that is variable and grows more rapidly than cartilage. Although smaller, being 60% the size of an adult ear, it appears relatively larger than other features of the head. In outline, it conforms more to a circle than an ellipse.

During infancy, the upper border of the ear is at about the same level as an upper eyelid. The lower border is on a level with the corner of the mouth. The ear width is about two-thirds its height. The slant of the ear resembles the slant of the forehead.

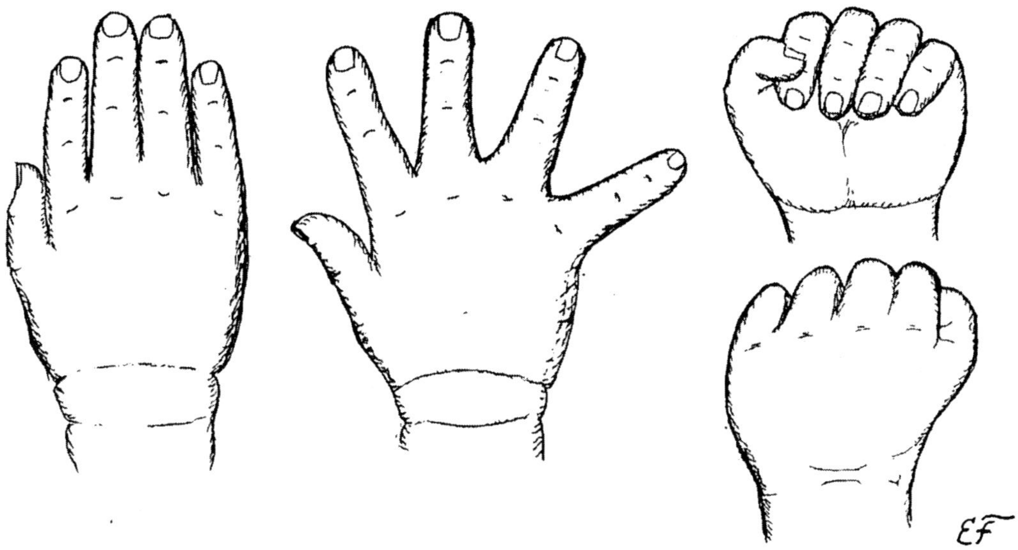

Infant Hands

The hand of a newborn infant, and for the first few months, is gently clenched at rest. The hands are plump and stubby. The breadth of the back or dorsum of the hand, including the thumb, is about twice the distance of wrist crease to knuckles. The palm is about one half the measure of wrist crease to fingertips, since the finger webs are more distant than the knuckles. When the fingers are extended, there are little dimples over the knuckles. The joints are flexible and often spread to a broad span. Parabolic curves can be drawn touching the tips of the extended fingers, when straight or when spread. The ratio of the segments of the fingers, like an adult, resembles the Fibonacci series, a natural factor in biological growth. The sum of the length of the farthest segment and the middle segment of a finger equals the segment nearest to the palm. The middle and nearest segment sum equals the back of the hand. The bones of the hands are long bones, and like leg and arm bones, they grow rapidly.

Infant Leg Development

There is some bowing of the legs in infancy that improves and become straighter by about three years in healthy individuals. The legs and feet grow more rapidly than the trunk during childhood. By two years of age the length a child's legs have doubled; by five years the length is two and one-half times; and by age seven the length has increased by three times. Average leg length is four times in both sexes at age thirteen. After mid teens, girls legs increase slightly, while boys legs increase, by age eighteen, to four and one-half times newborn leg length. Some male individuals continue to grow gradual from eighteen to twenty-five years and may have legs five times newborn leg length. Muscular development increases in a nearly parallel manner. As an individual nears adulthood, training and exercise increases muscle size and strength.[1]

1. Watson and Lowery, p. 116-118.

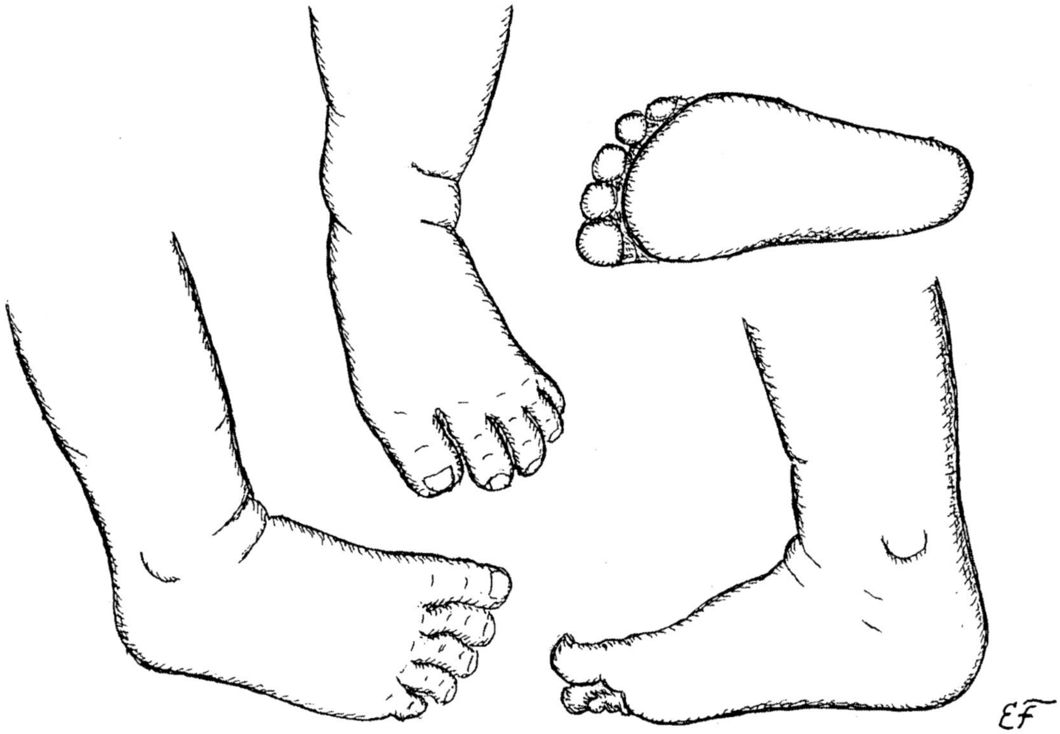

Infant Foot Development

The newborn infant has chubby feet with stubby toes. Like adults, on each foot a half parabola can be vaguely outlined on the toes. The feet do not begin to form an arch until the child has been walking for several years. Newborn infant's feet are proportionally more fat and wider than an adult. Fat pads in the foot obliterate the longitudinal arch and give the appearance of flat feet. When the infant starts to stand erect, holding onto an object, it will appear bowlegged and pigeon toed. The feet seem flat and continue in early life while toddling.

During the first two or three years, and after erect posture is achieved, infants' feet grow rapidly. The feet are broad across the toes and narrow at the heel. Many children prefer to go barefoot, when walking on sand or rugs. They like to take shoes off whenever they can, to the concern of parents. A well formed arch is not apparent until a child is about three or four years old. Feet grow more rapidly than the legs. At birth, they are about thirty percent of adult foot length; by one year they are forty-two percent; at three years they are fifty-seven percent; at five years they are seventy percent; at ten years they are eighty percent; and by early teens they have grown to nearly adult size.[1]

1. Lowery, G., p. 116-118.

ONE-YEAR-OLD INFANTS

There is considerable similarity of male and female from one-year-old infants to the age of five years. The ranges of twenty-fifth, fiftieth, and seventy-fifth percentiles are listed here for comparison.

Percentile	Twenty-fifth		Fiftieth		Seventy-fifth	
Boys:	**Short**		**Medium**		**Tall**	
Crown to Sole	74.3 cm,	29 1/4 in.	76 cm,	30 in.	78 cm,	30 1/2 in.
Weight	9.5 kg,	21 lb.	10.2 kg,	22 1/2 lb.	10.9 kg,	24 lb.
Girls:						
Crown to Sole	72.4 cm,	28 1/2 in.	74.3 cm,	29 1/4 in.	76.3 cm,	30 in.
Weight	8.8 kg,	19 1/2 lb.	9.5 kg,	21 lb.	10.2 kg,	22 1/2 lb.[1]

Male infants, on the average, are slightly longer and slightly heavier than are female infants. By ten-months, an infant may momentarily stand. By eleven-months they may stand alone well and during the next month may start walking without support or assistance. By age one year, most are walking as toddlers, often seeking a lateral support for balance. Feeding is less frequent and more time is spent at play. There is more social interaction. They like to play peek-a-boo and play with toys. They sit with adequate balance. They have a high center of gravity, and they walk with a wide base for adequate balance. They stand with toes turned out, feet far apart, and knees flexed for stability. They soon gain confidence and by fourteen months may start climbing stairs. The brain has continued rapid growth and is eighty-two percent of adult size. Stature has increased fifty percent over birth measurement. The head and face of these cute babies have assumed a round appearance, and they begin to say a word like da-da or ma-ma.

Body Measurements Compared To Height

	Feature Measure in cm	Feature Compared to Stature	Approx. Ratio of Feature	Compared to Median Adult
Stature: Crown to sole	75			42%
Head Height: Crown to chin	16.5	22 %	1/5	77%
Sitting Height: Crown to rump	47	63 %	2/3	50%
Trunk Measure: Sternal notch to symphysis pubis	27	36 %	1/3	53%
Leg Length: Greater trochanter of femur to sole	30	40 %	2/5	33%
Thigh Length: Greater trochanter to tibia at knee	13	17 %	1/6	30%
Tibia at Knee to Sole:	17	23 %	1/5	37 %
Arm Length: Shoulder to dependent fingertip	29	42 %	2/5	39%
Hand Length:	9	12 %	1/8	50 %
Foot Length:	11	15 %	1/6–	46 %

The deciduous incisor teeth erupt at about one year of age.[2]

1. Nelson, W.E., p. 38.
2. Lockhart, R.D., Hamilton, G.F., Fyfe, F.W., p. 57.

One-Year-Old Infant, anterior and lateral head

all measurements in centimeters

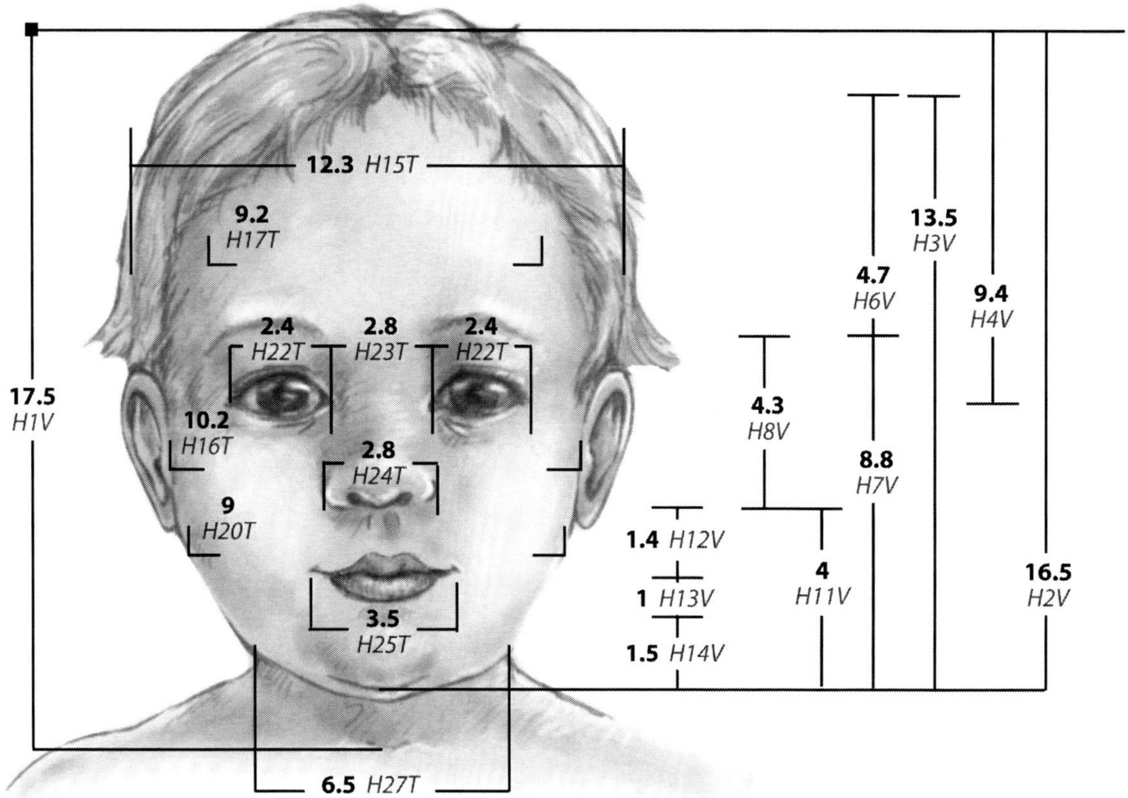

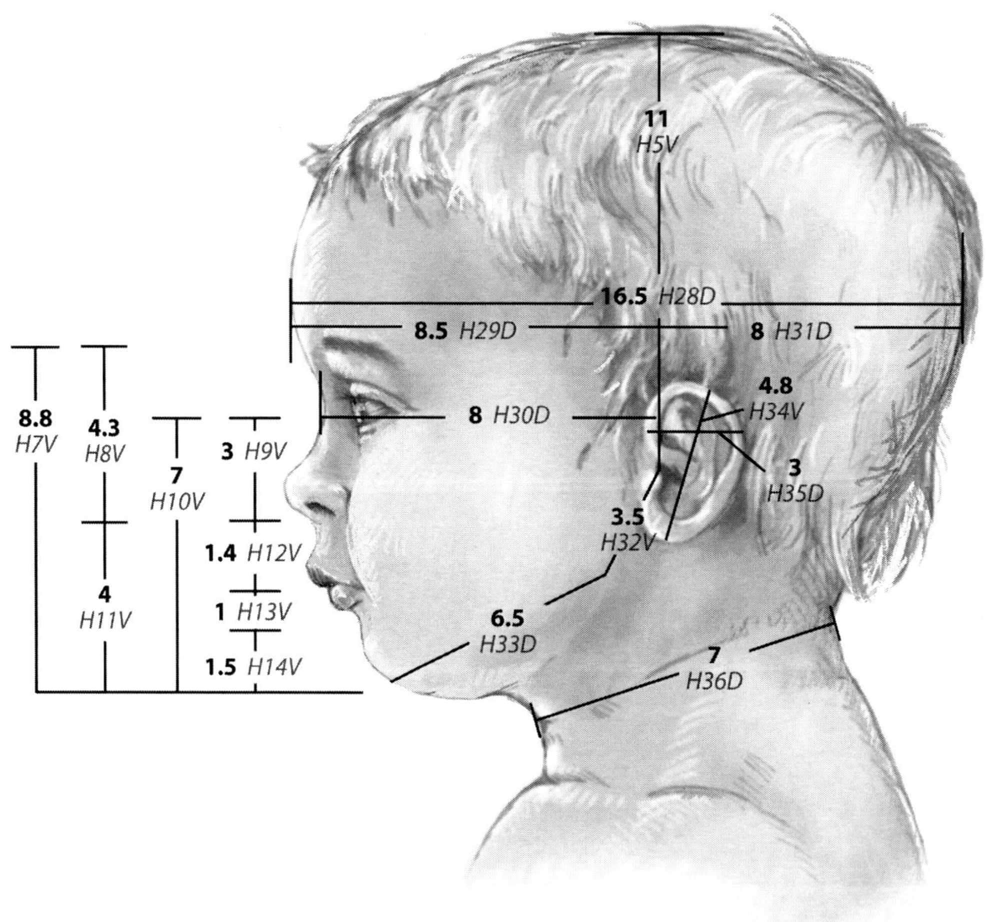

One-Year-Old Infant, anterior and lateral figure

all measurements in centimeters

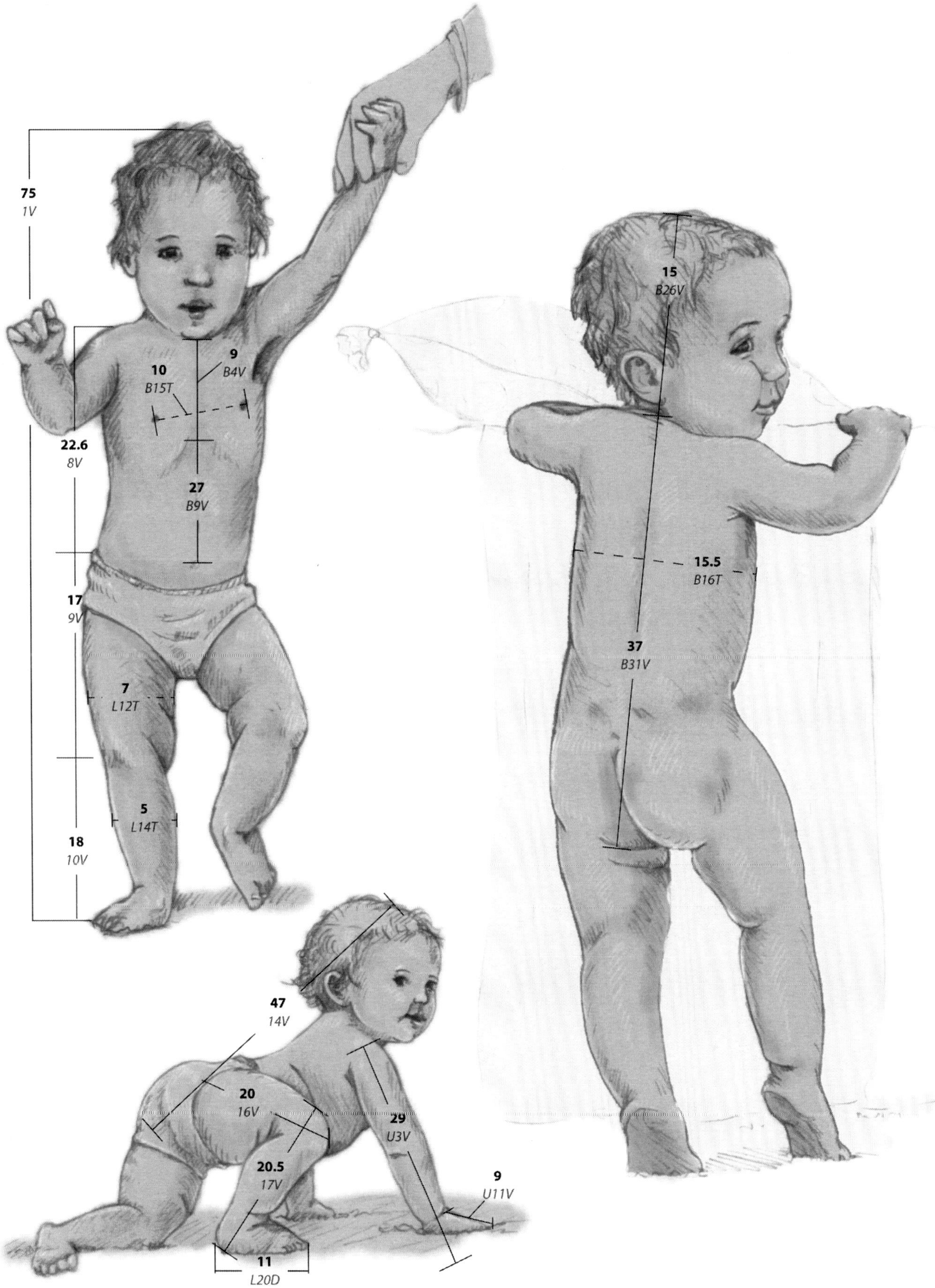

Life size			ONE-YEAR-OLD INFANT	Fractions of life size				
				3/4	2/3	1/2	1/3	1/4
Inches	cm	Code		cm	cm	cm	cm	cm
			VERTICAL MEASUREMENTS					
29.5	**75.0**	1 V	Stature, Height, Vertex to Sole	56.3	50.0	37.5	25.0	18.8
17.7	**45.0**	2 V	Vertex to Symphysis Pubis	33.8	30.0	22.5	15.0	11.3
11.8	**30.0**	* 3 V	Symphysis Pubis of Pelvis-Sole	22.5	20.0	15.0	10.0	7.5
22.6	**57.4**	4 V	Sternal Notch, Neck Pit, to Sole	43.1	38.3	28.7	19.1	14.4
6.9	**17.4**	5 V	Vertex to Acromion Process	13.1	11.6	8.7	5.8	4.4
22.7	**57.6**	6 V	Acromion Process to Sole	43.2	38.4	28.8	19.2	14.4
13.8	**35.0**	* 7 V	Ant.Sup.Spine of Ilium to Sole	26.3	23.3	17.5	11.7	8.8
8.9	**22.6**	8 V	Acrom. proc. to Ant. Sup. Spine	17.0	15.1	11.3	7.5	5.7
6.7	**17.0**	9 V	Ant.Sup.Spine of Ilium to Knee	12.8	11.3	8.5	5.7	4.3
7.1	**18.0**	10 V	Knee at Patella to Sole	13.5	12.0	9.0	6.0	4.5
15.4	**39.0**	11 V	Umbilicus to Sole	29.3	26.0	19.5	13.0	9.8
9.8	**25.0**	* 12 V	Crotch to Sole	18.8	16.7	12.5	8.3	6.3
21.7	**55.0**	13 V	Kneeling Height	41.3	36.7	27.5	18.3	13.8
18.5	**47.0**	14 V	Sitting Height - Crown Rump	35.3	31.3	23.5	15.7	11.8
13.8	**35.0**	15 V	Buttock, Straight Leg to Sole	26.3	23.3	17.5	11.7	8.8
7.9	**20.0**	16 V	Buttock to Flexed Knee	15.0	13.3	10.0	6.7	5.0
8.1	**20.5**	17 V	Flexed Knee to Sole	15.4	13.7	10.3	6.8	5.1
		** These are also listed with Lower Extremity Measurements.*						
			VERTICAL HEAD MEASUREMENTS					
6.9	**17.5**	H1 V	Vertex to Sternal Notch	13.1	11.7	8.8	5.8	4.4
6.5	**16.5**	H2 V	Vertex to Chin, Head Height	12.4	11.0	8.3	5.5	4.1
5.3	**13.5**	H3 V	Hairline to Chin, Face Height	10.1	9.0	6.8	4.5	3.4
3.7	**9.4**	H4 V	Vertex to Lateral Corner of Eye	7.1	6.3	4.7	3.1	2.4
4.3	**11.0**	H5 V	Vertex to Tragion of Ear	8.3	7.3	5.5	3.7	2.8
1.9	**4.7**	H6 V	Hairline to Brow, Glabella	3.5	3.1	2.4	1.6	1.2
3.5	**8.8**	H7 V	Brow, Glabella, to Chin, Menton	6.6	5.9	4.4	2.9	2.2
1.7	**4.3**	H8 V	Brow to Subnasal Point	3.2	2.9	2.2	1.4	1.1
1.2	**3.0**	H9 V	Sellion to Subnasal, Nose Length	2.3	2.0	1.5	1.0	0.8
2.8	**7.0**	H10 V	Sellion, Nose Saddle, to Chin	5.3	4.7	3.5	2.3	1.8
1.6	**4.0**	H11 V	Subnasal Point to Chin	3.0	2.7	2.0	1.3	1.0
0.6	**1.4**	H12 V	Subnasal to Lip Aperature	1.1	0.9	0.7	0.5	0.4
0.4	**1.0**	H13 V	Lip Aperat. to Sublabial Crease	0.8	0.7	0.5	0.3	0.3
0.6	**1.5**	H14 V	Sublabial Crease to Chin	1.1	1.0	0.8	0.5	0.4
			TRANSVERSE HEAD MEASUREMENTS					
5	**12.6**	H15 T	Bi-Parietal, Head Breadth	9.5	8.4	6.3	4.2	3.2
4.0	**10.2**	H16 T	Bi-Zygomatic, Cheekbones	7.7	6.8	5.1	3.4	2.6
3.6	**9.2**	H17 T	Bi-Fronto Temporal, Forehead	6.9	6.1	4.6	3.1	2.3

Life size			ONE-YEAR-OLD INFANT	Fractions of life size				
				3/4	2/3	1/2	1/3	1/4
Inches	cm	Code		cm	cm	cm	cm	cm
			TRANSVERSE HEAD MEASUREMENTS					
3.3	**8.3**	H18 T	Inter-Brow Ridge	6.2	5.5	4.2	2.8	2.1
4.3	**10.8**	H19 T	Inter-Tragion of Ears	8.1	7.2	5.4	3.6	2.7
3.5	**9.0**	H20 T	Inter-Angle of Mandible	6.8	6.0	4.5	3.0	2.3
3.0	**7.6**	H21 T	Inter-Lateral Corners of Eyes	5.7	5.1	3.8	2.5	1.9
0.9	**2.4**	H22 T	Width of Eye	1.8	1.6	1.2	0.8	0.6
1.1	**2.8**	H23 T	Inter-Inner Corner of Eyes	2.1	1.9	1.4	0.9	0.7
1.1	**2.8**	H24 T	Breadth of Nose	2.1	1.9	1.4	0.9	0.7
1.4	**3.5**	H25 T	Breadth of Mouth	2.6	2.3	1.8	1.2	0.9
1.0	**2.5**	H26 T	Breadth of Chin	1.9	1.7	1.3	0.8	0.6
2.6	**6.5**	H27 T	Breadth of Neck	4.9	4.3	3.3	2.2	1.6
			ANTERIOR-POSTERIOR HEAD MEASUREMENTS					
6.5	**16.5**	H28 D	Brow to Occiput, Posterior Head	12.4	11.0	8.3	5.5	4.1
3.3	**8.5**	H29 D	Brow to Tragus of Ear	6.4	5.7	4.3	2.8	2.1
3.1	**8.0**	H30 D	Sellion, Nose Saddle, to Tragus	6.0	5.3	4.0	2.7	2.0
3.1	**8.0**	H31 D	Tragus to Occiput, Back of Head	6.0	5.3	4.0	2.7	2.0
1.4	**3.5**	H32 V	Tragus to Angle Mandible	2.6	2.3	1.8	1.2	0.9
2.6	**6.5**	H33 D	Angle of Mandible to Chin	4.9	4.3	3.3	2.2	1.6
1.9	**4.8**	H34 V	Length of Ear, Vertical	3.6	3.2	2.4	1.6	1.2
1.2	**3.0**	H35 D	Width of Ear, Anterior-Posterior	2.3	2.0	1.5	1.0	0.8
2.8	**7.0**	H36 D	Anterior to Posterior of Neck	5.3	4.7	3.5	2.3	1.8
			VERTICAL BODY MEASUREMENTS					
0.8	**2.0**	B1 V	Chin to Sternal Notch, Neck Pit	1.5	1.3	1.0	0.7	0.5
1.2	**3.0**	B2 V	Sternal Notch to Axilla, Arm Pit	2.3	2.0	1.5	1.0	0.8
2.4	**6.0**	B3 V	Sternal Notch to Nipples	4.5	4.0	3.0	2.0	1.5
3.5	**9.0**	B4 V	Sternal Notch to Xyphoid Proc.	6.8	6.0	4.5	3.0	2.3
7.1	**18.0**	B5 V	Sternal Notch to Umbilicus	13.5	12.0	9.0	6.0	4.5
3.5	**9.0**	B6 V	Xyphoid Process to Umbilicus	6.8	6.0	4.5	3.0	2.3
8.3	**21.0**	B7 V	Sternal Notch to Crest Ilium	15.8	14.0	10.5	7.0	5.3
9.1	**23.0**	B8 V	Sternal Notch to Ant.S.Sp.Ilium	17.3	15.3	11.5	7.7	5.8
10.6	**27.0**	B9 V	Sternal Notch to Pubis	20.3	18.0	13.5	9.0	6.8
3.5	**9.0**	B10 V	Umbilicus to Pubis of Pelvis	6.8	6.0	4.5	3.0	2.3
			TRANSVERSE BODY MEASUREMENTS					
8.3	**21.0**	B11 T	Shoulder Breadth	15.8	14.0	10.5	7.0	5.3
3.1	**8.0**	B12 T	Clavicle to Acromion Process	6.0	5.3	4.0	2.7	2.0
6.7	**17.0**	B13 T	Bi-acromial Process Breadth	12.8	11.3	8.5	5.7	4.3
6.1	**15.5**	B14 T	Inter-Anterior Axillary Fold	11.6	10.3	7.8	5.2	3.9
3.9	**10.0**	B15 T	Inter-Nipples	7.5	6.7	5.0	3.3	2.5

Life size			ONE-YEAR-OLD INFANT	Fractions of life size				
				3/4	2/3	1/2	1/3	1/4
Inches	cm	Code		cm	cm	cm	cm	cm
			TRANSVERSE BODY MEASUREMENTS					
6.1	**15.5**	B16 T	Breadth of Thorax	11.6	10.3	7.8	5.2	3.9
6.1	**15.4**	B17 T	Inter-Costal. Rib Margin Breadth	11.6	10.3	7.7	5.1	3.9
5.7	**14.5**	B18 T	Inter-Crest of Ilium Breadth	10.9	9.7	7.3	4.8	3.6
4.3	**11.0**	B19 T	Inter-Anterior Superior Spine	8.3	7.3	5.5	3.7	2.8
6.7	**17.0**	B20 T	Inter-Greater Trochanter Femur	12.8	11.3	8.5	5.7	4.3
			ANTERIOR-POSTERIOR BODY MEASUREMENTS					
4.3	**11.0**	B21 D	Manubrium Dorsal Spine	8.3	7.3	5.5	3.7	2.8
4.3	**11.0**	B22 D	Xyphoid Process to Dorsum	8.3	7.3	5.5	3.7	2.8
4.9	**12.5**	B23 D	Umbilicus to Dorsum	9.4	8.3	6.3	4.2	3.1
4.3	**11.0**	B24 D	Pubis of Pelvis to Buttock	8.3	7.3	5.5	3.7	2.8
3.1	**8.0**	B25 D	Thigh to Gluteal Furrow	6.0	5.3	4.0	2.7	2.0
			VERTICAL BACK MEASUREMENTS					
5.9	**15.0**	B26 V	Vertex to 7th Cervical Spine	11.3	10.0	7.5	5.0	3.8
			7th Cervical Vertebra Spine to:					
2.1	**7.0**	B27 V	To Posterior Axillary Fold, Armpit	4.0	3.5	2.7	1.8	1.3
3.3	**8.3**	B28 V	To Inferior Tip of Scapula	6.2	5.5	4.2	2.8	2.1
8.3	**21.0**	B29 V	To Posterior Superior Spine Ilium	15.8	14.0	10.5	7.0	5.3
9.1	**23.0**	B30 V	To Midline Upper Buttock Crease	17.3	15.3	11.5	7.7	5.8
14.6	**37.0**	B31 V	To Gluteal Furrow	27.8	24.7	18.5	12.3	9.3
17.7	**45.0**	B32 V	To Popliteal Crease Behind Knee	33.8	30.0	22.5	15.0	11.3
			TRANSVERSE BACK MEASUREMENTS					
6.7	**17.0**	B33 T	Inter-Posterior Axillary Folds	12.8	11.3	8.5	5.7	4.3
3.5	**9.0**	B34 T	Inter-Scapulae Medial Borders	6.8	6.0	4.5	3.0	2.3
1.9	**4.8**	B35 T	Inter-Post. Sup. Spines Sacrum	3.6	3.2	2.4	1.6	1.2
			UPPER EXTREMITY MEASUREMENTS					
28.7	**73.0**	U1 V	Span of Arms	54.8	48.7	36.5	24.3	18.3
3.1	**8.0**	U2 V	Sternum to Acromion Process	6.0	5.3	4.0	2.7	2.0
11.4	**29.0**	U3 V	Acromion Process to Fingertip	21.8	19.3	14.5	9.7	7.3
5.7	**14.5**	U7 V	Shoulder to Flexed Elbow	10.9	9.7	7.3	4.8	3.6
4.7	**11.0**	U8 V	Acromion to Lat. Condyle Elbow	9.0	8.0	6.0	4.0	3.0
7.9	**20.0**	U9 D	Flexed Elbow to Fingertip	15.0	13.3	10.0	6.7	5.0
4.3	**11.0**	U10 D	Flexed Elbow Pivot to Wrist Pivot	8.3	7.3	5.5	3.7	2.8

Life size			ONE-YEAR-OLD INFANT	Fractions of life size				
				3/4	2/3	1/2	1/3	1/4
Inches	cm	Code		cm	cm	cm	cm	cm
			UPPER EXTREMITY MEASUREMENTS					
3.5	**9.0**	U11 V	Wrist to Fingertip	6.8	6.0	4.5	3.0	2.3
2.4	**6.0**	U12 V	Wrist to Thumb Tip	4.5	4.0	3.0	2.0	1.5
2.0	**5.2**	U13 V	Length of Palm	3.9	3.5	2.6	1.7	1.3
1.5	**3.8**	U14 V	Length 3rd Finger	2.9	2.5	1.9	1.3	1.0
2.2	**5.6**	U15 V	Thumb Crotch to Third Finger	4.2	3.7	2.8	1.9	1.4
1.5	**3.8**	U16 V	Thumb Crotch to Thumb Tip	2.9	2.5	1.9	1.3	1.0
1.7	**4.3**	U17 D	Anterior-Post. Arm at Biceps	3.2	2.9	2.2	1.4	1.1
1.8	**4.5**	U18 D	Anterior-Posterior at Forearm	3.4	3.0	2.3	1.5	1.1
1.0	**2.5**	U19 D	Depth of Wrist	1.9	1.7	1.3	0.8	0.6
1.2	**3.0**	U20 T	Breadth of Wrist	2.3	2.0	1.5	1.0	0.8
2.2	**5.5**	U21 T	Breadth of Hand	4.1	3.7	2.8	1.8	1.4
1.8	**4.5**	U22 T	Breadth of Palm	3.4	3.0	2.3	1.5	1.1
			LOWER EXTREMITY MEASUREMENTS					
15.0	**38.0**	L1 V	Crest of Ilium of Pelvis to Sole	28.5	25.3	19.0	12.7	9.5
13.8	**35.0**	* L2 V	Ant.Sup.Spine of Ilium to Sole	26.3	23.3	17.5	11.7	8.8
11.8	**30.0**	L3 V	Greater Trochanter Femur-Sole	22.5	20.0	15.0	10.0	7.5
11.8	**30.0**	* L4 V	Symphysis Pubis of Pelvis-Sole	22.5	20.0	15.0	10.0	7.5
9.8	**25.0**	* L5 V	Crotch to Sole	18.8	16.7	12.5	8.3	6.3
9.1	**23.0**	L6 V	Gluteal Furrow to Sole	17.3	15.3	11.5	7.7	5.8
5.9	**15.0**	L7 V	Posterior Knee Crease to Sole	11.3	10.0	7.5	5.0	3.8
6.7	**17.0**	L8 V	Tibia at Knee to Sole	12.8	11.3	8.5	5.7	4.3
1.2	**3.0**	L9 V	Med. Malleolus of Ankle to Sole	2.3	2.0	1.5	1.0	0.8
1.0	**2.5**	L10 V	Lat. Malleolus of Ankle to Sole	1.9	1.7	1.3	0.8	0.6
5.9	**15.0**	L11 T	Breadth of Hips	11.3	10.0	7.5	5.0	3.8
2.6	**7.0**	L12 T	Breadth Mid Thigh	4.9	4.3	3.3	2.2	1.6
2.2	**5.5**	L13 T	Breadth of Knee	4.1	3.7	2.8	1.8	1.4
2.0	**5.0**	L14 T	Breadth of Calf	3.8	3.3	2.5	1.7	1.3
1.6	**4.0**	L15 T	Breadth Ankle	3.0	2.7	2.0	1.3	1.0
2.8	**7.0**	L16 D	Mid Thigh, Anterior-Posterior	5.3	4.7	3.5	2.3	1.8
2.0	**5.0**	L17 D	Knee, Anterior-Posterior	3.8	3.3	2.5	1.7	1.3
1.9	**4.8**	L18 D	Mid Calf, Anterior-Posterior	3.6	3.2	2.4	1.6	1.2
1.6	**4.0**	L19 D	Ankle, Anterior-Posterior	3.0	2.7	2.0	1.3	1.0
4.3	**11.0**	L20 D	Length of Foot	8.3	7.3	5.5	3.7	2.8
3.3	**8.5**	L21 D	Heel to 5th Toe	6.4	5.7	4.3	2.8	2.1
1.8	**4.5**	L22 T	Breadth Foot	3.4	3.0	2.3	1.5	1.1
1.0	**2.5**	L23 T	Width of Heel	1.9	1.7	1.3	0.8	0.6
		** These are also listed with Vertical Measurements.*						

Superimposed Profiles of Infants and Children

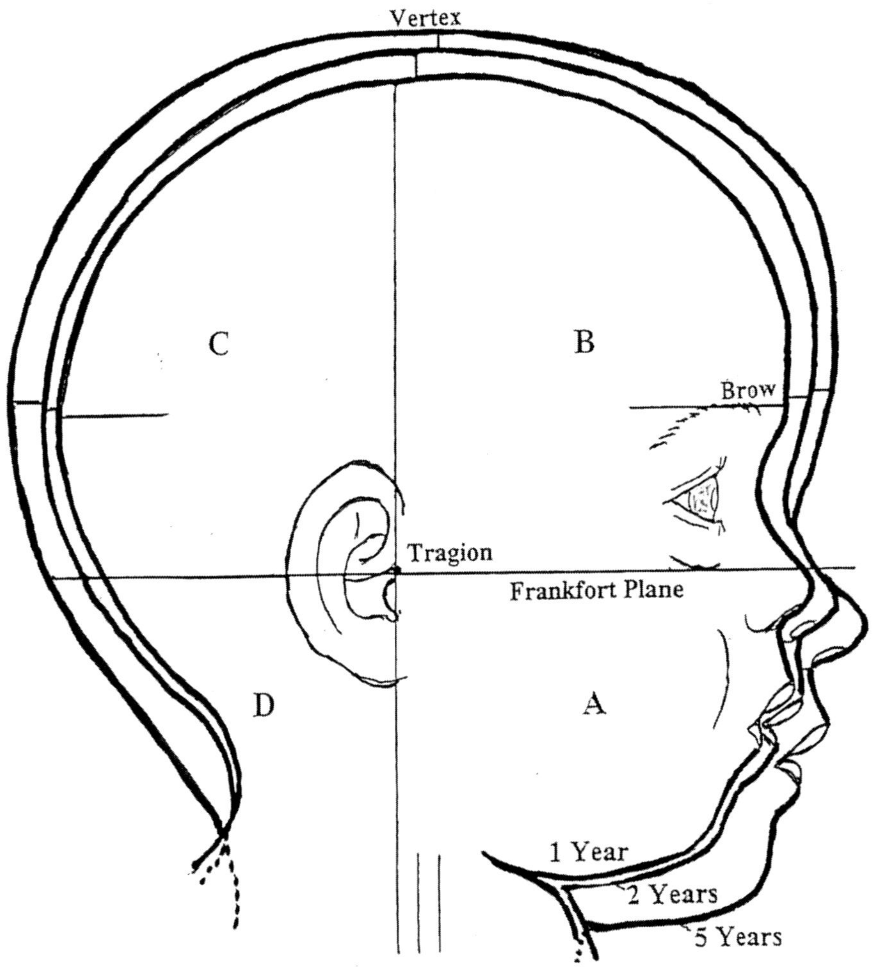

The Frankfort Plane

Anthropologists have described the Frankfort Plane as a baseline of the skull, a designation that is relatively constant during growth. The plane crosses two points, the inferior rim of the orbit, or eye socket, and the top of the ear, canal.[1] This plane, along the base of the cranium, changes very little during growth and maturation. The inferior orbital point is on an inverted arch that is manifest by a soft tissue arc shaped crease below an eye. This is present in infants but is subtle. The ear canal is a skeletal feature that may be located on the surface by the tragion, a notch on the upper border of the tragus, or ear flap, where the border of a helix curves into the concha.

Most cranial growth is above the Frankfort Plane and growth of the face is essentially below this plane.[2] A vertical line at right angles to the Frankfort Plane from the tragion crosses the cranium near the vertex. The head on profile may then be considered as having four quadrants. Most of the mid-face and lower face growth develops in A, the anterior lower quadrant. The cranium is in the upper two quadrants, B and C, and the lower posterior quadrant, D. The growth is most rapid in A, B, and C quadrants during the first five years.

1. Enlow, D.H., p. 164.
2. Enlow, D.H., p. 262.

TWO-YEAR-OLD CHILDREN

There is considerable similarity of measurements in male and female two-year-old infants. The ranges of twenty-fifth to fiftieth and to seventy-fifth percentiles are listed here for comparison.

Percentile	Twenty-fifth		Fiftieth		Seventy-fifth	
Boys:	**Short**		**Medium**		**Tall**	
Stature	85.6 cm,	33 3/4 in.	87.6 cm,	34 1/2 in.	90 cm,	35 1/2 in.
Weight	11.7 kg,	25 3/4 lb.	12.6 kg,	27 3/4 lb.	13.5 kg,	29 3/4 lb.
Girls:						
Stature	84 cm,	33 1/4 in.	86.5 cm,	34 in.	89 cm,	35 in.
Weight	11 kg,	24 1/2 lb.	12 kg,	26 1/4 lb.	12.7 kg,	26 lb.[1]

Male infants, on the average, are slightly longer and slightly heavier than female infants. By this age, most will be walking, but with a wide spread of their short legs for adequate balance. At about fifteen to eighteen months they want to explore, and may start climbing stairs. They can run and jump. They are very busy and need rest periods. They want socialization with older children and adults. During the fifteen to eighteen months, they learn more words and like to scribble with crayons. They like to play games, especially pushing or kicking balls. They can stack up to six blocks and knock them down. They enjoy picture books and learn illustrated new words. They develop a sense of independence as they try to put on their own cloths. They put words together and begin to make short sentences. They are curious and they often ask, "Why?"

BODY MEASUREMENTS COMPARED TO HEIGHT

	Feature Measure in cm	Feature Compared to Stature	Approx. Ratio of Feature	Compared to Median Adult
Stature: Crown to sole	87			49%
Head Height: Crown to chin	18	21%	1/5	82%
Sitting Height:	52	60%	2/3	56%
Trunk Measure: Sternal notch to symphysis pubis	29.5	34%	1/3	57%
Leg Length: Greater trochanter of femur to sole	38	44%	1/2–	44%
Thigh Length: Greater trochanter to tibia at knee	18.5	21%	1/5	43%
Tibia at Knee to Sole:	19.5	23%	1/4–	45%
Arm Length: Shoulder to dependent fingertip	36	41 %	2/5	48%
Hand Length:	9.8	11%	1/9	61%
Foot Length:	13	15%	1/6–	54%

Dentition: The deciduous canine (cuspid) and first molar have erupted.[2]

1. Nelson, W.E., p. 38.
2. Lockhart, p. 57.

Two-Year-Old Child, anterior and lateral head

all measurements in centimeters

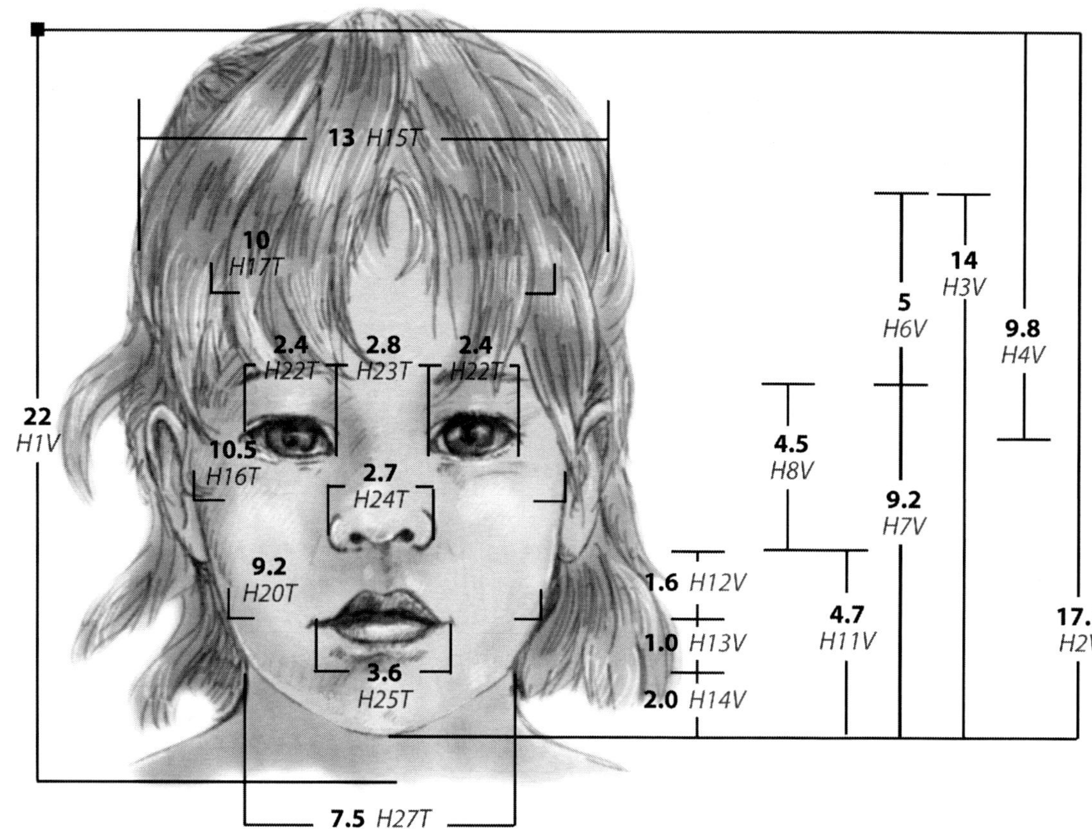

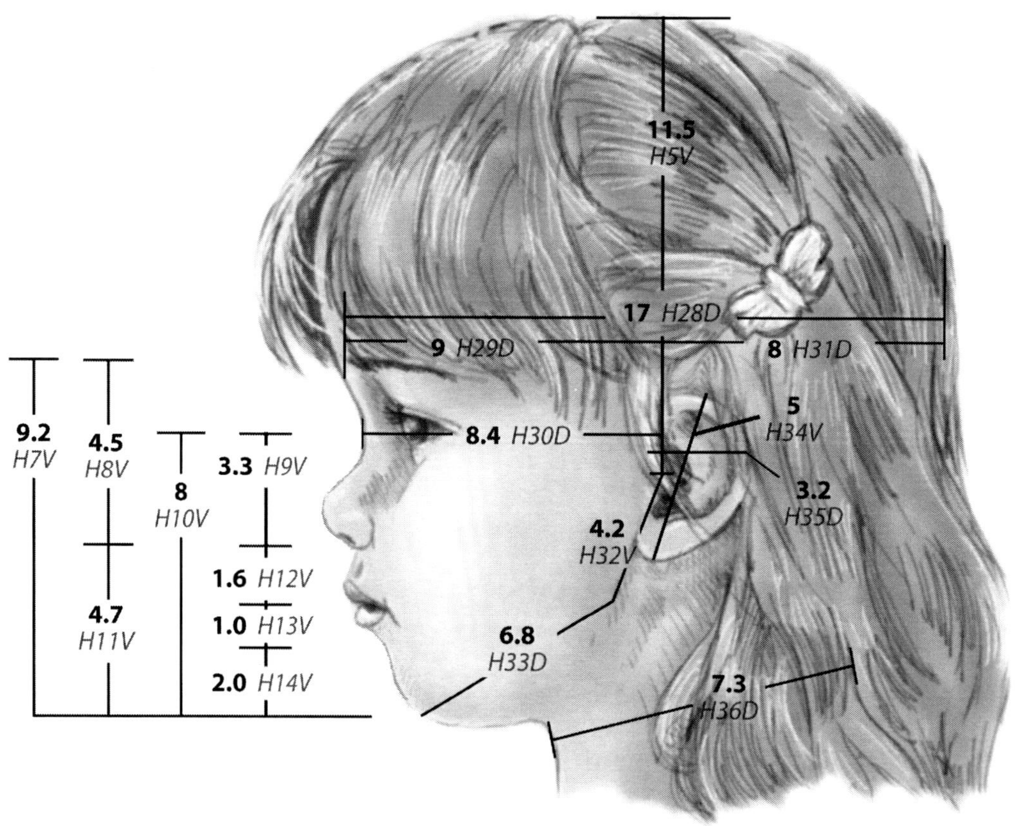

Two-Year-Old Child, anterior and lateral figure

all measurements in centimeters

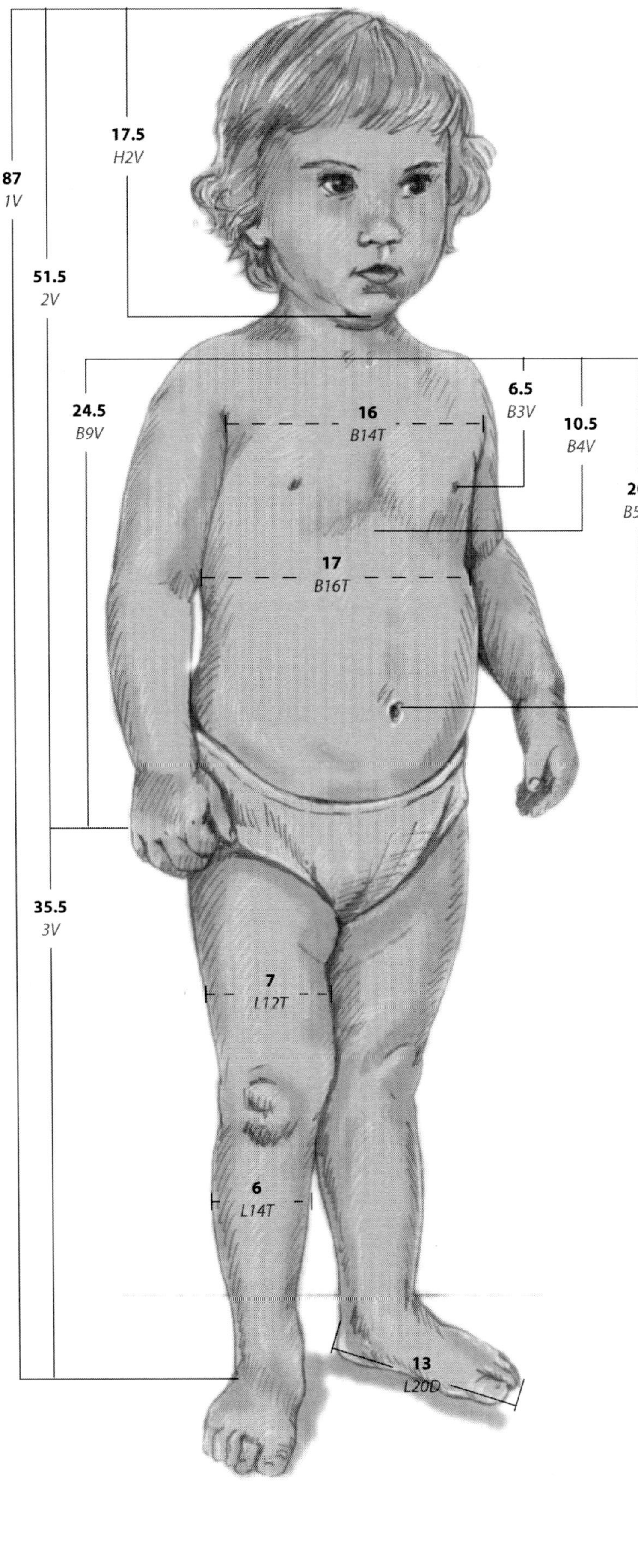

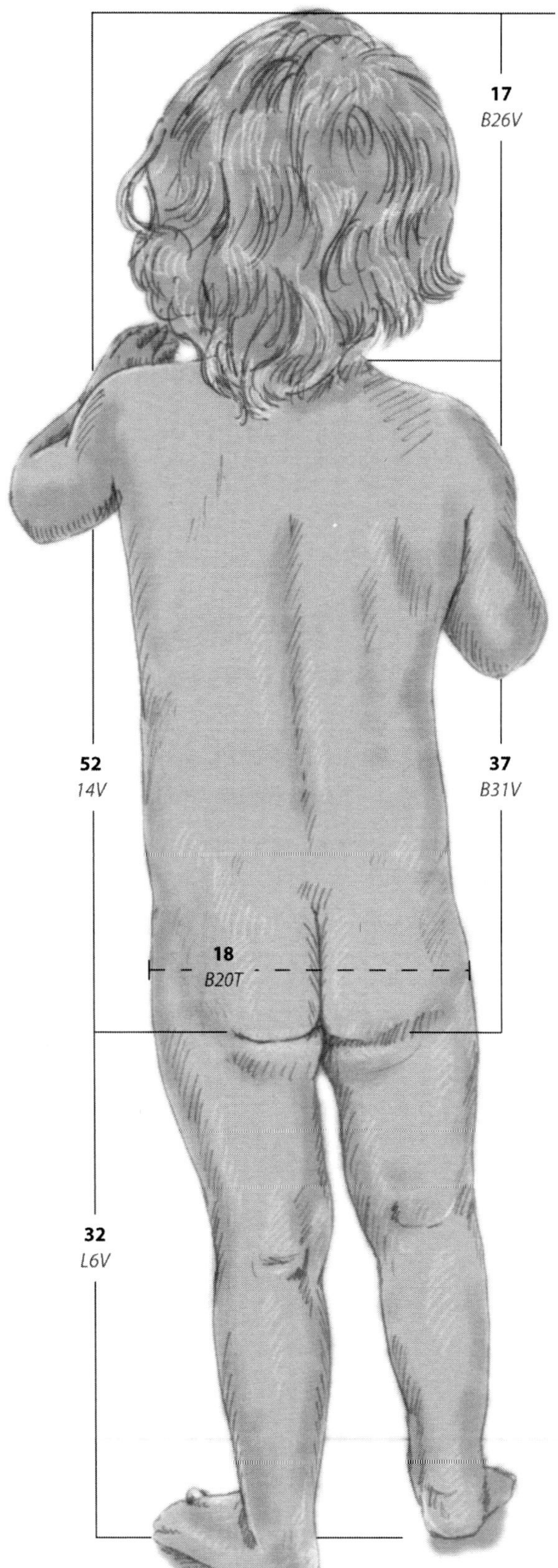

Life size			TWO-YEAR-OLD CHILD	Fractions of life size				
				3/4	2/3	1/2	1/3	1/4
Inches	cm	Code		cm	cm	cm	cm	cm
			VERTICAL MEASUREMENTS					
34.3	87.0	1 V	Stature, Height, Vertex to Sole	65.3	58.0	43.5	29.0	21.8
20.3	51.5	2 V	Vertex to Symphysis Pubis	38.6	34.3	25.8	17.2	12.9
14.0	35.5	* 3 V	Symphysis Pubis of Pelvis-Sole	26.6	23.7	17.8	11.8	8.9
25.6	65.0	4 V	Sternal Notch, Neck Pit, to Sole	48.8	43.3	32.5	21.7	16.3
8.3	21.0	5 V	Vertex to Acromion Process	15.8	14.0	10.5	7.0	5.3
26.0	66.0	6 V	Acromion Process to Sole	49.5	44.0	33.0	22.0	16.5
16.9	43.0	* 7 V	Ant.Sup.Spine of Ilium to Sole	32.3	28.7	21.5	14.3	10.8
9.1	23.0	8 V	Acrom. proc. to Ant. Sup. Spine	17.3	15.3	11.5	7.7	5.8
8.3	21.0	9 V	Ant.Sup.Spine of Ilium to Knee	15.8	14.0	10.5	7.0	5.3
8.7	22.0	10 V	Knee at Patella to Sole	16.5	14.7	11.0	7.3	5.5
17.7	45.0	11 V	Umbilicus to Sole	33.8	30.0	22.5	15.0	11.3
12.6	32.0	* 12 V	Crotch to Sole	24.0	21.3	16.0	10.7	8.0
25.6	65.0	13 V	Kneeling Height	48.8	43.3	32.5	21.7	16.3
20.5	52.0	14 V	Sitting Height - Crown Rump	39.0	34.7	26.0	17.3	13.0
16.9	43.0	15 V	Buttock, Straight Leg to Sole	32.3	28.7	21.5	14.3	10.8
9.8	25.0	16 V	Buttock to Flexed Knee	18.8	16.7	12.5	8.3	6.3
9.3	23.5	17 V	Flexed Knee to Sole	17.6	15.7	11.8	7.8	5.9
		These are also listed with Lower Extremity Measurements.						
			VERTICAL HEAD MEASUREMENTS					
8.7	22.0	H1 V	Vertex to Sternal Notch	16.5	14.7	11.0	7.3	5.5
6.9	17.5	H2 V	Vertex to Chin, Head Height	13.1	11.7	8.8	5.8	4.4
5.6	14.2	H3 V	Hairline to Chin, Face Height	10.7	9.5	7.1	4.7	3.6
3.9	9.8	H4 V	Vertex to Lateral Corner of Eye	7.4	6.5	4.9	3.3	2.5
4.5	11.5	H5 V	Vertex to Tragion of Ear	8.6	7.7	5.8	3.8	2.9
2.0	5.0	H6 V	Hairline to Brow, Glabella	3.8	3.3	2.5	1.7	1.3
3.6	9.2	H7 V	Brow, Glabella, to Chin, Menton	6.9	6.1	4.6	3.1	2.3
1.8	4.5	H8 V	Brow to Subnasal Point	3.4	3.0	2.3	1.5	1.1
1.3	3.3	H9 V	Sellion to Subnasal, Nose Length	2.5	2.2	1.7	1.1	0.8
3.1	8.0	H10 V	Sellion, Nose Saddle to Chin	6.0	5.3	4.0	2.7	2.0
1.9	4.7	H11 V	Subnasal Point to Chin	3.5	3.1	2.4	1.6	1.2
0.6	1.6	H12 V	Subnasal to Lip Aperature	1.2	1.1	0.8	0.5	0.4
0.5	1.1	H13 V	Lip Aperat. to Sublabial Crease	0.9	0.8	0.6	0.4	0.3
0.7	1.9	H14 V	Sublabial Crease to Chin	1.4	1.3	1.0	0.6	0.5
			TRANSVERSE HEAD MEASUREMENTS					
5.1	13.0	H15 T	Bi-Parietal, Head Breadth	9.8	8.7	6.5	4.3	3.3
4.1	10.5	H16 T	Bi-Zygomatic, Cheekbones	7.9	7.0	5.3	3.5	2.6
3.7	9.5	H17 T	Bi-Fronto Temporal, Forehead	7.1	6.3	4.8	3.2	2.4

Life size			TWO-YEAR-OLD CHILD	Fractions of life size				
				3/4	2/3	1/2	1/3	1/4
Inches	cm	Code		cm	cm	cm	cm	cm
			TRANSVERSE HEAD MEASUREMENTS					
3.3	**8.5**	H18 T	Inter-Brow Ridge	6.4	5.7	4.3	2.8	2.1
4.3	**11.0**	H19 T	Inter-Tragion of Ears	8.3	7.3	5.5	3.7	2.8
3.6	**9.2**	H20 T	Inter-Angle of Mandible	6.9	6.1	4.6	3.1	2.3
3.0	**7.6**	H21 T	Inter-Lateral Corners of Eyes	5.7	5.1	3.8	2.5	1.9
0.9	**2.4**	H22 T	Width of Eye	1.8	1.6	1.2	0.8	0.6
1.1	**2.8**	H23 T	Inter-Inner Corner of Eyes	2.1	1.9	1.4	0.9	0.7
1.1	**2.8**	H24 T	Breadth of Nose	2.1	1.9	1.4	0.9	0.7
1.4	**3.6**	H25 T	Breadth of Mouth	2.7	2.4	1.8	1.2	0.9
1.0	**2.5**	H26 T	Breadth of Chin	1.9	1.7	1.3	0.8	0.6
3.0	**7.5**	H27 T	Breadth of Neck	5.6	5.0	3.8	2.5	1.9
			ANTERIOR-POSTERIOR HEAD MEASUREMENTS					
6.7	**17.0**	H28 D	Brow to Occiput, Posterior Head	12.8	11.3	8.5	5.7	4.3
3.5	**9.0**	H29 D	Brow to Tragus of Ear	6.8	6.0	4.5	3.0	2.3
3.3	**8.4**	H30 D	Sellion, Nose Saddle, to Tragus	6.3	5.6	4.2	2.8	2.1
3.1	**8.0**	H31 D	Tragus to Occiput, Back of Head	6.0	5.3	4.0	2.7	2.0
1.7	**4.2**	H32 V	Tragus to Angle Mandible	3.2	2.8	2.1	1.4	1.1
2.7	**6.8**	H33 D	Angle of Mandible to Chin	5.1	4.5	3.4	2.3	1.7
2.0	**5.0**	H34 V	Length of Ear, Vertical	3.8	3.3	2.5	1.7	1.3
1.2	**3.0**	H35 D	Width of Ear, Anterior-Posterior	2.3	2.0	1.5	1.0	0.8
2.9	**7.3**	H36 D	Anterior to Posterior of Neck	5.5	4.9	3.7	2.4	1.8
			VERTICAL BODY MEASUREMENTS					
1.6	**4.0**	B1 V	Chin to Sternal Notch, Neck Pit	3.0	2.7	2.0	1.3	1.0
1.4	**3.5**	B2 V	Sternal Notch to Axilla, Arm Pit	2.6	2.3	1.8	1.2	0.9
2.6	**6.5**	B3 V	Sternal Notch to Nipples	4.9	4.3	3.3	2.2	1.6
4.1	**10.5**	B4 V	Sternal Notch to Xyphoid Proc.	7.9	7.0	5.3	3.5	2.6
7.9	**20.0**	B5 V	Sternal Notch to Umbilicus	15.0	13.3	10.0	6.7	5.0
3.7	**9.5**	B6 V	Xyphoid Process to Umbilicus	7.1	6.3	4.8	3.2	2.4
8.3	**21.0**	B7 V	Sternal Notch to Crest Ilium	15.8	14.0	10.5	7.0	5.3
9.4	**24.0**	B8 V	Sternal Notch to Ant.S.Sp.Ilium	18.0	16.0	12.0	8.0	6.0
11.6	**29.5**	B9 V	Sternal Notch to Pubis	22.1	19.7	14.8	9.8	7.4
3.7	**9.5**	B10 V	Umbilicus to Pubis of Pelvis	7.1	6.3	4.8	3.2	2.4
			TRANSVERSE BODY MEASUREMENTS					
9.1	**23.0**	B11 T	Shoulder Breadth	17.3	15.3	11.5	7.7	5.8
3.7	**9.5**	B12 T	Clavicle to Acromion Process	7.1	6.3	4.8	3.2	2.4
7.9	**20.0**	B13 T	Bi-acromial Process Breadth	15.0	13.3	10.0	6.7	5.0
6.3	**16.0**	B14 T	Inter-Anterior Axillary Fold	12.0	10.7	8.0	5.3	4.0
4.3	**11.0**	B15 T	Inter-Nipples	8.3	7.3	5.5	3.7	2.8

Life size			TWO-YEAR-OLD CHILD	Fractions of life size				
				3/4	2/3	1/2	1/3	1/4
Inches	cm	Code		cm	cm	cm	cm	cm
			TRANSVERSE BODY MEASUREMENTS					
6.7	**17.0**	B16 T	Breadth of Thorax	12.8	11.3	8.5	5.7	4.3
6.5	**16.5**	B17 T	Inter-Costal. Rib Margin Breadth	12.4	11.0	8.3	5.5	4.1
6.3	**16.0**	B18 T	Inter-Crest of Ilium Breadth	12.0	10.7	8.0	5.3	4.0
4.9	**12.5**	B19 T	Inter-Anterior Superior Spine	9.4	8.3	6.3	4.2	3.1
7.1	**18.0**	B20 T	Inter-Greater Trochanter Femur	13.5	12.0	9.0	6.0	4.5
			ANTERIOR-POSTERIOR BODY MEASUREMENTS					
4.7	**12.0**	B21 D	Manubrium to Dorsal Spine	9.0	8.0	6.0	4.0	3.0
4.7	**12.0**	B22 D	Xyphoid Process to Dorsum	9.0	8.0	6.0	4.0	3.0
5.3	**13.5**	B23 D	Umbilicus to Dorsum	10.1	9.0	6.8	4.5	3.4
4.5	**11.5**	B24 D	Pubis of Pelvis to Buttock	8.6	7.7	5.8	3.8	2.9
3.5	**9.0**	B25 D	Thigh to Gluteal Furrow	6.8	6.0	4.5	3.0	2.3
			VERTICAL BACK MEASUREMENTS					
6.7	**17.0**	B26 V	Vertex to 7th Cervical Spine	12.8	11.3	8.5	5.7	4.3
			7th Cervical Vertebra Spine to:					
3.1	**8.0**	B27 V	To Posterior Axillary Fold, Armpit	6.0	5.3	4.0	2.7	2.0
3.9	**10.0**	B28 V	To Inferior Tip of Scapula	7.5	6.7	5.0	3.3	2.5
9.4	**24.0**	B29 V	To Posterior Superior Spine Ilium	18.0	16.0	12.0	8.0	6.0
11.8	**30.0**	B30 V	To Midline Upper Buttock Crease	22.5	20.0	15.0	10.0	7.5
14.6	**37.0**	B31 V	To Gluteal Furrow	27.8	24.7	18.5	12.3	9.3
18.5	**47.0**	B32 V	To Popliteal Crease Behind Knee	35.3	31.3	23.5	15.7	11.8
			TRANSVERSE BACK MEASUREMENTS					
7.9	**20.0**	B33 T	Inter-Posterior Axillary Folds	15.0	13.3	10.0	6.7	5.0
3.9	**10.0**	B34 T	Inter-Scapulae Medial Borders	7.5	6.7	5.0	3.3	2.5
2.0	**5.0**	B35 T	Inter-Post. Sup. Spines Sacrum	3.8	3.3	2.5	1.7	1.3
			UPPER EXTREMITY MEASUREMENTS					
33.5	**85.0**	U1 V	Span of Arms	63.8	56.7	42.5	28.3	21.3
3.7	**9.5**	U2 V	Sternum to Acromion Process	7.1	6.3	4.8	3.2	2.4
14.2	**36.0**	U3 V	Acromion Process to Fingertip	27.0	24.0	18.0	12.0	9.0
40.9	**104.0**	U4 V	Vertical Reach to Fingertip	78.0	69.3	52.0	34.7	26.0
22.4	**57.0**	U5 T	Lateral Reach to Fingertip	42.8	38.0	28.5	19.0	14.3
16.7	**42.5**	U6 D	Forward Reach to Fingertip	31.9	28.3	21.3	14.2	10.6
5.9	**15.0**	U7 V	Shoulder to Flexed Elbow	11.3	10.0	7.5	5.0	3.8
5.3	**13.5**	U8 V	Acromion to Lat. Condyle Elbow	10.1	9.0	6.8	4.5	3.4
8.9	**22.5**	U9 D	Flexed Elbow to Fingertip	16.9	15.0	11.3	7.5	5.6
4.7	**12.5**	U10 D	Flexed Elbow Pivot to Wrist Pivot	9.4	8.3	6.3	4.2	3.1

Life size			TWO-YEAR-OLD CHILD	Fractions of life size				
				3/4	2/3	1/2	1/3	1/4
Inches	cm	Code		cm	cm	cm	cm	cm
			UPPER EXTREMITY MEASUREMENTS					
3.9	**9.8**	U11 V	Wrist to Fingertip	7.4	6.5	4.9	3.3	2.5
2.5	**6.3**	U12 V	Wrist to Thumb Tip	4.7	4.2	3.2	2.1	1.6
2.2	**5.5**	U13 V	Length of Palm	4.1	3.7	2.8	1.8	1.4
1.6	**4.0**	U14 V	Length 3rd Finger	3.0	2.7	2.0	1.3	1.0
2.5	**6.4**	U15 V	Thumb Crotch to 3rd Finger	4.8	4.3	3.2	2.1	1.6
1.6	**4.0**	U16 V	Thumb Crotch to Thumb Tip	3.0	2.7	2.0	1.3	1.0
1.6	**4.0**	U17 D	Anterior-Post. Arm at Biceps	3.0	2.7	2.0	1.3	1.0
1.8	**4.5**	U18 D	Anterior-Posterior at Forearm	3.4	3.0	2.3	1.5	1.1
1.0	**2.6**	U19 D	Depth of Wrist	2.0	1.7	1.3	0.9	0.7
1.6	**4.0**	U20 T	Breadth of Wrist	3.0	2.7	2.0	1.3	1.0
2.4	**6.0**	U21 T	Breadth of Hand	4.5	4.0	3.0	2.0	1.5
1.9	**4.8**	U22 T	Breadth of Palm	3.6	3.2	2.4	1.6	1.2
			LOWER EXTREMITY MEASUREMENTS					
18.5	**47.0**	L1 V	Crest of Ilium of Pelvis to Sole	35.3	31.3	23.5	15.7	11.8
16.9	**43.0**	* L2 V	Ant.Sup.Spine of Ilium to Sole	32.3	28.7	21.5	14.3	10.8
15.0	**38.0**	L3 V	Greater Trochanter Femur-Sole	28.5	25.3	19.0	12.7	9.5
13.8	**35.0**	* L4 V	Symphysis Pubis of Pelvis-Sole	26.3	23.3	17.5	11.7	8.8
12.2	**31.0**	* L5 V	Crotch to Sole	23.3	20.7	15.5	10.3	7.8
12.6	**32.0**	L6 V	Gluteal Furrow to Sole	24.0	21.3	16.0	10.7	8.0
8.3	**21.0**	L7 V	Posterior Knee Crease to Sole	15.8	14.0	10.5	7.0	5.3
7.9	**20.0**	L8 V	Tibia at Knee to Sole	15.0	13.3	10.0	6.7	5.0
1.5	**3.8**	L9 V	Med. Malleolus of Ankle to Sole	2.9	2.5	1.9	1.3	1.0
1.3	**3.2**	L10 V	Lat. Malleolus of Ankle to Sole	2.4	2.1	1.6	1.1	0.8
6.7	**17.0**	L11 T	Breadth of Hips, Sitting	12.8	11.3	8.5	5.7	4.3
2.8	**7.0**	L12 T	Breadth Mid Thigh	5.3	4.7	3.5	2.3	1.8
2.6	**6.5**	L13 T	Breadth of Knee	4.9	4.3	3.3	2.2	1.6
2.4	**6.0**	L14 T	Breadth of Calf	4.5	4.0	3.0	2.0	1.5
1.8	**4.5**	L15 T	Breadth Ankle	3.4	3.0	2.3	1.5	1.1
3.0	**7.5**	L16 D	Mid Thigh, Anterior-Posterior	5.6	5.0	3.8	2.5	1.9
2.4	**6.0**	L17 D	Knee, Anterior-Posterior	4.5	4.0	3.0	2.0	1.5
2.2	**5.5**	L18 D	Mid Calf, Anterior-Posterior	4.1	3.7	2.8	1.8	1.4
1.8	**4.5**	L19 D	Ankle, Anterior-Posterior	3.4	3.0	2.3	1.5	1.1
5.1	**13.0**	L20 D	Length of Foot	9.8	8.7	6.5	4.3	3.3
3.7	**9.5**	L21 D	Heel to 5th Toe	7.1	6.3	4.8	3.2	2.4
2.1	**5.4**	L22 T	Breadth Foot	4.1	3.6	2.7	1.8	1.4
1.2	**3.0**	L23 T	Width of Heel	2.3	2.0	1.5	1.0	0.8

* *These are also listed with Vertical Measurements.*

GROWTH OF THE FACE

Let us consider the face in three portions. The forehead represents the cranium. The mask of the face includes the mid-face and the lower face, from the brow to the chin. The mid-face is from the brow to the subnasal point, where the nose joins the upper lip. The lower face is from the subnasal point to the chin. The growth of the face is related to its different structures and their bony support. The mid-face includes the eyes and the nose. Its growth depends on the growth of the zygoma, or cheekbone, and the maxilla, or upper jawbone. The lower face growth occurs by an increase in the mandible, or lower jaw, and the growth of teeth in both the upper and the lower jaw.

The rapid growth of the facial bones, or viscerocranium, is necessary for respiration and ingestion of food. Yet the forehead, representing the neurocranium, enlarges more rapidly. Two major bones in the mid-face, the paired maxilla, grow to accommodate nasal air passages and the mastication of food by the arc of upper teeth. The mandible is also for mastication, and it grows to accommodate the arc of lower teeth. Forty-five percent of the maxillary growth, and forty percent of the mandible growth is complete by the fifth year.

This results in an apparent disproportion in five-year-old children, because they have a cranium that is ninety percent and a face that is forty percent adult size. By ten years, sixty-five percent of maxillary and mandibular growth is complete, compared to the cranium at ninety five percent. The remaining five percent of cranial growth develops between ten and twenty years. The remaining thirty-five percent maxillary and mandibular growth occurs during the same years.[1]

DENTITION

Growth of the lower face is related to the development of teeth. The deciduous or "milk teeth" are growing during the first two years. The first teeth to erupt are usually the central lower incisors at about six months. Approximate eruption times for others are next upper central incisors at eight months. Lateral incisors, canines, and first molars follow at fifteen to twenty months. The second molars erupt at twenty to twenty-four months. This results in a complement of twenty teeth, five on each side above and below. These teeth are small and muscles of mastication have not yet developed much strength. Firm food is difficult to chew so they prefer liquids or semi-liquid and soft food. Permanent teeth are growing during this time, which requires increase of both the maxilla, and mandible.

At age five, the deciduous teeth begin to be lost as their roots are reabsorbed and the permanent teeth push them out. The order is similar to the first eruption. The first permanent teeth erupt at age six. This is followed by eruption of central incisors at six to seven years. Lateral incisors erupt at eight year; first premolars at nine years, second premolars at ten years; canines at eleven years; second molars at twelve years and third molars may erupt from fifteen to twenty-five years. The adult complement of teeth is thirty-two, eight on each side above and below.[2]

1. Faulkner, F., p. 536.
2. Lockhart, R.D., Hamilton, O.F., Fyfe, F.W., p. 57.

THREE-YEAR-OLD CHILDREN

There is still considerable similarity of form in male and female three-year-old children. The range of twenty-fifth, fiftieth, and seventy-fifth percentile is shown here for comparison:

Percentile	Twenty-fifth		Fiftieth		Seventy-fifth	
Boys:	Short		Medium		Tall	
Stature	94 cm,	37 in.	95 cm,	38 in.	99 cm,	39 in.
Weight	13.6 kg,	30 lb.	14.7 kg,	32 1/2 lb.	15.6 kg,	36 3/4 lb.
Girls:						
Stature	93 cm,	36 3/4 in.	96 cm,	37 3/4 in.	98 cm,	38 1/2 in.
Weight	13 kg,	28 3/4 lb.	14 kg	30 1/4 lb.	15 kg,	33 1/4 lb.[1]

Male children, on the average, are slightly longer and slightly heavier than female children. Between two and three years their legs grow to be double the newborn leg length, and they achieve fifty percent of adult height. Social and intellectual development has progressed. By improved coordination, they climb, run, throw, jump, and enjoy outdoor games. They can pedal a tricycle, and enjoy games with others. Girls enjoy dressing up and make believe plays. Boys are more interested in mechanical toys. Most learn to feed themselves and begin to obey orders. They can put on their shoes but are not able to tie the laces. Language skills have improve as their vocabulary increases.

BODY MEASUREMENTS COMPARED TO HEIGHT

	Feature Measure in cm	Feature Compared to Stature	Approx. Ratio of Feature	Compared to Median Adult
Stature: Crown to sole	95			52%
Head Height: Crown to chin	18	19%	1/5	84%
Sitting Height:	58.4	57%	3/5–	58%
Trunk Measure: Sternal notch to symphysis pubis	32	33%	1/3	62%
Leg Length: Greater trochanter of femur to sole	43	45%	1/2–	48%
Thigh Length: Greater trochanter to tibia at knee	20	21%	1/5	44%
Tibia at Knee to Sole:	23.5	24 %	1/4–	53%
Arm Length, Shoulder to dependent fingertip	40	42%	2/5	51%
Hand Length:	10.5	11%	1/9	55%
Foot Length:	14.5	16%	1/6	65%

Dentition: all of the twenty deciduous teeth, including second molars, have erupted.[2]

1. Nelson, W.E., p. 38.

2. Lockhart, p. 57.

Three-Year-Old Child, anterior and lateral head

all measurements in centimeters

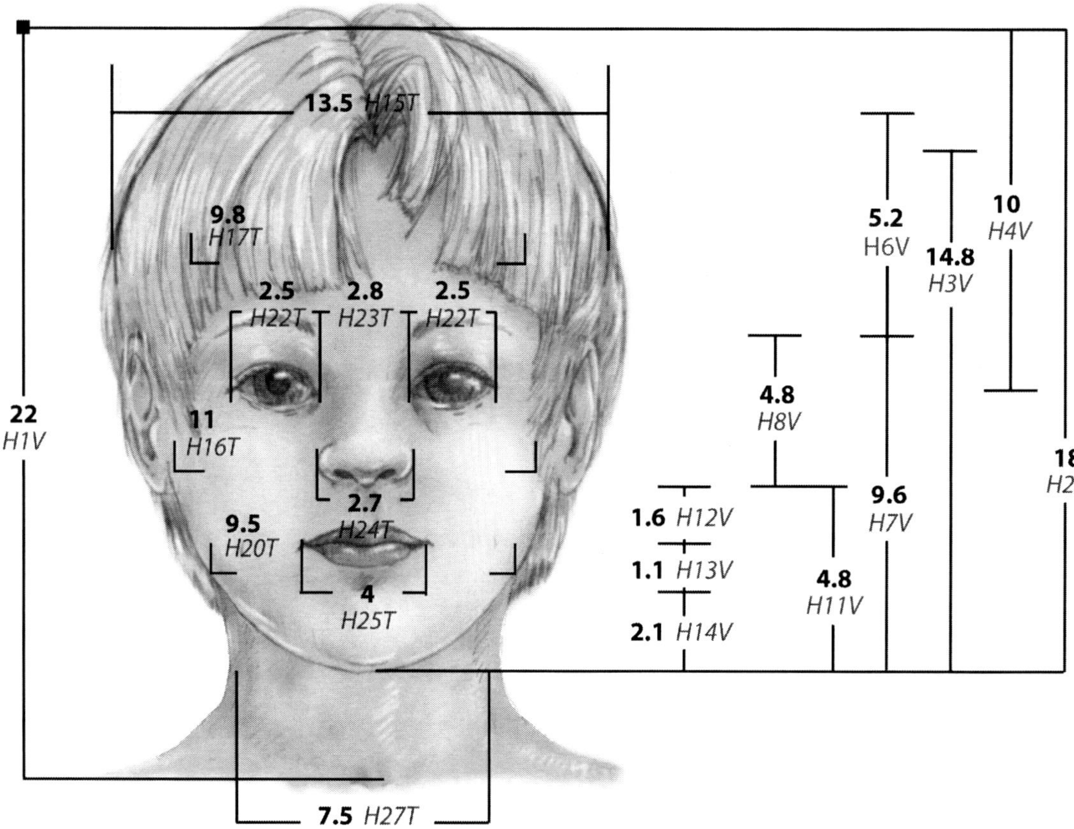

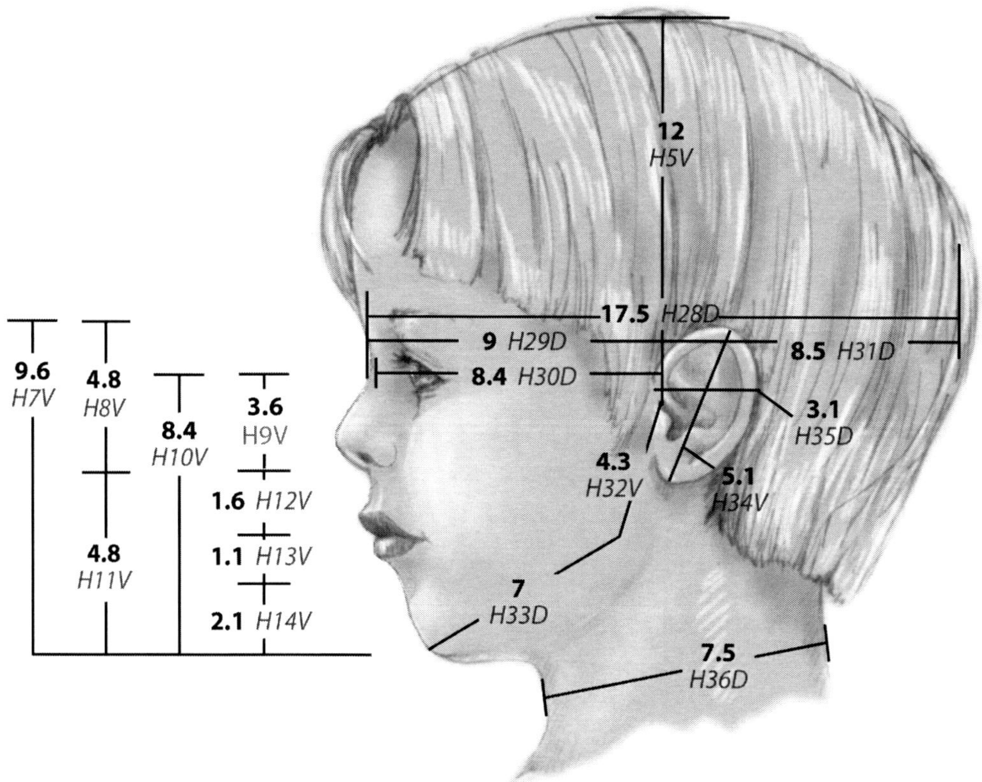

Three-Year-Old Child, anterior and lateral figure

all measurements in centimeters

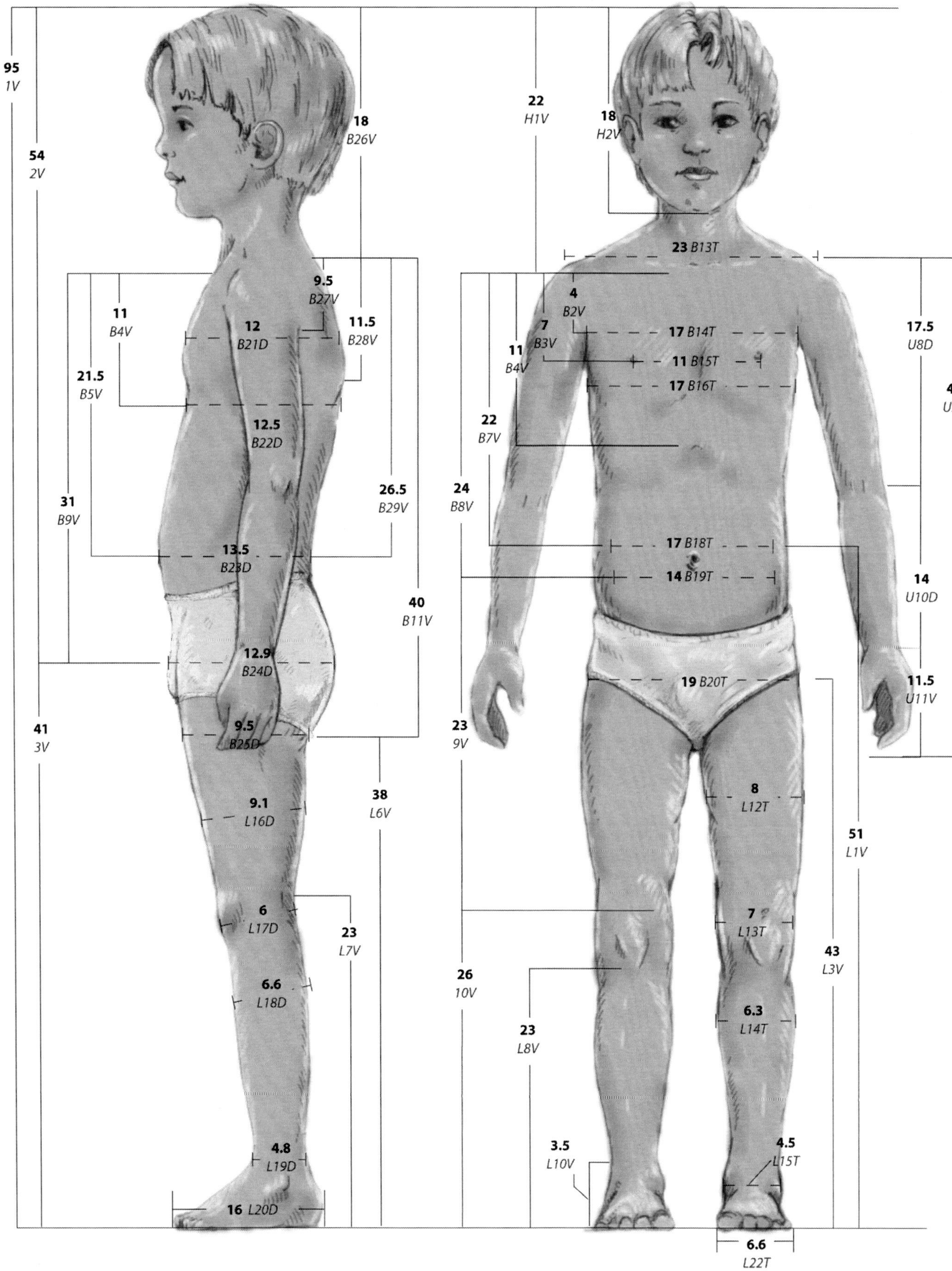

Life size			THREE-YEAR-OLD CHILD	Fractions of life size				
				3/4	2/3	1/2	1/3	1/4
Inches	cm	Code		cm	cm	cm	cm	cm
			VERTICAL MEASUREMENTS					
37.4	**95.0**	1 V	Stature, Height, Vertex to Sole	71.3	63.3	47.5	31.7	23.8
21.3	**54.0**	2 V	Vertex to Symphysis Pubis	40.5	36.0	27.0	18.0	13.5
16.1	**41.0**	* 3 V	Symphysis Pubis of Pelvis-Sole	30.8	27.3	20.5	13.7	10.3
28.7	**73.0**	4 V	Sternal Notch, Neck Pit, to Sole	54.8	48.7	36.5	24.3	18.3
8.6	**21.8**	5 V	Vertex to Acromion Process	16.4	14.5	10.9	7.3	5.5
28.9	**73.5**	6 V	Acromion Process to Sole	55.1	49.0	36.8	24.5	18.4
19.3	**49.0**	* 7 V	Ant.Sup.Spine of Ilium to Sole	36.8	32.7	24.5	16.3	12.3
9.6	**24.5**	8 V	Acrom. Proc. to Ant. Sup. Spine	18.4	16.3	12.3	8.2	6.1
9.1	**23.0**	9 V	Ant.Sup.Spine of Ilium to Knee	17.3	15.3	11.5	7.7	5.8
10.2	**26.0**	10 V	Knee at Patella to Sole	19.5	17.3	13.0	8.7	6.5
20.3	**51.5**	11 V	Umbilicus to Sole	38.6	34.3	25.8	17.2	12.9
14.8	**37.5**	* 12 V	Crotch to Sole	28.1	25.0	18.8	12.5	9.4
28.3	**72.0**	13 V	Kneeling Height	54.0	48.0	36.0	24.0	18.0
22.0	**56.0**	14 V	Sitting Height	42.0	37.3	28.0	18.7	14.0
19.5	**49.5**	15 V	Buttock, Straight Leg to Sole	37.1	33.0	24.8	16.5	12.4
11.4	**29.0**	16 V	Buttock to Flexed Knee	21.8	19.3	14.5	9.7	7.3
10.8	**27.5**	17 V	Flexed Knee to Sole	20.6	18.3	13.8	9.2	6.9
		** These are also listed with Lower Extremity Measurements.*						
			VERTICAL HEAD MEASUREMENTS					
8.7	**22.0**	H1 V	Vertex to Sternal Notch	16.5	14.7	11.0	7.3	5.5
7.1	**18.0**	H2 V	Vertex to Chin, Head Height	13.5	12.0	9.0	6.0	4.5
5.8	**14.8**	H3 V	Hairline to Chin, Face Height	11.1	9.9	7.4	4.9	3.7
3.9	**10.0**	H4 V	Vertex to Lateral Corner of Eye	7.5	6.7	5.0	3.3	2.5
4.7	**12.0**	H5 V	Vertex to Tragion of Ear	9.0	8.0	6.0	4.0	3.0
2.0	**5.0**	H6 V	Hairline to Brow, Glabella	3.9	3.5	2.6	1.7	1.3
3.8	**9.6**	H7 V	Brow, Glabella, to Chin, Menton	7.2	6.4	4.8	3.2	2.4
1.9	**4.8**	H8 V	Brow to Subnasal Point	3.6	3.2	2.4	1.6	1.2
1.4	**3.6**	H9 V	Sellion to Subnasal, Nose Length	2.7	2.4	1.8	1.2	0.9
3.3	**8.4**	H10 V	Sellion, Nose Saddle to Chin	6.3	5.6	4.2	2.8	2.1
1.9	**4.8**	H11 V	Subnasal Point to Chin	3.6	3.2	2.4	1.6	1.2
0.6	**1.6**	H12 V	Subnasal to Lip Aperature	1.2	1.1	0.8	0.5	0.4
0.4	**1.1**	H13 V	Lip Aperat. to Sublabial Crease	0.8	0.7	0.6	0.4	0.3
0.8	**2.1**	H14 V	Sublabial Crease to Chin	1.6	1.4	1.1	0.7	0.5
			TRANSVERSE HEAD MEASUREMENTS					
5.3	**13.5**	H15 T	Bi-Parietal, Head Breadth	10.1	9.0	6.8	4.5	3.4
4.3	**11.0**	H16 T	Bi-Zygomatic, Cheekbones	8.3	7.3	5.5	3.7	2.8
3.9	**9.8**	H17 T	Bi-Fronto Temporal, Forehead	7.4	6.5	4.9	3.3	2.5

Life size			THREE-YEAR-OLD CHILD	Fractions of life size				
				3/4	2/3	1/2	1/3	1/4
Inches	cm	Code		cm	cm	cm	cm	cm
			TRANSVERSE HEAD MEASUREMENTS					
3.3	8.5	H18 T	Inter-Brow Ridge	6.4	5.7	4.3	2.8	2.1
4.3	11.0	H19 T	Inter-Tragion of Ears	8.3	7.3	5.5	3.7	2.8
3.7	9.5	H20 T	Inter-Angle of Mandible	7.1	6.3	4.8	3.2	2.4
3.1	7.8	H21 T	Inter-Lateral Corners of Eyes	5.9	5.2	3.9	2.6	2.0
1.0	2.5	H22 T	Width of Eye	1.9	1.7	1.3	0.8	0.6
1.1	2.8	H23 T	Inter-Inner Corner of Eyes	2.1	1.9	1.4	0.9	0.7
1.1	2.8	H24 T	Breadth of Nose	2.1	1.9	1.4	0.9	0.7
1.6	4.0	H25 T	Breadth of Mouth	3.0	2.7	2.0	1.3	1.0
1.0	2.6	H26 T	Breadth of Chin	2.0	1.7	1.3	0.9	0.7
3.0	7.5	H27 T	Breadth of Neck	5.6	5.0	3.8	2.5	1.9
			ANTERIOR-POSTERIOR HEAD MEASUREMENTS					
6.9	17.5	H28 D	Brow to Occiput, Posterior Head	13.1	11.7	8.8	5.8	4.4
3.5	9.0	H29 D	Brow to Tragus of Ear	6.8	6.0	4.5	3.0	2.3
3.3	8.4	H30 D	Sellion, Nose Saddle, to Tragus	6.3	5.6	4.2	2.8	2.1
3.3	8.5	H31 D	Tragus to Occiput, Back of Head	6.4	5.7	4.3	2.8	2.1
1.7	4.3	H32 V	Tragus to Angle Mandible	3.2	2.9	2.2	1.4	1.1
2.8	7.0	H33 D	Angle of Mandible to Chin	5.3	4.7	3.5	2.3	1.8
2.0	5.0	H34 V	Length of Ear, Vertical	3.8	3.4	2.6	1.7	1.3
1.2	3.0	H35 D	Width of Ear, Anterior-Posterior	2.3	2.1	1.6	1.0	0.8
3.0	7.5	H36 D	Anterior to Posterior of Neck	5.6	5.0	3.8	2.5	1.9
			VERTICAL BODY MEASUREMENTS					
2.0	5.0	B1 V	Chin to Sternal Notch, Neck Pit	3.8	3.3	2.5	1.7	1.3
1.6	4.0	B2 V	Sternal Notch to Axilla, Arm Pit	3.0	2.7	2.0	1.3	1.0
2.8	7.0	B3 V	Sternal Notch to Nipples	5.3	4.7	3.5	2.3	1.8
4.3	11.0	B4 V	Sternal Notch to Xyphoid Proc.	8.3	7.3	5.5	3.7	2.8
8.5	21.5	B5 V	Sternal Notch to Umbilicus	16.1	14.3	10.8	7.2	5.4
4.1	10.5	B6 V	Xyphoid Process to Umbilicus	7.9	7.0	5.3	3.5	2.6
8.7	22.0	B7 V	Sternal Notch to Crest Ilium	16.5	14.7	11.0	7.3	5.5
9.4	24.0	B8 V	Sternal Notch to Ant.S.Sp.Ilium	18.0	16.0	12.0	8.0	6.0
12.2	31.0	B9 V	Sternal Notch to Pubis	23.3	20.7	15.5	10.3	7.8
3.9	10.0	B10 V	Umbilicus to Pubis of Pelvis	7.5	6.7	5.0	3.3	2.5
			TRANSVERSE BODY MEASUREMENTS					
9.8	25.0	B11 T	Shoulder Breadth	18.8	16.7	12.5	8.3	6.3
4.1	10.5	B12 T	Clavicle to Acromion Process	7.9	7.0	5.3	3.5	2.6
9.1	23.0	B13 T	Bi-acromial Process Breadth	17.3	15.3	11.5	7.7	5.8
6.7	17.0	B14 T	Inter-Anterior Axillary Fold	12.8	11.3	8.5	5.7	4.3
4.3	11.0	B15 T	Inter-Nipples	8.3	7.3	5.5	3.7	2.8

Life size			THREE-YEAR-OLD CHILD	Fractions of life size				
				3/4	2/3	1/2	1/3	1/4
Inches	cm	Code		cm	cm	cm	cm	cm
			TRANSVERSE BODY MEASUREMENTS					
6.7	**17.0**	B16 T	Breadth of Thorax	12.8	11.3	8.5	5.7	4.3
6.6	**16.8**	B17 T	Inter-Costal. Rib Margin Breadth	12.6	11.2	8.4	5.6	4.2
6.7	**17.0**	B18 T	Inter-Crest of Ilium Breadth	12.8	11.3	8.5	5.7	4.3
5.5	**14.0**	B19 T	Inter-Anterior Superior Spine	10.5	9.3	7.0	4.7	3.5
7.5	**19.0**	B20 T	Inter-Greater Trochanter Femur	14.3	12.7	9.5	6.3	4.8
			ANTERIOR-POSTERIOR BODY MEASUREMENTS					
4.7	**12.0**	B21 D	Manubrium to Dorsal Spine	9.0	8.0	6.0	4.0	3.0
4.9	**12.5**	B22 D	Xyphoid Process to Dorsum	9.4	8.3	6.3	4.2	3.1
5.3	**13.5**	B23 D	Umbilicus to Dorsum	10.1	9.0	6.8	4.5	3.4
5.1	**12.9**	B24 D	Pubis of Pelvis to Buttock	9.7	8.6	6.5	4.3	3.2
3.7	**9.5**	B25 D	Thigh to Gluteal Furrow	7.1	6.3	4.8	3.2	2.4
			VERTICAL BACK MEASUREMENTS					
7.1	**18.0**	B26 V	Vertex to 7th Cervical Spine	13.5	12.0	9.0	6.0	4.5
			7th Cervical Vertebra Spine to:					
3.7	**9.5**	B27 V	To Posterior Axillary Fold, Armpit	7.1	6.3	4.8	3.2	2.4
4.5	**11.5**	B28 V	To Inferior Tip of Scapula	8.6	7.7	5.8	3.8	2.9
10.4	**26.5**	B29 V	To Posterior Superior Spine Ilium	19.9	17.7	13.3	8.8	6.6
12.8	**32.5**	B30 V	To Midline Upper Buttock Crease	24.4	21.7	16.3	10.8	8.1
15.7	**40.0**	B31 V	To Gluteal Furrow	30.0	26.7	20.0	13.3	10.0
21.7	**55.0**	B32 V	To Popliteal Crease Behind Knee	41.3	36.7	27.5	18.3	13.8
			TRANSVERSE BACK MEASUREMENTS					
8.3	**21.0**	B33 T	Inter-Posterior Axillary Folds	15.8	14.0	10.5	7.0	5.3
4.3	**11.0**	B34 T	Inter-Scapulae Medial Borders	8.3	7.3	5.5	3.7	2.8
2.2	**5.5**	B35 T	Inter-Post. Sup. Spines Sacrum	4.1	3.7	2.8	1.8	1.4
			UPPER EXTREMITY MEASUREMENTS					
37.2	**94.5**	U1 V	Span of Arms	70.9	63.0	47.3	31.5	23.6
4.1	**10.5**	U2 V	Sternum to Acromion Process	7.9	7.0	5.3	3.5	2.6
15.7	**40.0**	U3 V	Acromion Process to Fingertip	30.0	26.7	20.0	13.3	10.0
44.9	**114.0**	U4 V	Vertical Reach to Fingertip	85.5	76.0	57.0	38.0	28.5
23.8	**60.5**	U5 T	Lateral Reach to Fingertip	45.4	40.3	30.3	20.2	15.1
18.9	**48.0**	U6 D	Forward Reach to Fingertip	36.0	32.0	24.0	16.0	12.0
7.5	**19.0**	U7 V	Shoulder to Flexed Elbow	14.3	12.7	9.5	6.3	4.8
6.9	**17.5**	U8 D	Acromion to Lat. Condyle Elbow	13.1	11.7	8.8	5.8	4.4
9.4	**24.0**	U9 D	Flexed Elbow to Fingertip	18.0	16.0	12.0	8.0	6.0
5.5	**14.0**	U10 D	Flexed Elbow Pivot to Wrist Pivot	10.5	9.3	7.0	4.7	3.5

Life size			THREE-YEAR-OLD CHILD	Fractions of life size				
				3/4	2/3	1/2	1/3	1/4
Inches	cm	Code		cm	cm	cm	cm	cm
			UPPER EXTREMITY MEASUREMENTS					
4.5	**11.5**	U11 V	Wrist to Fingertip	8.6	7.7	5.8	3.8	2.9
2.8	**7.0**	U12 V	Wrist to Thumb Tip	5.3	4.7	3.5	2.3	1.8
2.6	**6.7**	U13 V	Length of Palm	5.0	4.5	3.4	2.2	1.7
1.9	**4.8**	U14 V	Length 3rd Finger	3.6	3.2	2.4	1.6	1.2
3.1	**7.8**	U15 V	Thumb Crotch to 3rd Finger	5.9	5.2	3.9	2.6	2.0
1.6	**4.0**	U16 V	Thumb Crotch to Thumb Tip	3.0	2.7	2.0	1.3	1.0
2.0	**5.2**	U17 D	Anterior-Post. Arm at Biceps	3.9	3.5	2.6	1.7	1.3
1.8	**4.5**	U18 D	Anterior-Posterior at Forearm	3.4	3.0	2.3	1.5	1.1
1.1	**2.7**	U19 D	Depth of Wrist	2.0	1.8	1.4	0.9	0.7
1.6	**4.0**	U20 T	Breadth of Wrist	3.0	2.7	2.0	1.3	1.0
2.5	**6.4**	U21 T	Breadth of Hand	4.8	4.3	3.2	2.1	1.6
2.2	**5.5**	U22 T	Breadth of Palm	4.1	3.7	2.8	1.8	1.4
			LOWER EXTREMITY MEASUREMENTS					
20.1	**51.0**	L1 V	Crest of Ilium of Pelvis to Sole	38.3	34.0	25.5	17.0	12.8
19.1	**48.6**	* L2 V	Ant.Sup.Spine of Ilium to Sole	36.5	32.4	24.3	16.2	12.2
16.9	**43.0**	L3 V	Greater Trochanter Femur-Sole	32.3	28.7	21.5	14.3	10.8
16.1	**41.0**	* L4 V	Symphysis Pubis of Pelvis-Sole	30.8	27.3	20.5	13.7	10.3
15.2	**38.5**	* L5 V	Crotch to Sole	28.9	25.7	19.3	12.8	9.6
15.0	**38.0**	L6 V	Gluteal Furrow to Sole	28.5	25.3	19.0	12.7	9.5
9.1	**23.0**	L7 V	Posterior Knee Crease to Sole	17.3	15.3	11.5	7.7	5.8
9.1	**23.0**	L8 V	Tibia at Knee to Sole	17.3	15.3	11.5	7.7	5.8
1.6	**4.0**	L9 V	Med. Malleolus of Ankle to Sole	3.0	2.7	2.0	1.3	1.0
1.4	**3.5**	L10 V	Lat. Malleolus of Ankle to Sole	2.6	2.3	1.8	1.2	0.9
7.5	**19.0**	L11 T	Breadth of Hips, Sitting	14.3	12.7	9.5	6.3	4.8
3.1	**8.0**	L12 T	Breadth Mid Thigh	6.0	5.3	4.0	2.7	2.0
2.8	**7.0**	L13 T	Breadth of Knee	5.3	4.7	3.5	2.3	1.8
2.5	**6.3**	L14 T	Breadth of Calf	4.7	4.2	3.2	2.1	1.6
1.8	**4.5**	L15 T	Breadth Ankle	3.4	3.0	2.3	1.5	1.1
3.6	**9.1**	L16 D	Mid Thigh, Anterior-Posterior	6.8	6.1	4.6	3.0	2.3
2.4	**6.0**	L17 D	Knee, Anterior-Posterior	4.5	4.0	3.0	2.0	1.5
2.6	**6.6**	L18 D	Mid Calf, Anterior-Posterior	5.0	4.4	3.3	2.2	1.7
1.9	**4.8**	L19 D	Ankle, Anterior-Posterior	3.6	3.2	2.4	1.6	1.2
6.3	**16.0**	L20 D	Length of Foot	12.0	10.7	8.0	5.3	4.0
4.5	**11.5**	L21 D	Heel to 5th Toe	8.6	7.7	5.8	3.8	2.9
2.4	**6.0**	L22 T	Breadth Foot	4.5	4	3	2	1.5
1.5	**3.9**	L23 T	Width of Heel	2.9	2.6	2.0	1.3	1.0
		**These are also listed with Vertical Measurements.*						

Superimposed Profiles of Three-Week-Old Infant to Five-Year-Old Child

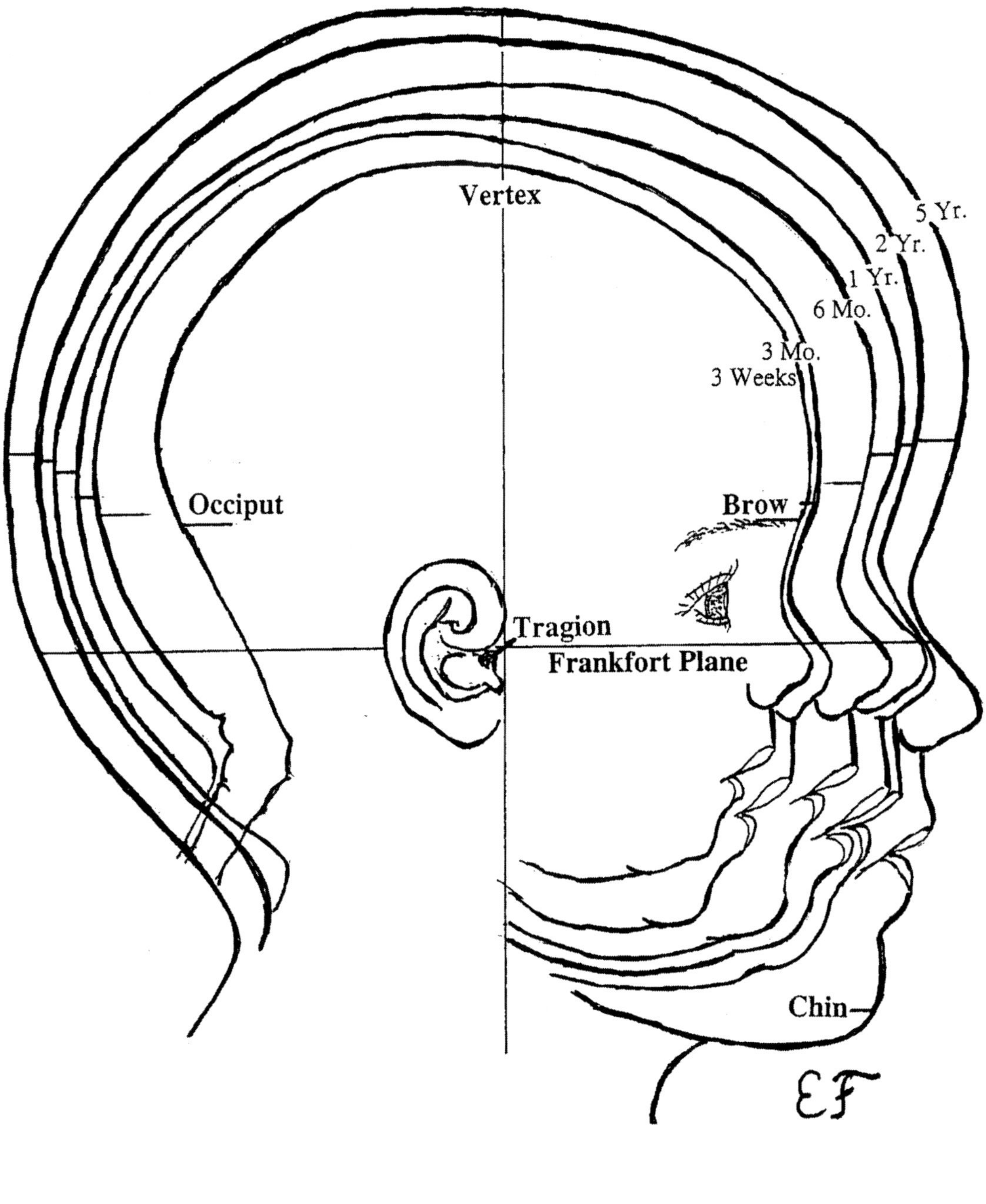

Illustrated is a composite of the profiles of infants, age three-weeks to five-years. The horizontal alignment is along the Frankfort Plane. The vertical alignment is along a ninety degree upright to this plane at the ear canal, or porion. By five years the cranial volume has increased to ninety-one percent of adult size. The mid face and lower face volume has increased only about forty-percent of adult size. This results in an apparent disproportion of cranium, compared to the mask of the face that is from the brow to the chin. This disparity improves as the mask of the face increases and the cranium's additional increase is gradual.

FOUR-YEAR-OLD CHILDREN

There is considerable similarity of male and female four-year-old children. The range of twenty-fifth, fiftieth, and seventy-fifth percentile is shown here for comparison.

Percentile	Twenty-fifth		Fiftieth		Seventy-fifth	
Boys:	**Short**		**Medium**		**Tall**	
Stature	100 cm,	39 1/4 in.	103 cm,	40 1/2 in.	106 cm,	41 1/2 in.
Weight	15.4 kg,	34 lb.	16.7 kg,	36 3/4 lb.	18 kg,	39 3/4 lb.
Girls:						
Stature	99 cm,	39 in.	102 cm,	40 in.	104 cm,	41 in.
Weight	14.9 kg,	32 3/4 lb.	16 kg,	35 1/4 lb.	17.6 kg,	38 3/4 lb.[1]

Male children, on the average, are slightly taller and heavier than female children. Females generally have more rounded contours. Families usually select the hair style and clothing appropriate for gender. Strength and coordination have improved so that the child can throw a ball overhand, hop on one foot, and walk down stairs alternating feet easily. The variety of social life increases as they play with more children As learning increases, artistic creativity advances with more recognizable figures in their crayon drawings. They enjoy having a parent read to them, and word usage increases. They develop a consciousness for colors. The boys prefer mechanical toys while the girls prefer domestic and home making objects, particularly dolls and dresses.

BODY MEASUREMENTS COMPARED TO HEIGHT

	Feature Measure in cm	Feature Compared to Stature	Approx. Ratio of Feature	Compared to Median Adult
Stature: Crown to sole	103			59%
Head Height: Crown to chin	18.5	18%	1/5	93%
Sitting Height:	58	56%	1/2	64%
Trunk Measure: Sternal notch to symphysis pubis	33	32%	1/3	67%
Leg Length: Greater trochanter of femur to sole	47	46%	1/2–	55%
Thigh Length: Greater trochanter to tibia at knee	23	23%	1/4-	53%
Tibia at Knee to Sole:	23.3	23%	1/4–	54%
Arm Length: Shoulder to dependent fingertip	44	43%	2/5	59%
Hand Length:	11.5	11%	1/9	60%
Foot Length:	16	16%	1/6	62%

1. Nelson, W.E., p. 38.

Four-Year-Old Child, anterior and lateral head

all measurements in centimeters

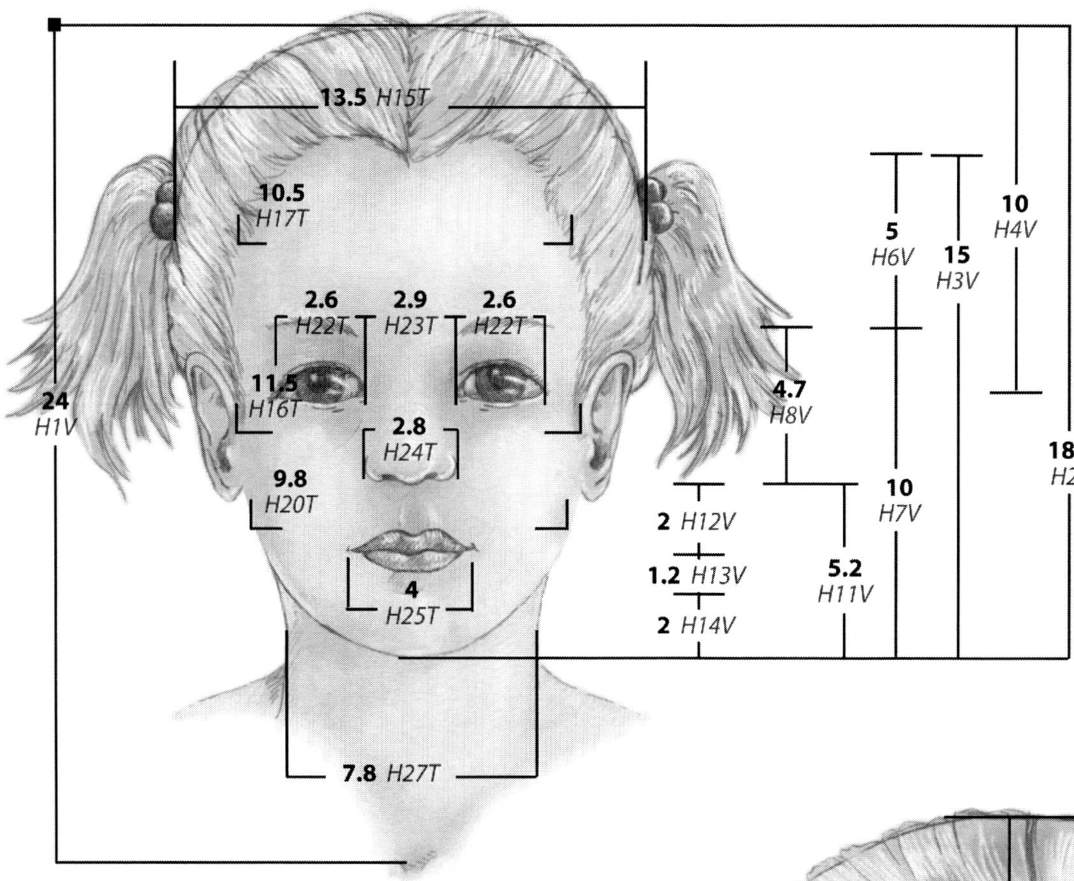

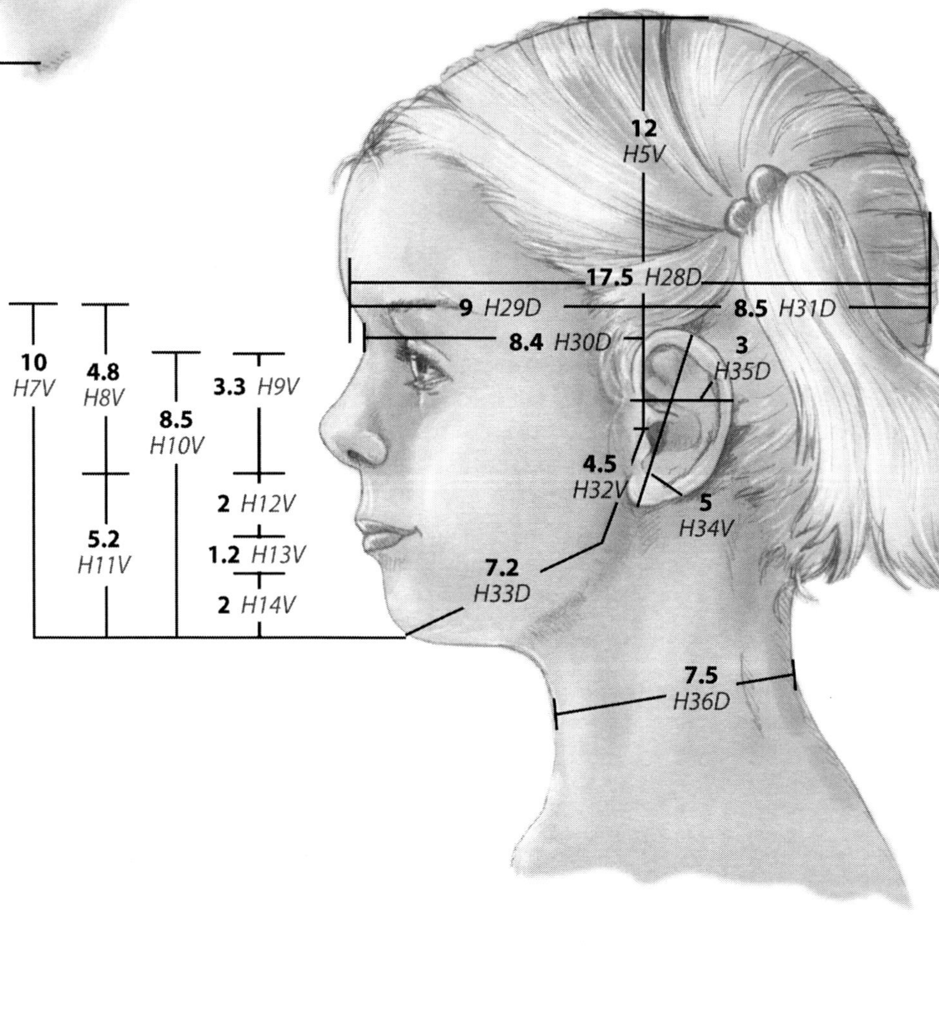

Four-Year-Old Child, anterior and lateral figure

all measurements in centimeters

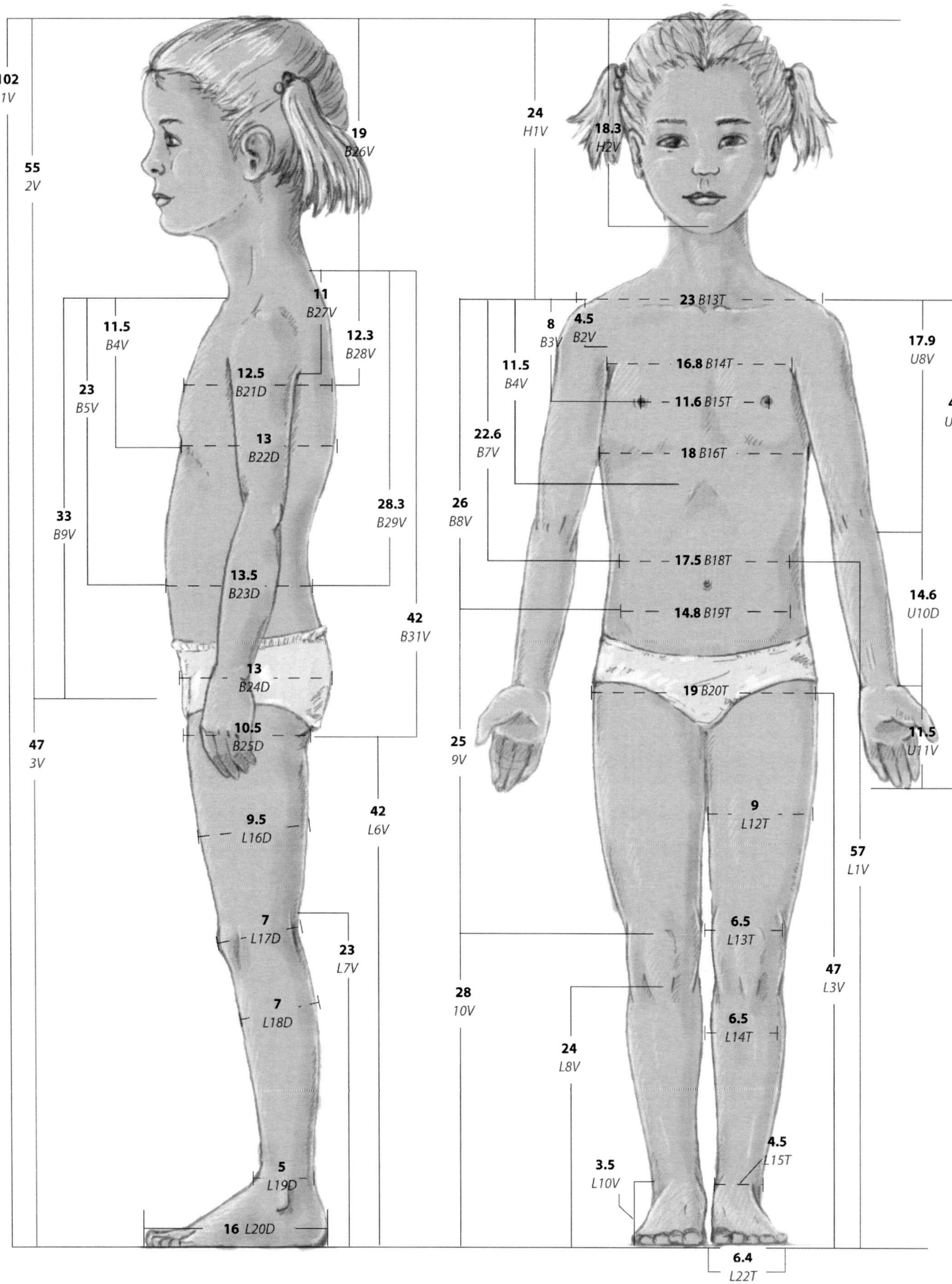

Life size			FOUR-YEAR-OLD CHILD	Fractions of life size				
				3/4	2/3	1/2	1/3	1/4
Inches	cm	Code		cm	cm	cm	cm	cm
			VERTICAL MEASUREMENTS					
40.2	**102.0**	1 V	Stature, Height, Vertex to Sole	76.5	68.0	51.0	34.0	25.5
21.7	**55.0**	2 V	Vertex to Symphysis Pubis	41.3	36.7	27.5	18.3	13.8
18.5	**47.0**	* 3 V	Symphysis Pubis of Pelvis-Sole	35.3	31.3	23.5	15.7	11.8
30.7	**78.0**	4 V	Sternal Notch, Neck Pit, to Sole	58.5	52.0	39.0	26.0	19.5
9.4	**24.0**	5 V	Vertex to Acromion Process	18.0	16.0	12.0	8.0	6.0
31.1	**79.0**	6 V	Acromion Process to Sole	59.3	52.7	39.5	26.3	19.8
20.5	**52.0**	* 7 V	Ant. Sup.Spine of Ilium to Sole	39.0	34.7	26.0	17.3	13.0
10.6	**27.0**	8 V	Acrom. proc. to Ant. Sup. Spine	20.3	18.0	13.5	9.0	6.8
9.8	**25.0**	9 V	Ant.Sup.Spine of Ilium to Knee	18.8	16.7	12.5	8.3	6.3
10.6	**27.0**	10 V	Knee at Patella to Sole	20.3	18.0	13.5	9.0	6.8
21.7	**55.0**	11 V	Umbilicus to Sole	41.3	36.7	27.5	18.3	13.8
16.9	**43.0**	* 12 V	Crotch to Sole	32.3	28.7	21.5	14.3	10.8
29.9	**76.0**	13 V	Kneeling Height	57.0	50.7	38.0	25.3	19.0
22.8	**58.0**	14 V	Sitting Height	43.5	38.7	29.0	19.3	14.5
21.3	**54.0**	15 V	Buttock, Straight Leg to Sole	40.5	36.0	27.0	18.0	13.5
12.2	**31.0**	16 V	Buttock to Flexed Knee	23.3	20.7	15.5	10.3	7.8
12.0	**30.5**	17 V	Flexed Knee to Sole	22.9	20.3	15.3	10.2	7.6
		**These are also listed with Lower Extremity Measurements.*						
			VERTICAL HEAD MEASUREMENTS					
9.4	**24.0**	H1 V	Vertex to Sternal Notch	18.0	16.0	12.0	8.0	6.0
7.2	**18.3**	H2 V	Vertex to Chin, Head Height	13.7	12.2	9.2	6.1	4.6
5.9	**15.0**	H3 V	Hairline to Chin, Face Height	11.3	10.0	7.5	5.0	3.8
3.9	**10.0**	H4 V	Vertex to Lateral Corner of Eye	7.5	6.7	5.0	3.3	2.5
4.7	**12.0**	H5 V	Vertex to Tragion of Ear	9.0	8.0	6.0	4.0	3.0
2.0	**5.0**	H6 V	Hairline to Brow, Glabella	3.8	3.3	2.5	1.7	1.3
3.9	**10.0**	H7 V	Brow, Glabella, to Chin, Menton	7.5	6.7	5.0	3.3	2.5
1.9	**4.8**	H8 V	Brow to Subnasal Point	3.6	3.2	2.4	1.6	1.2
1.3	**3.3**	H9 V	Sellion to Subnasal, Nose Length	2.5	2.2	1.7	1.1	0.8
3.3	**8.5**	H10 V	Sellion, Nose Saddle to Chin	6.4	5.7	4.3	2.8	2.1
2.0	**5.2**	H11 V	Subnasal Point to Chin	3.9	3.5	2.6	1.7	1.3
0.8	**2.0**	H12 V	Subnasal to Lip Aperature	1.5	1.3	1.0	0.7	0.5
0.5	**1.2**	H13 V	Lip Aperat. to Sublabial Crease	0.9	0.8	0.6	0.4	0.3
0.8	**2.0**	H14 V	Sublabial Crease to Chin	1.5	1.3	1.0	0.7	0.5
			TRANSVERSE HEAD MEASUREMENTS					
5.3	**13.5**	H15 T	Bi-Parietal, Head Breadth	10.1	9.0	6.8	4.5	3.4
4.5	**11.5**	H16 T	Bi-Zygomatic, Cheekbones	8.6	7.7	5.8	3.8	2.9
4.1	**10.5**	H17 T	Bi-Fronto Temporal, Forehead	7.9	7.0	5.3	3.5	2.6

Life size			FOUR-YEAR-OLD CHILD	Fractions of life size				
				3/4	2/3	1/2	1/3	1/4
Inches	cm	Code		cm	cm	cm	cm	cm
			TRANSVERSE HEAD MEASUREMENTS					
3.3	**8.5**	H18 T	Inter-Brow Ridge	6.4	5.7	4.3	2.8	2.1
4.5	**11.5**	H19 T	Inter-Tragion of Ears	8.6	7.7	5.8	3.8	2.9
3.9	**9.8**	H20 T	Inter-Angle of Mandible	7.4	6.5	4.9	3.3	2.5
3.1	**8.0**	H21 T	Inter-Lateral Corners of Eyes	6.0	5.3	4.0	2.7	2.0
1.0	**2.6**	H22 T	Width of Eye	2.0	1.7	1.3	0.9	0.7
1.1	**2.8**	H23 T	Inter-Inner Corner of Eyes	2.1	1.9	1.4	0.9	0.7
1.1	**2.8**	H24 T	Breadth of Nose	2.1	1.9	1.4	0.9	0.7
1.6	**4.0**	H25 T	Breadth of Mouth	3.0	2.7	2.0	1.3	1.0
1.0	**2.6**	H26 T	Breadth of Chin	2.0	1.7	1.3	0.9	0.7
3.1	**7.8**	H27 T	Breadth of Neck	5.9	5.2	3.9	2.6	2.0
			ANTERIOR-POSTERIOR HEAD MEASUREMENTS					
6.9	**17.5**	H28 D	Brow to Occiput, Posterior Head	13.1	11.7	8.8	5.8	4.4
3.5	**9.0**	H29 D	Brow to Tragus of Ear	6.8	6.0	4.5	3.0	2.3
3.3	**8.4**	H30 D	Sellion, Nose Saddle, to Tragus	6.3	5.6	4.2	2.8	2.1
3.3	**8.5**	H31 D	Tragus to Occiput, Back of Head	6.4	5.7	4.3	2.8	2.1
1.8	**4.5**	H32 V	Tragus to Angle Mandible	3.4	3.0	2.3	1.5	1.1
2.8	**7.2**	H33 D	Angle of Mandible to Chin	5.4	4.8	3.6	2.4	1.8
2.0	**5.0**	H34 V	Length of Ear, Vertical	3.8	3.3	2.5	1.7	1.3
1.2	**3.0**	H35 D	Width of Ear, Anterior-Posterior	2.3	2.0	1.5	1.0	0.8
3.0	**7.5**	H36 D	Anterior to Posterior of Neck	5.6	5.0	3.8	2.5	1.9
			VERTICAL BODY MEASUREMENTS					
1.8	**4.5**	B1 V	Chin to Sternal Notch, Neck Pit	3.4	3.0	2.3	1.5	1.1
1.8	**4.5**	B2 V	Sternal Notch to Axilla, Arm Pit	3.4	3.0	2.3	1.5	1.1
3.1	**8.0**	B3 V	Sternal Notch to Nipples	6.0	5.3	4.0	2.7	2.0
4.5	**11.5**	B4 V	Sternal Notch to Xyphoid Proc.	8.6	7.7	5.8	3.8	2.9
9.1	**23.0**	B5 V	Sternal Notch to Umbilicus	17.3	15.3	11.5	7.7	5.8
4.5	**11.5**	B6 V	Xyphoid Process to Umbilicus	8.6	7.7	5.8	3.8	2.9
8.9	**22.6**	B7 V	Sternal Notch to Crest Ilium	17.0	15.1	11.3	7.5	5.7
10.2	**26.0**	B8 V	Sternal Notch to Ant.S.Sp.Ilium	19.5	17.3	13.0	8.7	6.5
13.0	**33.0**	B9 V	Sternal Notch to Pubis	24.8	22.0	16.5	11.0	8.3
4.3	**11.0**	B10 V	Umbilicus to Pubis of Pelvis	8.3	7.3	5.5	3.7	2.8
			TRANSVERSE BODY MEASUREMENTS					
10.2	**26.0**	B11 T	Shoulder Breadth	19.5	17.3	13.0	8.7	6.5
4.4	**11.3**	B12 T	Clavicle to Acromion Process	8.5	7.5	5.7	3.8	2.8
9.1	**23.0**	B13 T	Bi-acromial Process Breadth	17.3	15.3	11.5	7.7	5.8
6.6	**16.8**	B14 T	Inter-Anterior Axillary Fold	12.6	11.2	8.4	5.6	4.2
4.6	**11.6**	B15 T	Inter-Nipples	8.7	7.7	5.8	3.9	2.9

Life size			FOUR-YEAR-OLD CHILD	Fractions of life size				
				3/4	2/3	1/2	1/3	1/4
Inches	cm	Code		cm	cm	cm	cm	cm
			TRANSVERSE BODY MEASUREMENTS					
7.1	**18.0**	B16 T	Breadth of Thorax	13.5	12.0	9.0	6.0	4.5
7.1	**18.0**	B17 T	Inter-Costal. Rib Margin Breadth	13.5	12.0	9.0	6.0	4.5
6.9	**17.5**	B18 T	Inter-Crest of Ilium Breadth	13.1	11.7	8.8	5.8	4.4
5.8	**14.8**	B19 T	Inter-Anterior Superior Spine	11.1	9.9	7.4	4.9	3.7
7.5	**19.0**	B20 T	Inter-Greater Trochanter Femur	14.3	12.7	9.5	6.3	4.8
			ANTERIOR-POSTERIOR BODY MEASUREMENTS					
4.9	**12.5**	B21 D	Manubrium to Dorsal Spine	9.4	8.3	6.3	4.2	3.1
5.1	**13.0**	B22 D	Xyphoid Process to Dorsum	9.8	8.7	6.5	4.3	3.3
5.3	**13.5**	B23 D	Umbilicus to Dorsum	10.1	9.0	6.8	4.5	3.4
5.1	**13.0**	B24 D	Pubis of Pelvis to Buttock	9.8	8.7	6.5	4.3	3.3
4.1	**10.5**	B25 D	Thigh to Gluteal Furrow	7.9	7.0	5.3	3.5	2.6
			VERTICAL BACK MEASUREMENTS					
7.5	**19.0**	B26 V	Vertex to 7th Cervical Spine	14.3	12.7	9.5	6.3	4.8
			7th Cervical Vertebra Spine to:		0.0	0.0	0.0	0.0
4.3	**11.0**	B27 V	To Posterior Axillary Fold, Armpit	8.3	7.3	5.5	3.7	2.8
4.8	**12.3**	B28 V	To Inferior Tip of Scapula	9.2	8.2	6.2	4.1	3.1
11.1	**28.3**	B29 V	To Posterior Superior Spine Ilium	21.2	18.9	14.2	9.4	7.1
13.0	**33.0**	B30 V	To Midline Upper Buttock Crease	24.8	22.0	16.5	11.0	8.3
16.5	**42.0**	B31 V	To Gluteal Furrow	31.5	28.0	21.0	14.0	10.5
22.8	**58.0**	B32 V	To Popliteal Crease Behind Knee	43.5	38.7	29.0	19.3	14.5
			TRANSVERSE BACK MEASUREMENTS					
8.3	**21.0**	B33 T	Inter-Posterior Axillary Folds	15.8	14.0	10.5	7.0	5.3
4.3	**11.0**	B34 T	Inter-Scapulae Medial Borders	8.3	7.3	5.5	3.7	2.8
2.2	**5.5**	B35 T	Inter-Post. Sup. Spines Sacrum	4.1	3.7	2.8	1.8	1.4
			UPPER EXTREMITY MEASUREMENTS					
39.0	**99.0**	U1 V	Span of Arms	74.3	66.0	49.5	33.0	24.8
4.3	**11.0**	U2 V	Sternum to Acromion Process	8.3	7.3	5.5	3.7	2.8
17.3	**44.0**	U3 V	Acromion Process to Fingertip	33.0	29.3	22.0	14.7	11.0
48.8	**124.0**	U4 V	Vertical Reach to Fingertip	93.0	82.7	62.0	41.3	31.0
25.2	**64.0**	U5 T	Lateral Reach to Fingertip	48.0	42.7	32.0	21.3	16.0
19.7	**50.0**	U6 D	Forward Reach to Fingertip	37.5	33.3	25.0	16.7	12.5
7.9	**20.0**	U7 V	Shoulder to Flexed Elbow	15.0	13.3	10.0	6.7	5.0
7.0	**17.9**	U8 V	Acromion to Lat. Condyle Elbow	13.4	11.9	9.0	6.0	4.5
10.5	**26.6**	U9 D	Flexed Elbow to Fingertip	20.0	17.7	13.3	8.9	6.7
5.7	**14.6**	U10 D	Flexed Elbow Pivot to Wrist Pivot	11.0	9.7	7.3	4.9	3.7

Life size			FOUR-YEAR-OLD CHILD	Fractions of life size				
				3/4	2/3	1/2	1/3	1/4
Inches	cm	Code		cm	cm	cm	cm	cm
			UPPER EXTREMITY MEASUREMENTS					
4.5	**11.5**	U11 V	Wrist to Fingertip	8.6	7.7	5.8	3.8	2.9
3.0	**7.5**	U12 V	Wrist to Thumb Tip	5.6	5.0	3.8	2.5	1.9
2.6	**6.6**	U13 V	Length of Palm	5.0	4.4	3.3	2.2	1.7
1.9	**4.8**	U14 V	Length 3rd Finger	3.6	3.2	2.4	1.6	1.2
3.1	**8.0**	U15 V	Thumb Crotch to 3rd Finger	6.0	5.3	4.0	2.7	2.0
1.6	**4.0**	U16 V	Thumb Crotch to Thumb Tip	3.0	2.7	2.0	1.3	1.0
2.2	**5.5**	U17 D	Anterior-Post. Arm at Biceps	4.1	3.7	2.8	1.8	1.4
2.0	**5.0**	U18 D	Anterior-Posterior at Forearm	3.8	3.3	2.5	1.7	1.3
1.2	**3.0**	U19 D	Depth of Wrist	2.3	2.0	1.5	1.0	0.8
1.7	**4.2**	U20 T	Breadth of Wrist	3.2	2.8	2.1	1.4	1.1
2.6	**6.6**	U21 T	Breadth of Hand	5.0	4.4	3.3	2.2	1.7
2.2	**5.5**	U22 T	Breadth of Palm	4.1	3.7	2.8	1.8	1.4
			LOWER EXTREMITY MEASUREMENTS					
22.4	**57.0**	L1 V	Crest of Ilium of Pelvis to Sole	42.8	38.0	28.5	19.0	14.3
20.7	**52.5**	* L2 V	Ant.Sup.Spine of Ilium to Sole	39.4	35.0	26.3	17.5	13.1
18.5	**47.0**	L3 V	Greater Trochanter Femur-Sole	35.3	31.3	23.5	15.7	11.8
18.5	**47.0**	* L4 V	Symphysis Pubis of Pelvis-Sole	35.3	31.3	23.5	15.7	11.8
16.9	**43.0**	* L5 V	Crotch to Sole	32.3	28.7	21.5	14.3	10.8
16.5	**42.0**	L6 V	Gluteal Furrow to Sole	31.5	28.0	21.0	14.0	10.5
9.1	**23.0**	L7 V	Posterior Knee Crease to Sole	17.3	15.3	11.5	7.7	5.8
9.4	**24.0**	L8 V	Tibia at Knee to Sole	18.0	16.0	12.0	8.0	6.0
1.8	**4.5**	L9 V	Med. Malleolus of Ankle to Sole	3.4	3.0	2.3	1.5	1.1
1.4	**3.5**	L10 V	Lat. Malleolus of Ankle to Sole	2.6	2.3	1.8	1.2	0.9
7.9	**20.0**	L11 T	Breadth of Hips, Sitting	15.0	13.3	10.0	6.7	5.0
3.5	**9.0**	L12 T	Breadth Mid Thigh	6.8	6.0	4.5	3.0	2.3
2.8	**7.2**	L13 T	Breadth of Knee	5.4	4.8	3.6	2.4	1.8
2.6	**6.5**	L14 T	Breadth of Calf	4.9	4.3	3.3	2.2	1.6
1.8	**4.5**	L15 T	Breadth Ankle	3.4	3.0	2.3	1.5	1.1
3.7	**9.5**	L16 D	Mid Thigh, Anterior-Posterior	7.1	6.3	4.8	3.2	2.4
2.8	**7.0**	L17 D	Knee, Anterior-Posterior	5.3	4.7	3.5	2.3	1.8
2.8	**7.0**	L18 D	Mid Calf, Anterior-Posterior	5.3	4.7	3.5	2.3	1.8
2.0	**5.0**	L19 D	Ankle, Anterior-Posterior	3.8	3.3	2.5	1.7	1.3
6.3	**16.0**	L20 D	Length of Foot	12.0	10.7	8.0	5.3	4.0
4.7	**12.0**	L21 D	Heel to 5th Toe	9.0	8.0	6.0	4.0	3.0
2.5	**6.4**	L22 T	Breadth Foot	4.8	4.3	3.2	2.1	1.6
1.5	**4.0**	L23 T	Width of Heel	2.9	2.5	1.9	1.3	1.0
		**These are also listed with Vertical Measurements.*						

FIVE-YEAR-OLD GIRLS

The lists of body proportions, for each sex, age five and older, are recorded separately in the tables of measurements. At this age most have achieved sixty-six percent of expected adult height. The ranges of twenty-fifth, fiftieth, and seventy-fifth percentile of stature are listed here for comparison.

Percentile	Twenty-fifth		Fiftieth		Seventy-fifth	
Girls:	Short		Medium		Tall	
Stature	105 cm,	41.5 in.	108 cm,	42.7 in.	113 cm,	44.7 in.
Weight	16.3 kg,	36 lb.	17.7 kg,	39 lb.	19.4 kg,	42.7 lb.[1]

At age five, girls are usually not as tall and weigh less than boys. The head has had considerable growth, particularly the cranium. The brain has increased to about ninety percent of adult size. The mask of the face, brow to chin, has not grown as rapidly. There may be an appearance of some disproportionate features. The most noticeable difference between boys and girls is the hair styling and clothing. However, on closer inspection, girls may have more rounded contours in face, shoulders, and hips. Their interests already are more in toys, and animals. Fancy dresses, play house, and dolls attract their attention. Hop Scotch is a popular game. Color appreciation has increased and creativity has advanced. School level is kindergarten.

BODY MEASUREMENTS COMPARED TO HEIGHT

	Feature Measure in cm	Feature Compared to Stature	Approx. Ratio of Feature	Compared to Median Adult
Stature: Crown to sole	109			65%
Head Height: Crown to chin	17.7	17%	1/6	87%
Sitting Height:	60	55%	1/2	67%
Trunk Measure: Sternal notch to symphysis pubis	34	32%	1/3	69%
Leg Length: Greater trochanter of femur to sole	52	48%	1/2	61%
Thigh Length: Greater trochanter to tibia at knee	26	24%	1/4	60%
Tibia at Knee to Sole:	26	24%	1/4	60%
Arm Length: Shoulder to dependent fingertip	46	42%	2/5+	61%
Hand Length:	12	11%	1/9	67%
Foot Length:	16.8	15%	1/6	70%

The deciduous incisor teeth are lost between five and six years of age.[2]

1. Nelson, W.E., p. 38.
2. Lockhart, R.D., p. 57.

FIVE-YEAR-OLD BOYS

The list of body proportions, for each sex, age five and older, are recorded separately in the tables of measurements. The range of twenty-fifth, fiftieth, and seventy-fifth percentile of stature is shown here for comparison.

Percentile	Twenty-fifth		Fiftieth		Seventy-fifth	
Boys:	Short		Medium		Tall	
Stature	106.5 cm,	42 in.	110 cm,	43.25 in.	113 cm,	44.5 in.
Weight	17 kg,	38 lb.	18.7 kg,	41.25 lb.	20 kg,	44.5 lb.[1]

On the average, boys are slightly taller and heavier than girls. The usual noticeable differences are in hair styling and clothing. By age five, the cranium has grown rapidly. The brain has increased to ninety percent of adult size. The face has not grown as rapidly. This results in the face looking too small or disproportionate for the rest of the head. It is more noticeable in boys since they have shorter haircuts. Body stem length has nearly doubled, and leg length has nearly tripled from newborn measurements. The arms, hands, and feet have grown to about 60% of adult size, and this is more growth than other bones of the leg. Little boys already identify with gun-toting cowboys, athletes, heroes performing daring deeds, monsters, dinosaurs, and mechanical toys. School level is kindergarten.

BODY MEASUREMENTS COMPARED TO HEIGHT

	Feature Measure in cm	Feature Compared to Stature	Approx. Ratio of Feature	Compared to Median Adult
Stature: Crown to sole	110			60%
Head Height: Crown to chin	18.5	17%	1/6	83%
Sitting Height:	61	55%	1/2+	64%
Trunk Measure: Sternal notch to symphysis pubis	35	32%	1/3	66%
Leg Length: Greater trochanter of femur to sole	52	47%	1/2	53%
Thigh Length: Greater trochanter to tibia at knee	25	23%	1/4	53%
Tibia at Knee to Sole:	27	25%	1/4	52%
Arm Length: Shoulder to dependant fingertip	48	44%	2/5+	59%
Hand Length:	12	11%	1/9	60%
Foot Length:	17.2	16%	1/6	61%

The deciduous incisor teeth are lost between five and six years of age.[2]

1. Nelson, W.E., p. 38.
2. Lockhart, R.D., p. 57.

Five-Year-Old Girl, anterior and lateral head

all measurements in centimeters

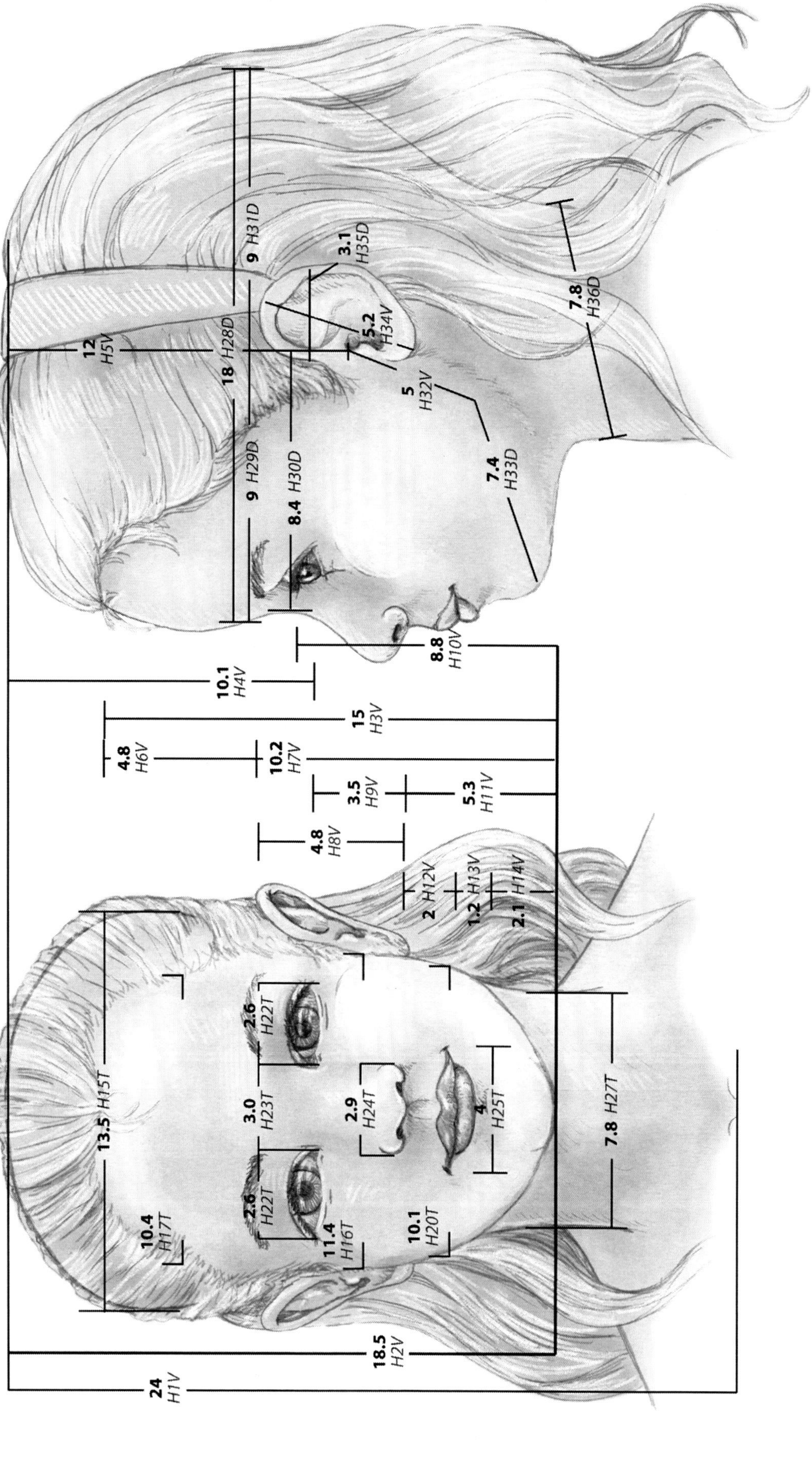

Five-Year-Old Girl, anterior and lateral figure

all measurements in centimeters

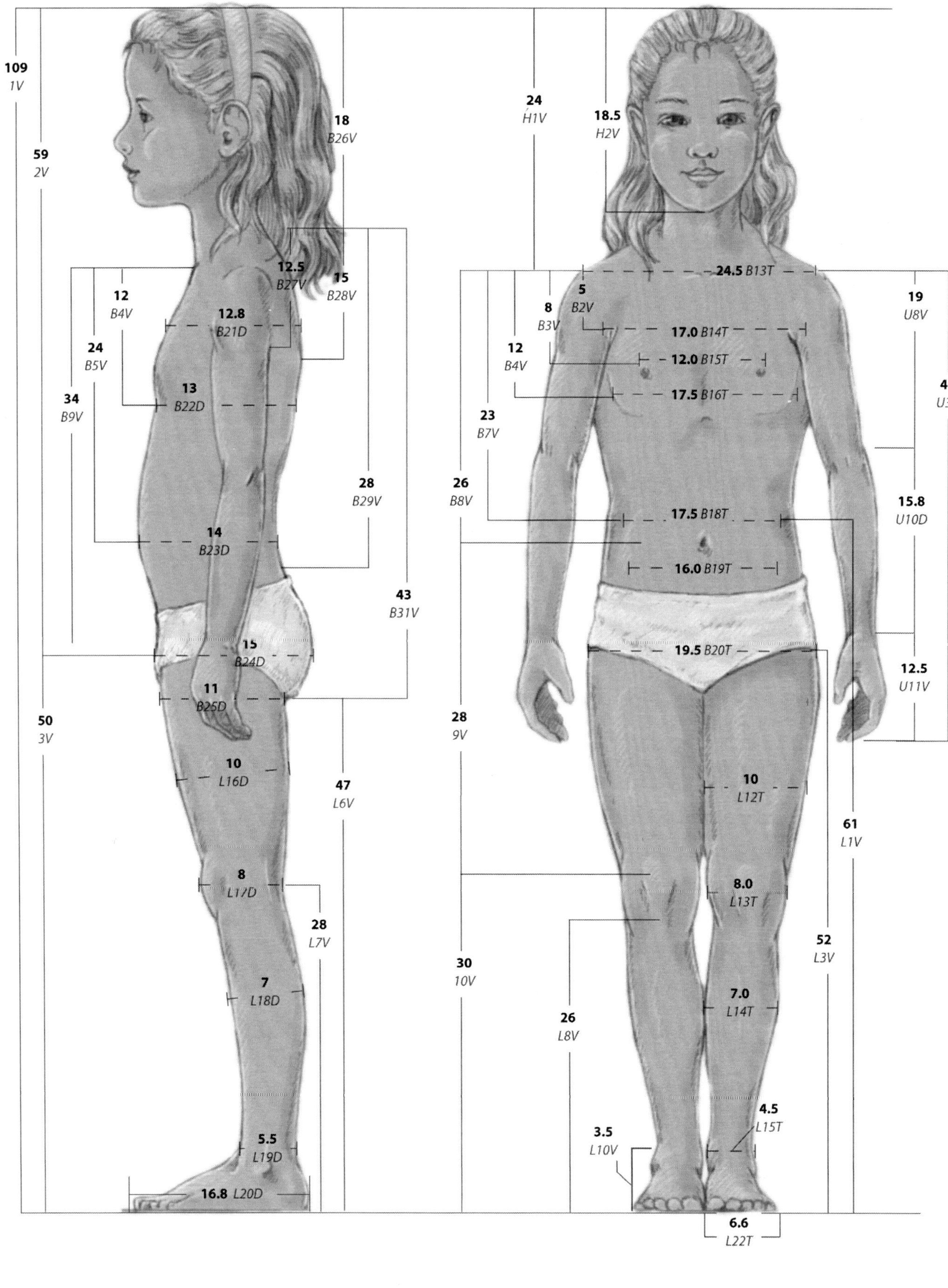

Life size			FIVE-YEAR-OLD GIRL	Fractions of life size				
Inches	cm	Code		3/4 cm	2/3 cm	1/2 cm	1/3 cm	1/4 cm
			VERTICAL MEASUREMENT					
42.9	**109.0**	1 V	Stature, Height, Vertex to Sole	81.8	72.7	54.5	36.3	27.3
23.2	**59.0**	2 V	Vertex to Symphysis Pubis	44.3	39.3	29.5	19.7	14.8
19.7	**50.0**	* 3 V	Symphysis Pubis of Pelvis-Sole	37.5	33.3	25.0	16.7	12.5
33.1	**84.0**	4 V	Sternal Notch, Neck Pit, to Sole	63.0	56.0	42.0	28.0	21.0
9.4	**24.0**	5 V	Vertex to Acromion Process	18.0	16.0	12.0	8.0	6.0
33.5	**85.0**	6 V	Acromion Process to Sole	63.8	56.7	42.5	28.3	21.3
22.8	**58.0**	* 7 V	Ant.Sup.Spine of Ilium to Sole	43.5	38.7	29.0	19.3	14.5
10.6	**27.0**	8 V	Acrom. proc. to Ant. Sup. Spine	20.3	18.0	13.5	9.0	6.8
11.0	**28.0**	9 V	Ant.Sup.Spine of Ilium to Knee	21.0	18.7	14.0	9.3	7.0
11.8	**30.0**	10 V	Knee at Patella to Sole	22.5	20.0	15.0	10.0	7.5
23.6	**60.0**	11 V	Umbilicus to Sole	45.0	40.0	30.0	20.0	15.0
18.5	**47.0**	* 12 V	Crotch to Sole	35.3	31.3	23.5	15.7	11.8
33.1	**84.0**	13 V	Kneeling Height	63.0	56.0	42.0	28.0	21.0
23.6	**60.0**	14 V	Sitting Height	45.0	40.0	30.0	20.0	15.0
21.7	**55.0**	15 V	Buttock, Straight Leg to Sole	41.3	36.7	27.5	18.3	13.8
13.6	**34.5**	16 V	Buttock to Flexed Knee	25.9	23.0	17.3	11.5	8.6
12.8	**32.5**	17 V	Flexed Knee to Sole	24.4	21.7	16.3	10.8	8.1
		These are also listed with Lower Extremity Measurements.						
			VERTICAL HEAD MEASUREMENTS					
9.4	**24.0**	H1 V	Vertex to Sternal Notch	18.0	16.0	12.0	8.0	6.0
7.3	**18.5**	H2 V	Vertex to Chin, Head Height	13.9	12.3	9.3	6.2	4.6
5.9	**15.0**	H3 V	Hairline to Chin, Face Height	11.3	10.0	7.5	5.0	3.8
4.0	**10.1**	H4 V	Vertex to Lateral Corner of Eye	7.6	6.7	5.1	3.4	2.5
4.7	**12.0**	H5 V	Vertex to Tragion of Ear	9.0	8.0	6.0	4.0	3.0
1.9	**4.8**	H6 V	Hairline to Brow, Glabella	3.6	3.2	2.4	1.6	1.2
4.0	**10.2**	H7 V	Brow, Glabella, to Chin, Menton	7.7	6.8	5.1	3.4	2.6
1.9	**4.8**	H8 V	Brow to Subnasal Point	3.6	3.2	2.4	1.6	1.2
1.4	**3.5**	H9 V	Sellion to Subnasal, Nose Length	2.6	2.3	1.8	1.2	0.9
3.5	**8.8**	H10 V	Sellion, Nose Saddle, to Chin	6.6	5.9	4.4	2.9	2.2
2.1	**5.3**	H11 V	Subnasal Point to Chin	4.0	3.5	2.7	1.8	1.3
0.8	**2.0**	H12 V	Subnasal to Lip Aperature	1.5	1.3	1.0	0.7	0.5
0.5	**1.2**	H13 V	Lip Aperat. to Sublabial Crease	0.9	0.8	0.6	0.4	0.3
0.8	**2.1**	H14 V	Sublabial Crease to Chin	1.6	1.4	1.1	0.7	0.5
			TRANSVERSE HEAD MEASUREMENTS					
5.3	**13.5**	H15 T	Bi-Parietal, Head Breadth	10.1	9.0	6.8	4.5	3.4
4.5	**11.4**	H16 T	Bi-Zygomatic, Cheekbones	8.6	7.6	5.7	3.8	2.9
4.1	**10.4**	H17 T	Bi-Fronto Temporal, Forehead	7.8	6.9	5.2	3.5	2.6

Life size			FIVE-YEAR-OLD GIRL	Fractions of life size				
				3/4	2/3	1/2	1/3	1/4
Inches	cm	Code		cm	cm	cm	cm	cm
			TRANSVERSE HEAD MEASUREMENTS					
3.4	8.7	H18 T	Inter-Brow Ridge	6.5	5.8	4.4	2.9	2.2
4.3	11.0	H19 T	Inter-Tragion of Ears	8.3	7.3	5.5	3.7	2.8
4.0	10.1	H20 T	Inter-Angle of Mandible	7.6	6.7	5.1	3.4	2.5
3.2	8.2	H21 T	Inter-Lateral Corners of Eyes	6.2	5.5	4.1	2.7	2.1
1.0	2.6	H22 T	Width of Eye	2.0	1.7	1.3	0.9	0.7
1.2	3.0	H23 T	Inter-Inner Corner of Eyes	2.3	2.0	1.5	1.0	0.8
1.2	3.0	H24 T	Breadth of Nose	2.3	2.0	1.5	1.0	0.8
1.6	4.0	H25 T	Breadth of Mouth	3.0	2.7	2.0	1.3	1.0
1.1	2.9	H26 T	Breadth of Chin	2.2	1.9	1.5	1.0	0.7
3.1	7.8	H27 T	Breadth of Neck	5.9	5.2	3.9	2.6	2.0
			ANTERIOR-POSTERIOR HEAD MEASUREMENTS					
7.1	18.0	H28 D	Brow to Occiput, Posterior Head	13.5	12.0	9.0	6.0	4.5
3.5	9.0	H29 D	Brow to Tragus of Ear	6.8	6.0	4.5	3.0	2.3
3.3	8.4	H30 D	Sellion, Nose Saddle, to Tragus	6.3	5.6	4.2	2.8	2.1
3.5	9.0	H31 D	Tragus to Occiput, Back of Head	6.8	6.0	4.5	3.0	2.3
2.0	5.0	H32 V	Tragus to Angle Mandible	3.8	3.3	2.5	1.7	1.3
2.9	7.4	H33 D	Angle of Mandible to Chin	5.6	4.9	3.7	2.5	1.9
2.0	5.2	H34 V	Length of Ear, Vertical	3.9	3.5	2.6	1.7	1.3
1.2	3.1	H35 D	Width of Ear, Anterior-Posterior	2.3	2.1	1.6	1.0	0.8
3.1	7.8	H36 D	Anterior to Posterior of Neck	5.9	5.2	3.9	2.6	2.0
			VERTICAL BODY MEASUREMENTS					
2.4	6.0	B1 V	Chin to Sternal Notch, Neck Pit	4.5	4.0	3.0	2.0	1.5
2.0	5.0	B2 V	Sternal Notch to Axilla, Arm Pit	3.8	3.3	2.5	1.7	1.3
3.1	8.0	B3 V	Sternal Notch to Nipples	6.0	5.3	4.0	2.7	2.0
4.7	12.0	B4 V	Sternal Notch to Xyphoid Proc.	9.0	8.0	6.0	4.0	3.0
9.4	24.0	B5 V	Sternal Notch to Umbilicus	18.0	16.0	12.0	8.0	6.0
4.7	12.0	B6 V	XyphoidProcess to Umbilicus	9.0	8.0	6.0	4.0	3.0
9.1	23.0	B7 V	Sternal Notch to Crest Ilium	17.3	15.3	11.5	7.7	5.8
10.2	26.0	B8 V	Sternal Notch to Ant.S.Sp.Ilium	19.5	17.3	13.0	8.7	6.5
13.4	34.0	B9 V	Sternal Notch to Pubis	25.5	22.7	17.0	11.3	8.5
4.3	11.0	B10 V	Umbilicus to Pubis of Pelvis	8.3	7.3	5.5	3.7	2.8
			TRANSVERSE BODY MEASUREMENTS					
10.6	27.0	B11 T	Shoulder Breadth	20.3	18.0	13.5	9.0	6.8
4.7	12.0	B12 T	Clavicle to Acromion Process	9.0	8.0	6.0	4.0	3.0
9.6	24.5	B13 T	Bi-acromial Process Breadth	18.4	16.3	12.3	8.2	6.1
6.7	17.0	B14 T	Inter-Anterior Axillary Fold	12.8	11.3	8.5	5.7	4.3
4.7	12.0	B15 T	Inter-Nipples	9.0	8.0	6.0	4.0	3.0

Life size			FIVE-YEAR-OLD GIRL	Fractions of life size				
				3/4	2/3	1/2	1/3	1/4
Inches	cm	Code		cm	cm	cm	cm	cm
			TRANSVERSE BODY MEASUREMENTS					
6.9	**17.5**	B16 T	Breadth of Thorax	13.1	11.7	8.8	5.8	4.4
6.7	**17.0**	B17 T	Inter-Costal. Rib Margin Breadth	12.8	11.3	8.5	5.7	4.3
6.9	**17.5**	B18 T	Inter-Crest of Ilium Breadth	13.1	11.7	8.8	5.8	4.4
6.3	**16.0**	B19 T	Inter-Anterior Superior Spine	12.0	10.7	8.0	5.3	4.0
7.7	**19.5**	B20 T	Inter-Greater Trochanter Femur	14.6	13.0	9.8	6.5	4.9
			ANTERIOR-POSTERIOR BODY MEASUREMENTS					
5.0	**12.8**	B21 D	Manubrium to Dorsal Spine	9.6	8.5	6.4	4.3	3.2
5.1	**13.0**	B22 D	Xyphoid Process to Dorsum	9.8	8.7	6.5	4.3	3.3
5.5	**14.0**	B23 D	Umbilicus to Dorsum	10.5	9.3	7.0	4.7	3.5
5.9	**15.0**	B24 D	Pubis of Pelvis to Buttock	11.3	10.0	7.5	5.0	3.8
4.3	**11.0**	B25 D	Thigh to Gluteal Furrow	8.3	7.3	5.5	3.7	2.8
			VERTICAL BACK MEASUREMENTS					
7.1	**18.0**	B26 V	Vertex to 7th Cervical Spine	13.5	12.0	9.0	6.0	4.5
			7th Cervical Vertebra Spine to:					
4.9	**12.5**	B27 V	To Posterior Axillary Fold, Armpit	9.4	8.3	6.3	4.2	3.1
5.9	**15.0**	B28 V	To Inferior Tip of Scapula	11.3	10.0	7.5	5.0	3.8
11.0	**28.0**	B29 V	To Posterior Superior Spine Ilium	21.0	18.7	14.0	9.3	7.0
13.8	**35.0**	B30 V	To Midline Upper Buttock Crease	26.3	23.3	17.5	11.7	8.8
16.9	**43.0**	B31 V	To Gluteal Furrow	32.3	28.7	21.5	14.3	10.8
22.8	**58.0**	B32 V	To Popliteal Crease Behind Knee	43.5	38.7	29.0	19.3	14.5
			TRANSVERSE BACK MEASUREMENTS					
9.1	**23.0**	B33 T	Inter-Posterior Axillary Folds	17.3	15.3	11.5	7.7	5.8
4.3	**11.0**	B34 T	Inter-Scapulae Medial Borders	8.3	7.3	5.5	3.7	2.8
2.2	**5.5**	B35 T	Inter-Post. Sup. Spines Sacrum	4.1	3.7	2.8	1.8	1.4
			UPPER EXTREMITY MEASUREMENTS					
42.1	**107.0**	U1 V	Span of Arms	80.3	71.3	53.5	35.7	26.8
4.6	**11.8**	U2 V	Sternum to Acromion Process	8.9	7.9	5.9	3.9	3.0
18.1	**46.0**	U3 V	Acromion Process to Fingertip	34.5	30.7	23.0	15.3	11.5
51.6	**131.0**	U4 V	Vertical Reach to Fingertip	98.3	87.3	65.5	43.7	32.8
26.4	**67.0**	U5 T	Lateral Reach to Fingertip	50.3	44.7	33.5	22.3	16.8
20.9	**53.0**	U6 D	Forward Reach to Fingertip	39.8	35.3	26.5	17.7	13.3
8.7	**22.0**	U7 V	Shoulder to Flexed Elbow	16.5	14.7	11.0	7.3	5.5
7.5	**19.0**	U8 V	Acromion to Lat. Condyle Elbow	14.3	12.7	9.5	6.3	4.8
11.1	**28.2**	U9 D	Flexed Elbow to Fingertip	21.2	18.8	14.1	9.4	7.1
6.2	**15.8**	U10 D	Flexed Elbow Pivot to Wrist Pivot	11.9	10.5	7.9	5.3	4.0

Life size			FIVE-YEAR-OLD GIRL	Fractions of life size				
				3/4	2/3	1/2	1/3	1/4
Inches	cm	Code		cm	cm	cm	cm	cm
			UPPER EXTREMITY MEASUREMENTS					
4.7	**12.0**	U11 V	Wrist to Fingertip	9.0	8.0	6.0	4.0	3.0
3.1	**8.0**	U12 V	Wrist to Thumb Tip	6.0	5.3	4.0	2.7	2.0
2.8	**7.0**	U13 V	Length of Palm	5.3	4.7	3.5	2.3	1.8
2.0	**5.0**	U14 V	Length 3rd Finger	3.8	3.3	2.5	1.7	1.3
3.3	**8.4**	U15 V	Thumb Crotch to 3rd Finger	6.3	5.6	4.2	2.8	2.1
1.6	**4.0**	U16 V	Thumb Crotch to Thumb Tip	3.0	2.7	2.0	1.3	1.0
2.1	**5.3**	U17 D	Anterior-Post. Arm at Biceps	4.0	3.5	2.7	1.8	1.3
2.0	**5.0**	U18 D	Anterior-Posterior at Forearm	3.8	3.3	2.5	1.7	1.3
1.2	**3.0**	U19 D	Depth of Wrist	2.3	2.0	1.5	1.0	0.8
1.6	**4.0**	U20 T	Breadth of Wrist	3.0	2.7	2.0	1.3	1.0
2.7	**6.8**	U21 T	Breadth of Hand	5.1	4.5	3.4	2.3	1.7
2.2	**5.5**	U22 T	Breadth of Palm	4.1	3.7	2.8	1.8	1.4
			LOWER EXTREMITY MEASUREMENTS					
24.0	**61.0**	L1 V	Crest of Ilium of Pelvis to Sole	45.8	40.7	30.5	20.3	15.3
22.8	**58.0**	* L2 V	Ant.Sup.Spine of Ilium to Sole	43.5	38.7	29.0	19.3	14.5
20.5	**52.0**	L3 V	Greater Trochanter Femur-Sole	39.0	34.7	26.0	17.3	13.0
19.7	**50.0**	* L4 V	Symphysis Pubis of Pelvis-Sole	37.5	33.3	25.0	16.7	12.5
18.5	**47.0**	* L5 V	Crotch to Sole	35.3	31.3	23.5	15.7	11.8
18.5	**47.0**	L6 V	Gluteal Furrow to Sole	35.3	31.3	23.5	15.7	11.8
11.0	**28.0**	L7 V	Posterior Knee Crease to Sole	21.0	18.7	14.0	9.3	7.0
10.2	**26.0**	L8 V	Tibia at Knee to Sole	19.5	17.3	13.0	8.7	6.5
2.0	**5.0**	L9 V	Med. Malleolus of Ankle to Sole	3.8	3.3	2.5	1.7	1.3
1.4	**3.5**	L10 V	Lat. Malleolus of Ankle to Sole	2.6	2.3	1.8	1.2	0.9
8.3	**21.0**	L11 T	Breadth of Hips, Sitting	15.8	14.0	10.5	7.0	5.3
3.9	**10.0**	L12 T	Breadth Mid Thigh	7.5	6.7	5.0	3.3	2.5
3.1	**8.0**	L13 T	Breadth of Knee	6.0	5.3	4.0	2.7	2.0
2.8	**7.0**	L14 T	Breadth of Calf	5.3	4.7	3.5	2.3	1.8
1.8	**4.5**	L15 T	Breadth Ankle	3.4	3.0	2.3	1.5	1.1
3.9	**10.0**	L16 D	Mid Thigh, Anterior-Posterior	7.5	6.7	5.0	3.3	2.5
3.1	**8.0**	L17 D	Knee, Anterior-Posterior	6.0	5.3	4.0	2.7	2.0
2.8	**7.0**	L18 D	Mid Calf, Anterior-Posterior	5.3	4.7	3.5	2.3	1.8
2.2	**5.5**	L19 D	Ankle, Anterior-Posterior	4.1	3.7	2.8	1.8	1.4
6.6	**16.8**	L20 D	Length of Foot	12.6	11.2	8.4	5.6	4.2
5.7	**14.5**	L21 D	Heel to 5th Toe	10.9	9.7	7.3	4.8	3.6
2.6	**6.6**	L22 T	Breadth Foot	5.0	4.4	3.3	2.2	1.7
1.6	**4.0**	L23 T	Width of Heel	3.0	2.7	2.0	1.3	1.0
		These are also listed with Vertical Measurements.						

Five-Year-Old Boy, anterior and lateral head

all measurements in centimeters

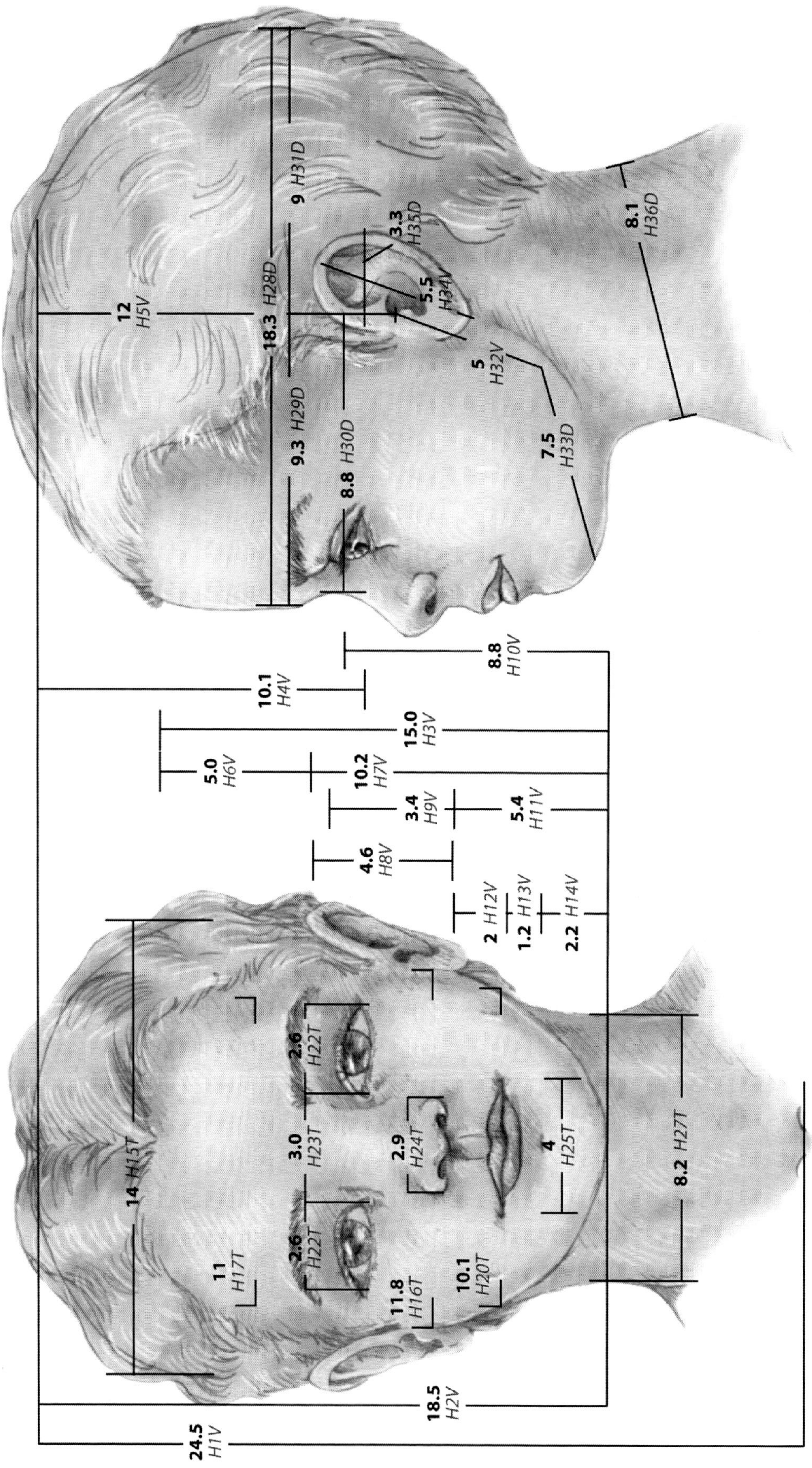

Five-Year-Old Boy, anterior and lateral figure

all measurements in centimeters

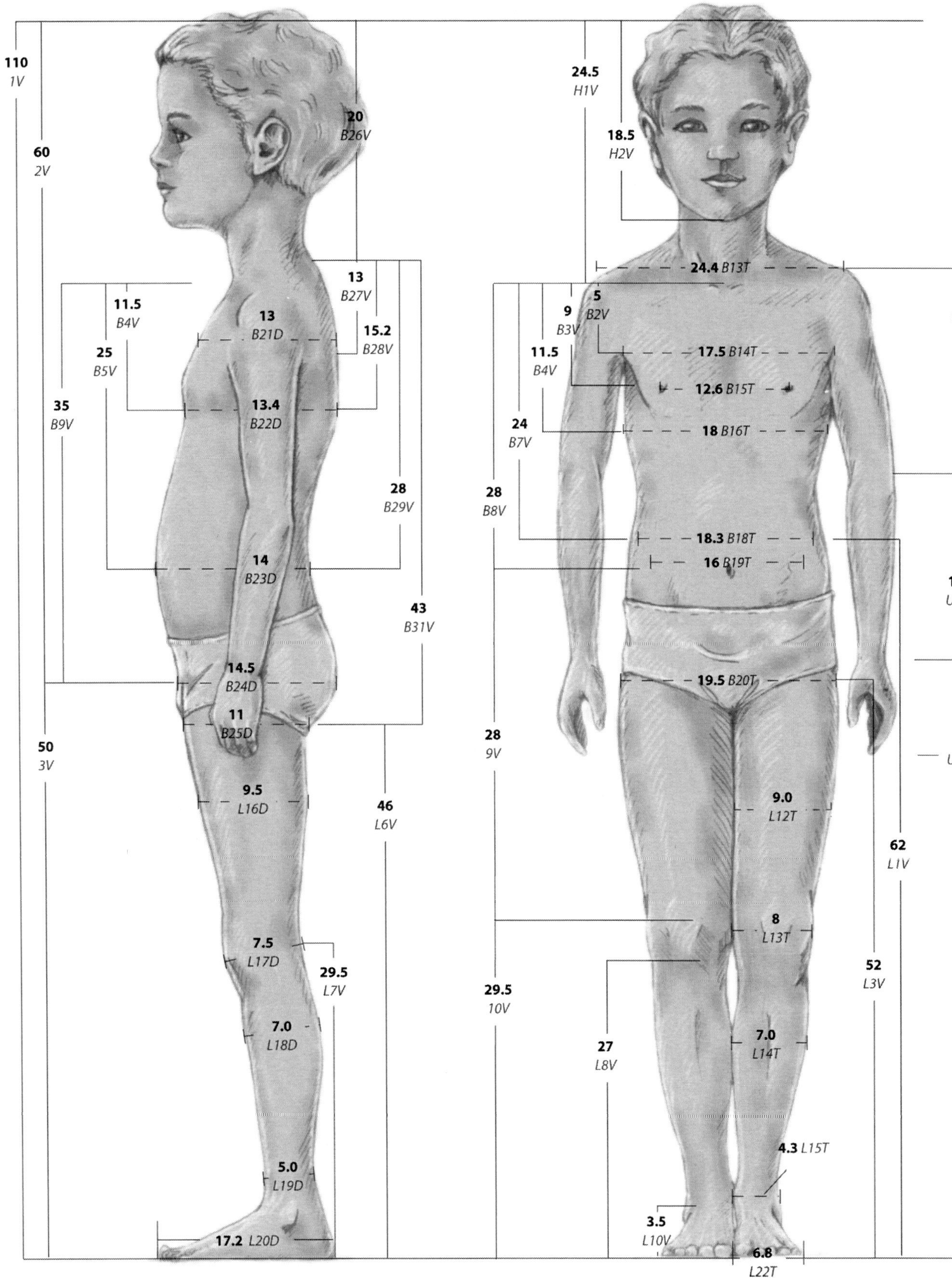

Life size			FIVE-YEAR-OLD BOY	Fractions of life size				
				3/4	2/3	1/2	1/3	1/4
Inches	cm	Code		cm	cm	cm	cm	cm
			VERTICAL MEASUREMENTS					
43.3	**110.0**	1 V	Stature, Height, Vertex to Sole	82.5	73.3	55.0	36.7	27.5
23.6	**60.0**	2 V	Vertex to Symphysis Pubis	45.0	40.0	30.0	20.0	15.0
19.7	**50.0**	* 3 V	Symphysis Pubis of Pelvis-Sole	37.5	33.3	25.0	16.7	12.5
33.5	**85.0**	4 V	Sternal Notch, Neck Pit, to Sole	63.8	56.7	42.5	28.3	21.3
9.4	**24.0**	5 V	Vertex to Acromion Process	18.0	16.0	12.0	8.0	6.0
33.9	**86.0**	6 V	Acromion Process to Sole	64.5	57.3	43.0	28.7	21.5
22.6	**57.5**	* 7 V	Ant.Sup.Spine of Ilium to Sole	43.1	38.3	28.8	19.2	14.4
11.2	**28.5**	8 V	Acrom. proc. to Ant. Sup. Spine	21.4	19.0	14.3	9.5	7.1
11.0	**28.0**	9 V	Ant.Sup.Spine of Ilium to Knee	21.0	18.7	14.0	9.3	7.0
11.6	**29.5**	10 V	Knee at Patella to Sole	22.1	19.7	14.8	9.8	7.4
24.0	**61.0**	11 V	Umbilicus to Sole	45.8	40.7	30.5	20.3	15.3
18.9	**48.0**	* 12 V	Crotch to Sole	36.0	32.0	24.0	16.0	12.0
33.5	**85.0**	13 V	Kneeling Height	63.8	56.7	42.5	28.3	21.3
24.4	**62.0**	14 V	Sitting Height	46.5	41.3	31.0	20.7	15.5
21.7	**55.0**	15 V	Buttock, Straight Leg to Sole	41.3	36.7	27.5	18.3	13.8
13.4	**34.0**	16 V	Buttock to Flexed Knee	25.5	22.7	17.0	11.3	8.5
12.8	**32.5**	17 V	Flexed Knee to Sole	24.4	21.7	16.3	10.8	8.1
		**These are also listed with Lower Extremity Measurements.*						
			VERTICAL HEAD MEASUREMENTS					
9.6	**24.5**	H1 V	Vertex to Sternal Notch	18.4	16.3	12.3	8.2	6.1
7.3	**18.5**	H2 V	Vertex to Chin, Head Height	13.9	12.3	9.3	6.2	4.6
5.9	**15.0**	H3 V	Hairline to Chin, Face Height	11.3	10.0	7.5	5.0	3.8
4.0	**10.1**	H4 V	Vertex to Lateral Corner of Eye	7.6	6.7	5.1	3.4	2.5
4.7	**12.0**	H5 V	Vertex to Tragion of Ear	9.0	8.0	6.0	4.0	3.0
2.0	**5.0**	H6 V	Hairline to Brow, Glabella	3.8	3.3	2.5	1.7	1.3
4.0	**10.2**	H7 V	Brow, Glabella, to Chin, Menton	7.7	6.8	5.1	3.4	2.6
2.0	**5.0**	H8 V	Brow to Subnasal Point	3.8	3.3	2.5	1.7	1.3
1.3	**3.4**	H9 V	Sellion to Subnasal, Nose Length	2.6	2.3	1.7	1.1	0.9
3.5	**8.8**	H10 V	Sellion, Nose Saddle to Chin	6.6	5.9	4.4	2.9	2.2
2.1	**5.4**	H11 V	Subnasal Point to Chin	4.1	3.6	2.7	1.8	1.4
0.8	**2.0**	H12 V	Subnasal to Lip Aperature	1.5	1.3	1.0	0.7	0.5
0.5	**1.2**	H13 V	Lip Aperat. to Sublabial Crease	0.9	0.8	0.6	0.4	0.3
0.9	**2.2**	H14 V	Sublabial Crease to Chin	1.7	1.5	1.1	0.7	0.6
			TRANSVERSE HEAD MEASUREMENTS					
5.5	**14.0**	H15 T	Bi-Parietal, Head Breadth	10.5	9.3	7.0	4.7	3.5
4.6	**11.8**	H16 T	Bi-Zygomatic, Cheekbones	8.9	7.9	5.9	3.9	3.0
4.4	**11.1**	H17 T	Bi-Fronto Temporal, Forehead	8.3	7.4	5.6	3.7	2.8

Life size			FIVE-YEAR-OLD BOY	Fractions of life size				
				3/4	2/3	1/2	1/3	1/4
Inches	cm	Code		cm	cm	cm	cm	cm
			TRANSVERSE HEAD MEASUREMENTS					
3.4	8.7	H18 T	Inter-Brow Ridge	6.5	5.8	4.4	2.9	2.2
4.5	11.5	H19 T	Inter-Tragion of Ears	8.6	7.7	5.8	3.8	2.9
4.0	10.1	H20 T	Inter-Angle of Mandible	7.6	6.7	5.1	3.4	2.5
3.2	8.2	H21 T	Inter-Lateral Corners of Eyes	6.2	5.5	4.1	2.7	2.1
1.0	2.6	H22 T	Width of Eye	2.0	1.7	1.3	0.9	0.7
1.2	3.0	H23 T	Inter-Inner Corner of Eyes	2.3	2.0	1.5	1.0	0.8
1.2	3.0	H24 T	Breadth of Nose	2.3	2.0	1.5	1.0	0.8
1.6	4.0	H25 T	Breadth of Mouth	3.0	2.7	2.0	1.3	1.0
1.2	3.0	H26 T	Breadth of Chin	2.3	2.0	1.5	1.0	0.8
3.2	8.2	H27 T	Breadth of Neck	6.2	5.5	4.1	2.7	2.1
			ANTERIOR-POSTERIOR HEAD MEASUREMENTS					
7.2	18.3	H28 D	Brow to Occiput, Posterior Head	13.7	12.2	9.2	6.1	4.6
3.7	9.3	H29 D	Brow to Tragus of Ear	7.0	6.2	4.7	3.1	2.3
3.5	8.8	H30 D	Sellion, Nose Saddle, to Tragus	6.6	5.9	4.4	2.9	2.2
3.5	9.0	H31 D	Tragus to Occiput, Back of Head	6.8	6.0	4.5	3.0	2.3
2.0	5.0	H32 V	Tragus to Angle Mandible	3.8	3.3	2.5	1.7	1.3
3.0	7.5	H33 D	Angle of Mandible to Chin	5.6	5.0	3.8	2.5	1.9
2.2	5.5	H34 V	Length of Ear, Vertical	4.1	3.7	2.8	1.8	1.4
1.3	3.3	H35 D	Width of Ear, Anterior-Posterior	2.5	2.2	1.7	1.1	0.8
3.2	8.1	H36 D	Anterior to Posterior of Neck	6.1	5.4	4.1	2.7	2.0
			VERTICAL BODY MEASUREMENTS					
2.6	6.5	B1 V	Chin to Sternal Notch, Neck Pit	4.9	4.3	3.3	2.2	1.6
2.0	5.0	B2 V	Sternal Notch to Axilla, Arm Pit	3.8	3.3	2.5	1.7	1.3
3.5	9.0	B3 V	Sternal Notch to Nipples	6.8	6.0	4.5	3.0	2.3
4.5	11.5	B4 V	Sternal Notch to Xyphoid Proc.	8.6	7.7	5.8	3.8	2.9
9.8	25.0	B5 V	Sternal Notch to Umbilicus	18.8	16.7	12.5	8.3	6.3
4.7	12.0	B6 V	Xyphoid Process to Umbilicus	9.0	8.0	6.0	4.0	3.0
9.4	24.0	B7 V	Sternal Notch to Crest Ilium	18.0	16.0	12.0	8.0	6.0
11.0	28.0	B8 V	Sternal Notch to Ant.S.Sp.Ilium	21.0	18.7	14.0	9.3	7.0
13.8	35.0	B9 V	Sternal Notch to Pubis	26.3	23.3	17.5	11.7	8.8
4.3	11.0	B10 V	Umbilicus to Pubis of Pelvis	8.3	7.3	5.5	3.7	2.8
			TRANSVERSE BODY MEASUREMENTS					
10.6	27.0	B11 T	Shoulder Breadth	20.3	18.0	13.5	9.0	6.8
4.8	12.2	B12 T	Clavicle to Acromion Process	9.2	8.1	6.1	4.1	3.1
9.6	24.4	B13 T	Bi-acromial Process Breadth	18.3	16.3	12.2	8.1	6.1
6.9	17.5	B14 T	Inter-Anterior Axillary Fold	13.1	11.7	8.8	5.8	4.4
5.0	12.6	B15 T	Inter-Nipples	9.5	8.4	6.3	4.2	3.2

Life size			FIVE-YEAR-OLD BOY	Fractions of life size				
				3/4	2/3	1/2	1/3	1/4
Inches	cm	Code		cm	cm	cm	cm	cm
			TRANSVERSE BODY MEASUREMENTS					
7.1	18.0	B16 T	Breadth of Thorax	13.5	12.0	9.0	6.0	4.5
7.1	18.0	B17 T	Inter-Costal. Rib Margin Breadth	13.5	12.0	9.0	6.0	4.5
7.2	18.3	B18 T	Inter-Crest of Ilium Breadth	13.7	12.2	9.2	6.1	4.6
6.3	16.0	B19 T	Inter-Anterior Superior Spine	12.0	10.7	8.0	5.3	4.0
7.7	19.5	B20 T	Inter-Greater Trochanter Femur	14.6	13.0	9.8	6.5	4.9
			ANTERIOR-POSTERIOR BODY MEASUREMENTS					
5.1	13.0	B21 D	Manubrium to Dorsal Spine	9.8	8.7	6.5	4.3	3.3
5.3	13.4	B22 D	Xyphoid Process to Dorsum	10.1	8.9	6.7	4.5	3.4
5.5	14.0	B23 D	Umbilicus to Dorsum	10.5	9.3	7.0	4.7	3.5
5.7	14.5	B24 D	Pubis of Pelvis to Buttock	10.9	9.7	7.3	4.8	3.6
4.3	11.0	B25 D	Thigh to Gluteal Furrow	8.3	7.3	5.5	3.7	2.8
			VERTICAL BACK MEASUREMENTS					
7.9	20.0	B26 V	Vertex to 7th Cervical Spine	15.0	13.3	10.0	6.7	5.0
			7th Cervical Vertebra Spine to:					
5.1	13.0	B27 V	To Posterior Axillary Fold, Armpit	9.8	8.7	6.5	4.3	3.3
6.0	15.0	B28 V	To Inferior Tip of Scapula	11.4	10.1	7.6	5.1	3.8
11.4	29.0	B29 V	To Posterior Superior Spine Ilium	21.8	19.3	14.5	9.7	7.3
13.8	35.0	B30 V	To Midline Upper Buttock Crease	26.3	23.3	17.5	11.7	8.8
16.9	43.0	B31 V	To Gluteal Furrow	32.3	28.7	21.5	14.3	10.8
23.6	60.0	B32 V	To Popliteal Crease Behind Knee	45.0	40.0	30.0	20.0	15.0
			TRANSVERSE BACK MEASUREMENTS					
9.1	23.0	B33 T	Inter-Posterior Axillary Folds	17.3	15.3	11.5	7.7	5.8
4.4	11.2	B34 T	Inter-Scapulae Medial Borders	8.4	7.5	5.6	3.7	2.8
2.2	5.6	B35 T	Inter-Post. Sup. Spines Sacrum	4.2	3.7	2.8	1.9	1.4
			UPPER EXTREMITY MEASUREMENTS					
43.3	110.0	U1 V	Span of Arms	82.5	73.3	55.0	36.7	27.5
4.7	12.0	U2 V	Sternum to Acromion Process	9.0	8.0	6.0	4.0	3.0
18.9	48.0	U3 V	Acromion Process to Fingertip	36.0	32.0	24.0	16.0	12.0
52.8	134.0	U4 V	Vertical Reach to Fingertip	100.5	89.3	67.0	44.7	33.5
27.0	68.5	U5 T	Lateral Reach to Fingertip	51.4	45.7	34.3	22.8	17.1
21.3	54.0	U6 D	Forward Reach to Fingertip	40.5	36.0	27.0	18.0	13.5
8.7	22.0	U7 V	Shoulder to Flexed Elbow	16.5	14.7	11.0	7.3	5.5
7.9	20.0	U8 V	Acromion to Lat. Condyle Elbow	15.0	13.3	10.0	6.7	5.0
11.0	28.0	U9 D	Flexed Elbow to Fingertip	21.0	18.7	14.0	9.3	7.0
6.2	15.8	U10 D	Flexed Elbow Pivot to Wrist Pivot	11.9	10.5	7.9	5.3	4.0

Life size			FIVE-YEAR-OLD BOY	Fractions of life size				
				3/4	2/3	1/2	1/3	1/4
Inches	cm	Code		cm	cm	cm	cm	cm
			UPPER EXTREMITY MEASUREMENTS					
4.7	12.0	U11 V	Wrist to Fingertip	9.0	8.0	6.0	4.0	3.0
3.1	8.0	U12 V	Wrist to Thumb Tip	6.0	5.3	4.0	2.7	2.0
2.8	7.0	U13 V	Length of Palm	5.3	4.7	3.5	2.3	1.8
2.0	5.2	U14 V	Length 3rd Finger	3.9	3.5	2.6	1.7	1.3
3.3	8.4	U15 V	Thumb Crotch to 3rd Finger	6.3	5.6	4.2	2.8	2.1
1.7	4.2	U16 V	Thumb Crotch to Thumb Tip	3.2	2.8	2.1	1.4	1.1
2.0	5.0	U17 D	Anterior-Post. Arm at Biceps	3.8	3.3	2.5	1.7	1.3
2.0	5.0	U18 D	Anterior-Posterior at Forearm	3.8	3.3	2.5	1.7	1.3
1.2	3.0	U19 D	Depth of Wrist	2.3	2.0	1.5	1.0	0.8
1.7	4.3	U20 T	Breadth of Wrist	3.2	2.9	2.2	1.4	1.1
2.8	7.2	U21 T	Breadth of Hand	5.4	4.8	3.6	2.4	1.8
2.3	5.8	U22 T	Breadth of Palm	4.4	3.9	2.9	1.9	1.5
			LOWER EXTREMITY MEASUREMENTS					
24.4	62.0	L1 V	Crest of Ilium of Pelvis to Sole	46.5	41.3	31.0	20.7	15.5
22.8	58.0	* L2 V	Ant.Sup.Spine of Ilium to Sole	43.5	38.7	29.0	19.3	14.5
20.5	52.0	L3 V	Greater Trochanter Femur-Sole	39.0	34.7	26.0	17.3	13.0
19.7	50.0	* L4 V	Symphysis Pubis of Pelvis-Sole	37.5	33.3	25.0	16.7	12.5
18.9	48.0	* L5 V	Crotch to Sole	36.0	32.0	24.0	16.0	12.0
18.1	46.0	L6 V	Gluteal Furrow to Sole	34.5	30.7	23.0	15.3	11.5
11.2	28.5	L7 V	Posterior Knee Crease to Sole	21.4	19	14.3	9.5	7.1
10.6	27.0	L8 V	Tibia at Knee to Sole	20.3	18.0	13.5	9.0	6.8
2.0	5.0	L9 V	Med. Malleolus of Ankle to Sole	3.8	3.3	2.5	1.7	1.3
1.4	3.5	L10 V	Lat. Malleolus of Ankle to Sole	2.6	2.3	1.8	1.2	0.9
8.1	20.5	L11 T	Breadth of Hips, Sitting	15.4	13.7	10.3	6.8	5.1
3.5	9.0	L12 T	Breadth Mid Thigh	6.8	6.0	4.5	3.0	2.3
3.1	8.0	L13 T	Breadth of Knee	6.0	5.3	4.0	2.7	2.0
2.8	7.0	L14 T	Breadth of Calf	5.3	4.7	3.5	2.3	1.8
1.7	4.3	L15 T	Breadth Ankle	3.2	2.9	2.2	1.4	1.1
3.7	9.5	L16 D	Mid Thigh, Anterior-Posterior	7.1	6.3	4.8	3.2	2.4
3.0	7.5	L17 D	Knee, Anterior- Posterior	5.6	5.0	3.8	2.5	1.9
2.8	7.0	L18 D	Mid Calf, Anterior-Posterior	5.3	4.7	3.5	2.3	1.8
2.0	5.0	L19 D	Ankle, Anterior-Posterior	3.8	3.3	2.5	1.7	1.3
6.8	17.2	L20 D	Length of Foot	12.9	11.5	8.6	5.7	4.3
4.9	12.5	L21 D	Heel to 5th Toe	9.4	8.3	6.3	4.2	3.1
2.7	6.8	L22 T	Breadth Foot	5.1	4.5	3.4	2.3	1.7
1.6	4.0	L23 T	Width of Heel	3.0	2.7	2.0	1.3	1.0
		**These are also listed with Vertical Measurements.*						

SIX-YEAR-OLD GIRLS

The body proportions are listed separately for gender in the tables of measurements. The range of twenty-fifth, fiftieth, and seventy-fifth percentile of stature is included here for comparison.

Percentile	Twenty-fifth		Fiftieth		Seventy-fifth	
Girls:	Short		Medium		Tall	
Stature	111 cm,	43.7 in.	114.6 cm,	45 in.	118 cm,	46.5 in.
Weight	18 kg,	39.3 lb.	19.5 kg,	43 lb.	21.5 kg,	47.3 lb.[1]

At this age girls are, on the average, not quite as tall and weigh slightly less boys. The most noticeable difference between boys and girls is the hair styling and the dress on special occasions. For every day play, the clothing is similar, but slightly more feminine. However, on closer inspection, the girls may already have more rounded contours in face, shoulders, and hips. Coordination continues to improve. During waking hours their activity is nearly constant. Interests in some toys and games may be similar to boys, but their interests still tend towards dresses, dress-up, dolls, and play house. The school level is first grade.

BODY MEASUREMENTS COMPARED TO HEIGHT

	Feature Measure in cm	Feature Compared to Stature	Approx. Ratio of Feature	Compared to Median Adult
Stature: Crown to sole	114.6			68%
Head Height: Crown to chin	18.2	16%	1/6	69%
Sitting Height:	63.5	55%	1/2+	71%
Trunk Measure: Sternal notch to symphysis pubis	36	31%	1/3–	73%
Leg Length: Greater trochanter of femur to sole	55	48%	1/2–	65%
Thigh Length: Greater trochanter to tibia at knee	27	24%	1/4–	64%
Tibia at Knee to Sole:	28	24%	1/4	67%
Arm Length: Shoulder to dependent fingertip	48	42 %	1/2–	64%
Hand Length:	12.5	11%	1/9	70%
Foot Length:	17.8	16%	1/6	74%

Permanent incisors erupt and these teeth may appear large for the face. The permanent first molars begin to erupt.[2]

1. Nelson, W.E., p. 39.
2. Lockhart , R.D., p. 57.

SIX-YEAR-OLD BOYS

The list of body proportions is shown in the tables of measurements. The range of twenty-fifth, to fiftieth, and to seventy-fifth percentile is listed here for comparison.

Percentile	Twenty-fifth		Fiftieth		Seventy-fifth	
Boys:	Short		Medium		Tall	
Stature	112.5 cm,	44.3 in.	116 cm,	45.8 in.	119 cm,	47 in.
Weight	19 kg,	47 lb.	20.7 kg,	45.5 lb.	22.4 kg,	49.5 lb.[1]

At this age boys are, on the average, slightly taller and slightly heavier than girls. Hair style is different, but clothing is similar for play and sometimes school, yet it is more masculine for special occasions. They are enthusiastic at starting projects but poor at completing them. Jealousy may occur between siblings. Boys show interest in tricycles, scooters, sports and mechanical toys. The school level is first grade.

BODY MEASUREMENTS COMPARED TO HEIGHT

	Feature Measure in cm	Feature Compared to Stature	Approx. Ratio of Feature	Compared to Median Adult
Stature: Crown to sole	116			63%
Head Height: Crown to chin	19	17%	1/6	83%
Sitting Height:	64	55%	1/2+	67%
Trunk Measure: Sternal notch to symphysis pubis	37	32%	1/3	70%
Leg Length: Greater trochanter of femur to sole	56	48%	1/2	57%
Thigh Length: Greater trochanter to tibia at knee	27	23%	1/4	58%
Tibia at Knee to Sole:	29	25%	1/4	57%
Arm Length: Acromion process at shoulder to dependant fingertip	50	43%	2/5	62%
Hand Length:	13	11%	1/9	65%
Foot Length:	18	16%	1/6	65%

Permanent incisors erupt and these teeth may appear too large for the face. The permanent first molars begin to erupt.[2]

1. Nelson, W.E., p. 39.
2. Lockhart , R.D., p. 57.

Six-Year-Old Girl, anterior and lateral head

all measurements in centimeters

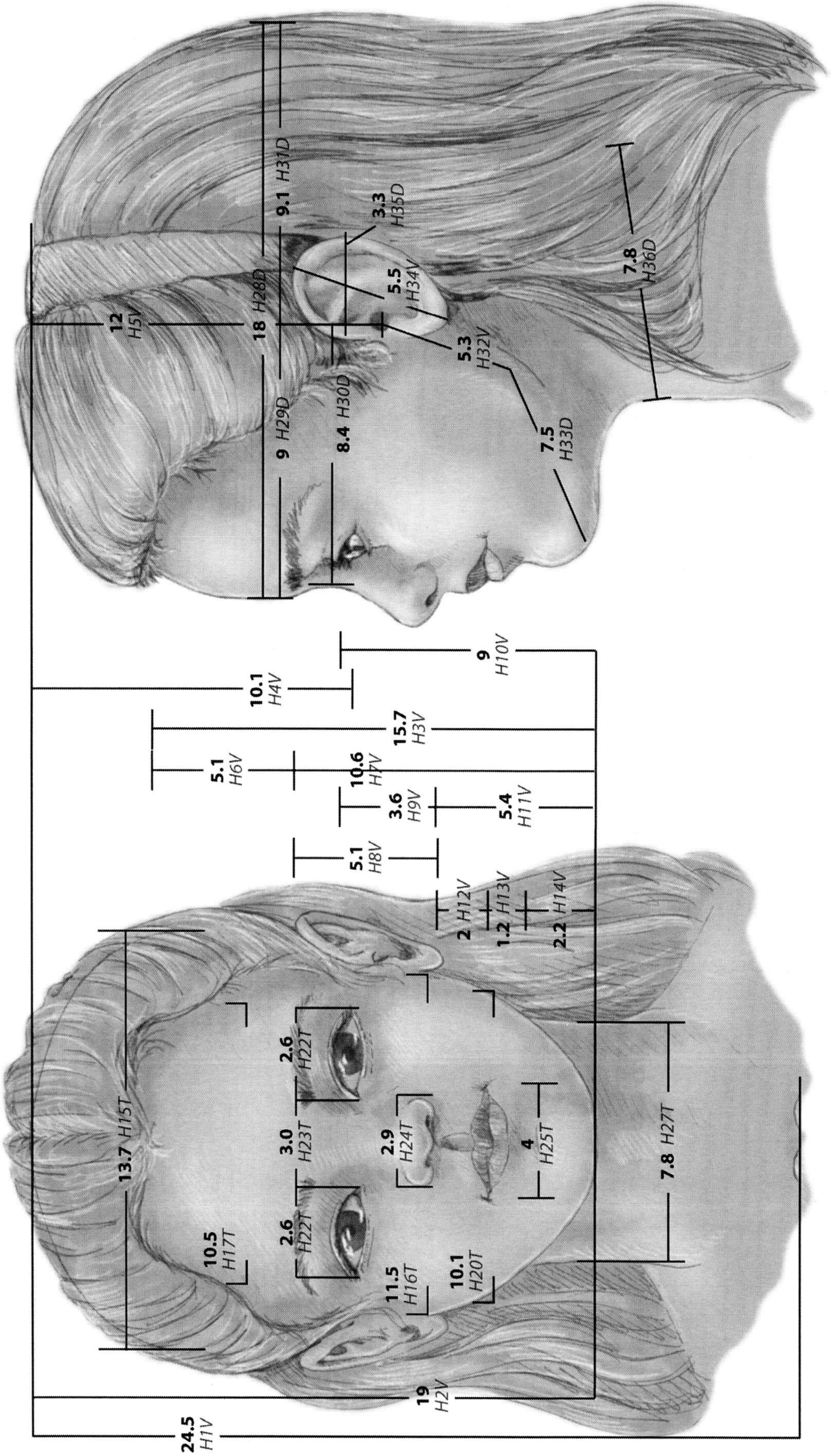

Six-Year-Old Girl, anterior and lateral figure

all measurements in centimeters

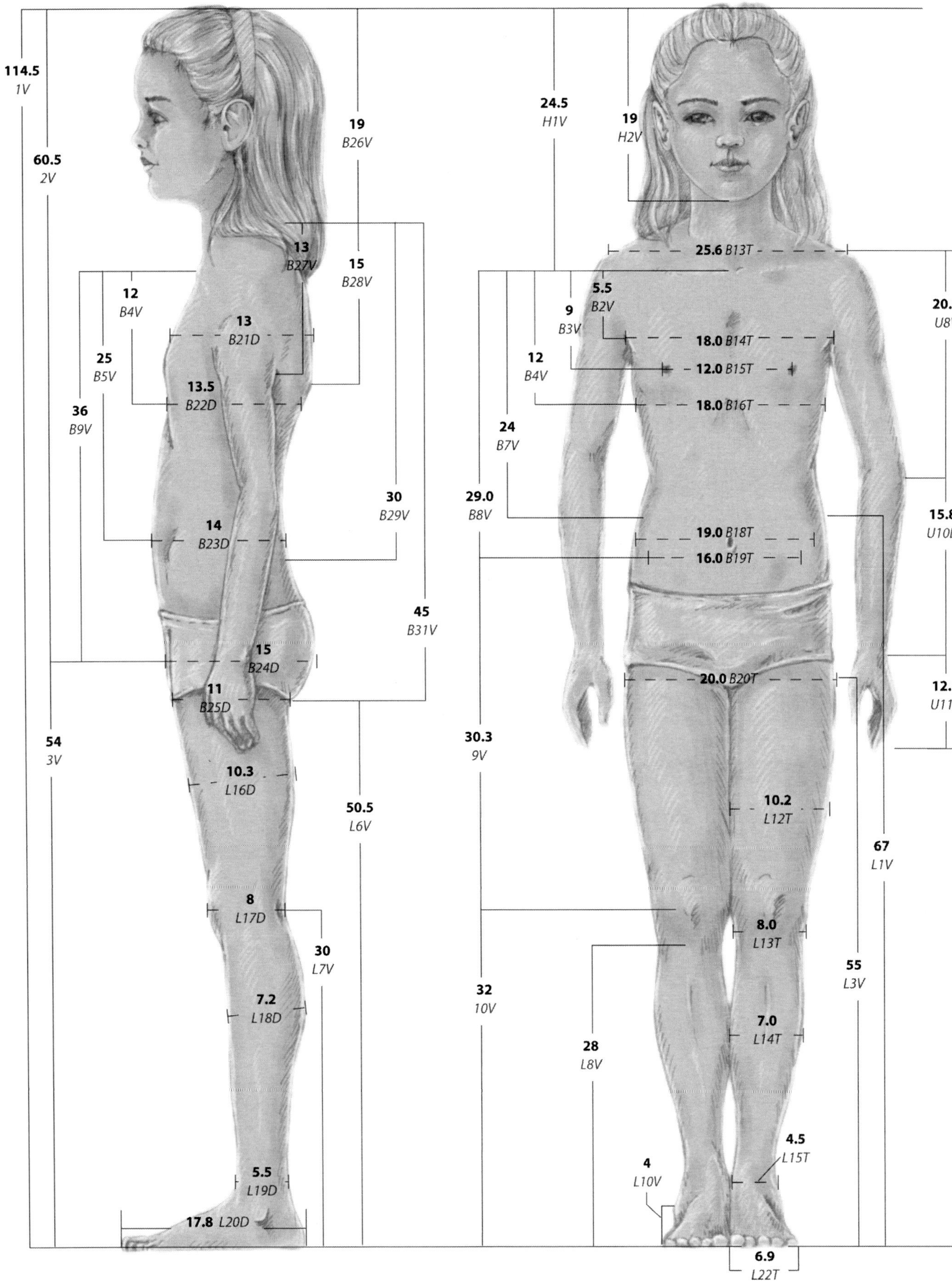

Life size			SIX-YEAR-OLD GIRL	Fractions of life size				
				3/4	2/3	1/2	1/3	1/4
Inches	cm	Code		cm	cm	cm	cm	cm
			VERTICAL MEASUREMENTS					
45.1	**114.5**	1 V	Stature, Height, Vertex to Sole	85.9	76.3	57.3	38.2	28.6
23.8	**60.5**	2 V	Vertex to Symphysis Pubis	45.4	40.3	30.3	20.2	15.1
21.3	**54.0**	* 3 V	Symphysis Pubis of Pelvis-Sole	40.5	36.0	27.0	18.0	13.5
35.4	**90.0**	4 V	Sternal Notch, Neck Pit, to Sole	67.5	60.0	45.0	30.0	22.5
9.4	**24.0**	5 V	Vertex to Acromion Process	18.0	16.0	12.0	8.0	6.0
36.2	**92.0**	6 V	Acromion Process to Sole	69.0	61.3	46.0	30.7	23.0
24.2	**61.5**	* 7 V	Ant.Sup.Spine of Ilium to Sole	46.1	41.0	30.8	20.5	15.4
11.6	**29.5**	8 V	Acrom. proc. to Ant. Sup. Spine	22.1	19.7	14.8	9.8	7.4
12.0	**30.5**	9 V	Ant.Sup.Spine of Ilium to Knee	22.9	20.3	15.3	10.2	7.6
12.6	**32.0**	10 V	Knee at Patella to Sole	24.0	21.3	16.0	10.7	8.0
25.6	**65.0**	11 V	Umbilicus to Sole	48.8	43.3	32.5	21.7	16.3
19.7	**50.0**	* 12 V	Crotch to Sole	37.5	33.3	25.0	16.7	12.5
34.4	**87.5**	13 V	Kneeling Height	65.6	58.3	43.8	29.2	21.9
25.0	**63.5**	14 V	Sitting Height	47.6	42.3	31.8	21.2	15.9
23.6	**60.0**	15 V	Buttock, Straight Leg to Sole	45.0	40.0	30.0	20.0	15.0
14.4	**36.5**	16 V	Buttock to Flexed Knee	27.4	24.3	18.3	12.2	9.1
13.8	**35.0**	17 V	Flexed Knee to Sole	26.3	23.3	17.5	11.7	8.8
			VERTICAL HEAD MEASUREMENTS					
9.6	**24.5**	H1 V	Vertex to Sternal Notch	18.4	16.3	12.3	8.2	6.1
7.5	**19.0**	H2 V	Vertex to Chin, Head Height	14.3	12.7	9.5	6.3	4.8
6.2	**15.7**	H3 V	Hairline to Chin, Face Height	11.8	10.5	7.9	5.2	3.9
4.0	**10.1**	H4 V	Vertex to Lateral Corner of Eye	7.6	6.7	5.1	3.4	2.5
4.7	**12.0**	H5 V	Vertex to Tragion of Ear	9.0	8.0	6.0	4.0	3.0
2.0	**5.1**	H6 V	Hairline to Brow, Glabella	3.8	3.4	2.6	1.7	1.3
4.2	**10.6**	H7 V	Brow, Glabella, to Chin, Menton	8.0	7.1	5.3	3.5	2.7
2.0	**5.1**	H8 V	Brow to Subnasal Point	3.8	3.4	2.6	1.7	1.3
1.4	**3.6**	H9 V	Sellion to Subnasal, Nose Length	2.7	2.4	1.8	1.2	0.9
3.5	**9.0**	H10 V	Sellion, Nose Saddle, to Chin	6.8	6.0	4.5	3.0	2.3
2.1	**5.4**	H11 V	Subnasal Point to Chin	4.1	3.6	2.7	1.8	1.4
0.8	**2.0**	H12 V	Subnasal to Lip Aperature	1.5	1.3	1.0	0.7	0.5
0.5	**1.2**	H13 V	Lip Aperat. to Sublabial Crease	0.9	0.8	0.6	0.4	0.3
1.0	**2.5**	H14 V	Sublabial Crease to Chin	1.9	1.7	1.3	0.8	0.6
		** These are also listed with Lower Extremity Measurements.*						
			TRANSVERSE HEAD MEASUREMENTS					
5.4	**13.7**	H15 T	Bi-Parietal, Head Breadth	10.3	9.1	6.9	4.6	3.4
4.5	**11.5**	H16 T	Bi-Zygomatic, Cheekbones	8.6	7.7	5.8	3.8	2.9
4.1	**10.4**	H17 T	Bi-Fronto Temporal, Forehead	7.8	6.9	5.2	3.5	2.6

Life size			SIX-YEAR-OLD GIRL	Fractions of life size				
				3/4	2/3	1/2	1/3	1/4
Inches	cm	Code		cm	cm	cm	cm	cm
			TRANSVERSE HEAD MEASUREMENTS					
3.5	**8.9**	H18 T	Inter-Brow Ridge	6.7	5.9	4.5	3.0	2.2
4.2	**11.0**	H19 T	Inter-Tragion of Ears	8.0	7.1	5.3	3.5	2.7
4.0	**10.1**	H20 T	Inter-Angle of Mandible	7.6	6.7	5.1	3.4	2.5
3.2	**8.2**	H21 T	Inter-Lateral Corners of Eyes	6.2	5.5	4.1	2.7	2.1
1.0	**2.6**	H22 T	Width of Eye	2.0	1.7	1.3	0.9	0.7
1.2	**3.0**	H23 T	Inter-Inner Corner of Eyes	2.3	2.0	1.5	1.0	0.8
1.2	**3.0**	H24 T	Breadth of Nose	2.3	2.0	1.5	1.0	0.8
1.6	**4.0**	H25 T	Breadth of Mouth	3.0	2.7	2.0	1.3	1.0
1.2	**3.0**	H26 T	Breadth of Chin	2.3	2.0	1.5	1.0	0.8
3.1	**7.8**	H27 T	Breadth of Neck	5.9	5.2	3.9	2.6	2.0
			ANTERIOR-POSTERIOR HEAD MEASUREMENTS					
7.2	**18.2**	H28 D	Brow to Occiput, Posterior Head	13.7	12.1	9.1	6.1	4.6
3.5	**9.0**	H29 D	Brow to Tragus of Ear	6.8	6.0	4.5	3.0	2.3
3.3	**8.4**	H30 D	Sellion, Nose Saddle, to Tragus	6.3	5.6	4.2	2.8	2.1
3.6	**9.1**	H31 D	Tragus to Occiput, Back of Head	6.8	6.1	4.6	3.0	2.3
2.1	**5.3**	H32 V	Tragus to Angle Mandible	4.0	3.5	2.7	1.8	1.3
3.0	**7.5**	H33 D	Angle of Mandible to Chin	5.6	5.0	3.8	2.5	1.9
2.2	**5.5**	H34 V	Length of Ear, Vertical	4.1	3.7	2.8	1.8	1.4
1.3	**3.3**	H35 D	Width of Ear, Anterior-Posterior	2.5	2.2	1.7	1.1	0.8
3.1	**7.8**	H36 D	Anterior to Posterior of Neck	5.9	5.2	3.9	2.6	2.0
			VERTICAL BODY MEASUREMENTS					
2.4	**6.0**	B1 V	Chin to Sternal Notch, Neck Pit	4.5	4.0	3.0	2.0	1.5
2.2	**5.5**	B2 V	Sternal Notch to Axilla, Arm Pit	4.1	3.7	2.8	1.8	1.4
3.5	**9.0**	B3 V	Sternal Notch to Nipples	6.8	6.0	4.5	3.0	2.3
4.7	**12.0**	B4 V	Sternal Notch to Xyphoid Proc.	9.0	8.0	6.0	4.0	3.0
9.8	**25.0**	B5 V	Sternal Notch to Umbilicus	18.8	16.7	12.5	8.3	6.3
5.1	**13.0**	B6 V	Xyphoid Process to Umbilicus	9.8	8.7	6.5	4.3	3.3
9.4	**24.0**	B7 V	Sternal Notch to Crest Ilium	18.0	16.0	12.0	8.0	6.0
11.4	**29.0**	B8 V	Sternal Notch to Ant.S.Sp.Ilium	21.8	19.3	14.5	9.7	7.3
14.2	**36.0**	B9 V	Sternal Notch to Pubis	27.0	24.0	18.0	12.0	9.0
4.3	**11.0**	B10 V	Umbilicus to Pubis of Pelvis	8.3	7.3	5.5	3.7	2.8
			TRANSVERSE BODY MEASUREMENTS					
11.0	**28.0**	B11 T	Shoulder Breadth	21.0	18.7	14.0	9.3	7.0
4.9	**12.4**	B12 T	Clavicle to Acromion Process	9.3	8.3	6.2	4.1	3.1
10.1	**25.6**	B13 T	Bi-acromial Process Breadth	19.2	17.1	12.8	8.5	6.4
7.1	**18.0**	B14 T	Inter-Anterior Axillary Fold	13.5	12.0	9.0	6.0	4.5
4.7	**12.0**	B15 T	Inter-Nipples	9.0	8.0	6.0	4.0	3.0

Life size			SIX-YEAR-OLD GIRL	Fractions of life size				
				3/4	2/3	1/2	1/3	1/4
Inches	cm	Code		cm	cm	cm	cm	cm
			TRANSVERSE BODY MEASUREMENTS					
7.1	**18.0**	B16 T	Breadth of Thorax	13.5	12.0	9.0	6.0	4.5
6.7	**17.0**	B17 T	Inter-Costal. Rib Margin Breadth	12.8	11.3	8.5	5.7	4.3
7.5	**19.0**	B18 T	Inter-Crest of Ilium Breadth	14.3	12.7	9.5	6.3	4.8
6.3	**16.0**	B19 T	Inter-Anterior Superior Spine	12.0	10.7	8.0	5.3	4.0
7.9	**20.0**	B20 T	Inter-Greater Trochanter Femur	15.0	13.3	10.0	6.7	5.0
			ANTERIOR-POSTERIOR BODY MEASUREMENTS					
5.1	**13.0**	B21 D	Manubrium to Dorsal Spine	9.8	8.7	6.5	4.3	3.3
5.3	**13.5**	B22 D	Xyphoid Process to Dorsum	10.1	9.0	6.8	4.5	3.4
5.5	**14.0**	B23 D	Umbilicus to Dorsum	10.5	9.3	7.0	4.7	3.5
5.9	**15.0**	B24 D	Pubis of Pelvis to Buttock	11.3	10.0	7.5	5.0	3.8
4.3	**11.0**	B25 D	Thigh to Gluteal Furrow	8.3	7.3	5.5	3.7	2.8
			VERTICAL BACK MEASUREMENTS					
7.5	**19.0**	B26 V	Vertex to 7th Cervical Spine	14.3	12.7	9.5	6.3	4.8
			7th Cervical Vertebra Spine to:					
5.1	**13.0**	B27 V	To Posterior Axillary Fold, Armpit	9.8	8.7	6.5	4.3	3.3
5.9	**15.5**	B28 V	To Inferior Tip of Scapula	11.3	10.0	7.5	5.0	3.8
11.8	**30.0**	B29 V	To Posterior Superior Spine Ilium	22.5	20.0	15.0	10.0	7.5
14.2	**36.0**	B30 V	To Midline Upper Buttock Crease	27.0	24.0	18.0	12.0	9.0
17.7	**45.0**	B31 V	To Gluteal Furrow	33.8	30.0	22.5	15.0	11.3
23.6	**60.0**	B32 V	To Popliteal Crease Behind Knee	45.0	40.0	30.0	20.0	15.0
			TRANSVERSE BACK MEASUREMENTS					
9.1	**23.0**	B33 T	Inter-Posterior Axillary Folds	17.3	15.3	11.5	7.7	5.8
4.3	**11.0**	B34 T	Inter-Scapulae Medial Borders	8.3	7.3	5.5	3.7	2.8
2.2	**5.5**	B35 T	Inter-Post. Sup. Spines Sacrum	4.1	3.7	2.8	1.8	1.4
			UPPER EXTREMITY MEASUREMENTS					
43.3	**110.0**	U1 V	Span of Arms	82.5	73.3	55.0	36.7	27.5
4.9	**12.5**	U2 V	Sternum to Acromion Process	9.4	8.3	6.3	4.2	3.1
18.9	**48.0**	U3 V	Acromion Process to Fingertip	36.0	32.0	24.0	16.0	12.0
56.3	**143.0**	U4 V	Vertical Reach to Fingertip	107.3	95.3	71.5	47.7	35.8
28.3	**72.0**	U5 T	Lateral Reach to Fingertip	54.0	48.0	36.0	24.0	18.0
21.7	**55.0**	U6 D	Forward Reach to Fingertip	41.3	36.7	27.5	18.3	13.8
9.1	**23.0**	U7 V	Shoulder to Flexed Elbow	17.3	15.3	11.5	7.7	5.8
7.9	**20.0**	U8 V	Acromion to Lat. Condyle Elbow	15.0	13.3	10.0	6.7	5.0
11.8	**30.0**	U9 D	Flexed Elbow to Fingertip	22.5	20.0	15.0	10.0	7.5
6.2	**15.8**	U10 D	Flexed Elbow Pivot to Wrist Pivot	11.9	10.5	7.9	5.3	4.0

Life size			SIX-YEAR-OLD GIRL	Fractions of life size				
				3/4	2/3	1/2	1/3	1/4
Inches	cm	Code		cm	cm	cm	cm	cm
			UPPER EXTREMITY MEASUREMENTS					
4.9	**12.5**	U11 V	Wrist to Fingertip	9.4	8.3	6.3	4.2	3.1
3.3	**8.5**	U12 V	Wrist to Thumb Tip	6.4	5.7	4.3	2.8	2.1
2.8	**7.2**	U13 V	Length of Palm	5.4	4.8	3.6	2.4	1.8
2.1	**5.3**	U14 V	Length 3rd Finger	4.0	3.5	2.7	1.8	1.3
3.3	**8.4**	U15 V	Thumb Crotch to 3rd Finger	6.3	5.6	4.2	2.8	2.1
1.7	**4.3**	U16 V	Thumb Crotch to Thumb Tip	3.2	2.9	2.2	1.4	1.1
2.1	**5.4**	U17 D	Anterior-Post. Arm at Biceps	4.1	3.6	2.7	1.8	1.4
2.0	**5.1**	U18 D	Anterior-Posterior at Forearm	3.8	3.4	2.6	1.7	1.3
1.2	**3.0**	U19 D	Depth of Wrist	2.3	2.0	1.5	1.0	0.8
1.7	**4.3**	U20 T	Breadth of Wrist	3.2	2.9	2.2	1.4	1.1
2.8	**7.2**	U21 T	Breadth of Hand	5.4	4.8	3.6	2.4	1.8
2.3	**5.8**	U22 T	Breadth of Palm	4.4	3.9	2.9	1.9	1.5
			LOWER EXTREMITY MEASUREMENTS					
26.4	**67.0**	L1 V	Crest of Ilium of Pelvis to Sole	50.3	44.7	33.5	22.3	16.8
24.4	**62.0**	* L2 V	Ant.Sup.Spine of Ilium to Sole	46.5	41.3	31.0	20.7	15.5
21.7	**55.0**	L3 V	Greater Trochanter Femur-Sole	41.3	36.7	27.5	18.3	13.8
21.3	**54.0**	* L4 V	Symphysis Pubis of Pelvis-Sole	40.5	36.0	27.0	18.0	13.5
19.7	**50.0**	* L5 V	Crotch to Sole	37.5	33.3	25.0	16.7	12.5
19.9	**50.5**	L6 V	Gluteal Furrow to Sole	37.9	33.7	25.3	16.8	12.6
11.8	**30.0**	L7 V	Posterior Knee Crease to Sole	22.5	20.0	15.0	10.0	7.5
11.0	**28.0**	L8 V	Tibia at Knee to Sole	21.0	18.7	14.0	9.3	7.0
2.0	**5.0**	L9 V	Med. Malleolus of Ankle to Sole	3.8	3.3	2.5	1.7	1.3
1.6	**4.0**	L10 V	Lat. Malleolus of Ankle to Sole	3.0	2.7	2.0	1.3	1.0
8.3	**21.0**	L11 T	Breadth of Hips Sitting	15.8	14.0	10.5	7.0	5.3
4.0	**10.2**	L12 T	Breadth Mid Thigh	7.7	6.8	5.1	3.4	2.6
3.1	**8.0**	L13 T	Breadth of Knee	6.0	5.3	4.0	2.7	2.0
2.8	**7.0**	L14 T	Breadth of Calf	5.3	4.7	3.5	2.3	1.8
1.8	**4.5**	L15 T	Breadth Ankle	3.4	3.0	2.3	1.5	1.1
4.1	**10.3**	L16 D	Mid Thigh, Anterior-Posterior	7.7	6.9	5.2	3.4	2.6
3.1	**8.0**	L17 D	Knee, Anterior-Posterior	6.0	5.3	4.0	2.7	2.0
2.8	**7.2**	L18 D	Mid Calf, Anterior-Posterior	5.4	4.8	3.6	2.4	1.8
2.2	**5.5**	L19 D	Ankle, Anterior-Posterior	4.1	3.7	2.8	1.8	1.4
7.0	**17.8**	L20 D	Length of Foot	13.4	11.9	8.9	5.9	4.5
5.9	**15.0**	L21 D	Heel to 5th Toe	11.3	10.0	7.5	5.0	3.8
2.7	**6.9**	L22 T	Breadth Foot	5.2	4.6	3.5	2.3	1.7
1.6	**4.0**	L23 T	Width of Heel	3.0	2.7	2.0	1.3	1.0
		* *These are also listed with Vertical Measurements.*						

Six-Year-Old Boy, anterior and lateral head

all measurements in centimeters

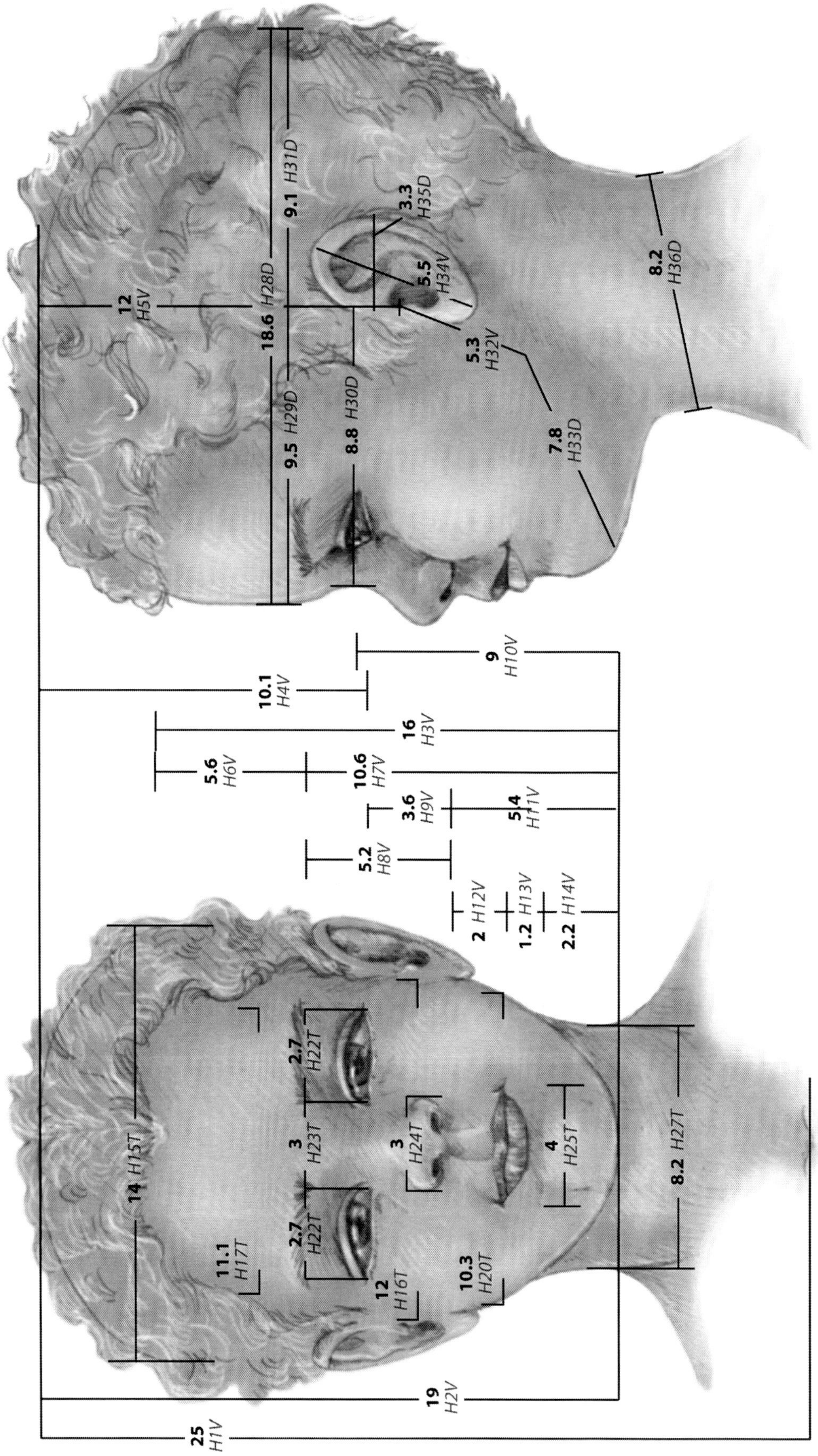

Six-Year-Old Boy, anterior and lateral figure

all measurements in centimeters

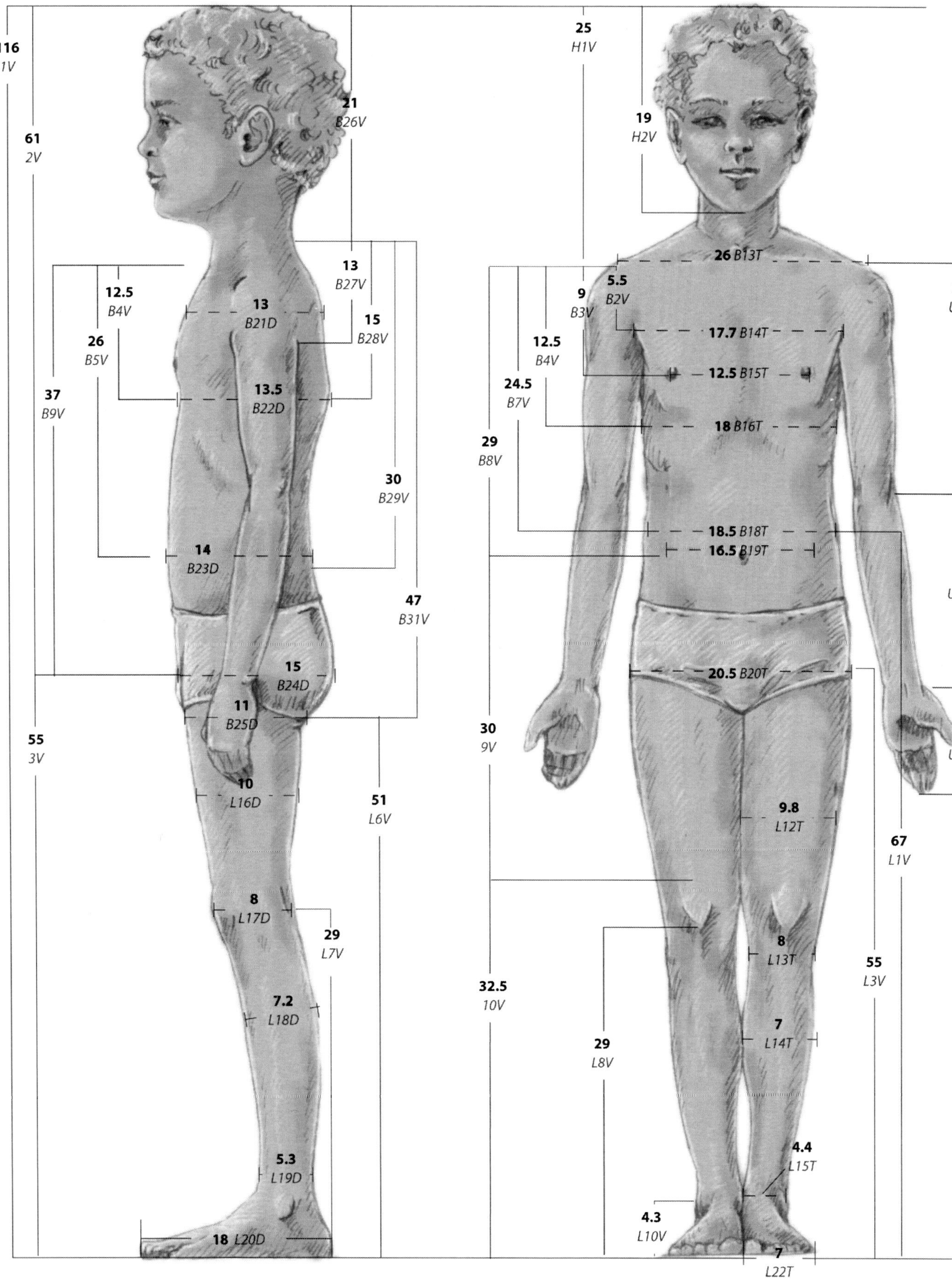

Life size			SIX-YEAR-OLD BOY	Fractions of life size				
				3/4	2/3	1/2	1/3	1/4
Inches	**cm**	**Code**		**cm**	**cm**	**cm**	**cm**	**cm**
			VERTICAL MEASUREMENTS					
45.7	**116.0**	1 V	Stature, Height, Vertex to Sole	87.0	77.3	58.0	38.7	29.0
24.0	**61.0**	2 V	Vertex to Symphysis Pubis	45.8	40.7	30.5	20.3	15.3
21.7	**55.0**	* 3 V	Symphysis Pubis of Pelvis-Sole	41.3	36.7	27.5	18.3	13.8
35.8	**91.0**	4 V	Sternal Notch, Neck Pit, to Sole	68.3	60.7	45.5	30.3	22.8
9.4	**24.0**	5 V	Vertex to Acromion Process	18.0	16.0	12.0	8.0	6.0
36.2	**92.0**	6 V	Acromion Process to Sole	69.0	61.3	46.0	30.7	23.0
24.4	**62.0**	* 7 V	Ant.Sup.Spine of Ilium to Sole	46.5	41.3	31.0	20.7	15.5
11.4	**29.0**	8 V	Acrom. proc. to Ant. Sup. Spine	21.8	19.3	14.5	9.7	7.3
11.8	**30.0**	9 V	Ant.Sup.Spine of Ilium to Knee	22.5	20.0	15.0	10.0	7.5
12.8	**32.5**	10 V	Knee at Patella to Sole	24.4	21.7	16.3	10.8	8.1
25.6	**65.0**	11 V	Umbilicus to Sole	48.8	43.3	32.5	21.7	16.3
20.1	**51.0**	* 12 V	Crotch to Sole	38.3	34.0	25.5	17.0	12.8
33.9	**86.0**	13 V	Kneeling Height	64.5	57.3	43.0	28.7	21.5
25.2	**64.0**	14 V	Sitting Height	48.0	42.7	32.0	21.3	16.0
23.6	**60.0**	15 V	Buttock, Straight Leg to Sole	45.0	40.0	30.0	20.0	15.0
14.2	**36.0**	16 V	Buttock to Flexed Knee	27.0	24.0	18.0	12.0	9.0
14.0	**35.5**	17 V	Flexed Knee to Sole	26.6	23.7	17.8	11.8	8.9
		** These are also listed with Lower Extremity Measurements.*						
			VERTICAL HEAD MEASUREMENTS					
9.8	**25.0**	H1 V	Vertex to Sternal Notch	18.8	16.7	12.5	8.3	6.3
7.5	**19.0**	H2 V	Vertex to Chin, Head Height	14.3	12.7	9.5	6.3	4.8
6.3	**16.0**	H3 V	Hairline to Chin, Face Height	12.0	10.7	8.0	5.3	4.0
4.0	**10.1**	H4 V	Vertex to Lateral Corner of Eye	7.6	6.7	5.1	3.4	2.5
4.7	**12.0**	H5 V	Vertex to Tragion of Ear	9.0	8.0	6.0	4.0	3.0
2.2	**5.6**	H6 V	Hairline to Brow, Glabella	4.2	3.7	2.8	1.9	1.4
4.2	**10.6**	H7 V	Brow, Glabella, to Chin, Menton	8.0	7.1	5.3	3.5	2.7
2.0	**5.2**	H8 V	Brow to Subnasal Point	3.9	3.5	2.6	1.7	1.3
1.4	**3.6**	H9 V	Sellion to Subnasal, Nose Length	2.7	2.4	1.8	1.2	0.9
3.5	**9.0**	H10 V	Sellion, Nose Saddle, to Chin	6.8	6.0	4.5	3.0	2.3
2.1	**5.4**	H11 V	Subnasal Point to Chin	4.1	3.6	2.7	1.8	1.4
0.8	**2.0**	H12 V	Subnasal to Lip Aperature	1.5	1.3	1.0	0.7	0.5
0.5	**1.2**	H13 V	Lip Aperat. to Sublabial Crease	0.9	0.8	0.6	0.4	0.3
0.9	**2.2**	H14 V	Sublabial Crease to Chin	1.7	1.5	1.1	0.7	0.6
			TRANSVERSE HEAD MEASUREMENTS					
5.5	**14.0**	H15 T	Bi-Parietal, Head Breadth	10.5	9.3	7.0	4.7	3.5
4.7	**12.0**	H16 T	Bi-Zygomatic, Cheekbones	9.0	8.0	6.0	4.0	3.0
4.4	**11.1**	H17 T	Bi-Fronto Temporal, Forehead	8.3	7.4	5.6	3.7	2.8

Life size			SIX-YEAR-OLD BOY	Fractions of life size				
				3/4	2/3	1/2	1/3	1/4
Inches	cm	Code		cm	cm	cm	cm	cm
			TRANSVERSE HEAD MEASUREMENTS					
3.5	**9.0**	H18 T	Inter-Brow Ridge	6.8	6.0	4.5	3.0	2.3
4.5	**11.5**	H19 T	Inter-Tragion of Ears	8.6	7.7	5.8	3.8	2.9
4.1	**10.3**	H20 T	Inter-Angle of Mandible	7.7	6.9	5.2	3.4	2.6
3.2	**8.2**	H21 T	Inter-Lateral Corners of Eyes	6.2	5.5	4.1	2.7	2.1
1.0	**2.6**	H22 T	Width of Eye	2.0	1.7	1.3	0.9	0.7
1.2	**3.0**	H23 T	Inter-Inner Corner of Eyes	2.3	2.0	1.5	1.0	0.8
1.2	**3.0**	H24 T	Breadth of Nose	2.3	2.0	1.5	1.0	0.8
1.6	**4.0**	H25 T	Breadth of Mouth	3.0	2.7	2.0	1.3	1.0
1.2	**3.0**	H26 T	Breadth of Chin	2.3	2.0	1.5	1.0	0.8
3.2	**8.2**	H27 T	Breadth of Neck	6.2	5.5	4.1	2.7	2.1
			ANTERIOR-POSTERIOR HEAD MEASUREMENTS					
7.3	**18.6**	H28 D	Brow to Occiput, Posterior Head	14.0	12.4	9.3	6.2	4.7
3.7	**9.5**	H29 D	Brow to Tragus of Ear	7.1	6.3	4.8	3.2	2.4
3.5	**8.8**	H30 D	Sellion, Nose Saddle, to Tragus	6.6	5.9	4.4	2.9	2.2
3.6	**9.1**	H31 D	Tragus to Occiput, Back of Head	6.8	6.1	4.6	3.0	2.3
2.1	**5.3**	H32 V	Tragus to Angle Mandible	4.0	3.5	2.7	1.8	1.3
3.1	**7.8**	H33 D	Angle of Mandible to Chin	5.9	5.2	3.9	2.6	2.0
2.2	**5.5**	H34 V	Length of Ear, Vertical	4.1	3.7	2.8	1.8	1.4
1.3	**3.3**	H35 D	Width of Ear, Anterior-Posterior	2.5	2.2	1.7	1.1	0.8
3.2	**8.2**	H36 D	Anterior to Posterior of Neck	6.2	5.5	4.1	2.7	2.1
			VERTICAL BODY MEASUREMENTS					
2.6	**6.5**	B1 V	Chin to Sternal Notch, Neck Pit	4.9	4.3	3.3	2.2	1.6
2.2	**5.5**	B2 V	Sternal Notch to Axilla, Arm Pit	4.1	3.7	2.8	1.8	1.4
3.5	**9.0**	B3 V	Sternal Notch to Nipples	6.8	6.0	4.5	3.0	2.3
4.9	**12.5**	B4 V	Sternal Notch to Xyphoid Proc.	9.4	8.3	6.3	4.2	3.1
10.2	**26.0**	B5 V	Sternal Notch to Umbilicus	19.5	17.3	13.0	8.7	6.5
5.3	**13.5**	B6 V	Xyphoid Process to Umbilicus	10.1	9.0	6.8	4.5	3.4
9.6	**24.5**	B7 V	Sternal Notch to Crest Ilium	18.4	16.3	12.3	8.2	6.1
11.4	**29.0**	B8 V	Sternal Notch to Ant.S.Sp.Ilium	21.8	19.3	14.5	9.7	7.3
14.6	**37.0**	B9 V	Sternal Notch to Pubis	27.8	24.7	18.5	12.3	9.3
4.3	**11.0**	B10 V	Umbilicus to Pubis of Pelvis	8.3	7.3	5.5	3.7	2.8
			TRANSVERSE BODY MEASUREMENTS					
11.2	**28.5**	B11 T	Shoulder Breadth	21.4	19.0	14.3	9.5	7.1
5.0	**12.7**	B12 T	Clavicle to Acromion Process	9.5	8.5	6.4	4.2	3.2
10.2	**26.0**	B13 T	Bi-acromial Process Breadth	19.5	17.3	13.0	8.7	6.5
7.0	**17.7**	B14 T	Inter-Anterior Axillary Fold	13.3	11.8	8.9	5.9	4.4
4.9	**12.5**	B15 T	Inter-Nipples	9.4	8.3	6.3	4.2	3.1

Life size			SIX-YEAR-OLD BOY	Fractions of life size				
				3/4	2/3	1/2	1/3	1/4
Inches	cm	Code		cm	cm	cm	cm	cm
			TRANSVERSE BODY MEASUREMENTS					
7.1	**18.0**	B16 T	Breadth of Thorax	13.5	12.0	9.0	6.0	4.5
7.2	**18.3**	B17 T	Inter-Costal. Rib Margin Breadth	13.7	12.2	9.2	6.1	4.6
7.3	**18.5**	B18 T	Inter-Crest of Ilium Breadth	13.9	12.3	9.3	6.2	4.6
6.5	**16.5**	B19 T	Inter-Anterior Superior Spine	12.4	11.0	8.3	5.5	4.1
8.1	**20.5**	B20 T	Inter-Greater Trochanter Femur	15.4	13.7	10.3	6.8	5.1
			ANTERIOR-POSTERIOR BODY MEASUREMENTS					
5.1	**13.0**	B21 D	Manubrium to Dorsal Spine	9.8	8.7	6.5	4.3	3.3
5.3	**13.5**	B22 D	Xyphoid Process to Dorsum	10.1	9.0	6.8	4.5	3.4
5.5	**14.0**	B23 D	Umbilicus to Dorsum	10.5	9.3	7.0	4.7	3.5
5.9	**15.0**	B24 D	Pubis of Pelvis to Buttock	11.3	10.0	7.5	5.0	3.8
4.3	**11.0**	B25 D	Thigh to Gluteal Furrow	8.3	7.3	5.5	3.7	2.8
			VERTICAL BACK MEASUREMENTS					
8.3	**21.0**	B26 V	Vertex to 7th Cervical Spine	15.8	14.0	10.5	7.0	5.3
			7th Cervical Vertebra Spine to:					
5.1	**13.0**	B27 V	To Posterior Axillary Fold, Armpit	9.8	8.7	6.5	4.3	3.3
6.1	**15.5**	B28 V	To Inferior Tip of Scapula	11.6	10.3	7.8	5.2	3.9
11.8	**30.0**	B29 V	To Posterior Superior Spine Ilium	22.5	20.0	15.0	10.0	7.5
14.2	**36.0**	B30 V	To Midline Upper Buttock Crease	27.0	24.0	18.0	12.0	9.0
18.5	**47.0**	B31 V	To Gluteal Furrow	35.3	31.3	23.5	15.7	11.8
25.6	**65.0**	B32 V	To Popliteal Crease Behind Knee	48.8	43.3	32.5	21.7	16.3
			TRANSVERSE BACK MEASUREMENTS					
9.1	**23.0**	B33 T	Inter-Posterior Axillary Folds	17.3	15.3	11.5	7.7	5.8
4.5	**11.5**	B34 T	Inter-Scapulae Medial Borders	8.6	7.7	5.8	3.8	2.9
2.4	**6.0**	B35 T	Inter-Post. Sup. Spines Sacrum	4.5	4.0	3.0	2.0	1.5
			UPPER EXTREMITY MEASUREMENTS					
44.9	**114.0**	U1 V	Span of Arms	85.5	76.0	57.0	38.0	28.5
5.0	**12.6**	U2 V	Sternum to Acromion Process	9.5	8.4	6.3	4.2	3.2
19.7	**50.0**	U3 V	Acromion Process to Fingertip	37.5	33.3	25.0	16.7	12.5
56.7	**144.0**	U4 V	Vertical Reach to Fingertip	108.0	96.0	72.0	48.0	36.0
28.7	**73.0**	U5 T	Lateral Reach to Fingertip	54.8	48.7	36.5	24.3	18.3
22.4	**57.0**	U6 D	Forward Reach to Fingertip	42.8	38.0	28.5	19.0	14.3
9.3	**23.5**	U7 V	Shoulder to Flexed Elbow	17.6	15.7	11.8	7.8	5.9
8.3	**21.0**	U8 V	Acromion to Lat. Condyle Elbow	15.8	14.0	10.5	7.0	5.3
12.1	**30.7**	U9 D	Flexed Elbow to Fingertip	23.0	20.5	15.4	10.2	7.7
6.3	**16.0**	U10 D	Flexed Elbow Pivot to Wrist Pivot	12.0	10.7	8.0	5.3	4.0

Life size			SIX-YEAR-OLD BOY	Fractions of life size				
				3/4	2/3	1/2	1/3	1/4
Inches	cm	Code		cm	cm	cm	cm	cm
			UPPER EXTREMITY MEASUREMENTS					
5.1	**13.0**	U11 V	Wrist to Fingertip	9.8	8.7	6.5	4.3	3.3
3.1	**8.0**	U12 V	Wrist to Thumb Tip	6.0	5.3	4.0	2.7	2.0
2.8	**7.0**	U13 V	Length of Palm	5.3	4.7	3.5	2.3	1.8
2.1	**5.4**	U14 V	Length 3rd Finger	4.1	3.6	2.7	1.8	1.4
3.4	**8.7**	U15 V	Thumb Crotch to 3rd Finger	6.5	5.8	4.4	2.9	2.2
1.7	**4.3**	U16 V	Thumb Crotch to Thumb Tip	3.2	2.9	2.2	1.4	1.1
2.0	**5.2**	U17 D	Anterior-Post. Arm at Biceps	3.9	3.5	2.6	1.7	1.3
2.0	**5.2**	U18 D	Anterior-Posterior at Forearm	3.9	3.5	2.6	1.7	1.3
1.2	**3.0**	U19 D	Depth of Wrist	2.3	2.0	1.5	1.0	0.8
1.7	**4.3**	U20 T	Breadth of Wrist	3.2	2.9	2.2	1.4	1.1
2.9	**7.4**	U21 T	Breadth of Hand	5.6	4.9	3.7	2.5	1.9
2.4	**6.0**	U22 T	Breadth of Palm	4.5	4.0	3.0	2.0	1.5
			LOWER EXTREMITY MEASUREMENTS					
26.4	**67.0**	L1 V	Crest of Ilium of Pelvis to Sole	50.3	44.7	33.5	22.3	16.8
24.4	**62.0**	* L2 V	Ant.Sup.Spine of Ilium to Sole	46.5	41.3	31.0	20.7	15.5
21.7	**55.0**	L3 V	Greater Trochanter Femur-Sole	41.3	36.7	27.5	18.3	13.8
21.7	**55.0**	* L4 V	Symphysis Pubis of Pelvis-Sole	41.3	36.7	27.5	18.3	13.8
20.1	**51.0**	* L5 V	Crotch to Sole	38.3	34.0	25.5	17.0	12.8
20.1	**51.0**	L6 V	Gluteal Furrow to Sole	38.3	34.0	25.5	17.0	12.8
12.0	**30.5**	L7 V	Posterior Knee Crease to Sole	22.9	20.3	15.3	10.2	7.6
11.4	**29.0**	L8 V	Tibia at Knee to Sole	21.8	19.3	14.5	9.7	7.3
2.0	**5.0**	L9 V	Med. Malleolus of Ankle to Sole	3.8	3.3	2.5	1.7	1.3
1.7	**4.3**	L10 V	Lat. Malleolus of Ankle to Sole	3.2	2.9	2.2	1.4	1.1
8.5	**21.5**	L11 T	Breadth of Hips, Sitting	16.1	14.3	10.8	7.2	5.4
3.9	**9.8**	L12 T	Breadth Mid Thigh	7.4	6.5	4.9	3.3	2.5
3.1	**8.0**	L13 T	Breadth of Knee	6.0	5.3	4.0	2.7	2.0
2.8	**7.0**	L14 T	Breadth of Calf	5.3	4.7	3.5	2.3	1.8
1.7	**4.4**	L15 T	Breadth Ankle	3.3	2.9	2.2	1.5	1.1
3.9	**10.0**	L16 D	Mid Thigh, Anterior-Posterior	7.5	6.7	5.0	3.3	2.5
3.1	**8.0**	L17 D	Knee, Anterior-Posterior	6.0	5.3	4.0	2.7	2.0
2.8	**7.2**	L18 D	Mid Calf, Anterior-Posterior	5.4	4.8	3.6	2.4	1.8
2.1	**5.3**	L19 D	Ankle, Anterior-Posterior	4.0	3.5	2.7	1.8	1.3
7.1	**18.0**	L20 D	Length of Foot	13.5	12.0	9.0	6.0	4.5
5.0	**12.7**	L21 D	Heel to 5th Toe	9.5	8.5	6.4	4.2	3.2
2.8	**7.0**	L22 T	Breadth Foot	5.3	4.7	3.5	2.3	1.8
1.6	**4.0**	L23 T	Width of Heel	3.0	2.7	2.0	1.3	1.0
		** These are also listed with Vertical Measurements.*						

DEVELOPMENT OF CHILDREN'S POSTURE

The preferred adult posture is described as having good balance at rest when a plumb line passes a little in front of the ankle, just in front of the knee joint, through the hip joint and just behind the ear. In this posture, the head is erect; there is an anterior curvature of neck vertebrae, a posterior curvature of thoracic, or chest, vertebrae and an anterior curvature or lordosis of the lumbar or low back vertebrae.

A child's posture has the same anterior and posterior curves, but the lower back lordosis is more pronounced. It changes with the years of development. Variables in body dynamics involve several factors: The force of gravity, the location of the center of gravity, the development of bones and muscle strength, the stamina of muscles or state of fatigue, the attitude of alertness or of lethargy; and the general health status are all factors.[1]

In early infancy, the head is proportionally larger and heavier than in later infancy and early childhood. The center of gravity is high on the body. An infant does not attempt to walk until there is increase in bone strength with growth, and increased muscle development with improved coordination. During the first year, after being able to stand erect, the legs are spread and the feet are widely apart to increase the base of support. There is a tendency to toe out, and there is considerable sway back.[2] The bones are flexible; especially the, spine and the muscles are weak. The abdomen is prominent and forward. In order to balance, the lumbar or lower back vertebrae have an anterior curve or lordosis. The center of gravity is near the xiphoid process, or at the lower tip of the breastbone, and it continues above the umbilicus in early childhood. At five to six years, the center of gravity moves below the umbilicus.[3]

The pelvis is tilted to about 30 degrees from horizontal in age two to six years. As muscles become stronger, balance improves. The lordosis is diminished, and the pelvis tilt progressively decreases to about 15 degrees in adulthood.[4] In the early years of development, an apparent suboptimal posture is common since the children have undeveloped muscles, and loose ligaments. They are very busy in play with frequent intermittent physical activity that can result in episodes of fatigue. Children may demonstrate happiness with a buoyant and vigorous approach to activities. Posture may also indicate an expression of sadness by sagging shoulders and lack of activity.[5]

During childhood years, posture usually improves and it indicates good health. The capacity for a military type posture with the head erect, shoulders back and the vertical plane of the abdomen similar to the thoracic plane begins to develop in adolescent years. Muscles and tendons have strengthened, while joints have become less flexible. The posture continues to improve with vigorous physical activity and with athletic participation. The protuberant abdomen decreases along with the improved lumbar lordosis, or swayback, as a child approaches puberty. A better posture is more often seen after mid-teen age.

1. Breckenridge, M.E., Murphy, M.N., p. 282.
2. Ibid, p. 274.
3. Lowery, G., p. 115.
4. Breckenridge, M.E., Murphy M.N., p. 282.
5. Ibid , p,. 284.

Posture of Children, One Through Nine Years Old

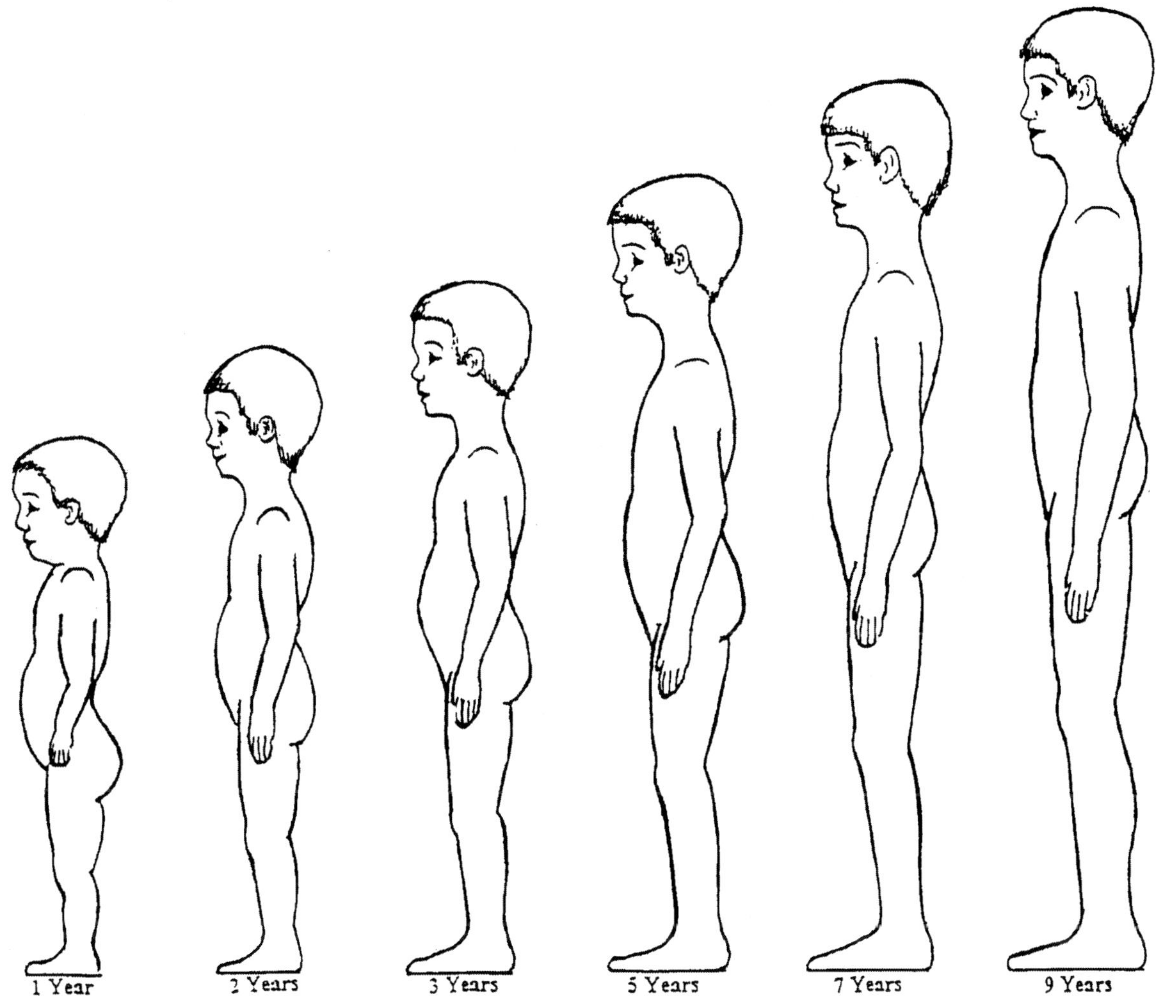

After an infant is able to stand erect and from one to four years of age, children's feet are relatively flat. The legs are bowed, especially at knees and ankles. The abdomen is usually protuberant or pot-bellied. The back exhibits marked lordosis or sway back. This suboptimal appearing posture is common in early years, since many are relatively thin. Ligaments are relatively loose and muscular development is limited. They have more mobility and flexibility of joints than adults. Muscular relaxation is common during fatigue and rest periods that occur between almost constant physical activities during their waking hours. Most have progressive improvement in muscle strength and tone during elementary school years as they approach puberty.[1]

1. Breckenridge, M.E., Murphy, M.N., p 285

SEVEN-YEAR-OLD GIRLS

The practical list of body proportions is listed separately in the tables of measurements. The range of height, twenty-fifth, to fiftieth, and to seventy-fifth percentile is shown here for comparison.

Percentile	Twenty-fifth		Fiftieth		Seventy-fifth	
Girls:	Short		Medium		Tall	
Stature	117 cm,	46 in.	121 cm,	47.5 in.	124.5 cm,	49 in.
Weight	20 kg,	43.5 lb.	22 kg,	48.5 lb.	24 kg,	53.3 lb.[1]

At this age, boys are on the average, slightly taller and slightly heavier than are girls. However, on closer inspection, the girls may have more rounded contours in face shoulders and hips. The most noticeable difference between boys and girls is the hair styling. For school and play clothing is often similar. For special occasions, girls enjoy dressing like little ladies. Their interest is in clothing, dolls, music, and costume jewelry. Play may be less active and emotions are better controlled. The school level is second grade.

BODY MEASUREMENTS COMPARED TO HEIGHT

	Feature Measure in cm	Feature Compared to Stature	Approx. Ratio of Feature	Compared to Median Adult
Stature: Crown to sole	121			72%
Head Height: Crown to chin	18.6	15%	1/6	91%
Sitting Height:	65.5	54%	1/2+	74%
Trunk Measure: Sternal notch to symphysis pubis	38	32%	1/3	78%
Leg Length: Greater trochanter of the femur to sole	59	49%	1/2	69%
Thigh Length: Greater trochanter of the femur to tibia at knee	28	23%	1/4–	69%
Tibia at Knee to Sole:	31	27%	1/4	68%
Arm Length: Acromion process at shoulder to dependent fingertip	50	41%	2/5	67%
Hand Length:	13.1	11%	1/9	73%
Foot Length:	18.8	16%	1/6	78%

Between six and seven years of age, deciduous incisors teeth are lost, and the lower central permanent incisor teeth appear.[2]

1. Nelson, W.E., p. 39.
2. Lockhart, R.D., p. 57.

SEVEN-YEAR-OLD BOYS

The practical list of body proportions, is shown in the tables of measurements. The range of twenty-fifth, fiftieth, and seventy-fifth percentile is shown here for comparison.

Percentile	Twenty-fifth		Fiftieth		Seventy-fifth	
Boys:	Short		Medium		Tall	
Stature	118 cm,	46.5 in.	122 cm,	48 in.	125 cm,	49.3 in.
Weight	21 kg,	46.3 lb.	23 kg,	50.3 lb.	25 kg,	55 lb.[1]

At this age, boys usually are slightly taller and slightly heavier than girls. Boys' interests include outdoor sports, ball games, mechanical toys, monsters, dinosaurs and super heroes.

Play is less exaggerated than the previous year and behavior is improved. They may be more concerned about parental and group approval. The school level is second grade.

BODY MEASUREMENTS COMPARED TO HEIGHT

	Feature Measure in cm	Feature Compared to Stature	Approx. Ratio of Feature	Compared to Median Adult
Stature: Crown to sole	123			67%
Head Height: Crown to chin	19.2	16%	1/6	84%
Sitting Height:	66	54%	1/2+	69%
Trunk Measure: Sternal notch to symphysis pubis	39	32%	1/3	74%
Leg Length: Greater trochanter of the femur to sole	60	49%	1/2	61%
Thigh Length: Greater trochanter of the femur to tibia at knee	27	22%	1/4–	58%
Tibia at Knee to Sole:	33	27%	1/4	61%
Arm Length: Acromion process at shoulder to dependent fingertip	52	43%	2/5+	64%
Hand Length:	13.3	11%	1/9	67%
Foot Length:	19	16%	1/6	68%

Between six and seven years of age, boys usually lose four deciduous incisor teeth, and the lower permanent central incisors erupt.[2]

1. Nelson, W.E., p. 38.
2. Lockhart, et al. p. 57.

Seven-Year-Old Girl, anterior and lateral head

all measurements in centimeters

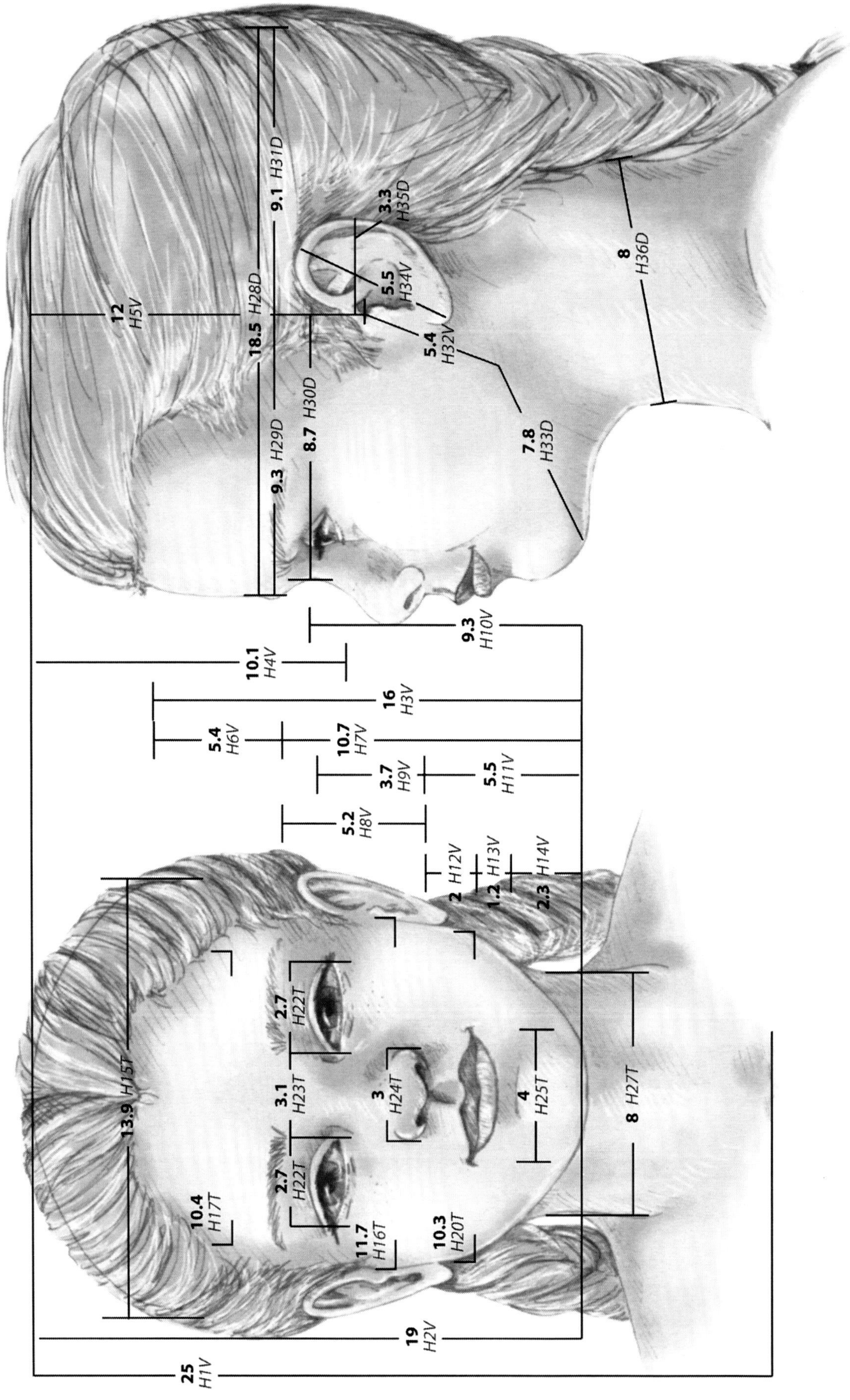

Seven-Year-Old Girl, anterior and lateral figure

all measurements in centimeters

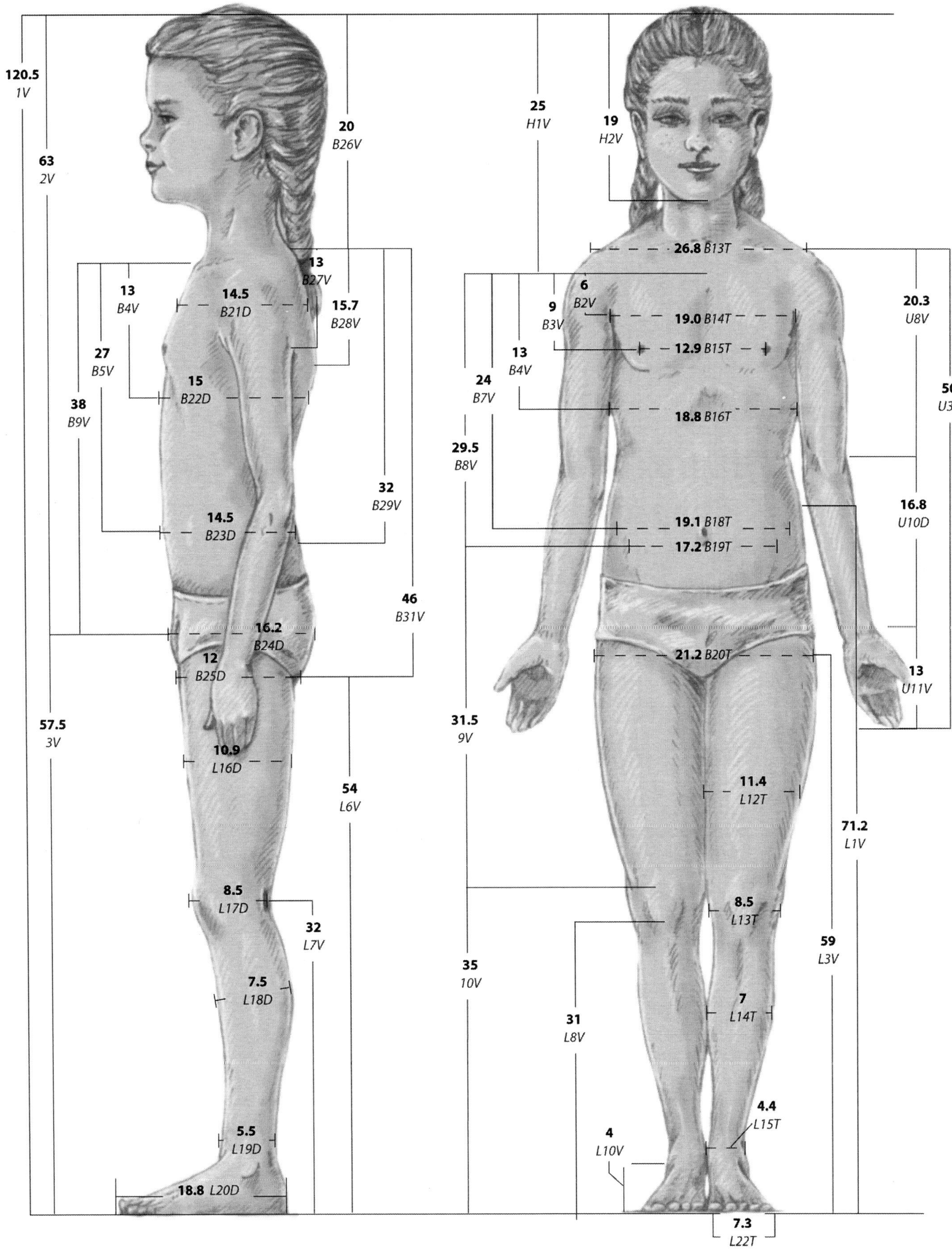

Life size			SEVEN-YEAR-OLD GIRL	Fractions of life size				
				3/4	2/3	1/2	1/3	1/4
Inches	cm	Code		cm	cm	cm	cm	cm
			VERTICAL MEASUREMENTS					
47.5	**120.5**	1 V	Stature, Height, Vertex to Sole	90.4	80.3	60.3	40.2	30.1
24.8	**63.0**	2 V	Vertex to Symphysis Pubis	47.3	42.0	31.5	21.0	15.8
22.7	**57.5**	* 3 V	Symphysis Pubis of Pelvis-Sole	43.1	38.3	28.8	19.2	14.4
37.6	**95.5**	4 V	Sternal Notch, Neck Pit, to Sole	71.6	63.7	47.8	31.8	23.9
9.5	**24.0**	5 V	Vertex to Acromion Process	18.0	16.0	12.0	8.0	6.0
38.0	**96.5**	6 V	Acromion Process to Sole	72.4	64.3	48.3	32.2	24.1
26.0	**66.0**	* 7 V	Ant.Sup.Spine of Ilium to Sole	49.5	44.0	33.0	22.0	16.5
12.0	**30.5**	8 V	Acrom. proc. to Ant. Sup. Spine	22.9	20.3	15.3	10.2	7.6
12.4	**31.5**	9 V	Ant.Sup.Spine of Ilium to Knee	23.6	21.0	15.8	10.5	7.9
13.8	**35.0**	10 V	Knee at Patella to Sole	26.3	23.3	17.5	11.7	8.8
27.0	**68.5**	11 V	Umbilicus to Sole	51.4	45.7	34.3	22.8	17.1
21.3	**54.0**	* 12 V	Crotch to Sole	40.5	36.0	27.0	18.0	13.5
35.9	**91.0**	13 V	Kneeling Height	68.3	60.7	45.5	30.3	22.8
25.8	**65.5**	14 V	Sitting Height	49.1	43.7	32.8	21.8	16.4
25.1	**63.8**	15 V	Buttock, Straight Leg to Sole	47.9	42.5	31.9	21.3	16.0
15.4	**39.0**	16 V	Buttock to Flexed Knee	29.3	26.0	19.5	13.0	9.8
14.4	**36.5**	17 V	Flexed Knee to Sole	27.4	24.3	18.3	12.2	9.1
		** These are also listed with Lower Extremity Measurements.*						
			VERTICAL HEAD MEASUREMENTS					
9.9	**25.0**	H1 V	Vertex to Sternal Notch	18.8	16.7	12.5	8.3	6.3
7.5	**19.0**	H2 V	Vertex to Chin, Head Height	14.3	12.7	9.5	6.3	4.8
6.3	**16.0**	H3 V	Hairline to Chin, Face Height	12.0	10.7	8.0	5.3	4.0
4.0	**10.1**	H4 V	Vertex to Lateral Corner of Eye	7.6	6.7	5.1	3.4	2.5
4.7	**12.0**	H5 V	Vertex to Tragion of Ear	9.0	8.0	6.0	4.0	3.0
2.1	**5.4**	H6 V	Hairline to Brow, Glabella	4.1	3.6	2.7	1.8	1.4
4.2	**10.7**	H7 V	Brow, Glabella, to Chin, Menton	8.0	7.1	5.4	3.6	2.7
2.0	**5.2**	H8 V	Brow to Subnasal Point	3.9	3.5	2.6	1.7	1.3
1.5	**3.7**	H9 V	Sellion to Subnasal, Nose Length	2.8	2.5	1.9	1.2	0.9
3.7	**9.3**	H10 V	Sellion, Nose Saddle, to Chin	7.0	6.2	4.7	3.1	2.3
2.2	**5.5**	H11 V	Subnasal Point to Chin	4.1	3.7	2.8	1.8	1.4
0.8	**2.0**	H12 V	Subnasal to Lip Aperature	1.5	1.3	1.0	0.7	0.5
0.5	**1.2**	H13 V	Lip Aperat. to Sublabial Crease	0.9	0.8	0.6	0.4	0.3
0.9	**2.3**	H14 V	Sublabial Crease to Chin	1.7	1.5	1.2	0.8	0.6
			TRANSVERSE HEAD MEASUREMENTS					
5.5	**13.9**	H15 T	Bi-Parietal, Head Breadth	10.4	9.3	7.0	4.6	3.5
4.6	**11.7**	H16 T	Bi-Zygomatic, Cheekbones	8.8	7.8	5.9	3.9	2.9
4.1	**10.4**	H17 T	Bi-Fronto Temporal, Forehead	7.8	6.9	5.2	3.5	2.6

Life size			SEVEN-YEAR-OLD GIRL	Fractions of life size				
				3/4	2/3	1/2	1/3	1/4
Inches	cm	Code		cm	cm	cm	cm	cm
			TRANSVERSE HEAD MEASUREMENTS					
3.6	9.2	H18 T	Inter-Brow Ridge	6.9	6.1	4.6	3.1	2.3
4.3	11.0	H19 T	Inter-Tragion of Ears	8.3	7.3	5.5	3.7	2.8
4.1	10.3	H20 T	Inter-Angle of Mandible	7.7	6.9	5.2	3.4	2.6
3.3	8.5	H21 T	Inter-Lateral Corners of Eyes	6.4	5.7	4.3	2.8	2.1
1.1	2.7	H22 T	Width of Eye	2.0	1.8	1.4	0.9	0.7
1.2	3.0	H23 T	Inter-Inner Corner of Eyes	2.3	2.0	1.5	1.0	0.8
1.2	3.0	H24 T	Breadth of Nose	2.3	2.0	1.5	1.0	0.8
1.6	4.0	H25 T	Breadth of Mouth	3.0	2.7	2.0	1.3	1.0
1.2	3.0	H26 T	Breadth of Chin	2.3	2.0	1.5	1.0	0.8
3.2	8.0	H27 T	Breadth of Neck	6.0	5.3	4.0	2.7	2.0
			ANTERIOR-POSTERIOR HEAD MEASUREMENTS					
7.3	18.5	H28 D	Brow to Occiput, Posterior Head	13.9	12.3	9.3	6.2	4.6
3.7	9.3	H29 D	Brow to Tragus of Ear	7.0	6.2	4.7	3.1	2.3
3.4	8.7	H30 D	Sellion, Nose Saddle, to Tragus	6.5	5.8	4.4	2.9	2.2
3.6	9.1	H31 D	Tragus to Occiput, Back of Head	6.8	6.1	4.6	3.0	2.3
2.1	5.4	H32 V	Tragus to Angle Mandible	4.1	3.6	2.7	1.8	1.4
3.1	7.8	H33 D	Angle of Mandible to Chin	5.9	5.2	3.9	2.6	2.0
2.2	5.5	H34 V	Length of Ear, Vertical	4.1	3.7	2.8	1.8	1.4
1.3	3.3	H35 D	Width of Ear, Anterior-Posterior	2.5	2.2	1.7	1.1	0.8
3.2	8.0	H36 D	Anterior to Posterior of Neck	6.0	5.3	4.0	2.7	2.0
			VERTICAL BODY MEASUREMENTS					
2.6	6.6	B1 V	Chin to Sternal Notch, Neck Pit	5.0	4.4	3.3	2.2	1.7
2.4	6.0	B2 V	Sternal Notch to Axilla, Arm Pit	4.5	4.0	3.0	2.0	1.5
3.5	9.0	B3 V	Sternal Notch to Nipples	6.8	6.0	4.5	3.0	2.3
5.1	13.0	B4 V	Sternal Notch to Xyphoid Proc.	9.8	8.7	6.5	4.3	3.3
10.6	27.0	B5 V	Sternal Notch to Umbilicus	20.3	18.0	13.5	9.0	6.8
5.1	13.0	B6 V	XyphoidProcess to Umbilicus	9.8	8.7	6.5	4.3	3.3
9.5	24.0	B7 V	Sternal Notch to Crest Ilium	18.0	16.0	12.0	8.0	6.0
11.6	29.5	B8 V	Sternal Notch to Ant.S.Sp.Ilium	22.1	19.7	14.8	9.8	7.4
14.6	37.0	B9 V	Sternal Notch to Pubis	27.8	24.7	18.5	12.3	9.3
4.3	11.0	B10 V	Umbilicus to Pubis of Pelvis	8.3	7.3	5.5	3.7	2.8
			TRANSVERSE BODY MEASUREMENTS					
11.4	29.0	B11 T	Shoulder Breadth	21.8	19.3	14.5	9.7	7.3
5.2	13.2	B12 T	Clavicle to Acromion Process	9.9	8.8	6.6	4.4	3.3
10.6	26.8	B13 T	Bi-acromial Process Breadth	20.1	17.9	13.4	8.9	6.7
7.5	19.0	B14 T	Inter-Anterior Axillary Fold	14.3	12.7	9.5	6.3	4.8
5.1	12.9	B15 T	Inter-Nipples	9.7	8.6	6.5	4.3	3.2

Life size			SEVEN-YEAR-OLD GIRL	Fractions of life size				
				3/4	2/3	1/2	1/3	1/4
Inches	cm	Code		cm	cm	cm	cm	cm
			TRANSVERSE BODY MEASUREMENTS					
7.4	**18.8**	B16 T	Breadth of Thorax	14.1	12.5	9.4	6.3	4.7
7.1	**18.0**	B17 T	Inter-Costal. Rib Margin Breadth	13.5	12.0	9.0	6.0	4.5
7.5	**19.1**	B18 T	Inter-Crest of Ilium Breadth	14.3	12.7	9.6	6.4	4.8
6.8	**17.2**	B19 T	Inter-Anterior Superior Spine	12.9	11.5	8.6	5.7	4.3
8.4	**21.2**	B20 T	Inter-Greater Trochanter Femur	15.9	14.1	10.6	7.1	5.3
			ANTERIOR-POSTERIOR BODY MEASUREMENTS					
5.7	**14.5**	B21 D	Manubrium to Dorsal Spine	10.9	9.7	7.3	4.8	3.6
5.9	**15.0**	B22 D	Xyphoid Process to Dorsum	11.3	10.0	7.5	5.0	3.8
5.7	**14.5**	B23 D	Umbilicus to Dorsum	10.9	9.7	7.3	4.8	3.6
6.4	**16.2**	B24 D	Pubis of Pelvis to Buttock	12.2	10.8	8.1	5.4	4.1
4.7	**12.0**	B25 D	Thigh to Gluteal Furrow	9.0	8.0	6.0	4.0	3.0
			VERTICAL BACK MEASUREMENTS					
7.9	**20.0**	B26 V	Vertex to 7th Cervical Spine	15.0	13.3	10.0	6.7	5.0
			7th Cervical Vertebra Spine to:					
5.1	**13.0**	B27 V	To Posterior Axillary Fold, Armpit	9.8	8.7	6.5	4.3	3.3
6.2	**15.7**	B28 V	To Inferior Tip of Scapula	11.8	10.5	7.9	5.2	3.9
12.6	**32.0**	B29 V	To Posterior Superior Spine Ilium	24.0	21.3	16.0	10.7	8.0
14.6	**37.0**	B30 V	To Midline Upper Buttock Crease	27.8	24.7	18.5	12.3	9.3
18.1	**46.0**	B31 V	To Gluteal Furrow	34.5	30.7	23.0	15.3	11.5
26.8	**68.0**	B32 V	To Popliteal Crease Behind Knee	51.0	45.3	34.0	22.7	17.0
			TRANSVERSE BACK MEASUREMENTS					
9.7	**24.5**	B33 T	Inter-Posterior Axillary Folds	18.4	16.3	12.3	8.2	6.1
4.6	**11.7**	B34 T	Inter-Scapulae Medial Borders	8.8	7.8	5.9	3.9	2.9
2.6	**6.5**	B35 T	Inter-Post. Sup. Spines Sacrum	4.9	4.3	3.3	2.2	1.6
			UPPER EXTREMITY MEASUREMENTS					
45.7	**116.0**	U1 V	Span of Arms	87.0	77.3	58.0	38.7	29.0
5.1	**13.0**	U2 V	Sternum to Acromion Process	9.8	8.7	6.5	4.3	3.3
19.7	**50.0**	U3 V	Acromion Process to Fingertip	37.5	33.3	25.0	16.7	12.5
60.3	**153.0**	U4 V	Vertical Reach to Fingertip	114.8	102.0	76.5	51.0	38.3
29.6	**75.0**	U5 T	Lateral Reach to Fingertip	56.3	50.0	37.5	25.0	18.8
23.2	**59.0**	U6 D	Forward Reach to Fingertip	44.3	39.3	29.5	19.7	14.8
9.6	**24.3**	U7 V	Shoulder to Flexed Elbow	18.2	16.2	12.2	8.1	6.1
8.0	**20.3**	U8 V	Acromion to Lat. Condyle Elbow	15.2	13.5	10.2	6.8	5.1
12.3	**31.2**	U9 D	Flexed Elbow to Fingertip	23.4	20.8	15.6	10.4	7.8
6.6	**16.8**	U10 D	Flexed Elbow Pivot to Wrist Pivot	12.6	11.2	8.4	5.6	4.2

Life size			SEVEN-YEAR-OLD GIRL	Fractions of life size				
				3/4	2/3	1/2	1/3	1/4
Inches	cm	Code		cm	cm	cm	cm	cm
			UPPER EXTREMITY MEASUREMENTS					
5.1	**13.0**	U11 V	Wrist to Fingertip	9.8	8.7	6.5	4.3	3.3
3.3	**8.5**	U12 V	Wrist to Thumb Tip	6.4	5.7	4.3	2.8	2.1
2.9	**7.4**	U13 V	Length of Palm	5.6	4.9	3.7	2.5	1.9
2.2	**5.7**	U14 V	Length 3rd Finger	4.3	3.8	2.9	1.9	1.4
3.6	**9.2**	U15 V	Thumb Crotch to 3rd Finger	6.9	6.1	4.6	3.1	2.3
1.8	**4.6**	U16 V	Thumb Crotch to Thumb Tip	3.5	3.1	2.3	1.5	1.2
2.2	**5.6**	U17 D	Anterior-Post. Arm at Biceps	4.2	3.7	2.8	1.9	1.4
2.0	**5.2**	U18 D	Anterior-Posterior at Forearm	3.9	3.5	2.6	1.7	1.3
1.2	**3.1**	U19 D	Depth of Wrist	2.3	2.1	1.6	1.0	0.8
1.8	**4.5**	U20 T	Breadth of Wrist	3.4	3.0	2.3	1.5	1.1
2.9	**7.4**	U21 T	Breadth of Hand	5.6	4.9	3.7	2.5	1.9
2.4	**6.0**	U22 T	Breadth of Palm	4.5	4.0	3.0	2.0	1.5
			LOWER EXTREMITY MEASUREMENTS					
28.1	**71.2**	L1 V	Crest of Ilium of Pelvis to Sole	53.4	47.5	35.6	23.7	17.8
26.0	**66.0**	* L2 V	Ant.Sup.Spine of Ilium to Sole	49.5	44.0	33.0	22.0	16.5
23.2	**59.0**	L3 V	Greater Trochanter Femur-Sole	44.3	39.3	29.5	19.7	14.8
22.7	**57.5**	* L4 V	Symphysis Pubis of Pelvis-Sole	43.1	38.3	28.8	19.2	14.4
21.3	**54.0**	* L5 V	Crotch to Sole	40.5	36.0	27.0	18.0	13.5
21.3	**54.0**	L6 V	Gluteal Furrow to Sole	40.5	36.0	27.0	18.0	13.5
12.6	**32.0**	L7 V	Posterior Knee Crease to Sole	24.0	21.3	16.0	10.7	8.0
12.2	**31.0**	L8 V	Tibia at Knee to Sole	23.3	20.7	15.5	10.3	7.8
2.0	**5.0**	L9 V	Med. Malleolus of Ankle to Sole	3.8	3.3	2.5	1.7	1.3
1.6	**4.0**	L10 V	Lat. Malleolus of Ankle to Sole	3.0	2.7	2.0	1.3	1.0
8.9	**22.5**	L11 T	Breadth of Hips	16.9	15.0	11.3	7.5	5.6
4.3	**11.0**	L12 T	Breadth Mid Thigh	8.3	7.3	5.5	3.7	2.8
3.3	**8.5**	L13 T	Breadth of Knee	6.4	5.7	4.3	2.8	2.1
2.8	**7.0**	L14 T	Breadth of Calf	5.3	4.7	3.5	2.3	1.8
1.8	**4.5**	L15 T	Breadth Ankle	3.4	3.0	2.3	1.5	1.1
4.3	**10.9**	L16 D	Mid Thigh, Anterior-Posterior	8.2	7.3	5.5	3.6	2.7
3.3	**8.5**	L17 D	Knee, Anterior-Posterior	6.4	5.7	4.3	2.8	2.1
3.0	**7.5**	L18 D	Mid Calf, Anterior-Posterior	5.6	5.0	3.8	2.5	1.9
2.2	**5.5**	L19 D	Ankle, Anterior-Posterior	4.1	3.7	2.8	1.8	1.4
7.4	**18.8**	L20 D	Length of Foot	14.1	12.5	9.4	6.3	4.7
6.0	**15.3**	L21 D	Heel to 5th Toe	11.5	10.2	7.7	5.1	3.8
2.9	**7.3**	L22 T	Breadth Foot	5.5	4.9	3.7	2.4	1.8
1.7	**4.3**	L23 T	Width of Heel	3.2	2.9	2.2	1.4	1.1
		** These are also listed with Vertical Measurements.*						

Seven-Year-Old Boy, anterior and lateral head

all measurements in centimeters

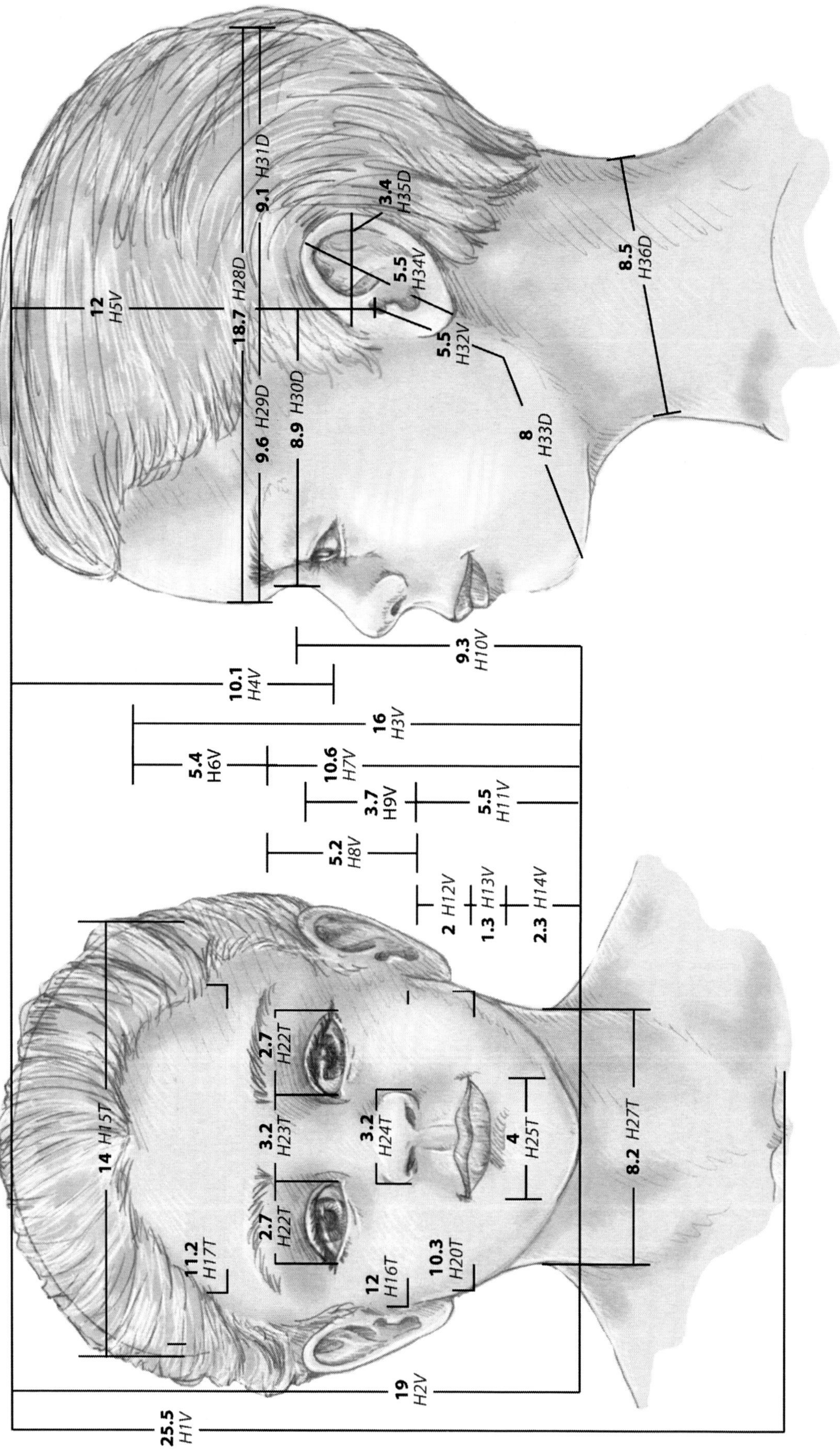

Seven-Year-Old Boy, anterior and lateral figure

all measurements in centimeters

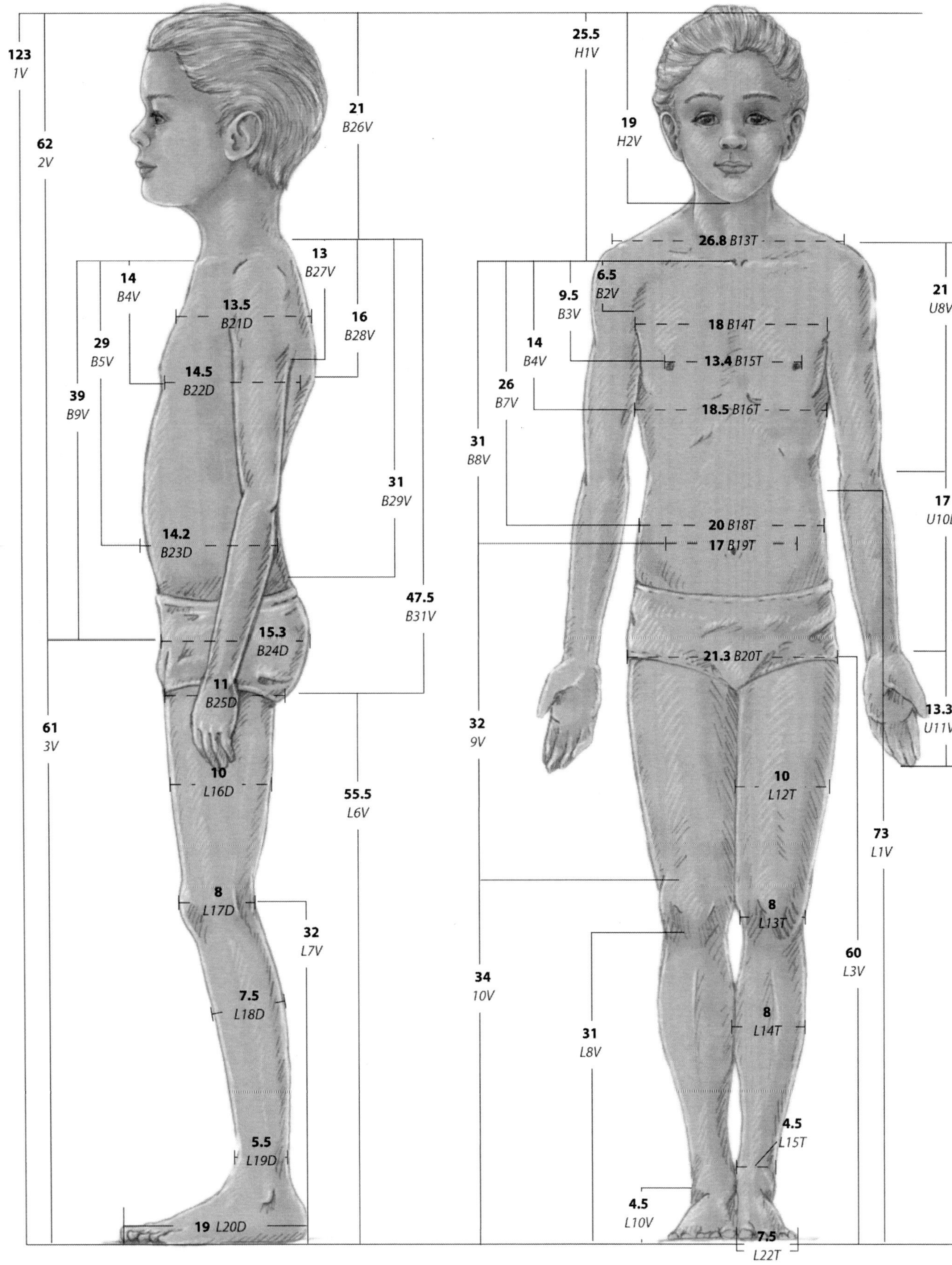

Life size			SEVEN-YEAR-OLD BOY	Fractions of life size				
				3/4	2/3	1/2	1/3	1/4
Inches	cm	Code		cm	cm	cm	cm	cm
			VERTICAL MEASUREMENTS					
48.4	**123.0**	1 V	Stature, Height, Vertex to Sole	92.3	82.0	61.5	41.0	30.8
24.4	**62.0**	2 V	Vertex to Symphysis Pubis	46.5	41.3	31.0	20.7	15.5
24.0	**61.0**	* 3 V	Symphysis Pubis of Pelvis-Sole	45.8	40.7	30.5	20.3	15.3
38.6	**98.0**	4 V	Sternal Notch, Neck Pit, to Sole	73.5	65.3	49.0	32.7	24.5
9.4	**24.0**	5 V	Vertex to Acromion Process	18.0	16.0	12.0	8.0	6.0
39.0	**99.0**	6 V	Acromion Process to Sole	74.3	66.0	49.5	33.0	24.8
26.0	**66.0**	* 7 V	Ant.Sup.Spine of Ilium to Sole	49.5	44.0	33.0	22.0	16.5
11.8	**30.0**	8 V	Acrom. proc. to Ant. Sup. Spine	22.5	20.0	15.0	10.0	7.5
12.6	**32.0**	9 V	Ant.Sup.Spine of Ilium to Knee	24.0	21.3	16.0	10.7	8.0
13.4	**34.0**	10 V	Knee at Patella to Sole	25.5	22.7	17.0	11.3	8.5
27.2	**69.0**	11 V	Umbilicus to Sole	51.8	46.0	34.5	23.0	17.3
21.7	**55.0**	* 12 V	Crotch to Sole	41.3	36.7	27.5	18.3	13.8
36.2	**92.0**	13 V	Kneeling Height	69.0	61.3	46.0	30.7	23.0
26.4	**67.0**	14 V	Sitting Height	50.3	44.7	33.5	22.3	16.8
25.2	**64.0**	15 V	Buttock, Straight Leg to Sole	48.0	42.7	32.0	21.3	16.0
15.4	**39.0**	16 V	Buttock to Flexed Knee	29.3	26.0	19.5	13.0	9.8
15.2	**38.5**	17 V	Flexed Knee to Sole	28.9	25.7	19.3	12.8	9.6
		** These are also listed with Lower Extremity Measurements.*						
			VERTICAL HEAD MEASUREMENTS					
10.0	**25.5**	H1 V	Vertex to Sternal Notch	19.1	17.0	12.8	8.5	6.4
7.5	**19.0**	H2 V	Vertex to Chin, Head Height	14.3	12.7	9.5	6.3	4.8
6.3	**16.0**	H3 V	Hairline to Chin, Face Height	12.0	10.7	8.0	5.3	4.0
4.0	**10.1**	H4 V	Vertex to Lateral Corner of Eye	7.6	6.7	5.1	3.4	2.5
4.7	**12.0**	H5 V	Vertex to Tragion of Ear	9.0	8.0	6.0	4.0	3.0
2.1	**5.6**	H6 V	Hairline to Brow, Glabella	4.1	3.6	2.7	1.8	1.4
4.2	**10.6**	H7 V	Brow, Glabella, to Chin, Menton	8.0	7.1	5.3	3.5	2.7
2.0	**5.2**	H8 V	Brow to Subnasal Point	3.9	3.5	2.6	1.7	1.3
1.5	**3.7**	H9 V	Sellion to Subnasal, Nose Length	2.8	2.5	1.9	1.2	0.9
3.7	**9.3**	H10 V	Sellion, Nose Saddle, to Chin	7.0	6.2	4.7	3.1	2.3
2.2	**5.5**	H11 V	Subnasal Point to Chin	4.1	3.7	2.8	1.8	1.4
0.8	**2.0**	H12 V	Subnasal to Lip Aperature	1.5	1.3	1.0	0.7	0.5
0.5	**1.3**	H13 V	Lip Aperat. to Sublabial Crease	1.0	0.9	0.7	0.4	0.3
0.9	**2.3**	H14 V	Sublabial Crease to Chin	1.7	1.5	1.2	0.8	0.6
			TRANSVERSE HEAD MEASUREMENTS					
5.5	**14.0**	H15 T	Bi-Parietal, Head Breadth	10.5	9.3	7.0	4.7	3.5
4.7	**12.0**	H16 T	Bi-Zygomatic, Cheekbones	9.0	8.0	6.0	4.0	3.0
4.4	**11.2**	H17 T	Bi-Fronto Temporal, Forehead	8.4	7.5	5.6	3.7	2.8

Life size			SEVEN-YEAR-OLD BOY	Fractions of life size				
				3/4	2/3	1/2	1/3	1/4
Inches	cm	Code		cm	cm	cm	cm	cm
			TRANSVERSE HEAD MEASUREMENTS					
3.5	9.0	H18 T	Inter-Brow Ridge	6.8	6.0	4.5	3.0	2.3
4.6	11.7	H19 T	Inter-Tragion of Ears	8.8	7.8	5.9	3.9	2.9
4.1	10.3	H20 T	Inter-Angle of Mandible	7.7	6.9	5.2	3.4	2.6
3.3	8.5	H21 T	Inter-Lateral Corners of Eyes	6.4	5.7	4.3	2.8	2.1
1.1	2.7	H22 T	Width of Eye	2.0	1.8	1.4	0.9	0.7
1.2	3.0	H23 T	Inter-Inner Corner of Eyes	2.3	2.0	1.5	1.0	0.8
1.2	3.0	H24 T	Breadth of Nose	2.3	2.0	1.5	1.0	0.8
1.6	4.0	H25 T	Breadth of Mouth	3.0	2.7	2.0	1.3	1.0
1.2	3.0	H26 T	Breadth of Chin	2.3	2.0	1.5	1.0	0.8
3.2	8.2	H27 T	Breadth of Neck	6.2	5.5	4.1	2.7	2.1
			ANTERIOR-POSTERIOR HEAD MEASUREMENTS					
7.4	18.7	H28 D	Brow to Occiput, Posterior Head	14.0	12.5	9.4	6.2	4.7
3.8	9.6	H29 D	Brow to Tragus of Ear	7.2	6.4	4.8	3.2	2.4
3.5	8.9	H30 D	Sellion, Nose Saddle, to Tragus	6.7	5.9	4.5	3.0	2.2
3.6	9.1	H31 D	Tragus to Occiput, Back of Head	6.8	6.1	4.6	3.0	2.3
2.2	5.5	H32 V	Tragus to Angle Mandible	4.1	3.7	2.8	1.8	1.4
3.1	8.0	H33 D	Angle of Mandible to Chin	6.0	5.3	4.0	2.7	2.0
2.2	5.5	H34 V	Length of Ear, Vertical	4.1	3.7	2.8	1.8	1.4
1.3	3.4	H35 D	Width of Ear, Anterior-Posterior	2.6	2.3	1.7	1.1	0.9
3.3	8.5	H36 D	Anterior to Posterior of Neck	6.4	5.7	4.3	2.8	2.1
			VERTICAL BODY MEASUREMENTS					
2.7	6.8	B1 V	Chin to Sternal Notch, Neck Pit	5.1	4.5	3.4	2.3	1.7
2.6	6.5	B2 V	Sternal Notch to Axilla, Arm Pit	4.9	4.3	3.3	2.2	1.6
3.7	9.5	B3 V	Sternal Notch to Nipples	7.1	6.3	4.8	3.2	2.4
5.5	14.0	B4 V	Sternal Notch to Xyphoid Proc.	10.5	9.3	7.0	4.7	3.5
11.4	29.0	B5 V	Sternal Notch to Umbilicus	21.8	19.3	14.5	9.7	7.3
5.9	15.0	B6 V	XyphoidProcess to Umbilicus	11.3	10.0	7.5	5.0	3.8
10.2	26.0	B7 V	Sternal Notch to Crest Ilium	19.5	17.3	13.0	8.7	6.5
12.2	31.0	B8 V	Sternal Notch to Ant.S.Sp.Ilium	23.3	20.7	15.5	10.3	7.8
15.4	39.0	B9 V	Sternal Notch to Pubis	29.3	26.0	19.5	13.0	9.8
4.3	11.0	B10 V	Umbilicus to Pubis of Pelvis	8.3	7.3	5.5	3.7	2.8
			TRANSVERSE BODY MEASUREMENTS					
11.5	29.3	B11 T	Shoulder Breadth	22.0	19.5	14.7	9.8	7.3
5.2	13.2	B12 T	Clavicle to Acromion Process	9.9	8.8	6.6	4.4	3.3
10.6	26.8	B13 T	Bi-acromial Process Breadth	20.1	17.9	13.4	8.9	6.7
7.1	18.0	B14 T	Inter-Anterior Axillary Fold	13.5	12.0	9.0	6.0	4.5
5.3	13.4	B15 T	Inter-Nipples	10.1	8.9	6.7	4.5	3.4

Life size			SEVEN-YEAR-OLD BOY	Fractions of life size				
				3/4	2/3	1/2	1/3	1/4
Inches	cm	Code		cm	cm	cm	cm	cm
			TRANSVERSE BODY MEASUREMENTS					
7.3	**18.5**	B16 T	Breadth of Thorax	13.9	12.3	9.3	6.2	4.6
7.1	**18.0**	B17 T	Inter-Costal. Rib Margin Breadth	13.5	12.0	9.0	6.0	4.5
7.9	**20.0**	B18 T	Inter-Crest of Ilium Breadth	15.0	13.3	10.0	6.7	5.0
6.7	**17.0**	B19 T	Inter-Anterior Superior Spine	12.8	11.3	8.5	5.7	4.3
8.4	**21.3**	B20 T	Inter-Greater Trochanter Femur	16.0	14.2	10.7	7.1	5.3
			ANTERIOR-POSTERIOR BODY MEASUREMENTS					
5.3	**13.5**	B21 D	Manubrium to Dorsal Spine	10.1	9.0	6.8	4.5	3.4
5.7	**14.5**	B22 D	Xyphoid Process to Dorsum	10.9	9.7	7.3	4.8	3.6
5.6	**14.2**	B23 D	Umbilicus to Dorsum	10.7	9.5	7.1	4.7	3.6
6.0	**15.3**	B24 D	Pubis of Pelvis to Buttock	11.5	10.2	7.7	5.1	3.8
4.3	**11.0**	B25 D	Thigh to Gluteal Furrow	8.3	7.3	5.5	3.7	2.8
			VERTICAL BACK MEASUREMENTS					
8.3	**21.0**	B26 V	Vertex to 7th Cervical Spine	15.8	14.0	10.5	7.0	5.3
			7th Cervical Vertebra Spine to:					
5.1	**13.0**	B27 V	To Posterior Axillary Fold, Armpit	9.8	8.7	6.5	4.3	3.3
6.3	**16.0**	B28 V	To Inferior Tip of Scapula	12.0	10.7	8.0	5.3	4.0
12.2	**31.0**	B29 V	To Posterior Superior Spine Ilium	23.3	20.7	15.5	10.3	7.8
16.1	**41.0**	B30 V	To Midline Upper Buttock Crease	30.8	27.3	20.5	13.7	10.3
18.7	**47.5**	B31 V	To Gluteal Furrow	35.6	31.7	23.8	15.8	11.9
27.6	**70.0**	B32 V	To Popliteal Crease Behind Knee	52.5	46.7	35.0	23.3	17.5
			TRANSVERSE BACK MEASUREMENTS					
9.6	**24.5**	B33 T	Inter-Posterior Axillary Folds	18.4	16.3	12.3	8.2	6.1
4.7	**12.0**	B34 T	Inter-Scapulae Medial Borders	9.0	8.0	6.0	4.0	3.0
2.4	**6.0**	B35 T	Inter-Post. Sup. Spines Sacrum	4.5	4.0	3.0	2.0	1.5
			UPPER EXTREMITY MEASUREMENTS					
47.6	**121.0**	U1 V	Span of Arms	90.8	80.7	60.5	40.3	30.3
5.1	**13.0**	U2 V	Sternum to Acromion Process	9.8	8.7	6.5	4.3	3.3
20.5	**52.0**	U3 V	Acromion Process to Fingertip	39.0	34.7	26.0	17.3	13.0
60.6	**154.0**	U4 V	Vertical Reach to Fingertip	115.5	102.7	77.0	51.3	38.5
29.5	**75.0**	U5 T	Lateral Reach to Fingertip	56.3	50.0	37.5	25.0	18.8
23.0	**58.5**	U6 D	Forward Reach to Fingertip	43.9	39.0	29.3	19.5	14.6
9.8	**25.0**	U7 V	Shoulder to Flexed Elbow	18.8	16.7	12.5	8.3	6.3
8.3	**21.0**	U8 V	Acromion to Lat. Condyle Elbow	15.8	14.0	10.5	7.0	5.3
12.6	**32.0**	U9 D	Flexed Elbow to Fingertip	24.0	21.3	16.0	10.7	8.0
6.7	**17.0**	U10 D	Flexed Elbow Pivot to Wrist Pivot	12.8	11.3	8.5	5.7	4.3

Life size			SEVEN-YEAR-OLD BOY	Fractions of life size				
				3/4	2/3	1/2	1/3	1/4
Inches	cm	Code		cm	cm	cm	cm	cm
			UPPER EXTREMITY MEASUREMENTS					
5.2	13.3	U11 V	Wrist to Fingertip	10.0	8.9	6.7	4.4	3.3
3.3	8.5	U12 V	Wrist to Thumb Tip	6.4	5.7	4.3	2.8	2.1
3.0	7.5	U13 V	Length of Palm	5.6	5.0	3.8	2.5	1.9
2.3	5.8	U14 V	Length 3rd Finger	4.4	3.9	2.9	1.9	1.5
3.7	9.4	U15 V	Thumb Crotch to 3rd Finger	7.1	6.3	4.7	3.1	2.4
1.8	4.6	U16 V	Thumb Crotch to Thumb Tip	3.5	3.1	2.3	1.5	1.2
2.0	5.2	U17 D	Anterior-Post. Arm at Biceps	3.9	3.5	2.6	1.7	1.3
2.1	5.4	U18 D	Anterior-Posterior at Forearm	4.1	3.6	2.7	1.8	1.4
1.3	3.2	U19 D	Depth of Wrist	2.4	2.1	1.6	1.1	0.8
1.8	4.5	U20 T	Breadth of Wrist	3.4	3.0	2.3	1.5	1.1
3.0	7.7	U21 T	Breadth of Hand	5.8	5.1	3.9	2.6	1.9
2.4	6.2	U22 T	Breadth of Palm	4.7	4.1	3.1	2.1	1.6
			LOWER EXTREMITY MEASUREMENTS					
28.7	73.0	L1 V	Crest of Ilium of Pelvis to Sole	54.8	48.7	36.5	24.3	18.3
26.4	67.0	* L2 V	Ant.Sup.Spine of Ilium to Solc	50.3	44.7	33.5	22.3	16.8
23.6	60.0	L3 V	Greater Trochanter Femur-Sole	45.0	40.0	30.0	20.0	15.0
24.0	61.0	* L4 V	Symphysis Pubis of Pelvis-Sole	45.8	40.7	30.5	20.3	15.3
21.7	55.0	* L5 V	Crotch to Sole	41.3	36.7	27.5	18.3	13.8
21.9	55.5	L6 V	Gluteal Furrow to Sole	41.6	37.0	27.8	18.5	13.9
12.6	32.0	L7 V	Posterior Knee Crease to Sole	24.0	21.3	16.0	10.7	8.0
12.2	31.0	L8 V	Tibia at Knee to Sole	23.3	20.7	15.5	10.3	7.8
2.0	5.0	L9 V	Med. Malleolus of Ankle to Sole	3.8	3.3	2.5	1.7	1.3
1.8	4.5	L10 V	Lat. Malleolus of Ankle to Sole	3.4	3.0	2.3	1.5	1.1
8.7	22.0	L11 T	Breadth of Hips, Sitting	16.5	14.7	11.0	7.3	5.5
3.9	10.0	L12 T	Breadth Mid Thigh	7.5	6.7	5.0	3.3	2.5
3.1	8.0	L13 T	Breadth of Knee	6.0	5.3	4.0	2.7	2.0
3.1	8.0	L14 T	Breadth of Calf	6.0	5.3	4.0	2.7	2.0
1.8	4.5	L15 T	Breadth Ankle	3.4	3.0	2.3	1.5	1.1
3.9	10.0	L16 D	Mid Thigh, Anterior-Posterior	7.5	6.7	5.0	3.3	2.5
3.1	8.0	L17 D	Knee, Anterior-Posterior	6.0	5.3	4.0	2.7	2.0
3.0	7.5	L18 D	Mid Calf, Anterior-Posterior	5.6	5.0	3.8	2.5	1.9
2.2	5.5	L19 D	Ankle, Anterior-Posterior	4.1	3.7	2.8	1.8	1.4
7.5	19.0	L20 D	Length of Foot	14.3	12.7	9.5	6.3	4.8
5.5	14.0	L21 D	Heel to 5th Toe	10.5	9.3	7.0	4.7	3.5
3.0	7.5	L22 T	Breadth Foot	5.6	5.0	3.8	2.5	1.9
1.8	4.5	L23 T	Width of Heel	3.4	3.0	2.3	1.5	1.1
		These are also listed with Vertical Measurements.						

EIGHT-YEAR-OLD GIRLS

The lists of body proportions are shown separately in the tables of measurements. The range of twenty-fifth to fiftieth and to seventy-fifth percentile is listed here for comparison:

Percentile	Twenty-fifth		Fiftieth		Seventy-fifth	
Girls:	Short		Medium		Tall	
Stature	122 cm,	48 in.	126.5 cm,	49.3 in.	131 cm,	51.5 in.
Weight	22 kg,	49 lb.	25 kg,	54.8 lb.	28 kg,	61.5 lb.[1]

The most noticeable difference between boys and girls is the hair styling and dress. However, on closer inspection, the girls may have more rounded contours in face shoulders and hips. Their interests are in doll houses, dresses, music, and costume jewelry. They may wear clothing like boys in school and play, especially if there is an opportunity to be a tom-boy. For special occasions they dress like little ladies. They begin to choose playmates and groups with their own sex. School level is third grade.

BODY MEASUREMENTS COMPARED TO HEIGHT

	Feature Measure in cm	Feature Compared to Stature	Approx. Ratio of Feature	Compared to Median Adult
Stature: Crown to sole	**126.4**			**75%**
Head Height: Crown to chin	**18.8**	**15%**	**1/6–**	**92%**
Sitting Height:	**68**	**54%**	**1/2+**	**77%**
Trunk Measure: Sternal notch to symphysis pubis	**38**	**30%**	**1/3**	**78%**
Leg Length: Greater trochanter of the femur to sole	**63**	**50%**	**1/2**	**73%**
Thigh Length: Greater trochanter to the tibia at knee	**31.5**	**25%**	**1/4**	**73%**
Tibia at Knee to Sole:	**31.5**	**25%**	**1/4**	**73%**
Arm Length: Acromion process at shoulder to dependent fingertip	**53**	**42%**	**2/5+**	**71%**
Hand Length:	**13.7**	**11%**	**1/9**	**76%**
Foot Length:	**19.5**	**15%**	**1/6–**	**81%**

The permanent upper and lower lateral incisors erupt at seven to eight years of age.[2]

1. Nelson, W.E., p. 38.
2. Lockhart, R.D., p. 57.

EIGHT-YEAR-OLD BOYS

The list of average body proportions is shown separately in the tables of measurements. The range of twenty-fifth, to fiftieth, and to seventy-fifth percentile is listed here for comparison:

Percentile	Twenty-fifth		Fiftieth		Seventy-fifth	
Boys:	Short		Medium		Tall	
Stature	123 cm,	48.5 in.	127 cm,	50 in.	130.5 cm,	51.5 in.
Weight	23 kg,	51 lb.	25.3 kg,	55.3 lb.	28 kg,	61.5 lb.[1]

Boys are, on average, are slightly taller and slightly heavier than are girls. Hair styling is different, and boys are not as attentive to hair care as girls. Athletics, ball games, outdoor activities, mechanical toys, segmental constructive toys, dinosaurs, and monsters are their greater interests. They try to emulate adult male images, particularly athletic stars. The school level is third grade.

BODY MEASUREMENTS COMPARED TO HEIGHT

	Feature Measure in cm	Feature Compared to Stature	Approx. Ratio of Feature	Compared to Median Adult
Stature: Crown to sole	127			72%
Head Height: Crown to chin	19.2	15%	1/6–	81%
Sitting Height:	69	54%	1/2+	71%
Trunk Measure: Sternal notch to symphysis pubis	40.5	32%	1/3	72%
Leg Length: Greater trochanter of the femur to sole	64	50%	1/2	64%
Thigh Length: Greater trochanter of the femur to tibia at knee	30	24%	1/4	67%
Tibia at Knee to Sole:	34	27%	1/4+	62%
Arm Length: Acromion process at shoulder to dependent fingertip	55	43%	2/5+	65%
Hand Length:	14	11%	1/9	68%
Foot Length:	20	16%	1/6	70%

The permanent upper and lower lateral incisors erupt at seven to eight years of age.[2]

1. Nelson, W.E., p. 39.
2. Lockhart, R.D., p. 57.

Eight-Year-Old Girl, anterior and lateral head

all measurements in centimeters

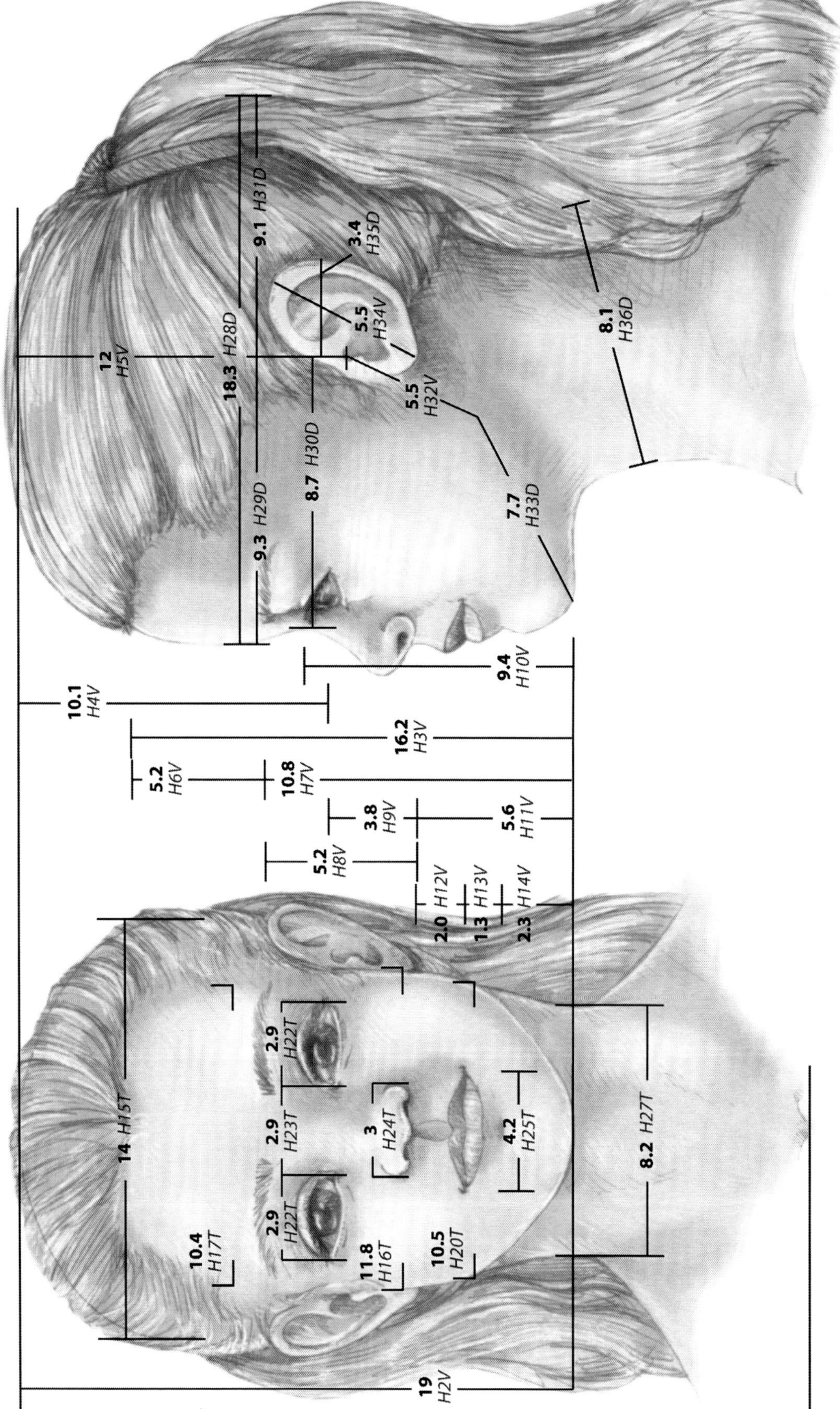

Eight-Year-Old Girl, anterior and lateral figure

all measurements in centimeters

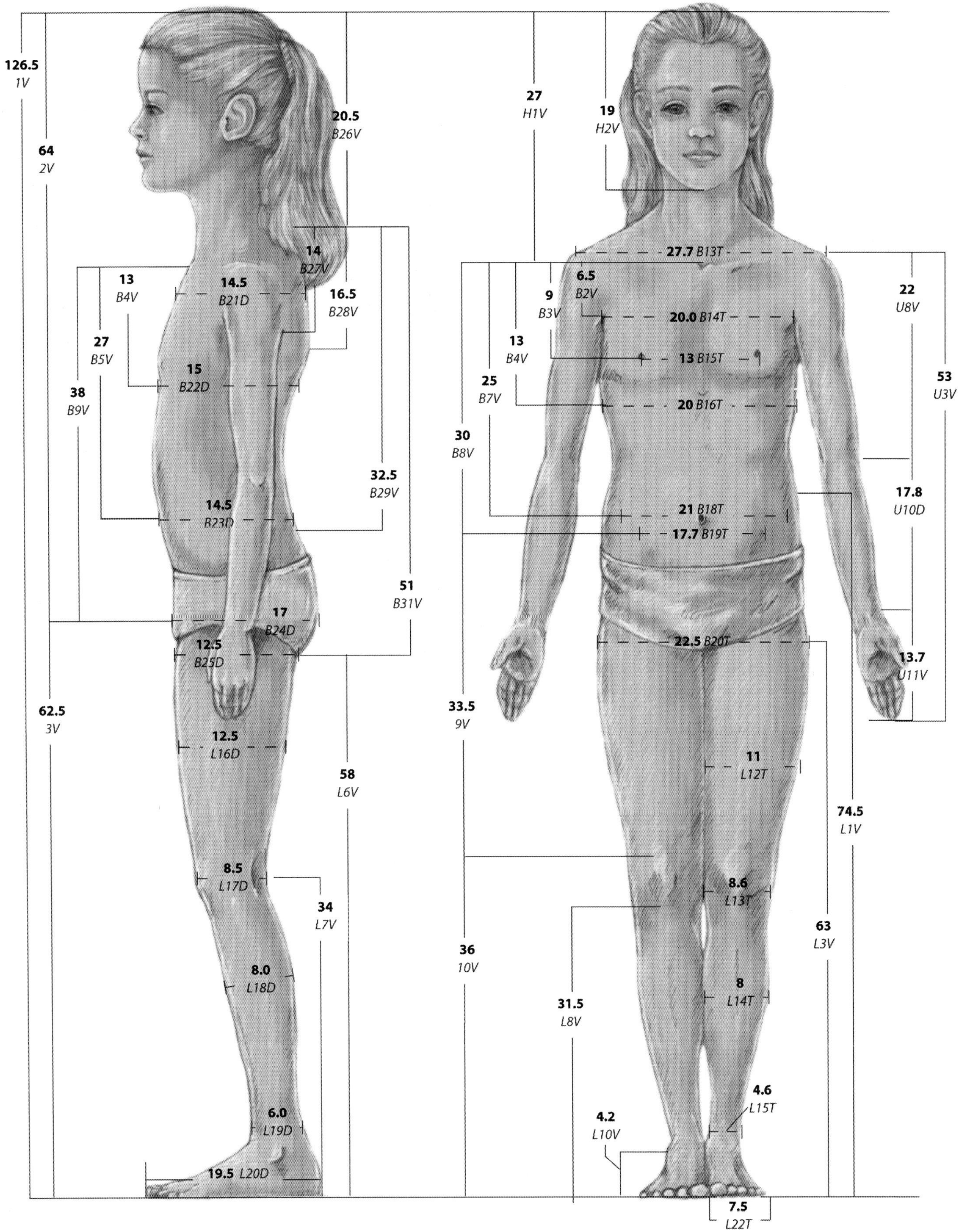

Life size			EIGHT-YEAR-OLD GIRL	Fractions of life size				
				3/4	2/3	1/2	1/3	1/4
Inches	cm	Code		cm	cm	cm	cm	cm
			VERTICAL MEASUREMENTS					
49.8	**126.5**	1 V	Stature, Height, Vertex to Sole	94.9	84.3	63.3	42.2	31.6
25.2	**64.0**	2 V	Vertex to Symphysis Pubis	48.0	42.7	32.0	21.3	16.0
24.6	**62.5**	* 3 V	Symphysis Pubis of Pelvis-Sole	46.9	41.7	31.3	20.8	15.6
39.0	**99.0**	4 V	Sternal Notch, Neck Pit, to Sole	74.3	66.0	49.5	33.0	24.8
9.4	**24.0**	5 V	Vertex to Acromion Process	18.0	16.0	12.0	8.0	6.0
39.8	**101.0**	6 V	Acromion Process to Sole	75.8	67.3	50.5	33.7	25.3
27.2	**69.0**	* 7 V	Ant.Sup.Spine of Ilium to Sole	51.8	46.0	34.5	23.0	17.3
12.6	**32.0**	8 V	Acrom. proc. to Ant. Sup. Spine	24.0	21.3	16.0	10.7	8.0
13.0	**33.0**	9 V	Ant.Sup.Spine of Ilium to Knee	24.8	22.0	16.5	11.0	8.3
14.2	**36.0**	10 V	Knee at Patella to Sole	27.0	24.0	18.0	12.0	9.0
28.7	**73.0**	11 V	Umbilicus to Sole	54.8	48.7	36.5	24.3	18.3
23.2	**59.0**	* 12 V	Crotch to Sole	44.3	39.3	29.5	19.7	14.8
37.8	**96.0**	13 V	Kneeling Height	72.0	64.0	48.0	32.0	24.0
26.8	**68.0**	14 V	Sitting Height	51.0	45.3	34.0	22.7	17.0
27.6	**70.0**	15 V	Buttock, Straight Leg to Sole	52.5	46.7	35.0	23.3	17.5
16.3	**41.5**	16 V	Buttock to Flexed Knee	31.1	27.7	20.8	13.8	10.4
15.4	**39.0**	17 V	Flexed Knee to Sole	29.3	26.0	19.5	13.0	9.8
		** These are also listed with Lower Extremity Measurements.*						
			VERTICAL HEAD MEASUREMENTS					
10.6	**27.0**	H1 V	Vertex to Sternal Notch	20.3	18.0	13.5	9.0	6.8
7.5	**19.0**	H2 V	Vertex to Chin, Head Height	14.3	12.7	9.5	6.3	4.8
6.4	**16.2**	H3 V	Hairline to Chin, Face Height	12.2	10.8	8.1	5.4	4.1
4.0	**10.1**	H4 V	Vertex to Lateral Corner of Eye	7.6	6.7	5.1	3.4	2.5
4.7	**12.0**	H5 V	Vertex to Tragion of Ear	9.0	8.0	6.0	4.0	3.0
2.0	**5.3**	H6 V	Hairline to Brow, Glabella	3.9	3.5	2.6	1.7	1.3
4.3	**10.8**	H7 V	Brow, Glabella, to Chin, Menton	8.1	7.2	5.4	3.6	2.7
2.0	**5.2**	H8 V	Brow to Subnasal Point	3.9	3.5	2.6	1.7	1.3
1.5	**3.8**	H9 V	Sellion to Subnasal, Nose Length	2.9	2.5	1.9	1.3	1.0
3.7	**9.4**	H10 V	Sellion, Nose Saddle, D87 to Chin	7.1	6.3	4.7	3.1	2.4
2.2	**5.6**	H11 V	Subnasal Point to Chin	4.2	3.7	2.8	1.9	1.4
0.8	**2.0**	H12 V	Subnasal to Lip Aperature	1.5	1.3	1.0	0.7	0.5
0.5	**1.3**	H13 V	Lip Aperat. to Sublabial Crease	1.0	0.9	0.7	0.4	0.3
0.9	**2.3**	H14 V	Sublabial Crease to Chin	1.7	1.5	1.2	0.8	0.6
			TRANSVERSE HEAD MEASUREMENTS					
5.5	**14.0**	H15 T	Bi-Parietal, Head Breadth	10.5	9.3	7.0	4.7	3.5
4.6	**11.8**	H16 T	Bi-Zygomatic, Cheekbones	8.9	7.9	5.9	3.9	3.0
4.1	**10.4**	H17 T	Bi-Fronto Temporal, Forehead	7.8	6.9	5.2	3.5	2.6

Life size			EIGHT-YEAR-OLD GIRL	Fractions of life size				
				3/4	2/3	1/2	1/3	1/4
Inches	cm	Code		cm	cm	cm	cm	cm
			TRANSVERSE HEAD MEASUREMENTS					
3.7	**9.3**	H18 T	Inter-Brow Ridge	7.0	6.2	4.7	3.1	2.3
4.5	**11.5**	H19 T	Inter-Tragion of Ears	8.6	7.7	5.8	3.8	2.9
4.1	**10.5**	H20 T	Inter-Angle of Mandible	7.9	7.0	5.3	3.5	2.6
3.4	**8.6**	H21 T	Inter-Lateral Corners of Eyes	6.5	5.7	4.3	2.9	2.2
1.1	**2.9**	H22 T	Width of Eye	2.2	1.9	1.5	1.0	0.7
1.2	**3.0**	H23 T	Inter-Inner Corner of Eyes	2.3	2	1.5	1	0.8
1.2	**3.0**	H24 T	Breadth of Nose	2.3	2.0	1.5	1.0	0.8
1.7	**4.2**	H25 T	Breadth of Mouth	3.2	2.8	2.1	1.4	1.1
1.2	**3.0**	H26 T	Breadth of Chin	2.3	2.0	1.5	1.0	0.8
3.2	**8.2**	H27 T	Breadth of Neck	6.2	5.5	4.1	2.7	2.1
			ANTERIOR-POSTERIOR HEAD MEASUREMENTS					
7.2	**18.3**	H28 D	Brow to Occiput, Poserior Head	13.7	12.2	9.2	6.1	4.6
3.7	**9.3**	H29 D	Brow to Tragus of Ear	7.0	6.2	4.7	3.1	2.3
3.4	**8.7**	H30 D	Sellion, Nose Saddle, to Tragus	6.5	5.8	4.4	2.9	2.2
3.6	**9.1**	H31 D	Tragus to Occiput, Back of Head	6.8	6.1	4.6	3.0	2.3
2.2	**5.5**	H32 V	Tragus to Angle Mandible	4.1	3.7	2.8	1.8	1.4
3.0	**7.7**	H33 D	Angle of Mandible to Chin	5.8	5.1	3.9	2.6	1.9
2.2	**5.5**	H34 V	Length of Ear, Vertical	4.1	3.7	2.8	1.8	1.4
1.3	**3.4**	H35 D	Width of Ear, Anterior-Posterior	2.6	2.3	1.7	1.1	0.9
3.2	**8.1**	H36 D	Anterior to Posterior of Neck	6.1	5.4	4.1	2.7	2.0
			VERTICAL BODY MEASUREMENTS					
2.8	**7.0**	B1 V	Chin to Sternal Notch, Neck Pit	5.3	4.7	3.5	2.3	1.8
2.6	**6.5**	B2 V	Sternal Notch to Axilla, Arm Pit	4.9	4.3	3.3	2.2	1.6
3.5	**9.0**	B3 V	Sternal Notch to Nipples	6.8	6.0	4.5	3.0	2.3
5.1	**13.0**	B4 V	Sternal Notch to Xyphoid Proc.	9.8	8.7	6.5	4.3	3.3
10.6	**27.0**	B5 V	Sternal Notch to Umbilicus	20.3	18.0	13.5	9.0	6.8
5.5	**14.0**	B6 V	Xyphoid Process to Umbilicus	10.5	9.3	7.0	4.7	3.5
9.8	**25.0**	B7 V	Sternal Notch to Crest Ilium	18.8	16.7	12.5	8.3	6.3
11.8	**30.0**	B8 V	Sternal Notch to Ant.S.Sp.Ilium	22.5	20.0	15.0	10.0	7.5
15.0	**38.0**	B9 V	Sternal Notch to Pubis	28.5	25.3	19.0	12.7	9.5
4.7	**12.0**	B10 V	Umbilicus to Pubis of Pelvis	9.0	8.0	6.0	4.0	3.0
			TRANSVERSE BODY MEASUREMENTS					
11.8	**30.0**	B11 T	Shoulder Breadth	22.5	20.0	15.0	10.0	7.5
5.2	**13.2**	B12 T	Clavicle to Acromion Process	9.9	8.8	6.6	4.4	3.3
10.9	**27.7**	B13 T	Bi-acromial Process Breadth	20.8	18.5	13.9	9.2	6.9
7.9	**20.0**	B14 T	Inter-Anterior Axillary Fold	15.0	13.3	10.0	6.7	5.0
5.1	**13.0**	B15 T	Inter-Nipples	9.8	8.7	6.5	4.3	3.3

Life size			EIGHT-YEAR-OLD GIRL	Fractions of life size				
				3/4	2/3	1/2	1/3	1/4
Inches	cm	Code		cm	cm	cm	cm	cm
			TRANSVERSE BODY MEASUREMENTS					
7.9	**20.0**	B16 T	Breadth of Thorax	15.0	13.3	10.0	6.7	5.0
7.6	**19.3**	B17 T	Inter-Costal. Rib Margin Breadth	14.5	12.9	9.7	6.4	4.8
8.3	**21.0**	B18 T	Inter-Crest of Ilium Breadth	15.8	14.0	10.5	7.0	5.3
7.0	**17.7**	B19 T	Inter-Anterior Superior Spine	13.3	11.8	8.9	5.9	4.4
8.8	**22.4**	B20 T	Inter-Greater Trochanter Femur	16.8	14.9	11.2	7.5	5.6
			ANTERIOR-POSTERIOR BODY MEASUREMENTS					
5.7	**14.5**	B21 D	Manubrium to Dorsal Spine	10.9	9.7	7.3	4.8	3.6
5.9	**15.0**	B22 D	Xyphoid Process to Dorsum	11.3	10.0	7.5	5.0	3.8
5.7	**14.5**	B23 D	Umbilicus to Dorsum	10.9	9.7	7.3	4.8	3.6
6.7	**17.0**	B24 D	Pubis of Pelvis to Buttock	12.8	11.3	8.5	5.7	4.3
4.9	**12.5**	B25 D	Thigh to Gluteal Furrow	9.4	8.3	6.3	4.2	3.1
			VERTICAL BACK MEASUREMENTS					
8.1	**20.5**	B26 V	Vertex to 7th Cervical Spine	15.4	13.7	10.3	6.8	5.1
			7th Cervical Vertebra Spine to:					
5.3	**13.5**	B27 V	To Posterior Axillary Fold, Armpit	10.1	9	6.8	4.5	3.4
6.5	**16.5**	B28 V	To Inferior Tip of Scapula	12.4	11.0	8.3	5.5	4.1
12.8	**32.5**	B29 V	To Posterior Superior Spine Ilium	24.4	21.7	16.3	10.8	8.1
15.0	**38.0**	B30 V	To Midline Upper Buttock Crease	28.5	25.3	19.0	12.7	9.5
20.1	**51.0**	B31 V	To Gluteal Furrow	38.3	34.0	25.5	17.0	12.8
28.3	**72.0**	B32 V	To Popliteal Crease Behind Knee	54.0	48.0	36.0	24.0	18.0
			TRANSVERSE BACK MEASUREMENTS					
9.8	**25.0**	B33 T	Inter-Posterior Axillary Folds	18.8	16.7	12.5	8.3	6.3
4.7	**12.0**	B34 T	Inter-Scapulae Medial Borders	9.0	8.0	6.0	4.0	3.0
2.6	**6.6**	B35 T	Inter-Post. Sup. Spines Sacrum	5.0	4.4	3.3	2.2	1.7
			UPPER EXTREMITY MEASUREMENTS					
48.4	**123.0**	U1 V	Span of Arms	92.3	82.0	61.5	41.0	30.8
5.2	**13.2**	U2 V	Sternum to Acromion Process	9.9	8.8	6.6	4.4	3.3
20.9	**53.0**	U3 V	Acromion Process to Fingertip	39.8	35.3	26.5	17.7	13.3
62.6	**159.0**	U4 V	Vertical Reach to Fingertip	119.3	106.0	79.5	53.0	39.8
30.7	**78.0**	U5 T	Lateral Reach to Fingertip	58.5	52.0	39.0	26.0	19.5
24.0	**61.0**	U6 D	Forward Reach to Fingertip	45.8	40.7	30.5	20.3	15.3
10.1	**25.7**	U7 V	Shoulder to Flexed Elbow	19.3	17.1	12.9	8.6	6.4
8.7	**22.0**	U8 V	Acromion to Lat. Condyle Elbow	16.5	14.7	11.0	7.3	5.5
13.0	**33.0**	U9 D	Flexed Elbow to Fingertip	24.8	22.0	16.5	11.0	8.3
7.0	**17.8**	U10 D	Flexed Elbow Pivot to Wrist Pivot	13.4	11.9	8.9	5.9	4.5

Life size			EIGHT-YEAR-OLD GIRL	Fractions of life size				
				3/4	2/3	1/2	1/3	1/4
Inches	cm	Code		cm	cm	cm	cm	cm
			UPPER EXTREMITY MEASUREMENTS					
5.4	**13.7**	U11 V	Wrist to Fingertip	10.3	9.1	6.9	4.6	3.4
3.5	**9.0**	U12 V	Wrist to Thumb Tip	6.8	6.0	4.5	3.0	2.3
3.1	**8.0**	U13 V	Length of Palm	6.0	5.3	4.0	2.7	2.0
2.2	**5.7**	U14 V	Length 3rd Finger	4.3	3.8	2.9	1.9	1.4
3.6	**9.2**	U15 V	Thumb Crotch to 3rd Finger	6.9	6.1	4.6	3.1	2.3
1.9	**4.8**	U16 V	Thumb Crotch to Thumb Tip	3.6	3.2	2.4	1.6	1.2
2.4	**6.2**	U17 D	Anterior-Post. Arm at Biceps	4.7	4.1	3.1	2.1	1.6
2.4	**6.0**	U18 D	Anterior-Posterior at Forearm	4.5	4.0	3.0	2.0	1.5
1.3	**3.2**	U19 D	Depth of Wrist	2.4	2.1	1.6	1.1	0.8
1.8	**4.6**	U20 T	Breadth of Wrist	3.5	3.1	2.3	1.5	1.2
3.0	**7.7**	U21 T	Breadth of Hand	5.8	5.1	3.9	2.6	1.9
2.4	**6.2**	U22 T	Breadth of Palm	4.7	4.1	3.1	2.1	1.6
			LOWER EXTREMITY MEASUREMENTS					
29.3	**74.5**	L1 V	Crest of Ilium of Pelvis to Sole	55.9	49.7	37.3	24.8	18.6
27.2	**69.0**	* L2 V	Ant.Sup.Spine of Ilium to Sole	51.8	46.0	34.5	23.0	17.3
24.8	**63.0**	L3 V	Greater Trochanter Femur-Sole	47.3	42.0	31.5	21.0	15.8
24.6	**62.5**	* L4 V	Symphysis Pubis of Pelvis-Sole	46.9	41.7	31.3	20.8	15.6
23.2	**59.0**	* L5 V	Crotch to Sole	44.3	39.3	29.5	19.7	14.8
22.8	**58.0**	L6 V	Gluteal Furrow to Sole	43.5	38.7	29.0	19.3	14.5
13.4	**34.0**	L7 V	Posterior Knee Crease to Sole	25.5	22.7	17.0	11.3	8.5
12.4	**31.5**	L8 V	Tibia at Knee to Sole	23.6	21.0	15.8	10.5	7.9
2.0	**5.0**	L9 V	Med. Malleolus of Ankle to Sole	3.8	3.3	2.5	1.7	1.3
1.7	**4.2**	L10 V	Lat. Malleolus of Ankle to Sole	3.2	2.8	2.1	1.4	1.1
9.4	**24.0**	L11 T	Breadth of Hips, Sitting	18.0	16.0	12.0	8.0	6.0
4.3	**11.0**	L12 T	Breadth Mid Thigh	8.3	7.3	5.5	3.7	2.8
3.4	**8.6**	L13 T	Breadth of Knee	6.5	5.7	4.3	2.9	2.2
3.1	**8.0**	L14 T	Breadth of Calf	6.0	5.3	4.0	2.7	2.0
1.8	**4.6**	L15 T	Breadth Ankle	3.5	3.1	2.3	1.5	1.2
4.9	**12.5**	L16 D	Mid Thigh, Anterior-Posterior	9.4	8.3	6.3	4.2	3.1
3.3	**8.5**	L17 D	Knee, Anterior-Posterior	6.4	5.7	4.3	2.8	2.1
3.1	**8.0**	L18 D	Mid Calf, Anterior-Posterior	6.0	5.3	4.0	2.7	2.0
2.4	**6.0**	L19 D	Ankle, Anterior-Posterior	4.5	4.0	3.0	2.0	1.5
7.7	**19.5**	L20 D	Length of Foot	14.6	13.0	9.8	6.5	4.9
6.5	**16.5**	L21 D	Heel to 5th Toe	12.4	11.0	8.3	5.5	4.1
3.0	**7.5**	L22 T	Breadth Foot	5.6	5.0	3.8	2.5	1.9
1.8	**4.5**	L23 T	Width of Heel	3.4	3.0	2.3	1.5	1.1
		** These are also listed with Vertical Measurements.*						

Eight-Year-Old Boy, anterior and lateral head

all measurements in centimeters

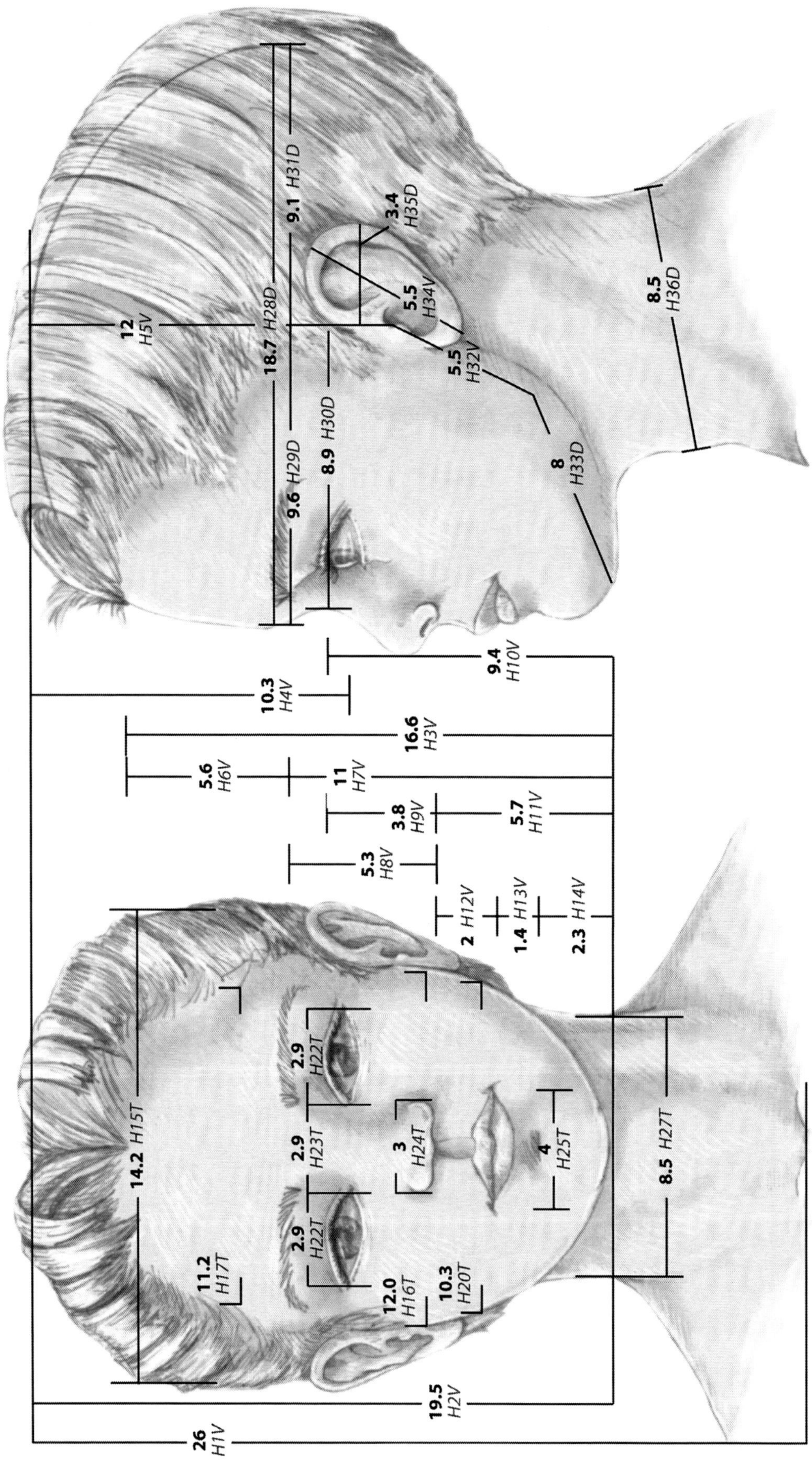

Eight-Year-Old Boy, anterior and lateral figure

all measurements in centimeters

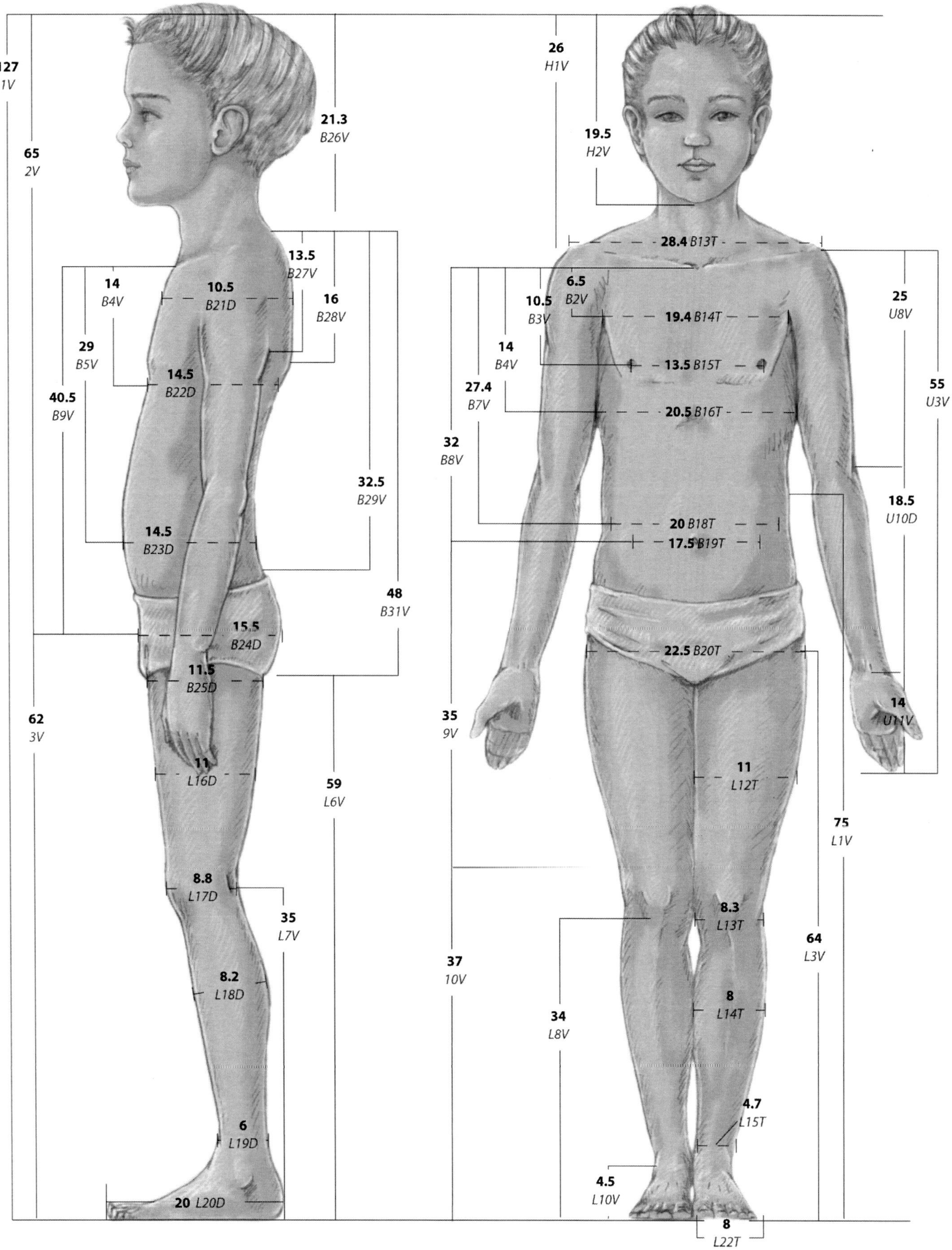

Life size			EIGHT-YEAR-OLD BOY	Fractions of life size				
				3/4	2/3	1/2	1/3	1/4
Inches	cm	Code		cm	cm	cm	cm	cm
			VERTICAL MEASUREMENTS					
50.0	**127.0**	1 V	Stature, Height, Vertex to Sole	95.3	84.7	63.5	42.3	31.8
25.6	**65.0**	2 V	Vertex to Symphysis Pubis	48.8	43.3	32.5	21.7	16.3
24.4	**62.0**	* 3 V	Symphysis Pubis of Pelvis-Sole	46.5	41.3	31.0	20.7	15.5
40.2	**102.0**	4 V	Sternal Notch, Neck Pit, to Sole	76.5	68.0	51.0	34.0	25.5
9.4	**24.0**	5 V	Vertex to Acromion Process	18.0	16.0	12.0	8.0	6.0
40.6	**103.0**	6 V	Acromion Process to Sole	77.3	68.7	51.5	34.3	25.8
28.3	**72.0**	* 7 V	Ant.Sup.Spine of Ilium to Sole	54.0	48.0	36.0	24.0	18.0
12.6	**32.0**	8 V	Acrom. proc. to Ant. Sup. Spine	24.0	21.3	16.0	10.7	8.0
13.8	**35.0**	9 V	Ant.Sup.Spine of Ilium to Knee	26.3	23.3	17.5	11.7	8.8
14.2	**36.0**	10 V	Knee at Patella to Sole	27.0	24.0	18.0	12.0	9.0
28.7	**73.0**	11 V	Umbilicus to Sole	54.8	48.7	36.5	24.3	18.3
22.2	**56.5**	* 12 V	Crotch to Sole	42.4	37.7	28.3	18.8	14.1
38.2	**97.0**	13 V	Kneeling Height	72.8	64.7	48.5	32.3	24.3
27.2	**69.0**	14 V	Sitting Height	51.8	46.0	34.5	23.0	17.3
27.8	**70.5**	15 V	Buttock, Straight Leg to Sole	52.9	47.0	35.3	23.5	17.6
16.3	**41.5**	16 V	Buttock to Flexed Knee	31.1	27.7	20.8	13.8	10.4
15.9	**40.5**	17 V	Flexed Knee to Sole	30.4	27.0	20.3	13.5	10.1
		** These are also listed with Lower Extremity Measurements.*						
			VERTICAL HEAD MEASUREMENTS					
10.2	**26.0**	H1 V	Vertex to Sternal Notch	19.5	17.3	13.0	8.7	6.5
7.7	**19.5**	H2 V	Vertex to Chin, Head Height	14.6	13.0	9.8	6.5	4.9
6.5	**16.6**	H3 V	Hairline to Chin, Face Height	12.5	11.1	8.3	5.5	4.2
4.1	**10.3**	H4 V	Vertex to Lateral Corner of Eye	7.7	6.9	5.2	3.4	2.6
4.7	**12.0**	H5 V	Vertex to Tragion of Ear	9.0	8.0	6.0	4.0	3.0
2.2	**5.6**	H6 V	Hairline to Brow, Glabella	4.2	3.7	2.8	1.9	1.4
4.3	**11.0**	H7 V	Brow, Glabella, to Chin, Menton	8.3	7.3	5.5	3.7	2.8
2.1	**5.3**	H8 V	Brow to Subnasal Point	4.0	3.5	2.7	1.8	1.3
1.5	**3.8**	H9 V	Sellion to Subnasal, Nose Length	2.9	2.5	1.9	1.3	1.0
3.7	**9.4**	H10 V	Sellion, Nose Saddle, to Chin	7.1	6.3	4.7	3.1	2.4
2.2	**5.7**	H11 V	Subnasal point to Chin	4.3	3.8	2.9	1.9	1.4
0.8	**2.0**	H12 V	Subnasal to Lip Aperature	1.5	1.3	1.0	0.7	0.5
0.5	**1.3**	H13 V	Lip Aperat. to Sublabial Crease	1.0	0.9	0.7	0.4	0.3
0.9	**2.3**	H14 V	Sublabial Crease to Chin	1.7	1.5	1.2	0.8	0.6
			TRANSVERSE HEAD MEASUREMENTS					
5.6	**14.2**	H15 T	Bi-Parietal, Head Breadth	10.7	9.5	7.1	4.7	3.6
4.7	**12.0**	H16 T	Bi-Zygomatic, Cheekbones	9.0	8.0	6.0	4.0	3.0
4.4	**11.2**	H17 T	Bi-Fronto Temporal, Forehead	8.4	7.5	5.6	3.7	2.8

Life size			EIGHT-YEAR-OLD BOY	Fractions of life size				
				3/4	2/3	1/2	1/3	1/4
Inches	cm	Code		cm	cm	cm	cm	cm
			TRANSVERSE HEAD MEASUREMENTS					
3.5	**9.0**	H18 T	Inter-Brow Ridge	6.8	6.0	4.5	3.0	2.3
4.6	**11.8**	H19 T	Inter-Tragion of Ears	8.9	7.9	5.9	3.9	3.0
4.1	**10.5**	H20 T	Inter-Angle of Mandible	7.9	7.0	5.3	3.5	2.6
3.4	**8.7**	H21 T	Inter-Lateral Corners of Eyes	6.5	5.8	4.4	2.9	2.2
1.1	**2.9**	H22 T	Width of Eye	2.2	1.9	1.5	1.0	0.7
1.2	**3.0**	H23 T	Inter-Inner Corner of eyes	2.3	2.0	1.5	1.0	0.8
1.2	**3.0**	H24 T	Breadth of Nose	2.3	2.0	1.5	1.0	0.8
1.6	**4.0**	H25 T	Breadth of Mouth	3.0	2.7	2.0	1.3	1.0
1.3	**3.2**	H26 T	Breadth of Chin	2.4	2.1	1.6	1.1	0.8
3.3	**8.5**	H27 T	Breadth of Neck	6.4	5.7	4.3	2.8	2.1
			ANTERIOR-POSTERIOR HEAD MEASUREMENTS					
7.4	**18.7**	H28 D	Brow to Occiput, Poserior Head	14.0	12.5	9.4	6.2	4.7
3.8	**9.6**	H29 D	Brow to Tragus of Ear	7.2	6.4	4.8	3.2	2.4
3.5	**8.9**	H30 D	Sellion, Nose Saddle, to Tragus	6.7	5.9	4.5	3.0	2.2
3.6	**9.1**	H31 D	Tragus to Occiput, Back of Head	6.8	6.1	4.6	3.0	2.3
2.2	**5.5**	H32 V	Tragus to Angle Mandible	4.1	3.7	2.8	1.8	1.4
3.1	**8.0**	H33 D	Angle of Mandible to Chin	6.0	5.3	4.0	2.7	2.0
2.2	**5.5**	H34 V	Length of Ear, Vertical	4.1	3.7	2.8	1.8	1.4
1.3	**3.4**	H35 D	Width of Ear, Anterior-Posterior	2.6	2.3	1.7	1.1	0.9
3.3	**8.5**	H36 D	Anterior to Posterior of Neck	6.4	5.7	4.3	2.8	2.1
			VERTICAL BODY MEASUREMENTS					
2.8	**7.0**	B1 V	Chin to Sternal Notch, Neck Pit	5.3	4.7	3.5	2.3	1.8
2.6	**6.5**	B2 V	Sternal Notch to Axilla, Arm Pit	4.9	4.3	3.3	2.2	1.6
4.1	**10.5**	B3 V	Sternal Notch to Nipples	7.9	7.0	5.3	3.5	2.6
5.5	**14.0**	B4 V	Sternal Notch to Xyphoid Proc.	10.5	9.3	7.0	4.7	3.5
11.4	**29.0**	B5 V	Sternal Notch to Umbilicus	21.8	19.3	14.5	9.7	7.3
5.9	**15.0**	B6 V	Xyphoid Process to Umbilicus	11.3	10.0	7.5	5.0	3.8
10.8	**27.4**	B7 V	Sternal Notch to Crest Ilium	20.6	18.3	13.7	9.1	6.9
12.6	**32.0**	B8 V	Sternal Notch to Ant.S.Sp.Ilium	24.0	21.3	16.0	10.7	8.0
15.9	**40.5**	B9 V	Sternal Notch to Pubis	30.4	27.0	20.3	13.5	10.1
4.5	**11.5**	B10 V	Umbilicus to Pubis of Pelvis	8.6	7.7	5.8	3.8	2.9
			TRANSVERSE BODY MEASUREMENTS					
11.8	**30.0**	B11 T	Shoulder Breadth	22.5	20.0	15.0	10.0	7.5
5.5	**13.9**	B12 T	Clavicle to Acromion Process	10.4	9.3	7.0	4.6	3.5
11.2	**28.4**	B13 T	Bi-acromial Process Breadth	21.3	18.9	14.2	9.5	7.1
7.6	**19.4**	B14 T	Inter-Anterior Axillary Fold	14.6	12.9	9.7	6.5	4.9
5.3	**13.5**	B15 T	Inter-Nipples	10.1	9.0	6.8	4.5	3.4

Life size			EIGHT-YEAR-OLD BOY	Fractions of life size				
				3/4	2/3	1/2	1/3	1/4
Inches	cm	Code		cm	cm	cm	cm	cm
			TRANSVERSE BODY MEASUREMENTS					
8.1	**20.5**	B16 T	Breadth of Thorax	15.4	13.7	10.3	6.8	5.1
7.9	**20.0**	B17 T	Inter-Costal. Rib Margin Breadth	15.0	13.3	10.0	6.7	5.0
7.9	**20.0**	B18 T	Inter-Crest of Ilium Breadth	15.0	13.3	10.0	6.7	5.0
6.9	**17.5**	B19 T	Inter-Anterior Superior Spine	13.1	11.7	8.8	5.8	4.4
8.9	**22.5**	B20 T	Inter-Greater Trochanter Femur	16.9	15.0	11.3	7.5	5.6
			ANTERIOR-POSTERIOR BODY MEASUREMENTS					
5.5	**14.0**	B21 D	Manubrium to Dorsal Spine	10.5	9.3	7.0	4.7	3.5
5.7	**14.5**	B22 D	Xyphoid Process to Dorsum	10.9	9.7	7.3	4.8	3.6
5.7	**14.5**	B23 D	Umbilicus to Dorsum	10.9	9.7	7.3	4.8	3.6
6.1	**15.5**	B24 D	Pubis of Pelvis to Buttock	11.6	10.3	7.8	5.2	3.9
4.5	**11.5**	B25 D	Thigh to Gluteal Furrow	8.6	7.7	5.8	3.8	2.9
			VERTICAL BACK MEASUREMENTS					
8.4	**21.3**	B26 V	Vertex to 7th Cervical Spine	16.0	14.2	10.7	7.1	5.3
			7th Cervical Vertebra Spine to:					
5.3	**13.5**	B27 V	To Posterior Axillary Fold, Armpit	10.1	9.0	6.8	4.5	3.4
6.3	**16.0**	B28 V	To Inferior Tip of Scapula	12.0	10.7	8.0	5.3	4.0
12.8	**32.5**	B29 V	To Posterior Superior Spine Ilium	24.4	21.7	16.3	10.8	8.1
16.1	**41.0**	B30 V	To Midline Upper Buttock Crease	30.8	27.3	20.5	13.7	10.3
18.9	**48.0**	B31 V	To Gluteal Furrow	36.0	32.0	24.0	16.0	12.0
28.0	**71.0**	B32 V	To Popliteal Crease Behind Knee	53.3	47.3	35.5	23.7	17.8
			TRANSVERSE BACK MEASUREMENTS					
9.8	**25.0**	B33 T	Inter-Posterior Axillary Folds	18.8	16.7	12.5	8.3	6.3
4.7	**12.0**	B34 T	Inter-Scapulae Medial Borders	9.0	8.0	6.0	4.0	3.0
2.4	**6.0**	B35 T	Inter-Post. Sup. Spines Sacrum	4.5	4.0	3.0	2.0	1.5
			UPPER EXTREMITY MEASUREMENTS					
50.0	**127.0**	U1 V	Span of Arms	95.3	84.7	63.5	42.3	31.8
5.5	**13.9**	U2 V	Sternum to Acromion Process	10.4	9.3	7.0	4.6	3.5
21.7	**55.0**	U3 V	Acromion Process to Fingertip	41.3	36.7	27.5	18.3	13.8
68.1	**173.0**	U4 V	Vertical Reach to Fingertip	129.8	115.3	86.5	57.7	43.3
32.7	**83.0**	U5 T	Lateral Reach to Fingertip	62.3	55.3	41.5	27.7	20.8
25.2	**64.0**	U6 D	Forward Reach to Fingertip	48.0	42.7	32.0	21.3	16.0
10.4	**26.3**	U7 V	Shoulder to Flexed Elbow	19.7	17.5	13.2	8.8	6.6
9.8	**25.0**	U8 V	Acromion to Lat. Condyle Elbow	18.8	16.7	12.5	8.3	6.3
13.3	**33.8**	U9 D	Flexed Elbow to Fingertip	25.4	22.5	16.9	11.3	8.5
7.3	**18.5**	U10 D	Flexed Elbow Pivot to Wrist Pivot	13.9	12.3	9.3	6.2	4.6

Life size			EIGHT-YEAR-OLD BOY	Fractions of life size				
				3/4	2/3	1/2	1/3	1/4
Inches	cm	Code		cm	cm	cm	cm	cm
			UPPER EXTREMITY MEASUREMENTS					
5.5	**14.0**	U11 V	Wrist to Fingertip	10.5	9.3	7.0	4.7	3.5
3.5	**9.0**	U12 V	Wrist to Thumb Tip	6.8	6.0	4.5	3.0	2.3
3.1	**8.0**	U13 V	Length of Palm	6.0	5.3	4.0	2.7	2.0
2.4	**6.0**	U14 V	Length 3rd Finger	4.5	4.0	3.0	2.0	1.5
3.8	**9.6**	U15 V	Thumb Crotch to 3rd Finger	7.2	6.4	4.8	3.2	2.4
1.9	**4.7**	U16 V	Thumb Crotch to Thumb Tip	3.5	3.1	2.4	1.6	1.2
2.5	**6.4**	U17 D	Anterior-Post. Arm at Biceps	4.8	4.3	3.2	2.1	1.6
2.4	**6.0**	U18 D	Anterior-Posterior at Forearm	4.5	4.0	3.0	2.0	1.5
1.3	**3.2**	U19 D	Depth of Wrist	2.4	2.1	1.6	1.1	0.8
1.8	**4.6**	U20 T	Breadth of Wrist	3.5	3.1	2.3	1.5	1.2
3.1	**8.0**	U21 T	Breadth of Hand	6.0	5.3	4.0	2.7	2.0
2.6	**6.5**	U22 T	Breadth of Palm	4.9	4.3	3.3	2.2	1.6
			LOWER EXTREMITY MEASUREMENTS					
29.5	**75.0**	L1 V	Crest of Ilium of Pelvis to Sole	56.3	50.0	37.5	25.0	18.8
28.3	**72.0**	* L2 V	Ant.Sup.Spine of Ilium to Sole	54.0	48.0	36.0	24.0	18.0
25.2	**64.0**	L3 V	Greater Trochanter Femur-Sole	48.0	42.7	32.0	21.3	16.0
24.4	**62.0**	* L4 V	Symphysis Pubis of Pelvis-Sole	46.5	41.3	31.0	20.7	15.5
22.4	**57.0**	* L5 V	Crotch to Sole	42.8	38.0	28.5	19.0	14.3
23.2	**59.0**	L6 V	Gluteal Furrow to Sole	44.3	39.3	29.5	19.7	14.8
13.8	**35.0**	L7 V	Posterior Knee Crease to Sole	26.3	23.3	17.5	11.7	8.8
13.4	**34.0**	L8 V	Tibia at Knee to Sole	25.5	22.7	17.0	11.3	8.5
2.2	**5.5**	L9 V	Med. Malleolus of Ankle to Sole	4.1	3.7	2.8	1.8	1.4
1.8	**4.5**	L10 V	Lat. Malleolus of Ankle to Sole	3.4	3.0	2.3	1.5	1.1
9.1	**23.0**	L11 T	Breadth of Hips, Sitting	17.3	15.3	11.5	7.7	5.8
4.3	**11.0**	L12 T	Breadth Mid Thigh	8.3	7.3	5.5	3.7	2.8
3.3	**8.3**	L13 T	Breadth of Knee	6.2	5.5	4.2	2.8	2.1
3.1	**8.0**	L14 T	Breadth of Calf	6.0	5.3	4.0	2.7	2.0
1.9	**4.7**	L15 T	Breadth Ankle	3.5	3.1	2.4	1.6	1.2
4.3	**11.0**	L16 D	Mid Thigh, Anterior-Posterior	8.3	7.3	5.5	3.7	2.8
3.5	**8.8**	L17 D	Knee, Anterior-Posterior	6.6	5.9	4.4	2.9	2.2
3.2	**8.2**	L18 D	Mid Calf, Anterior-Posterior	6.2	5.5	4.1	2.7	2.1
2.4	**6.0**	L19 D	Ankle, Anterior-Posterior	4.5	4.0	3.0	2.0	1.5
7.9	**20.0**	L20 D	Length of Foot	15.0	13.3	10.0	6.7	5.0
6.1	**15.4**	L21 D	Heel to 5th Toe	11.6	10.3	7.7	5.1	3.9
3.1	**8.0**	L22 T	Breadth Foot	6.0	5.3	4.0	2.7	2.0
1.8	**4.6**	L23 T	Width of Heel	3.5	3.1	2.3	1.5	1.2
		These are also listed with Vertical Measurements.						

NINE-YEAR-OLD GIRLS

There is still considerable physical similarity of boys and girls at this age, although differences are beginning to be manifest. The range of twenty-fifth to fiftieth and to seventy-fifth percentile is listed here for comparison:

Percentile	Twenty-fifth		Fiftieth		Seventy-fifth	
Girls:	Short		Medium		Tall	
Stature	128 cm,	50.3 in.	132 cm,	52 in.	137 cm,	55.5 in.
Weight	26 kg,	56 lb.	28.5 kg,	63 lb.	32.5 kg,	71.5 lb.[1]

At this age, girls, on the average, are about the same height and weight as boys. On closer inspection, the girls may have more rounded contours in face shoulders and hips. The most noticeable difference between boys and girls is the hair styling. For school and play dress is often similar. For special occasions, girls like to be little ladies. They participate in outdoor and competitive athletics, but not as vigorously as boys. Their interest is in music, dancing, clothing, and already in costume jewelry. They are more concerned about their hair and appearance than boys. The school level is fourth grade.

BODY MEASUREMENTS COMPARED TO HEIGHT

	Feature Measure in cm	Feature Compared to Stature	Approx. Ratio of Feature	Compared to Median Adult
Stature: Crown to sole	132			80%
Head Height: Crown to chin	18.9	14%	1/7	93%
Sitting Height:	70	53%	1/2+	79%
Trunk Measure: Sternal notch to symphysis pubis	38	29%	1/3–	78%
Leg Length: Greater trochanter of the femur to sole	67.4	51%	1/2	78%
Thigh Length: Greater trochanter of the femur to tibia at knee	32	24%	1/4	80%
Tibia at Knee to Sole:	35	26%	1/4+	80%
Arm Length: Acromion process at shoulder to dependant fingertip	56.5	43%	2/5+	74%
Hand Length:	14.5	11%	1/9	79%
Foot Length:	21	16%	1/6	88%

The mask of the face, brow to chin, begins to grow more as dentition increases. The upper lateral incisors and lower canine (cuspid) teeth erupt at eight to nine years of age.[2]

1. Nelson, W.E., p. 39.
2. Lockhart, R.D. p. 57.

NINE-YEAR-OLD BOYS

The lists of body proportions are shown separately in the tables of measurements. The range of twenty-fifth to fiftieth and to seventy-fifth percentile is listed here for comparison:

Percentile	Twenty-fifth		Fiftieth		Seventy-fifth	
Boys:	Short		Medium		Tall	
Stature	128 cm,	50.5 in.	132 cm,	52 in.	136 cm,	53.5 in.
Weight	25.4 kg,	56 lb.	28 kg,	62 lb.	31.5 kg,	69.3 lb.[1]

At this age boys and girls are still similar in height and weight. Hair styling is different, and boys are not as attentive to hair care as are girls. Athletics, ball games, outdoor activities, mechanical toys, constructive toys, monsters, and dinosaurs are their greater interests. They try to emulate adult male images, especially if one is interested in camping and fishing. The school level is fourth grade.

BODY MEASUREMENTS COMPARED TO HEIGHT

	Feature Measure in cm	Feature Compared to Stature	Approx. Ratio of Feature	Compared to Median Adult
Stature: Crown to sole	132			72%
Head Height: Crown to chin	19.4	15%	1/6–	84%
Sitting Height:	71	54%	1/2+	74%
Trunk Measure: Sternal notch to symphysis pubis	42	32%	1/3	79%
Leg Length: Greater trochanter of the femur to sole	69	52%	1/2	70%
Thigh Length: Greater trochanter of the femur to tibia at knee	33.7	26%	1/4	66%
Tibia at Knee to Sole:	35.3	27%	1/4+	69%
Arm Length: Acromion process at shoulder to dependent fingertip	60	45%	2/5+	74%
Hand Length:	14.3	11%	1/9	72%
Foot Length:	21	16%	1/6	75%

The mask of the face, brow to chin, begins to grow more as dentition increases. The permanent upper lateral incisors and lower canine (cuspids) erupt at eight to nine years of age.[2]

1. Nelson, W.E., p. 38.
2. Lockhart, R.D., p. 57.

Nine-Year-Old Girl, anterior and lateral head

all measurements in centimeters

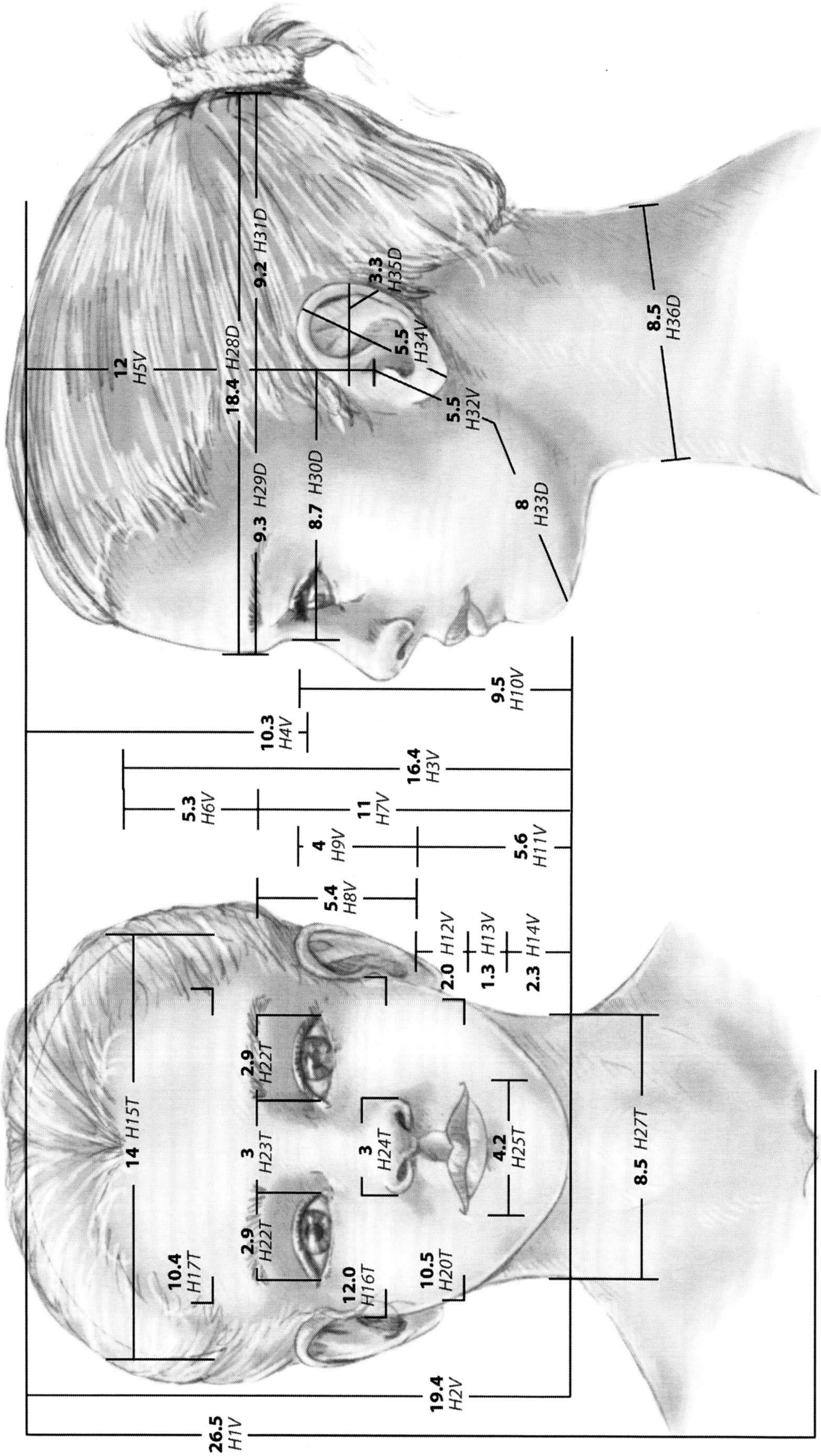

Nine-Year-Old Girl, anterior and lateral figure

all measurements in centimeters

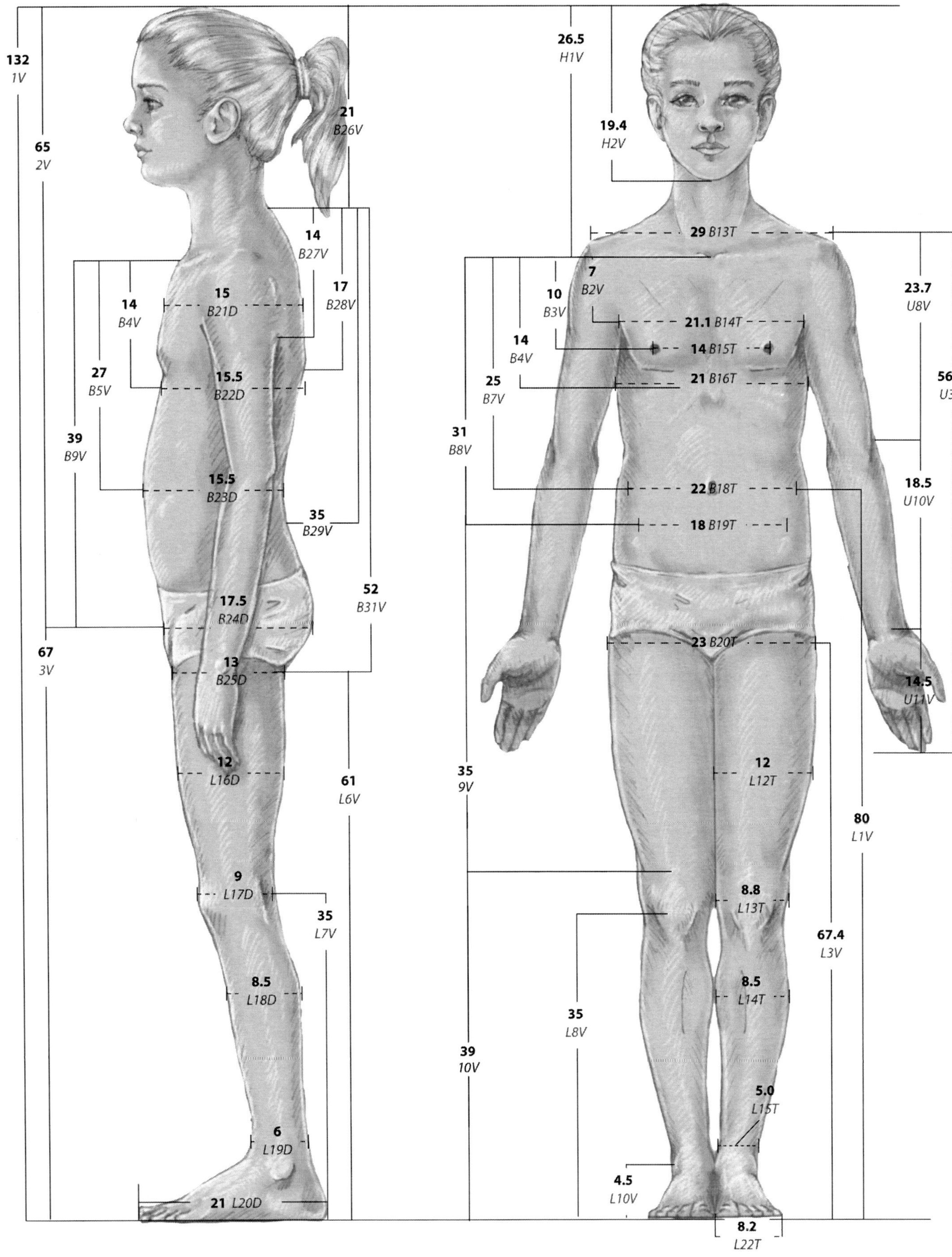

Life size			NINE-YEAR-OLD GIRL	Fractions of life size				
				3/4	2/3	1/2	1/3	1/4
Inches	cm	Code		cm	cm	cm	cm	cm
			VERTICAL MEASUREMENTS					
52.0	**132.0**	1 V	Stature, Height, Vertex to Sole	99.0	88.0	66.0	44.0	33.0
25.6	**65.0**	2 V	Vertex to Symphysis Pubis	48.8	43.3	32.5	21.7	16.3
26.4	**67.0**	* 3 V	Symphysis Pubis of Pelvis-Sole	50.3	44.7	33.5	22.3	16.8
41.3	**105.0**	4 V	Sternal Notch, Neck Pit, to Sole	78.8	70.0	52.5	35.0	26.3
10.2	**26.0**	5 V	Vertex to Acromion Process	19.5	17.3	13.0	8.7	6.5
41.7	**106.0**	6 V	Acromion Process to Sole	79.5	70.7	53.0	35.3	26.5
29.1	**74.0**	* 7 V	Ant.Sup.Spine of Ilium to Sole	55.5	49.3	37.0	24.7	18.5
12.6	**32.0**	8 V	Acrom. proc. to Ant. Sup. Spine	24.0	21.3	16.0	10.7	8.0
13.8	**35.0**	9 V	Ant.Sup.Spine of Ilium to Knee	26.3	23.3	17.5	11.7	8.8
15.4	**39.0**	10 V	Knee at Patella to Sole	29.3	26.0	19.5	13.0	9.8
31.5	**80.0**	11 V	Umbilicus to Sole	60.0	53.3	40.0	26.7	20.0
24.0	**61.0**	* 12 V	Crotch to Sole	45.8	40.7	30.5	20.3	15.3
39.8	**101.0**	13 V	Kneeling Height	75.8	67.3	50.5	33.7	25.3
27.6	**70.0**	14 V	Sitting Height	52.5	46.7	35.0	23.3	17.5
29.1	**74.0**	15 V	Buttock, Straight Leg to Sole	55.5	49.3	37.0	24.7	18.5
17.3	**44.0**	16 V	Buttock to Flexed Knee	33.0	29.3	22.0	14.7	11.0
16.5	**42.0**	17 V	Flexed Knee to Sole	31.5	28.0	21.0	14.0	10.5
		** These are also listed with Lower Extremity Measurements.*						
			VERTICAL HEAD MEASUREMENTS					
10.6	**27.0**	H1 V	Vertex to Sternal Notch	20.3	18	13.5	9.0	6.8
7.6	**19.4**	H2 V	Vertex to Chin, Head Height	14.6	12.9	9.7	6.5	4.9
6.5	**16.4**	H3 V	Hairline to Chin, Face Height	12.3	10.9	8.2	5.5	4.1
4.1	**10.3**	H4 V	Vertex to Lateral Corner of Eye	7.7	6.9	5.2	3.4	2.6
4.7	**12.0**	H5 V	Vertex to Tragion of Ear	9.0	8.0	6.0	4.0	3.0
2.1	**5.3**	H6 V	Hairline to Brow, Glabella	4.0	3.5	2.7	1.8	1.3
4.3	**11.0**	H7 V	Brow, Glabella, to Chin, Menton	8.3	7.3	5.5	3.7	2.8
2.1	**5.4**	H8 V	Brow to Subnasal Point	4.1	3.6	2.7	1.8	1.4
1.6	**4.0**	H9 V	Sellion to Subnasal, Nose Length	3.0	2.7	2.0	1.3	1.0
3.7	**9.5**	H10 V	Sellion, Nose Saddle, to Chin	7.1	6.3	4.8	3.2	2.4
2.2	**5.6**	H11 V	Subnasal Point to Chin	4.2	3.7	2.8	1.9	1.4
0.8	**2.0**	H12 V	Subnasal to Lip Aperature	1.5	1.3	1.0	0.7	0.5
0.5	**1.3**	H13 V	Lip Aperat. to Sublabial Crease	1.0	0.9	0.7	0.4	0.3
0.9	**2.3**	H14 T	Sublabial Crease to Chin	1.7	1.5	1.2	0.8	0.6
			TRANSVERSE HEAD MEASUREMENTS					
5.5	**14.0**	H15 T	Bi-Parietal, Head Breadth	10.5	9.3	7.0	4.7	3.5
4.7	**12.0**	H16 T	Bi-Zygomatic, Cheekbones	9.0	8.0	6.0	4.0	3.0
4.1	**10.4**	H17 T	Bi-Fronto Temporal, Forehead	7.8	6.9	5.2	3.5	2.6

Life size			NINE-YEAR-OLD GIRL	Fractions of life size				
				3/4	2/3	1/2	1/3	1/4
Inches	cm	Code		cm	cm	cm	cm	cm
			TRANSVERSE HEAD MEASUREMENTS					
3.7	9.4	H18 T	Inter-Brow Ridge	7.1	6.3	4.7	3.1	2.4
4.4	11.3	H19 T	Inter-Tragion of Ears	8.5	7.5	5.7	3.8	2.8
4.1	10.5	H20 T	Inter Angle of Mandible	7.9	7.0	5.3	3.5	2.6
3.5	8.8	H21 T	Inter Lateral Corners of Eyes	6.6	5.9	4.4	2.9	2.2
1.1	2.9	H22 T	Width of Eye	2.2	1.9	1.5	1.0	0.7
1.2	3.0	H23 T	Inter Inner Corner of eyes	2.3	2.0	1.5	1.0	0.8
1.2	3.0	H24 T	Breadth of Nose	2.3	2.0	1.5	1.0	0.8
1.7	4.2	H25 T	Breadth of Mouth	3.2	2.8	2.1	1.4	1.1
1.2	3.0	H26 T	Breadth of Chin	2.3	2.0	1.5	1.0	0.8
3.3	8.5	H27 T	Breadth of Neck	6.4	5.7	4.3	2.8	2.1
			ANTERIOR-POSTERIOR HEAD MEASUREMENTS					
7.2	18.4	H28 D	Brow to Occiput, Poserior Head	13.8	12.3	9.2	6.1	4.6
3.7	9.3	H29 D	Brow to Tragus of Ear	7.0	6.2	4.7	3.1	2.3
3.4	8.7	H30 D	Sellion, Nose Saddle, to Tragus	6.5	5.8	4.4	2.9	2.2
3.6	9.2	H31 D	Tragus to Occiput, Back of Head	6.9	6.1	4.6	3.1	2.3
2.2	5.5	H32 V	Tragus to Angle Mandible	4.1	3.7	2.8	1.8	1.4
3.1	8.0	H33 D	Angle of Mandible to Chin	6.0	5.3	4.0	2.7	2.0
2.2	5.5	H34 V	Length of Ear, Vertical	4.1	3.7	2.8	1.8	1.4
1.3	3.3	H35 D	Width of Ear, Anterior-Posterior	2.5	2.2	1.7	1.1	0.8
3.3	8.5	H36.D	Anterior to Posterior of Neck	6.4	5.7	4.3	2.8	2.1
			VERTICAL BODY MEASUREMENTS					
3.0	7.5	B1 V	Chin to Sternal Notch, Neck Pit	5.6	5.0	3.8	2.5	1.9
2.8	7.0	B2 V	Sternal Notch to Axilla, Arm Pit	5.3	4.7	3.5	2.3	1.8
3.9	10.0	B3 V	Sternal Notch to Nipples	7.5	6.7	5.0	3.3	2.5
5.5	14.0	B4 V	Sternal Notch to Xyphoid Proc.	10.5	9.3	7.0	4.7	3.5
10.6	27.0	B5 V	Sternal Notch to Umbilicus	20.3	18.0	13.5	9.0	6.8
5.5	14.0	B6 V	Xyphoid Process to Umbilicus	10.5	9.3	7.0	4.7	3.5
9.8	25.0	B7 V	Sternal Notch to Crest Ilium	18.8	16.7	12.5	8.3	6.3
12.2	31.0	B8 V	Sternal Notch to Ant.S.Sp.Ilium	23.3	20.7	15.5	10.3	7.8
15.4	39.0	B9 V	Sternal Notch to Pubis	29.3	26.0	19.5	13.0	9.8
5.1	13.0	B10 V	Umbilicus to Pubis	9.8	8.7	6.5	4.3	3.3
			TRANSVERSE BODY MEASUREMENTS					
12.4	31.5	B11 T	Shoulder Breadth	23.6	21.0	15.8	10.5	7.9
5.6	14.3	B12 T	Clavicle to Acromion Process	10.7	9.5	7.2	4.8	3.6
11.4	29.0	B13 T	Biacromial Process Breadth	21.8	19.3	14.5	9.7	7.3
8.3	21.1	B14 T	Inter Anterior Axillary Fold	15.8	14.1	10.6	7.0	5.3
5.5	14.0	B15 T	Inter Nipples	10.5	9.3	7.0	4.7	3.5

Life size			NINE-YEAR-OLD GIRL	Fractions of life size				
				3/4	2/3	1/2	1/3	1/4
Inches	cm	Code		cm	cm	cm	cm	cm
			TRANSVERSE BODY MEASUREMENTS					
8.3	**21.2**	B16 T	Breadth of Thorax	15.9	14.1	10.6	7.1	5.3
8.7	**22.0**	B17 T	Inter Costal Margin Breadth	16.5	14.7	11.0	7.3	5.5
8.7	**22.0**	B18 T	Inter Crest of Ilium Breadth	16.5	14.7	11.0	7.3	5.5
7.1	**18.0**	B19 T	Inter Anterior Superior Spine	13.5	12.0	9.0	6.0	4.5
9.1	**23.0**	B20 T	Inter Greater Trochanter Femur	17.3	15.3	11.5	7.7	5.8
			ANTERIOR-POSTERIOR BODY MEASUREMENTS					
5.9	**15.0**	B21 D	Manubrium to Dorsal Spine	11.3	10.0	7.5	5.0	3.8
6.1	**15.5**	B22 D	Xyphoid Process to Dorsum	11.6	10.3	7.8	5.2	3.9
6.1	**15.5**	B23 D	Umbilicus to Dorsum	11.6	10.3	7.8	5.2	3.9
6.9	**17.5**	B24 D	Pubis of Pelvis to Buttock	13.1	11.7	8.8	5.8	4.4
5.1	**13.0**	B25 D	Thigh to Gluteal Furrow	9.8	8.7	6.5	4.3	3.3
			VERTICAL BACK MEASUREMENTS					
8.3	**21.0**	B26 V	Vertex to 7th Cervical Spine	15.8	14.0	10.5	7.0	5.3
			7th Cervical Vertebra Spine to:					
5.5	**14.0**	B27 V	To Posterior Axillary Fold, Armpit	10.5	9.3	7.0	4.7	3.5
6.7	**17.0**	B28 V	To Inferior Tip of Scapula	12.8	11.3	8.5	5.7	4.3
13.8	**35.0**	B29 V	To Posterior Superior Spine Ilium	26.3	23.3	17.5	11.7	8.8
16.5	**42.0**	B30 V	To Midline Upper Buttock Crease	31.5	28.0	21.0	14.0	10.5
20.5	**52.0**	B31 V	To Gluteal Furrow	39.0	34.7	26.0	17.3	13.0
29.1	**74.0**	B32 V	To Popliteal Crease Behind Knee	55.5	49.3	37.0	24.7	18.5
			TRANSVERSE BACK MEASUREMENTS					
10.0	**25.4**	B33 T	Inter Posterior Axillary Folds	19.1	16.9	12.7	8.5	6.4
4.7	**12.0**	B34 T	Inter Scapulae Medial Borders	9.0	8.0	6.0	4.0	3.0
2.7	**6.9**	B35 T	Inter Post. Sup. Spines Sacrum	5.2	4.6	3.5	2.3	1.7
			UPPER EXTREMITY MEASUREMENTS					
51.2	**130.0**	U1 V	Span of Arms	97.5	86.7	65.0	43.3	32.5
5.6	**14.3**	U2 V	Sternum to Acromion Process	10.7	9.5	7.2	4.8	3.6
22.2	**56.5**	U3 V	Acromion Process to Fingertip	42.4	37.7	28.3	18.8	14.1
67.1	**170.4**	U4 V	Vertical Reach to Fingertip	127.8	113.6	85.2	56.8	42.6
31.5	**80.0**	U5 T	Lateral Reach to Fingertip	60.0	53.3	40.0	26.7	20.0
26.0	**66.0**	U6 D	Forward Reach to Fingertip	49.5	44.0	33.0	22.0	16.5
10.6	**27.0**	U7 V	Shoulder to Elbow Flexed	20.3	18.0	13.5	9.0	6.8
9.3	**23.7**	U8 V	Acromion to Lat. Condyle Elbow	17.8	15.8	11.9	7.9	5.9
13.6	**34.6**	U9 D	Flexed Elbow to Fingertip	26.0	23.1	17.3	11.5	8.7
7.3	**18.5**	U10 D	Flexed Elbow Pivot to Wrist Pivot	13.9	12.3	9.3	6.2	4.6

Life size			NINE-YEAR-OLD GIRL	Fractions of life size				
				3/4	2/3	1/2	1/3	1/4
Inches	cm	Code		cm	cm	cm	cm	cm
			UPPER EXTREMITY MEASUREMENTS					
5.7	**14.5**	U11 V	Wrist to Fingertip	10.9	9.7	7.3	4.8	3.6
3.7	**9.5**	U12 V	Wrist to Thumb Tip	7.1	6.3	4.8	3.2	2.4
3.1	**8.0**	U13 V	Length of Palm	6.0	5.3	4.0	2.7	2.0
2.4	**6.2**	U14 V	Length 3rd Finger	4.7	4.1	3.1	2.1	1.6
3.8	**9.6**	U15 V	Thumb Crotch to 3rd Finger	7.2	6.4	4.8	3.2	2.4
1.9	**4.9**	U16 V	Thumb Crotch to Thumb Tip	3.7	3.3	2.5	1.6	1.2
2.6	**6.5**	U17 D	Anterior-Post. Arm at Biceps	4.9	4.3	3.3	2.2	1.6
2.4	**6.0**	U18 D	Anterior-Posterior at Forearm	4.5	4.0	3.0	2.0	1.5
1.3	**3.4**	U19 D	Depth of Wrist	2.6	2.3	1.7	1.1	0.9
1.9	**4.9**	U20 T	Breadth of Wrist	3.7	3.3	2.5	1.6	1.2
3.1	**8.0**	U21 T	Breadth of Hand	6.0	5.3	4.0	2.7	2.0
2.5	**6.4**	U22 T	Breadth of Palm	4.8	4.3	3.2	2.1	1.6
			LOWER EXTREMITY MEASUREMENTS					
31.5	**80.0**	L1 V	Crest of Ilium to Sole	60.0	53.3	40.0	26.7	20.0
29.1	**74.0**	* L2 V	Ant.Sup.Spine of Ilium to Sole	55.5	49.3	37.0	24.7	18.5
26.5	**67.4**	L3 V	Greater Trochanter Femur-Sole	50.6	44.9	33.7	22.5	16.9
26.4	**67.0**	* L4 V	Symphysis Pubis of Pelvis-Sole	50.3	44.7	33.5	22.3	16.8
24.0	**61.0**	* L5 V	Crotch to Sole	45.8	40.7	30.5	20.3	15.3
24.0	**61.0**	L6 V	Gluteal Furrow to Sole	45.8	40.7	30.5	20.3	15.3
13.8	**35.0**	L7 V	Posterior Knee Crease to Sole	26.3	23.3	17.5	11.7	8.8
13.8	**35.0**	L8 V	Tibia at Knee to Sole	26.3	23.3	17.5	11.7	8.8
2.2	**5.5**	L9 V	Med. Malleolus of Ankle to Sole	4.1	3.7	2.8	1.8	1.4
1.8	**4.5**	L10 V	Lat. Malleolus of Ankle to Sole	3.4	3.0	2.3	1.5	1.1
9.8	**25.0**	L11 T	Breadth of Hips, Sitting	18.8	16.7	12.5	8.3	6.3
4.7	**12.0**	L12 T	Breadth Mid Thigh	9.0	8.0	6.0	4.0	3.0
3.5	**8.8**	L13 T	Breadth of Knee	6.6	5.9	4.4	2.9	2.2
3.3	**8.5**	L14 T	Breadth of Calf	6.4	5.7	4.3	2.8	2.1
2.0	**5.0**	L15 T	Breadth Ankle	3.8	3.3	2.5	1.7	1.3
4.7	**12.0**	L16 D	Mid Thigh, Anterior-Posterior	9.0	8.0	6.0	4.0	3.0
3.5	**9.0**	L17 D	Knee, Anterior-Posterior	6.8	6.0	4.5	3.0	2.3
3.3	**8.5**	L18 D	Mid Calf, Anterior-Posterior	6.4	5.7	4.3	2.8	2.1
2.4	**6.0**	L19 D	Ankle, Anterior-Posterior	4.5	4.0	3.0	2.0	1.5
8.3	**21.0**	L20 D	Length of Foot	15.8	14.0	10.5	7.0	5.3
6.5	**16.5**	L21 D	Heel to 5th Toe	12.4	11.0	8.3	5.5	4.1
3.2	**8.2**	L22 T	Breadth Foot	6.2	5.5	4.1	2.7	2.1
1.8	**4.5**	L23 T	Width of Heel	3.4	3.0	2.3	1.5	1.1
		These are also listed with Vertical Measurements.						

Nine-Year-Old Boy, anterior and lateral head

all measurements in centimeters

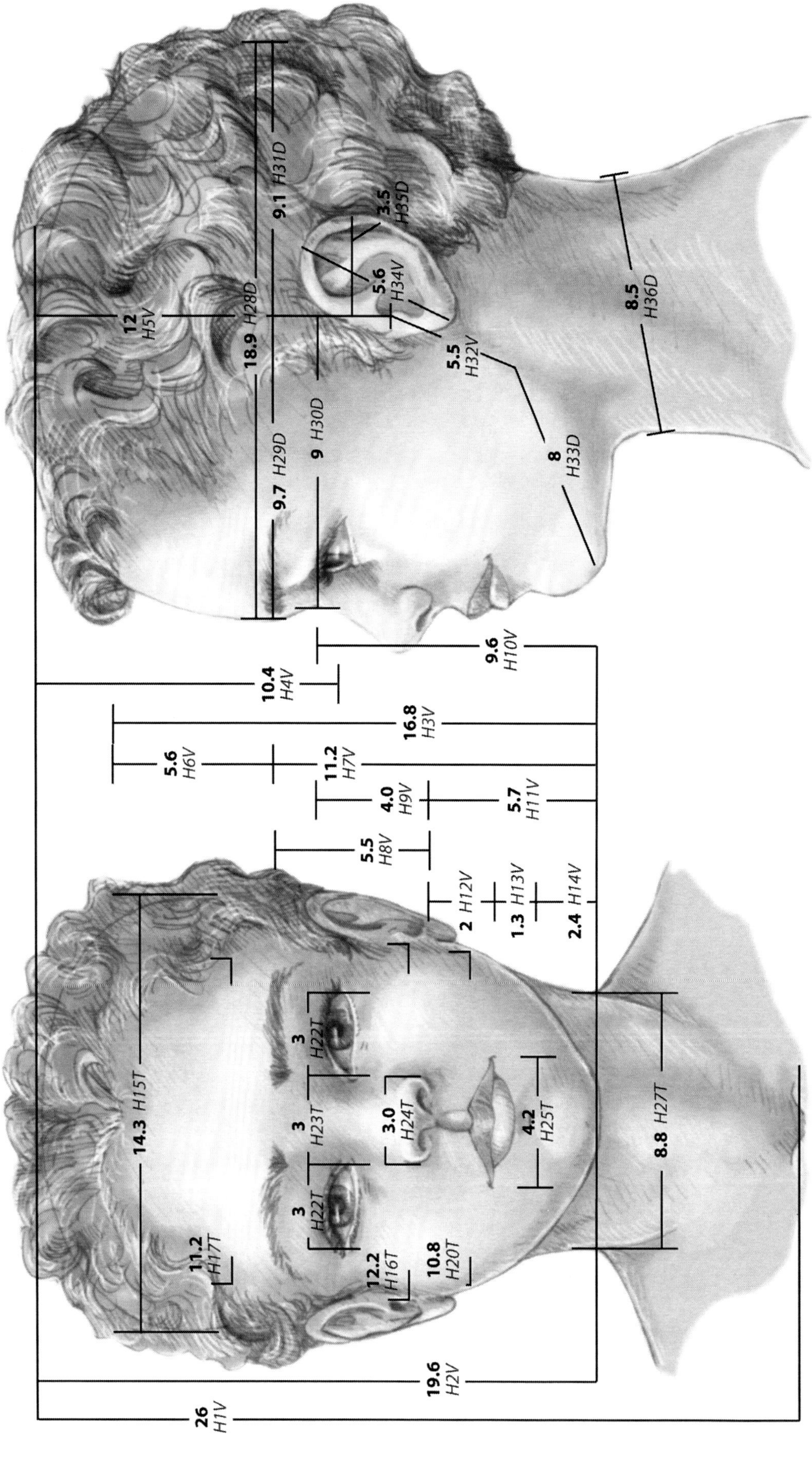

Nine-Year-Old Boy, anterior and lateral figure

all measurements in centimeters

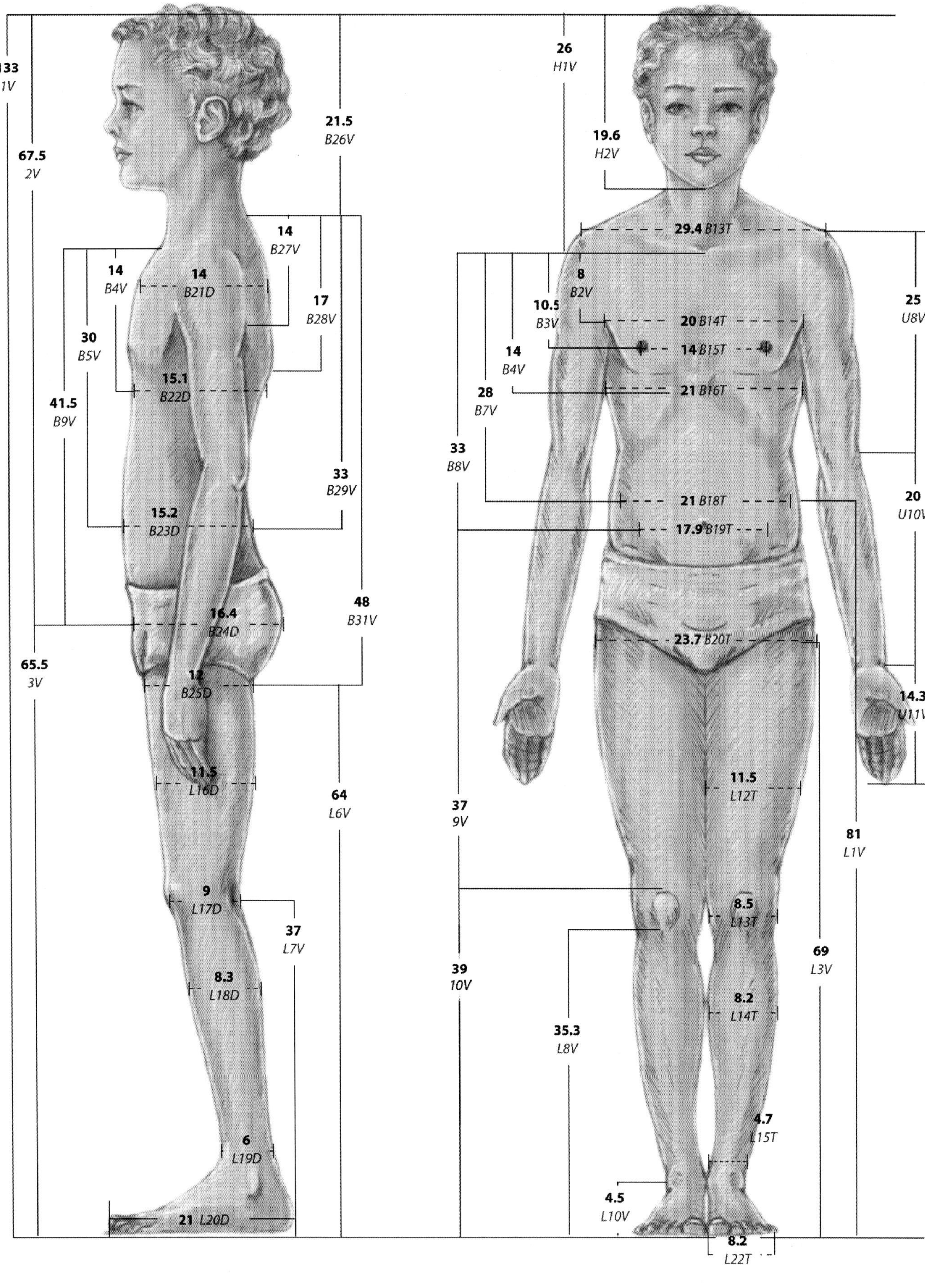

Life size			NINE-YEAR-OLD BOY	Fractions of life size				
				3/4	2/3	1/2	1/3	1/4
Inches	cm	Code		cm	cm	cm	cm	cm
			VERTICAL MEASUREMENTS					
52.4	**133.0**	1 V	Stature, Height, Vertex to Sole	99.8	88.7	66.5	44.3	33.3
26.6	**67.5**	2 V	Vertex to Symphysis Pubis	50.6	45.0	33.8	22.5	16.9
25.8	**65.5**	* 3 V	Symphysis Pubis of Pelvis-Sole	49.1	43.7	32.8	21.8	16.4
42.9	**109.0**	4 V	Sternal Notch, Neck Pit, to Sole	81.8	72.7	54.5	36.3	27.3
9.8	**25.0**	5 V	Vertex to Acromion Process	18.8	16.7	12.5	8.3	6.3
43.3	**110.0**	6 V	Acromion Process to Sole	82.5	73.3	55.0	36.7	27.5
29.9	**76.0**	* 7 V	Ant.Sup.Spine of Ilium to Sole	57.0	50.7	38.0	25.3	19.0
13.4	**34.0**	8 V	Acrom. proc. to Ant. Sup. Spine	25.5	22.7	17.0	11.3	8.5
14.6	**37.0**	9 V	Ant.Sup.Spine of Ilium to Knee	27.8	24.7	18.5	12.3	9.3
15.4	**39.0**	10 V	Knee at Patella to Sole	29.3	26.0	19.5	13.0	9.8
30.7	**78.0**	11 V	Umbilicus to Sole	58.5	52.0	39.0	26.0	19.5
23.8	**60.5**	* 12 V	Crotch to Sole	45.4	40.3	30.3	20.2	15.1
39.4	**100.0**	13 V	Kneeling Height	75.0	66.7	50.0	33.3	25.0
28.0	**71.0**	14 V	Sitting Height	53.3	47.3	35.5	23.7	17.8
29.1	**74.0**	15 V	Buttock, Straight Leg to Sole	55.5	49.3	37.0	24.7	18.5
17.1	**43.5**	16 V	Buttock to Flexed Knee	32.6	29.0	21.8	14.5	10.9
16.5	**42.0**	17 V	Flexed Knee to Sole	31.5	28.0	21.0	14.0	10.5
		** These are also listed with Lower Extremity Measurements.*						
			VERTICAL HEAD MEASUREMENTS					
10.2	**26.0**	H1 V	Vertex to Sternal Notch	19.5	17.3	13.0	8.7	6.5
7.7	**19.6**	H2 V	Vertex to Chin, Head Height	14.7	13.1	9.8	6.5	4.9
6.6	**16.8**	H3 V	Hairline to Chin, Face Height	12.6	11.2	8.4	5.6	4.2
4.1	**10.4**	H4 V	Vertex to Lateral Corner of Eye	7.8	6.9	5.2	3.5	2.6
4.7	**12.0**	H5 V	Vertex to Tragion of Ear	9.0	8.0	6.0	4.0	3.0
2.2	**5.6**	H6 V	Hairline to Brow, Glabella	4.2	3.7	2.8	1.9	1.4
4.4	**11.2**	H7 V	Brow, Glabella, to Chin, Menton	8.4	7.5	5.6	3.7	2.8
2.2	**5.5**	H8 V	Brow to Subnasal Point	4.1	3.7	2.8	1.8	1.4
1.6	**4.0**	H9 V	Sellion to Subnasal, Nose Length	3.0	2.7	2.0	1.3	1.0
3.8	**9.6**	H10 V	Sellion, Nose Saddle, to Chin	7.2	6.4	4.8	3.2	2.4
2.2	**5.7**	H11 V	Subnasal point to Chin	4.3	3.8	2.9	1.9	1.4
0.8	**2.0**	H12 V	Subnasal to Lip Aperature	1.5	1.3	1.0	0.7	0.5
0.5	**1.3**	H13 V	Lip Aperat. to Sublabial Crease	1.0	0.9	0.7	0.4	0.3
0.9	**2.4**	H14 V	Sublabial Crease to Chin	1.8	1.6	1.2	0.8	0.6
			TRANSVERSE HEAD MEASUREMENTS					
5.6	**14.3**	H15 T	Bi-Parietal, Head Breadth	10.7	9.5	7.2	4.8	3.6
4.8	**12.2**	H16 T	Bi-Zygomatic, Cheekbones	9.2	8.1	6.1	4.1	3.1
4.4	**11.2**	H17 T	Bi-Fronto Temporal, Forehead	8.4	7.5	5.6	3.7	2.8

Life size			NINE-YEAR-OLD BOY	Fractions of life size				
				3/4	2/3	1/2	1/3	1/4
Inches	cm	Code		cm	cm	cm	cm	cm
			TRANSVERSE HEAD MEASUREMENTS					
3.6	**9.1**	H18 T	Inter-Brow Ridge	6.8	6.1	4.6	3.0	2.3
4.7	**12.0**	H19 T	Inter-Tragion of Ears	9.0	8.0	6.0	4.0	3.0
4.3	**10.8**	H20 T	Inter-Angle of Mandible	8.1	7.2	5.4	3.6	2.7
3.5	**9.0**	H21 T	Inter-Lateral Corners of Eyes	6.8	6.0	4.5	3.0	2.3
1.2	**3.0**	H22 T	Width of Eye	2.3	2.0	1.5	1.0	0.8
1.2	**3.0**	H23 T	Inter-Inner Corner of Eyes	2.3	2.0	1.5	1.0	0.8
1.2	**3.0**	H24 T	Breadth of Nose	2.3	2.0	1.5	1.0	0.8
1.7	**4.2**	H25 T	Breadth of Mouth	3.2	2.8	2.1	1.4	1.1
1.3	**3.2**	H26 T	Breadth of Chin	2.4	2.1	1.6	1.1	0.8
3.5	**8.8**	H27 T	Breadth of Neck	6.6	5.9	4.4	2.9	2.2
			ANTERIOR-POSTERIOR HEAD MEASUREMENTS					
7.4	**18.9**	H28 D	Brow to Occiput, Poserior Head	14.2	12.6	9.5	6.3	4.7
3.8	**9.7**	H29 D	Brow to Tragus of Ear	7.3	6.5	4.9	3.2	2.4
3.5	**9.0**	H30 D	Sellion, Nose Saddle, to Tragus	6.8	6.0	4.5	3.0	2.3
3.6	**9.1**	H31 D	Tragus to Occiput, Back of Head	6.8	6.1	4.6	3.0	2.3
2.2	**5.5**	H32 V	Tragus to Angle Mandible	4.1	3.7	2.8	1.8	1.4
3.1	**8.0**	H33 D	Angle of Mandible to Chin	6.0	5.3	4.0	2.7	2.0
2.2	**5.6**	H34 V	Length of Ear, Vertical	4.2	3.7	2.8	1.9	1.4
1.4	**3.5**	H35 D	Width of Ear, Anterior-Posterior	2.6	2.3	1.8	1.2	0.9
3.3	**8.5**	H36 D	Anterior to Posterior of Neck	6.4	5.7	4.3	2.8	2.1
			VERTICAL BODY MEASUREMENTS					
2.8	**7.2**	B1 V	Chin to Sternal Notch, Neck Pit	5.4	4.8	3.6	2.4	1.8
3.1	**8.0**	B2 V	Sternal Notch to Axilla, Arm Pit	6.0	5.3	4.0	2.7	2.0
4.1	**10.5**	B3 V	Sternal Notch to Nipples	7.9	7.0	5.3	3.5	2.6
5.5	**14.0**	B4 V	Sternal Notch to Xyphoid Proc.	10.5	9.3	7.0	4.7	3.5
11.8	**30.0**	B5 V	Sternal Notch to Umbilicus	22.5	20.0	15.0	10.0	7.5
5.9	**15.0**	B6 V	Xyphoid Process to Umbilicus	11.3	10.0	7.5	5.0	3.8
11.0	**28.0**	B7 V	Sternal Notch to Crest Ilium	21.0	18.7	14.0	9.3	7.0
13.0	**33.0**	B8 V	Sternal Notch to Ant.S.Sp.Ilium	24.8	22.0	16.5	11.0	8.3
16.3	**41.5**	B9 V	Sternal Notch to Pubis	31.1	27.7	20.8	13.8	10.4
4.9	**12.5**	B10 V	Umbilicus to Pubis of Pelvis	9.4	8.3	6.3	4.2	3.1
			TRANSVERSE BODY MEASUREMENTS					
12.6	**32.0**	B11 T	Shoulder Breadth	24.0	21.3	16.0	10.7	8.0
5.6	**14.3**	B12 T	Clavicle to Acromion Process	10.7	9.5	7.2	4.8	3.6
11.6	**29.4**	B13 T	Bi-acromial Process Breadth	22.1	19.6	14.7	9.8	7.4
7.9	**20.0**	B14 T	Inter-Anterior Axillary Fold	15.0	13.3	10.0	6.7	5.0
5.5	**14.0**	B15 T	Inter-Nipples	10.5	9.3	7.0	4.7	3.5

Life size			NINE-YEAR-OLD BOY	Fractions of life size				
				3/4	2/3	1/2	1/3	1/4
Inches	cm	Code		cm	cm	cm	cm	cm
			TRANSVERSE BODY MEASUREMENTS					
8.3	21.0	B16 T	Breadth of Thorax	15.8	14.0	10.5	7.0	5.3
7.9	20.0	B17 T	Inter-Costal. Rib Margin Breadth	15.0	13.3	10.0	6.7	5.0
8.3	21.0	B18 T	Inter-Crest of Ilium Breadth	15.8	14.0	10.5	7.0	5.3
7.0	17.9	B19 T	Inter-Anterior Superior Spine	13.4	11.9	9.0	6.0	4.5
9.3	23.7	B20 T	Inter-Greater Trochanter Femur	17.8	15.8	11.9	7.9	5.9
			ANTERIOR-POSTERIOR BODY MEASUREMENTS					
5.5	14.0	B21 D	Manubrium to Dorsal Spine	10.5	9.3	7.0	4.7	3.5
5.9	15.1	B22 D	Xyphoid Process to Dorsum	11.3	10.1	7.6	5.0	3.8
6.0	15.2	B23 D	Umbilicus to Dorsum	11.4	10.1	7.6	5.1	3.8
6.5	16.4	B24 D	Pubis of Pelvis to Buttock	12.3	10.9	8.2	5.5	4.1
4.7	12.0	B25 D	Thigh to Gluteal Furrow	9.0	8.0	6.0	4.0	3.0
			VERTICAL BACK MEASUREMENTS					
8.5	21.5	B26 V	Vertex to 7th Cervical Spine	16.1	14.3	10.8	7.2	5.4
			7th Cervical Vertebra Spine to:					
5.5	14.0	B27 V	To Posterior Axillary Fold, Armpit	10.5	9.3	7.0	4.7	3.5
6.7	17.0	B28 V	To Inferior Tip of Scapula	12.8	11.3	8.5	5.7	4.3
13.0	33.0	B29 V	To Posterior Superior Spine Ilium	24.8	22.0	16.5	11.0	8.3
16.1	41.0	B30 V	To Midline Upper Buttock Crease	30.8	27.3	20.5	13.7	10.3
18.9	48.0	B31 V	To Gluteal Furrow	36.0	32.0	24.0	16.0	12.0
28.3	72.0	B32 V	To Popliteal Crease Behind Knee	54.0	48.0	36.0	24.0	18.0
			TRANSVERSE BACK MEASUREMENTS					
9.8	25.0	B33 T	Inter-Posterior Axillary Folds	18.8	16.7	12.5	8.3	6.3
4.7	12.0	B34 T	Inter-Scapulae Medial Borders	9.0	8.0	6.0	4.0	3.0
2.6	6.5	B35 T	Inter-Post. Sup. Spines Sacrum	4.9	4.3	3.3	2.2	1.6
			UPPER EXTREMITY MEASUREMENTS					
52.4	133.0	U1 V	Span of Arms	99.8	88.7	66.5	44.3	33.3
5.6	14.3	U2 V	Sternum to Acromion Process	10.7	9.5	7.2	4.8	3.6
23.6	60.0	U3 V	Acromion Process to Fingertip	45.0	40.0	30.0	20.0	15.0
68.1	173.0	U4 V	Vertical Reach to Fingertip	129.8	115.3	86.5	57.7	43.3
32.7	83.0	U5 T	Lateral Reach to Fingertip	62.3	55.3	41.5	27.7	20.8
25.2	64.0	U6 D	Forward Reach to Fingertip	48.0	42.7	32.0	21.3	16.0
10.9	27.6	U7 V	Shoulder to Flexed Elbow	20.7	18.4	13.8	9.2	6.9
9.8	25.0	U8 V	Acromion to Lat. Condyle Elbow	18.8	16.7	12.5	8.3	6.3
13.9	35.2	U9 D	Flexed Elbow to Fingertip	26.4	23.5	17.6	11.7	8.8
7.9	20.0	U10 D	Flexed Elbow Pivot to Wrist Pivot	15.0	13.3	10.0	6.7	5.0

Life size			NINE-YEAR-OLD BOY	Fractions of life size				
				3/4	2/3	1/2	1/3	1/4
Inches	cm	Code		cm	cm	cm	cm	cm
			UPPER EXTREMITY MEASUREMENTS					
5.6	**14.3**	U11 V	Wrist to Fingertip	10.7	9.5	7.2	4.8	3.6
3.7	**9.3**	U12 V	Wrist to Thumb Tip	7.0	6.2	4.7	3.1	2.3
3.3	**8.3**	U13 V	Length of Palm	6.2	5.5	4.2	2.8	2.1
2.4	**6.0**	U14 V	Length 3rd Finger	4.5	4.0	3.0	2.0	1.5
4.0	**10.2**	U15 V	Thumb Crotch to 3rd Finger	7.7	6.8	5.1	3.4	2.6
2.0	**5.0**	U16 V	Thumb Crotch to Thumb Tip	3.8	3.3	2.5	1.7	1.3
2.5	**6.4**	U17 D	Anterior-Post. Arm at Biceps	4.8	4.3	3.2	2.1	1.6
2.4	**6.0**	U18 D	Anterior-Posterior at Forearm	4.5	4.0	3.0	2.0	1.5
1.3	**3.2**	U19 D	Depth of Wrist	2.4	2.1	1.6	1.1	0.8
2.0	**5.0**	U20 T	Breadth of Wrist	3.8	3.3	2.5	1.7	1.3
3.2	**8.2**	U21 T	Breadth of Hand	6.2	5.5	4.1	2.7	2.1
2.6	**6.7**	U22 T	Breadth of Palm	5.0	4.5	3.4	2.2	1.7
			LOWER EXTREMITY MEASUREMENTS					
31.9	**81.0**	L1 V	Crest of Ilium of Pelvis to Sole	60.8	54.0	40.5	27.0	20.3
29.5	**75.0**	* L2 V	Ant.Sup.Spine of Ilium to Sole	56.3	50.0	37.5	25.0	18.8
27.2	**69.0**	L3 V	Greater Trochanter Femur-Sole	51.8	46.0	34.5	23.0	17.3
25.8	**65.5**	* L4 V	Symphysis Pubis of Pelvis-Sole	49.1	43.7	32.8	21.8	16.4
23.8	**60.5**	* L5 V	Crotch to Sole	45.4	40.3	30.3	20.2	15.1
25.2	**64.0**	L6 V	Gluteal Furrow to Sole	48.0	42.7	32.0	21.3	16.0
14.6	**37.0**	L7 V	Posterior Knee Crease to Sole	27.8	24.7	18.5	12.3	9.3
13.9	**35.3**	L8 V	Tibia at Knee to Sole	26.5	23.5	17.7	11.8	8.8
2.4	**6.0**	L9 V	Med. Malleolus of Ankle to Sole	4.5	4.0	3.0	2.0	1.5
1.8	**4.5**	L10 V	Lat. Malleolus of Ankle to Sole	3.4	3.0	2.3	1.5	1.1
9.4	**24.0**	L11 T	Breadth of Hips, Sitting	18.0	16.0	12.0	8.0	6.0
4.5	**11.5**	L12 T	Breadth Mid Thigh	8.6	7.7	5.8	3.8	2.9
3.3	**8.5**	L13 T	Breadth of Knee	6.4	5.7	4.3	2.8	2.1
3.2	**8.2**	L14 T	Breadth of Calf	6.2	5.5	4.1	2.7	2.1
1.9	**4.7**	L15 T	Breadth Ankle	3.5	3.1	2.4	1.6	1.2
4.5	**11.5**	L16 D	Mid Thigh, Anterior-Posterior	8.6	7.7	5.8	3.8	2.9
3.5	**9.0**	L17 D	Knee, Anterior-Posterior	6.8	6.0	4.5	3.0	2.3
3.3	**8.3**	L18 D	Mid Calf, Anterior-Posterior	6.2	5.5	4.2	2.8	2.1
2.4	**6.0**	L19 D	Ankle, Anterior-Posterior	4.5	4.0	3.0	2.0	1.5
8.3	**21.0**	L20 D	Length of Foot	15.8	14.0	10.5	7.0	5.3
6.2	**15.8**	L21 D	Heel to 5th Toe	11.9	10.5	7.9	5.3	4.0
3.2	**8.2**	L22 T	Breadth Foot	6.2	5.5	4.1	2.7	2.1
1.8	**4.6**	L23 T	Width of Heel	3.5	3.1	2.3	1.5	1.2
		** These are also listed with Vertical Measurements.*						

TEN-YEAR-OLD GIRLS

The lists of body proportions for girls are shown in the tables of measurements. The stature range of twenty-fifth, to fiftieth, and to seventy-fifth percentile is listed here for comparison:

Percentile	Twenty-fifth		Fiftieth		Seventy-fifth	
Girls:	Short		Medium		Tall	
Stature	134 cm,	52.5 in.	138 cm,	54.5 in.	143 cm,	56.3 in.
Weight	29 kg,	63.3 lb.	32.5 kg,	71.8 lb.	37.5 kg,	82.8 lb.[1]

Girls at this age are approaching the onset of menstruation, and begin to grow taller than boys. Feminine characteristics may begin to be evident. Hair styling and dress receive more attention. Many athletic interests are similar to those in which boys participate. Some girls develop a special interest in horses, both in care and riding and as a competitive sport. This is a combination of psychological pride in the domination of a larger animal and the sense of security they feel while mounted. This fondness of horses may continue for a lifetime or it may be substituted for an automobile at the age of obtaining a driver's license. School level is fifth grade.

BODY MEASUREMENTS COMPARED TO HEIGHT

	Feature Measure in cm	Feature Compared to Stature	Approx. Ratio of Feature	Compared to Median Adult
Stature: Crown to sole	138			82%
Head Height: Crown to chin	19.5	14%	1/7	96%
Sitting Height:	72.5	53%	1/2+	82%
Trunk Measure: Sternal notch to symphysis pubis	41	30%	1/3−	84%
Leg Length: Greater trochanter of the femur to sole	70.4	51%	1/2	82%
Thigh Length: Greater trochanter of the femur to tibia at knee	33.6	24%	1/4	81%
Tibia at Knee to Sole:	36	26%	1/4	83%
Arm Length: Acromion process at shoulder to dependent fingertip	58	42%	2/5	77%
Hand Length:	15	11%	1/9	84%
Foot Length:	21.5	15%	1/6	90%

The mask of the face, brow to chin, continues to grow more as dentition increases. When more deciduous teeth are lost and replaced by permanent teeth, the face gradually grows down and forward. The upper first permanent premolars erupt between ten and eleven years. The lower first premolars and the upper second premolars erupt between ten and twelve years of age.[2]

1. Nelson, W.E., p. 38.
2. Lockhart, R.D., p. 57.

TEN-YEAR-OLD BOYS

The list of body proportions, for each sex, is shown separately in the tables of measurements. The range of twenty-fifth to fiftieth and to seventy-fifth percentile is listed here for comparison:

Percentile	Twenty-fifth		Fiftieth		Seventy-fifth	
Boys:	Short		Medium		Tall	
Stature	133.4 cm,	52.5.5 in.	138 cm,	54.3 in.	142 cm,	55.8 in.
Weight	28. kg,	62 lb.	31.5 kg,	69.3 lb.	35.6 kg,	78.5 lb.[1]

At this age, boys are, on the average, about the same height and weight as most girls. Athletics, ball games, bicycles, and outdoor activities, like hiking, camping and fishing are their greater interests. Pet dogs receive attention. They still have interest in dinosaurs, monsters, and many become interested in space aliens. They try to emulate adult male images. Attention to hair is casual and dress varies with style of the times and the neighborhood. School level is fifth grade.

BODY MEASUREMENTS COMPARED TO HEIGHT

	Feature Measure in cm	Feature Compared to Stature	Approx. Ratio of Feature	Compared to Median Adult
Stature: Crown to sole	138			78%
Head Height: Crown to chin	19.6	14%	1/7	85%
Sitting Height:	73	53%	1/2+	76%
Trunk Measure: Sternal notch to symphysis pubis	42	31%	1/3–	79%
Leg Length: Greater trochanter of the femur to sole	71	52%	1/2	72%
Thigh Length: Greater trochanter of the femur to tibia at knee	34	25%	1/4	72%
Tibia at Knee to Sole:	37	27%	1/4+	73%
Arm Length: Acromion process at shoulder to dependent fingertip	62	52%	1/2+	85%
Hand Length:	15	11%	1/9	75%
Foot Length:	21.6	16%	1/6	77%

The mask of the face, brow to chin, continues to grow more as dentition increases. When more deciduous teeth are lost and replaced by permanent teeth, the face gradually grows down and forward. The upper first permanent premolars erupt between ten and eleven years. The lower first premolars and the upper second premolars erupt between ten and twelve years of age.[2]

1. Nelson, W.E., p. 39.
2. Lockhart, R.D., p. 57.

Ten-Year-Old Girl, anterior and lateral head

all measurements in centimeters

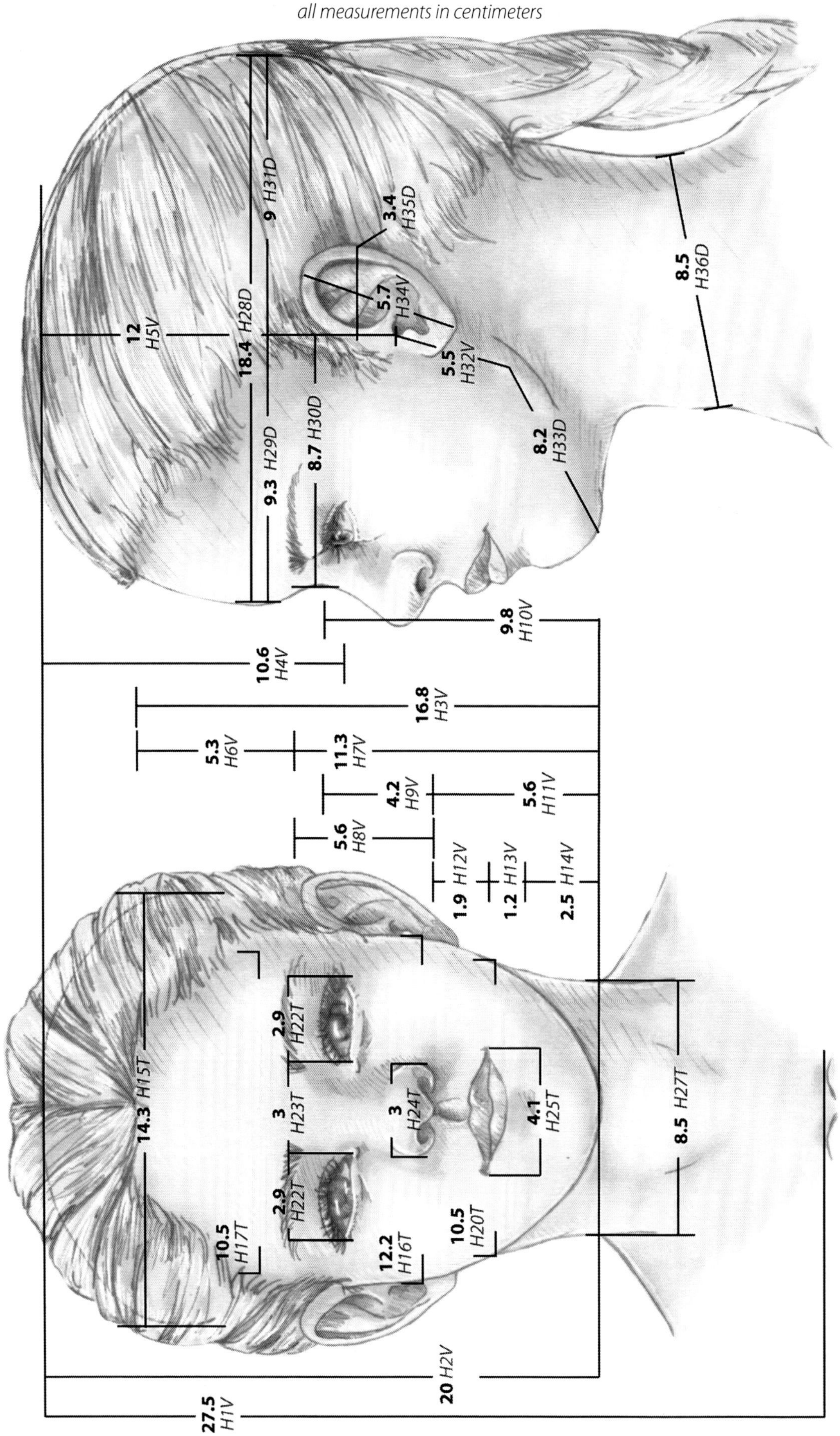

Ten-Year-Old Girl, anterior and lateral figure

all measurements in centimeters

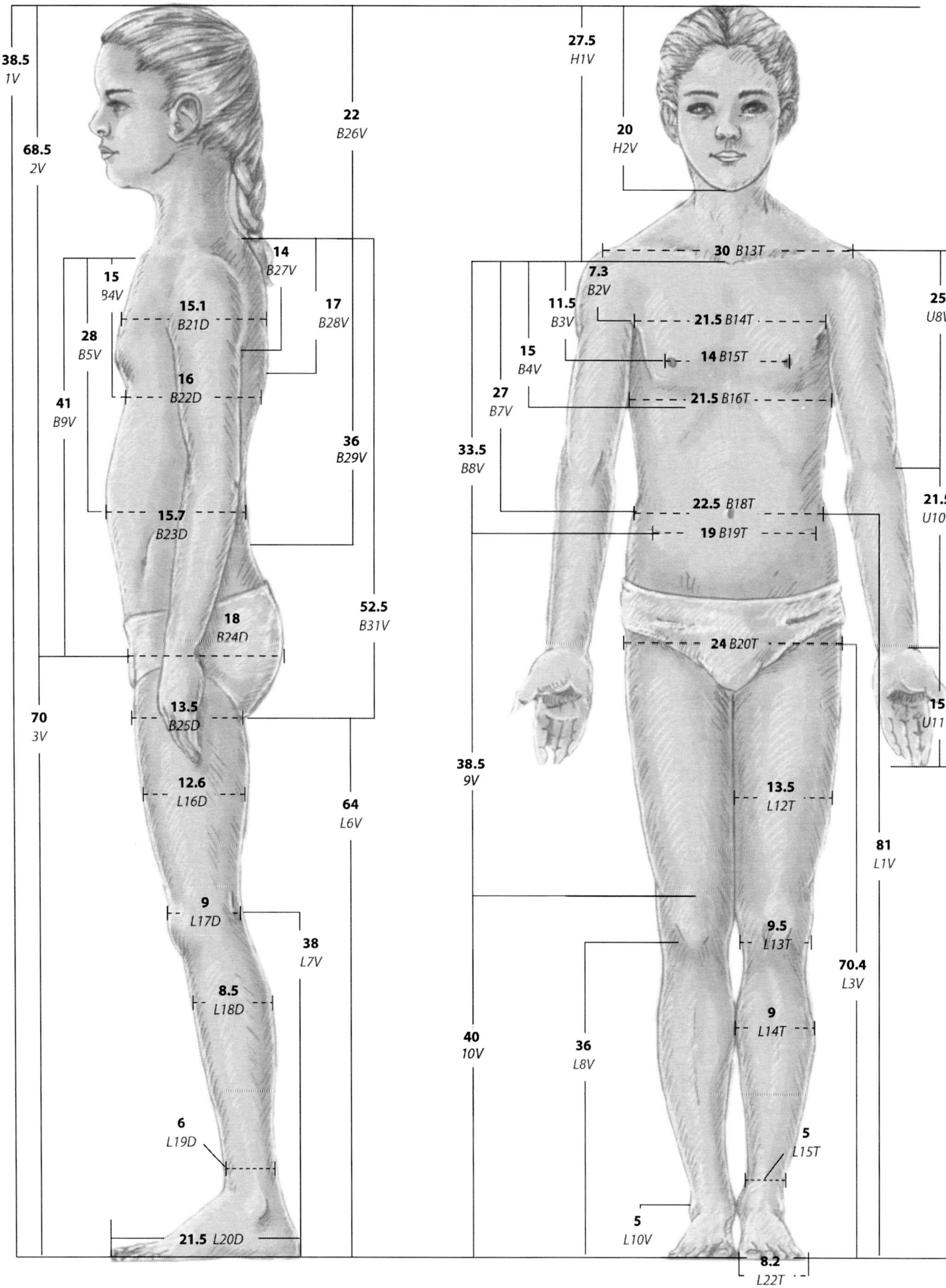

Life size			TEN-YEAR-OLD GIRL	Fractions of life size				
				3/4	2/3	1/2	1/3	1/4
Inches	cm	Code		cm	cm	cm	cm	cm
			VERTICAL MEASUREMENTS					
54.5	**138.5**	1 V	Stature, Height, Vertex to Sole	103.9	92.3	69.3	46.2	34.6
27.0	**68.5**	2 V	Vertex to Symphysis Pubis	51.4	45.7	34.3	22.8	17.1
27.6	**70.0**	* 3 V	Symphysis Pubis of Pelvis-Sole	52.5	46.7	35.0	23.3	17.5
43.7	**111.0**	4 V	Sternal Notch, Neck Pit, to Sole	83.3	74.0	55.5	37.0	27.8
10.4	**26.5**	5 V	Vertex to Acromion Process	19.9	17.7	13.3	8.8	6.6
44.1	**112.0**	6 V	Acromion Process to Sole	84.0	74.7	56.0	37.3	28.0
30.9	**78.5**	* 7 V	Ant.Sup.Spine of Ilium to Sole	58.9	52.3	39.3	26.2	19.6
13.2	**33.5**	8 V	Acrom. proc. to Ant. Sup. Spine	25.1	22.3	16.8	11.2	8.4
15.2	**38.5**	9 V	Ant.Sup.Spine of Ilium to Knee	28.9	25.7	19.3	12.8	9.6
15.7	**40.0**	10 V	Knee at Patella to Sole	30.0	26.7	20.0	13.3	10.0
32.7	**83.0**	11 V	Umbilicus to Sole	62.3	55.3	41.5	27.7	20.8
25.0	**63.5**	* 12 V	Crotch to Sole	47.6	42.3	31.8	21.2	15.9
40.2	**102.0**	13 V	Kneeling Height	76.5	68.0	51.0	34.0	25.5
28.5	**72.5**	14 V	Sitting Height	54.4	48.3	36.3	24.2	18.1
29.9	**76.0**	15 V	Buttock, Straight Leg to Sole	57.0	50.7	38.0	25.3	19.0
18.1	**46.0**	16 V	Buttock to Flexed Knee	34.5	30.7	23.0	15.3	11.5
17.3	**44.0**	17 V	Flexed Knee to Sole	33.0	29.3	22.0	14.7	11.0
		** These are also listed with Lower Extremity Measurements.*						
			VERTICAL HEAD MEASUREMENTS					
10.8	**27.5**	H1 V	Vertex to Sternal Notch	20.6	18.3	13.8	9.2	6.9
7.9	**20.0**	H2 V	Vertex to Chin, Head Height	15.0	13.3	10.0	6.7	5.0
6.6	**16.8**	H3 V	Hairline to Chin, Face Height	12.6	11.2	8.4	5.6	4.2
4.2	**10.6**	H4 V	Vertex to Lateral Corner of Eye	8.0	7.1	5.3	3.5	2.7
4.7	**12.0**	H5 V	Vertex to Tragion of Ear	9.0	8.0	6.0	4.0	3.0
2.1	**5.3**	H6 V	Hairline to Brow, Glabella	4.0	3.5	2.7	1.8	1.3
4.4	**11.3**	H7 V	Brow, Glabella, to Chin, Menton	8.5	7.5	5.7	3.8	2.8
2.2	**5.6**	H8 V	Brow to Subnasal Point	4.2	3.7	2.8	1.9	1.4
1.7	**4.2**	H9 V	Sellion to Subnasal, Nose Length	3.2	2.8	2.1	1.4	1.1
3.9	**9.8**	H10 V	Sellion, Nose Saddle, to Chin	7.4	6.5	4.9	3.3	2.5
2.2	**5.7**	H11 V	Subnasal Point to Chin	4.3	3.8	2.9	1.9	1.4
0.8	**2.0**	H12 V	Subnasal to Lip Aperature	1.5	1.3	1.0	0.7	0.5
0.5	**1.3**	H13 V	Lip Aperat. to Sublabial Crease	1.0	0.9	0.7	0.4	0.3
0.9	**2.4**	H14 V	Sublabial Crease to Chin	1.8	1.6	1.2	0.8	0.6
			TRANSVERSE HEAD MEASUREMENTS					
5.6	**14.3**	H15 T	Bi-Parietal, Head Breadth	10.7	9.5	7.2	4.8	3.6
4.8	**12.2**	H16 T	Bi-Zygomatic, Cheekbones	9.2	8.1	6.1	4.1	3.1
4.1	**10.5**	H17 T	Bi-Fronto Temporal, Forehead	7.9	7.0	5.3	3.5	2.6

Life size			TEN-YEAR-OLD GIRL	Fractions of life size				
				3/4	2/3	1/2	1/3	1/4
Inches	cm	Code		cm	cm	cm	cm	cm
			TRANSVERSE HEAD MEASUREMENTS					
3.7	**9.5**	H18 T	Inter-Brow Ridge	7.1	6.3	4.8	3.2	2.4
4.6	**11.6**	H19 T	Inter-Tragion of Ears	8.7	7.7	5.8	3.9	2.9
4.1	**10.5**	H20 T	Inter Angle of Mandible	7.9	7.0	5.3	3.5	2.6
3.5	**8.8**	H21 T	Inter Lateral Corners of Eyes	6.6	5.9	4.4	2.9	2.2
1.1	**2.9**	H22 T	Width of Eye	2.2	1.9	1.5	1.0	0.7
1.2	**3.0**	H23 T	Inter Inner Corner of Eyes	2.3	2.0	1.5	1.0	0.8
1.2	**3.0**	H24 T	Breadth of Nose	2.3	2.0	1.5	1.0	0.8
1.6	**4.1**	H25 T	Breadth of Mouth	3.1	2.7	2.1	1.4	1.0
1.2	**3.0**	H26 T	Breadth of Chin	2.3	2.0	1.5	1.0	0.8
3.3	**8.5**	H27 T	Breadth of Neck	6.4	5.7	4.3	2.8	2.1
			ANTERIOR-POSTERIOR HEAD MEASUREMENTS					
7.2	**18.4**	H28 D	Brow to Occiput, Posterior Head	13.8	12.3	9.2	6.1	4.6
3.7	**9.3**	H29 D	Brow to Tragus of Ear	7.0	6.2	4.7	3.1	2.3
3.4	**8.7**	H30 D	Sellion, Nose Saddle, to Tragus	6.5	5.8	4.4	2.9	2.2
3.6	**9.2**	H31 D	Tragus to Occiput, Back of Head	6.9	6.0	4.5	3.0	2.3
2.2	**5.5**	H32 V	Tragus to Angle Mandible	4.1	3.7	2.8	1.8	1.4
3.2	**8.2**	H33 D	Angle of Mandible to Chin	6.2	5.5	4.1	2.7	2.1
2.2	**5.7**	H34 V	Length of Ear, Vertical	4.3	3.8	2.9	1.9	1.4
1.3	**3.4**	H35 D	Width of Ear, Anterior-Posterior	2.6	2.3	1.7	1.1	0.9
3.4	**8.6**	H36 D	Anterior to Posterior of Neck	6.5	5.7	4.3	2.9	2.2
			VERTICAL BODY MEASUREMENTS					
3.0	**7.5**	B1 V	Chin to Sternal Notch, Neck Pit	5.6	5.0	3.8	2.5	1.9
2.9	**7.3**	B2 V	Sternal Notch to Axilla, Arm Pit	5.5	4.9	3.7	2.4	1.8
4.5	**11.5**	B3 V	Sternal Notch to Nipples	8.6	7.7	5.8	3.8	2.9
5.9	**15.0**	B4 V	Sternal Notch to Xyphoid Proc.	11.3	10.0	7.5	5.0	3.8
11.0	**28.0**	B5 V	Sternal Notch to Umbilicus	21.0	18.7	14.0	9.3	7.0
5.5	**14.0**	B6 V	Xyphoid Process to Umbilicus	10.5	9.3	7.0	4.7	3.5
10.6	**27.0**	B7 V	Sternal Notch to Crest Ilium	20.3	18.0	13.5	9.0	6.8
13.2	**33.5**	B8 V	Sternal Notch to Ant.S.Sp.Ilium	25.1	22.3	16.8	11.2	8.4
16.1	**41.0**	B9 V	Sternal Notch to Pubis	30.8	27.3	20.5	13.7	10.3
5.1	**13.0**	B10 V	Umbilicus to Pubis of Pelvis	9.8	8.7	6.5	4.3	3.3
			TRANSVERSE BODY MEASUREMENTS					
13.0	**33.0**	B11 T	Shoulder Breadth	24.8	22.0	16.5	11.0	8.3
5.9	**15.0**	B12 T	Clavicle to Acromion Process	11.3	10.0	7.5	5.0	3.8
11.8	**30.0**	B13 T	Biacromial Process Breadth	22.5	20.0	15.0	10.0	7.5
8.5	**21.5**	B14 T	Inter Anterior Axillary Fold	16.1	14.3	10.8	7.2	5.4
5.5	**14.0**	B15 T	Inter Nipples	10.5	9.3	7.0	4.7	3.5

Life size			TEN-YEAR-OLD GIRL	Fractions of life size				
				3/4	2/3	1/2	1/3	1/4
Inches	cm	Code		cm	cm	cm	cm	cm
			TRANSVERSE BODY MEASUREMENTS					
8.5	**21.5**	B16 T	Breadth of Thorax	16.1	14.3	10.8	7.2	5.4
8.3	**21.0**	B17 T	Inter-Costal. Rib Margin Breadth	15.8	14.0	10.5	7.0	5.3
8.9	**22.5**	B18 T	Inter Crest of Ilium Breadth	16.9	15.0	11.3	7.5	5.6
7.5	**19.0**	B19 T	Inter Anterior Superior Spine	14.3	12.7	9.5	6.3	4.8
9.4	**24.0**	B20 T	Inter Greater Trochanter Femur	18.0	16.0	12.0	8.0	6.0
			ANTERIOR-POSTERIOR BODY MEASUREMENTS					
5.9	**15.1**	B21 D	Manubrium to Dorsal Spine	11.3	10.1	7.6	5.0	3.8
6.3	**16.0**	B22 D	Xyphoid Process to Dorsum	12.0	10.7	8.0	5.3	4.0
6.2	**15.7**	B23 D	Umbilicus to Dorsum	11.8	10.5	7.9	5.2	3.9
7.1	**18.0**	B24 D	Pubis to Buttock	13.5	12.0	9.0	6.0	4.5
5.3	**13.5**	B25 D	Thigh to Gluteal Furrow	10.1	9.0	6.8	4.5	3.4
			VERTICAL BACK MEASUREMENTS					
8.7	**22.0**	B26 V	Vertex to 7th Cervical Spine	16.5	14.7	11.0	7.3	5.5
			7th Cervical Vertebra Spine to:					
5.5	**15.0**	B27 V	To Posterior Axillary Fold	10.5	9.3	7.0	4.7	3.5
6.7	**17.0**	B28 V	To Inferior Tip of the Scapula	12.8	11.3	8.5	5.7	4.3
14.2	**36.0**	B29 V	To Posterior Superior Spine Ilium	27.0	24.0	18.0	12.0	9.0
17.3	**44.0**	B30 V	To Midline Upper Buttock Crease	33.0	29.3	22.0	14.7	11.0
20.7	**52.5**	B31 V	To Gluteal Furrow	39.4	35.0	26.3	17.5	13.1
30.3	**77.0**	B32 V	To Popliteal Crease of Knee	57.8	51.3	38.5	25.7	19.3
			TRANSVERSE BACK MEASUREMENTS					
10.6	**27.0**	B33 T	Inter Posterior Axillary Folds	20.3	18.0	13.5	9.0	6.8
4.7	**12.0**	B34 T	Inter Scapulae Medial Borders	9.0	8.0	6.0	4.0	3.0
2.8	**7.0**	B35 T	Inter Post. Sup. Spines Sacrum	5.3	4.7	3.5	2.3	1.8
			UPPER EXTREMITY MEASUREMENTS					
53.9	**137.0**	U1 V	Span of Arms	102.8	91.3	68.5	45.7	34.3
5.8	**14.8**	U2 V	Sternum to Acromion Process	11.1	9.9	7.4	4.9	3.7
22.8	**58.0**	U3 V	Acromion Process to Fingertip	43.5	38.7	29.0	19.3	14.5
70.5	**179.0**	U4 V	Vertical Reach to Fingertip	134.3	119.3	89.5	59.7	44.8
33.5	**85.0**	U5 T	Lateral Reach to Fingertip	63.8	56.7	42.5	28.3	21.3
26.4	**67.0**	U6 D	Forward Reach to Fingertip	50.3	44.7	33.5	22.3	16.8
11.2	**28.5**	U7 V	Shoulder to Elbow Flexed	21.4	19.0	14.3	9.5	7.1
9.8	**25.0**	U8 V	Acromion to Lat. Condyle Elbow	18.8	16.7	12.5	8.3	6.3
14.4	**36.5**	U9 D	Flexed Elbow to Fingertip	27.4	24.3	18.3	12.2	9.1
8.5	**21.5**	U10 D	Flexed Elbow Pivot to Wrist Pivot	16.1	14.3	10.8	7.2	5.4

Life size			TEN-YEAR-OLD GIRL	Fractions of life size				
				3/4	2/3	1/2	1/3	1/4
Inches	cm	Code		cm	cm	cm	cm	cm
			UPPER EXTREMITY MEASUREMENTS					
5.9	**15.0**	U11 V	Wrist to Fingertip	11.3	10.0	7.5	5.0	3.8
3.9	**10.0**	U12 V	Wrist to Thumb Tip	7.5	6.7	5.0	3.3	2.5
3.3	**8.5**	U13 V	Length of Palm	6.4	5.7	4.3	2.8	2.1
2.6	**6.5**	U14 V	Length 3rd Finger	4.9	4.3	3.3	2.2	1.6
4.0	**10.2**	U15 V	Thumb Crotch to 3rd Finger	7.7	6.8	5.1	3.4	2.6
2.1	**5.3**	U16 V	Thumb Crotch to Thumb Tip	4.0	3.5	2.7	1.8	1.3
2.6	**6.5**	U17 D	Anterior-Post. Arm at Biceps	4.9	4.3	3.3	2.2	1.6
2.4	**6.0**	U18 D	Anterior-Posterior at Forearm	4.5	4.0	3.0	2.0	1.5
1.4	**3.5**	U19 D	Depth of Wrist	2.6	2.3	1.8	1.2	0.9
2.0	**5.0**	U20 T	Breadth of Wrist	3.8	3.3	2.5	1.7	1.3
3.1	**8.0**	U21 T	Breadth of Hand	6.0	5.3	4.0	2.7	2.0
2.6	**6.6**	U22 T	Breadth of Palm	5.0	4.4	3.3	2.2	1.7
			LOWER EXTREMITY MEASUREMENTS					
31.9	**81.0**	L1 V	Crest of Ilium of Pelvis to Sole	60.8	54.0	40.5	27.0	20.3
30.9	**78.5**	* L2 V	Ant.Sup.Spine of Ilium to Sole	58.9	52.3	39.3	26.2	19.6
27.7	**70.4**	L3 V	Greater Trochanter Femur-Sole	52.8	46.9	35.2	23.5	17.6
27.6	**70.0**	* L4 V	Symphysis Pubis of Pelvis-Sole	52.5	46.7	35.0	23.3	17.5
25.0	**63.5**	* L5 V	Crotch to Sole	47.6	42.3	31.8	21.2	15.9
25.2	**64.0**	L6 V	Gluteal Furrow to Sole	48.0	42.7	32.0	21.3	16.0
15.0	**38.0**	L7 V	Posterior Knee Crease to Sole	28.5	25.3	19.0	12.7	9.5
14.2	**36.0**	L8 V	Tibia at Knee to Sole	27.0	24.0	18.0	12.0	9.0
2.4	**6.0**	L9 V	Med. Malleolus of Ankle to Sole	4.5	4.0	3.0	2.0	1.5
2.0	**5.0**	L10 V	Lat. Malleolus of Ankle to Sole	3.8	3.3	2.5	1.7	1.3
10.4	**26.5**	L11 T	Breadth of Hips, Sitting	19.9	17.7	13.3	8.8	6.6
5.3	**13.5**	L12 T	Breadth Mid Thigh	10.1	9.0	6.8	4.5	3.4
3.7	**9.5**	L13 T	Breadth of Knee	7.1	6.3	4.8	3.2	2.4
3.5	**9.0**	L14 T	Breadth of Calf	6.8	6.0	4.5	3.0	2.3
2.0	**5.0**	L15 T	Breadth Ankle	3.8	3.3	2.5	1.7	1.3
5.0	**12.6**	L16 D	Mid Thigh, Anterior-Posterior	9.5	8.4	6.3	4.2	3.2
3.5	**9.0**	L17 D	Knee, Anterior-Posterior	6.8	6.0	4.5	3.0	2.3
3.3	**8.5**	L18 D	Mid Calf, Anterior-Posterior	6.4	5.7	4.3	2.8	2.1
2.4	**6.0**	L19 D	Ankle, Anterior-Posterior	4.5	4.0	3.0	2.0	1.5
8.5	**21.5**	L20 D	Length of Foot	16.1	14.3	10.8	7.2	5.4
6.3	**16.0**	L21 D	Heel to 5th Toe	12.0	10.7	8.0	5.3	4.0
3.2	**8.2**	L22 T	Breadth Foot	6.2	5.5	4.1	2.7	2.1
1.8	**4.5**	L23 T	Width of Heel	3.4	3.0	2.3	1.5	1.1
		** These are also listed with Vertical Measurements.*						

Ten-Year-Old Boy, anterior and lateral head

all measurements in centimeters

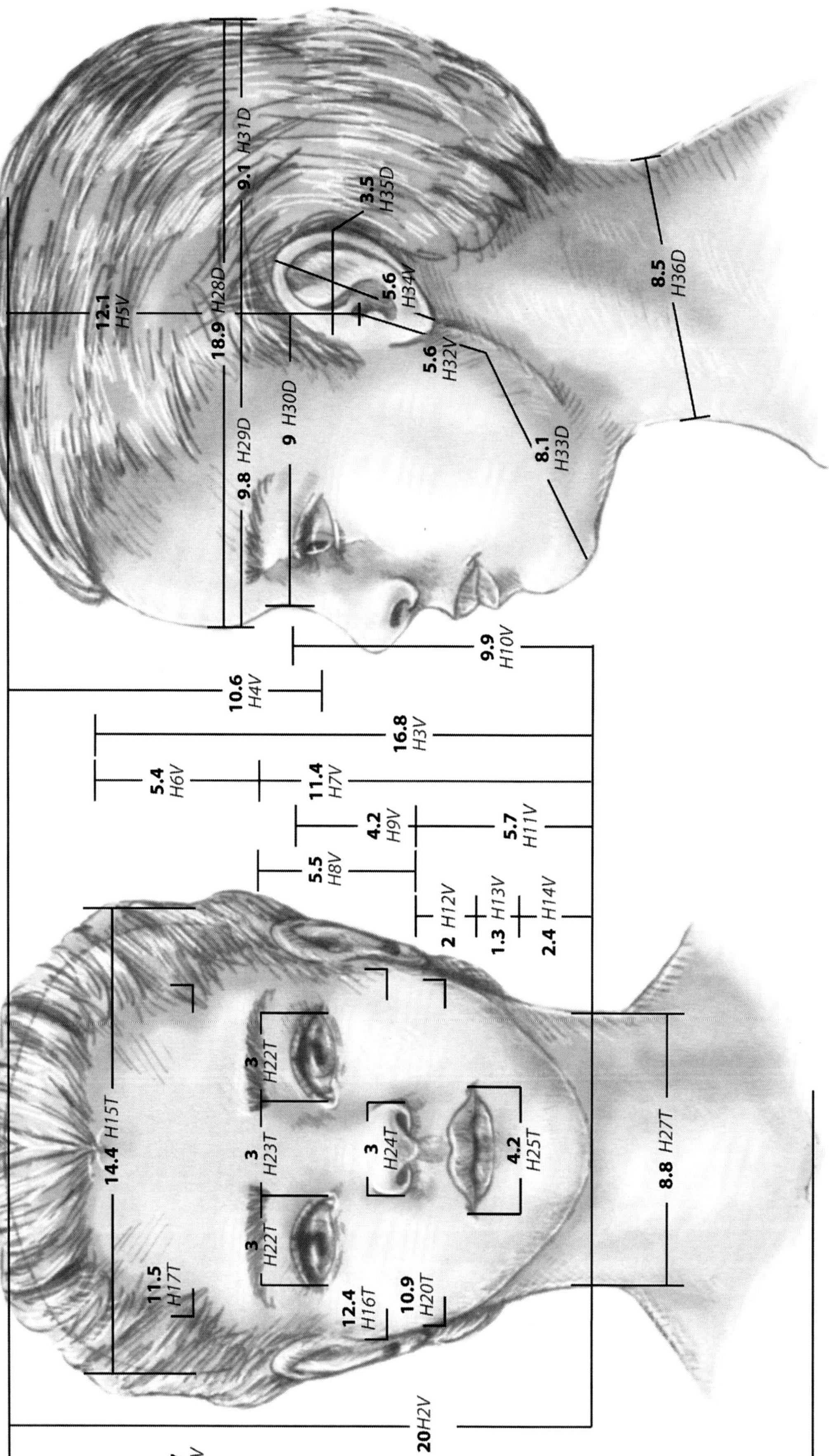

Ten-Year-Old Boy, anterior and lateral figure

all measurements in centimeters

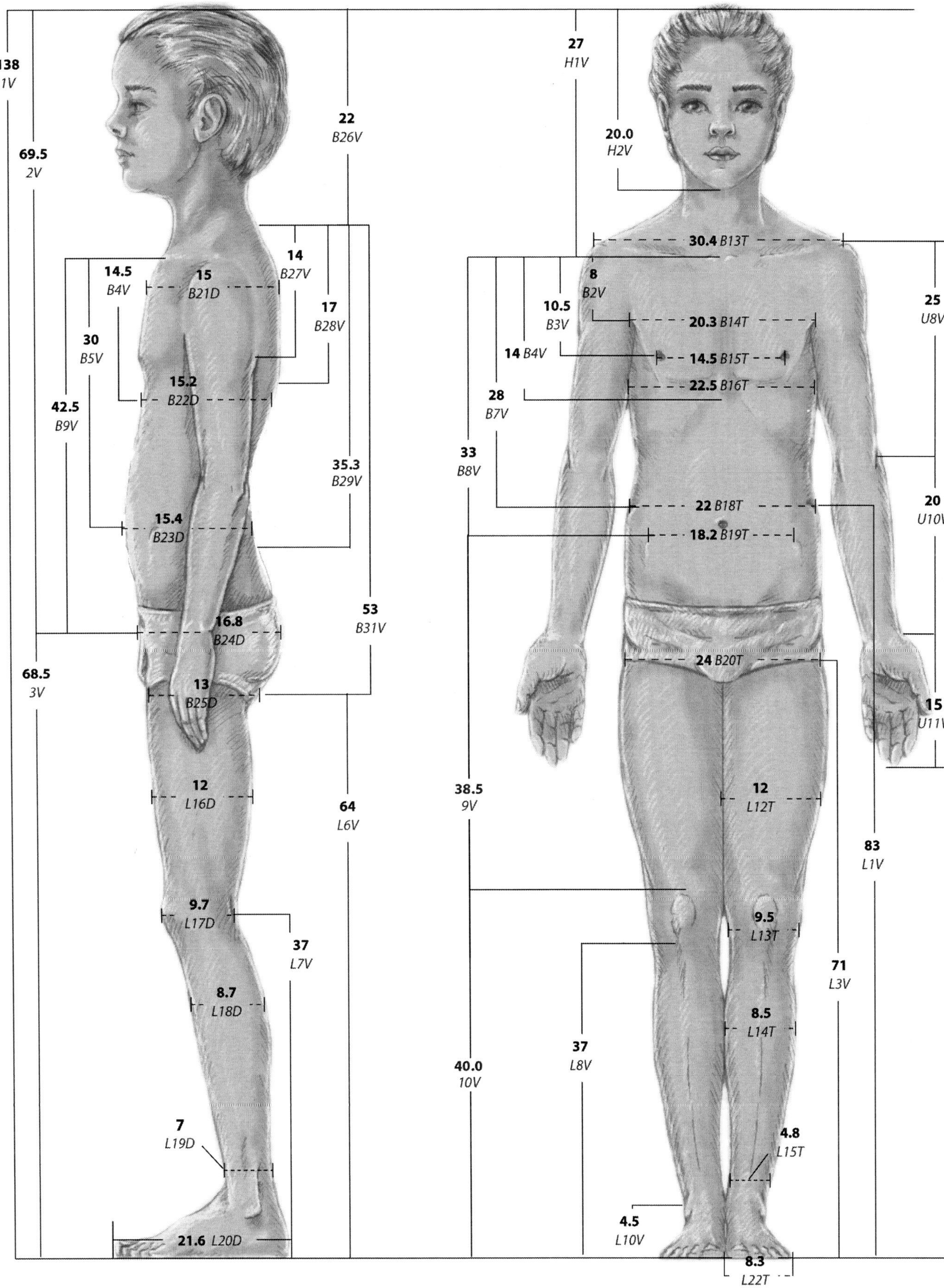

Life size			TEN-YEAR-OLD BOY	Fractions of life size				
				3/4	2/3	1/2	1/3	1/4
Inches	cm	Code		cm	cm	cm	cm	cm
			VERTICAL MEASUREMENTS					
54.3	**138.0**	1 V	Stature, Height, Vertex to Sole	103.5	92.0	69.0	46.0	34.5
27.4	**69.5**	2 V	Vertex to Symphysis Pubis	52.1	46.3	34.8	23.2	17.4
27.0	**68.5**	* 3 V	Symphysis Pubis of Pelvis-Sole	51.4	45.7	34.3	22.8	17.1
43.7	**111.0**	4 V	Sternal Notch, Neck Pit, to Sole	83.3	74.0	55.5	37.0	27.8
10.2	**26.0**	5 V	Vertex to Acromion Process	19.5	17.3	13.0	8.7	6.5
44.1	**112.0**	6 V	Acromion Process to Sole	84.0	74.7	56.0	37.3	28.0
30.7	**78.0**	* 7 V	Ant.Sup.Spine of Ilium to Sole	58.5	52.0	39.0	26.0	19.5
13.4	**34.0**	8 V	Acrom. proc. to Ant. Sup. Spine	25.5	22.7	17.0	11.3	8.5
15.2	**38.5**	9 V	Ant.Sup.Spine of Ilium to Knee	28.9	25.7	19.3	12.8	9.6
15.7	**40.0**	10 V	Knee at Patella to Sole	30.0	26.7	20.0	13.3	10.0
31.9	**81.0**	11 V	Umbilicus to Sole	60.8	54.0	40.5	27.0	20.3
25.0	**63.5**	* 12 V	Crotch to Sole	47.6	42.3	31.8	21.2	15.9
40.6	**103.0**	13 V	Kneeling Height	77.3	68.7	51.5	34.3	25.8
28.5	**72.3**	14 V	Sitting Height	54.2	48.2	36.2	24.1	18.1
29.1	**74.0**	15 V	Buttock, Straight Leg to Sole	55.5	49.3	37.0	24.7	18.5
17.9	**45.4**	16 V	Buttock to Flexed Knee	34.1	30.3	22.7	15.1	11.4
17.2	**43.6**	17 V	Flexed Knee to Sole	32.7	29.1	21.8	14.5	10.9
		** These are also listed with Lower Extremity Measurements.*						
			VERTICAL HEAD MEASUREMENTS					
10.6	**27.0**	H1 V	Vertex to Sternal Notch	20.3	18.0	13.5	9.0	6.8
7.9	**20.0**	H2 V	Vertex to Chin, Head Height	15.0	13.3	10.0	6.7	5.0
6.6	**16.8**	H3 V	Hairline to Chin, Face Height	12.6	11.2	8.4	5.6	4.2
4.2	**10.6**	H4 V	Vertex to Lateral Corner of Eye	8.0	7.1	5.3	3.5	2.7
4.8	**12.1**	H5 V	Vertex to Tragion of Ear	9.1	8.1	6.1	4.0	3.0
2.1	**5.4**	H6 V	Hairline to Brow, Glabella	4.1	3.6	2.7	1.8	1.4
4.5	**11.4**	H7 V	Brow, Glabella, to Chin, Menton	8.6	7.6	5.7	3.8	2.9
2.2	**5.5**	H8 V	Brow to Subnasal Point	4.1	3.7	2.8	1.8	1.4
1.7	**4.2**	H9 V	Sellion to Subnasal, Nose Length	3.2	2.8	2.1	1.4	1.1
3.9	**9.9**	H10 V	Sellion, Nose Saddle, to Chin	7.4	6.6	5.0	3.3	2.5
2.2	**5.7**	H11 V	Subnasal point to Chin	4.3	3.8	2.9	1.9	1.4
0.8	**2.0**	H12 V	Subnasal to Lip Aperature	1.5	1.3	1.0	0.7	0.5
0.5	**1.3**	H13 V	Lip Aperat. to Sublabial Crease	1.0	0.9	0.7	0.4	0.3
0.9	**2.4**	H14 V	Sublabial Crease to Chin	1.8	1.6	1.2	0.8	0.6
			TRANSVERSE HEAD MEASUREMENTS					
5.7	**14.4**	H15 T	Bi-Parietal, Head Breadth	10.8	9.6	7.2	4.8	3.6
4.9	**12.4**	H16 T	Bi-Zygomatic, Cheekbones	9.3	8.3	6.2	4.1	3.1
4.4	**11.2**	H17 T	Bi-Fronto Temporal, Forehead	8.4	7.5	5.6	3.7	2.8

Life size			TEN-YEAR-OLD BOY	Fractions of life size				
				3/4	2/3	1/2	1/3	1/4
Inches	cm	Code		cm	cm	cm	cm	cm
			TRANSVERSE HEAD MEASUREMENTS					
3.6	**9.2**	H18 T	Inter-Brow Ridge	6.9	6.1	4.6	3.1	2.3
4.7	**12.0**	H19 T	Inter-Tragion of Ears	9.0	8.0	6.0	4.0	3.0
4.3	**10.9**	H20 T	Inter-Angle of Mandible	8.2	7.3	5.5	3.6	2.7
3.5	**9.0**	H21 T	Inter-Lateral Corners of Eyes	6.8	6.0	4.5	3.0	2.3
1.2	**3.0**	H22 T	Width of Eye	2.3	2.0	1.5	1.0	0.8
1.2	**3.0**	H23 T	Inter-Inner Corner of Eyes	2.3	2.0	1.5	1.0	0.8
1.2	**3.0**	H24 T	Breadth of Nose	2.3	2.0	1.5	1.0	0.8
1.7	**4.2**	H25 T	Breadth of Mouth	3.2	2.8	2.1	1.4	1.1
1.3	**3.2**	H26 T	Breadth of Chin	2.4	2.1	1.6	1.1	0.8
3.5	**8.8**	H27 T	Breadth of Neck	6.6	5.9	4.4	2.9	2.2
			ANTERIOR-POSTERIOR HEAD MEASUREMENTS					
7.4	**18.9**	H28 D	Brow to Occiput, Poserior Head	14.2	12.6	9.5	6.3	4.7
3.9	**9.8**	H29 D	Brow to Tragus of Ear	7.4	6.5	4.9	3.3	2.5
3.5	**9.0**	H30 D	Sellion, Nose Saddle, to Tragus	6.8	6.0	4.5	3.0	2.3
3.6	**9.1**	H31 D	Tragus to Occiput, Back of Head	6.8	6.1	4.6	3.0	2.3
2.2	**5.6**	H32 V	Tragus to Angle Mandible	4.2	3.7	2.8	1.9	1.4
3.2	**8.1**	H33 D	Angle of Mandible to Chin	6.1	5.4	4.1	2.7	2.0
2.2	**5.6**	H34 V	Length of Ear, Vertical	4.2	3.7	2.8	1.9	1.4
1.4	**3.5**	H35 D	Width of Ear, Anterior-Posterior	2.6	2.3	1.8	1.2	0.9
3.3	**8.5**	H36 D	Anterior to Posterior of Neck	6.4	5.7	4.3	2.8	2.1
			VERTICAL BODY MEASUREMENTS					
3.0	**7.5**	B1 V	Chin to Sternal Notch, Neck Pit	5.6	5.0	3.8	2.5	1.9
3.1	**8.0**	B2 V	Sternal Notch to Axilla, Arm Pit	6.0	5.3	4.0	2.7	2.0
4.3	**11.0**	B3 V	Sternal Notch to Nipples	8.3	7.3	5.5	3.7	2.8
5.7	**14.5**	B4 V	Sternal Notch to Xyphoid Proc.	10.9	9.7	7.3	4.8	3.6
11.8	**30.0**	B5 V	Sternal Notch to Umbilicus	22.5	20.0	15.0	10.0	7.5
5.9	**15.0**	B6 V	Xyphoid Process to Umbilicus	11.3	10.0	7.5	5.0	3.8
11.0	**28.0**	B7 V	Sternal Notch to Crest Ilium	21.0	18.7	14.0	9.3	7.0
13.0	**33.0**	B8 V	Sternal Notch to Ant.S.Sp.Ilium	24.8	22.0	16.5	11.0	8.3
16.7	**42.5**	B9 V	Sternal Notch to Pubis	31.9	28.3	21.3	14.2	10.6
4.7	**12.0**	B10 V	Umbilicus to Pubis of Pelvis	9.0	8.0	6.0	4.0	3.0
			TRANSVERSE BODY MEASUREMENTS					
13.0	**33.0**	B11 T	Shoulder Breadth	24.8	22.0	16.5	11.0	8.3
5.8	**14.7**	B12 T	Clavicle to Acromion Process	11.0	9.8	7.4	4.9	3.7
12.0	**30.4**	B13 T	Bi-acromial Process Breadth	22.8	20.3	15.2	10.1	7.6
8.0	**20.3**	B14 T	Inter-Anterior Axillary Fold	15.2	13.5	10.2	6.8	5.1
5.7	**14.5**	B15 T	Inter-Nipples	10.9	9.7	7.3	4.8	3.6

Life size			TEN-YEAR-OLD BOY	Fractions of life size				
				3/4	2/3	1/2	1/3	1/4
Inches	cm	Code		cm	cm	cm	cm	cm
			TRANSVERSE BODY MEASUREMENTS					
8.9	**22.5**	B16 T	Breadth of Thorax	16.9	15.0	11.3	7.5	5.6
8.5	**21.5**	B17 T	Inter-Costal. Rib Margin Breadth	16.1	14.3	10.8	7.2	5.4
8.7	**22.0**	B18 T	Inter-Crest of Ilium Breadth	16.5	14.7	11.0	7.3	5.5
7.2	**18.2**	B19 T	Inter-Anterior Superior Spine	13.7	12.1	9.1	6.1	4.6
9.4	**24.0**	B20 T	Inter-Greater Trochanter Femur	18.0	16.0	12.0	8.0	6.0
			ANTERIOR-POSTERIOR BODY MEASUREMENTS					
5.9	**15.0**	B21 D	Manubrium to Dorsal Spine	11.3	10.0	7.5	5.0	3.8
6.0	**15.2**	B22 D	Xyphoid Process to Dorsum	11.4	10.1	7.6	5.1	3.8
6.1	**15.4**	B23 D	Umbilicus to Dorsum	11.6	10.3	7.7	5.1	3.9
6.6	**16.8**	B24 D	Pubis of Pelvis to Buttock	12.6	11.2	8.4	5.6	4.2
5.1	**13.0**	B25 D	Thigh to Gluteal Furrow	9.8	8.7	6.5	4.3	3.3
			VERTICAL BACK MEASUREMENTS					
8.7	**22.0**	B26 V	Vertex to 7th Cervical Spine	16.5	14.7	11.0	7.3	5.5
			7th Cervical Vertebra Spine to:					
5.5	**14.0**	B27 V	To Posterior Axillary Fold, Armpit	10.5	9.3	7.0	4.7	3.5
6.7	**17.0**	B28 V	To Inferior Tip of Scapula	12.8	11.3	8.5	5.7	4.3
13.9	**35.3**	B29 V	To Posterior Superior Spine Ilium	26.5	23.5	17.7	11.8	8.8
16.5	**42.0**	B30 V	To Midline Upper Buttock Crease	31.5	28.0	21.0	14.0	10.5
20.9	**53.0**	B31 V	To Gluteal Furrow	39.8	35.3	26.5	17.7	13.3
31.1	**79.0**	B32 V	To Popliteal Crease Behind Knee	59.3	52.7	39.5	26.3	19.8
			TRANSVERSE BACK MEASUREMENTS					
10.9	**27.8**	B33 T	Inter-Posterior Axillary Folds	20.9	18.5	13.9	9.3	7.0
4.7	**12.0**	B34 T	Inter-Scapulae Medial Borders	9.0	8.0	6.0	4.0	3.0
2.8	**7.0**	B35 T	Inter-Post. Sup. Spines Sacrum	5.3	4.7	3.5	2.3	1.8
			UPPER EXTREMITY MEASUREMENTS					
53.5	**136.0**	U1 V	Span of Arms	102.0	90.7	68.0	45.3	34.0
5.8	**14.7**	U2 V	Sternum to Acromion Process	11.0	9.8	7.4	4.9	3.7
24.4	**62.0**	U3 V	Acromion Process to Fingertip	46.5	41.3	31.0	20.7	15.5
69.3	**176.0**	U4 V	Vertical Reach to Fingertip	132.0	117.3	88.0	58.7	44.0
34.2	**86.9**	U5 T	Lateral Reach to Fingertip	65.2	57.9	43.5	29.0	21.7
26.4	**67.0**	U6 D	Forward Reach to Fingertip	50.3	44.7	33.5	22.3	16.8
11.2	**28.5**	U7 V	Shoulder to Flexed Elbow	21.4	19.0	14.3	9.5	7.1
10.4	**26.5**	U8 V	Acromion to Lat. Condyle Elbow	19.9	17.7	13.3	8.8	6.6
14.4	**36.7**	U9 D	Flexed Elbow to Fingertip	27.5	24.5	18.4	12.2	9.2
8.1	**20.7**	U10 D	Flexed Elbow Pivot to Wrist Pivot	15.5	13.8	10.4	6.9	5.2

Life size			TEN-YEAR-OLD BOY	Fractions of life size				
				3/4	2/3	1/2	1/3	1/4
Inches	cm	Code		cm	cm	cm	cm	cm
			UPPER EXTREMITY MEASUREMENTS					
5.9	**15.0**	U11 V	Wrist to Fingertip	11.3	10.0	7.5	5.0	3.8
3.9	**9.8**	U12 V	Wrist to Thumb Tip	7.4	6.5	4.9	3.3	2.5
3.4	**8.6**	U13 V	Length of Palm	6.5	5.7	4.3	2.9	2.2
2.5	**6.4**	U14 V	Length 3rd Finger	4.8	4.3	3.2	2.1	1.6
4.3	**10.9**	U15 V	Thumb Crotch to 3rd Finger	8.2	7.3	5.5	3.6	2.7
2.1	**5.3**	U16 V	Thumb Crotch to Thumb Tip	4.0	3.5	2.7	1.8	1.3
2.6	**6.5**	U17 D	Anterior-Post. Arm at Biceps	4.9	4.3	3.3	2.2	1.6
2.4	**6.0**	U18 D	Anterior-Posterior at Forearm	4.5	4.0	3.0	2.0	1.5
1.3	**3.3**	U19 D	Depth of Wrist	2.5	2.2	1.7	1.1	0.8
2.0	**5.0**	U20 T	Breadth of Wrist	3.8	3.3	2.5	1.7	1.3
3.3	**8.5**	U21 T	Breadth of Hand	6.4	5.7	4.3	2.8	2.1
2.7	**6.9**	U22 T	Breadth of Palm	5.2	4.6	3.5	2.3	1.7
			LOWER EXTREMITY MEASUREMENTS					
32.7	**83.0**	L1 V	Crest of Ilium of Pelvis to Sole	62.3	55.3	41.5	27.7	20.8
30.7	**78.0**	* L2 V	Ant.Sup.Spine of Ilium to Sole	58.5	52.0	39.0	26.0	19.5
28.0	**71.0**	L3 V	Greater Trochanter Femur-Sole	53.3	47.3	35.5	23.7	17.8
27.4	**69.5**	* L4 V	Symphysis Pubis of Pelvis-Sole	52.1	46.3	34.8	23.2	17.4
25.0	**63.5**	* L5 V	Crotch to Sole	47.6	42.3	31.8	21.2	15.9
25.8	**65.5**	L6 V	Gluteal Furrow to Sole	49.1	43.7	32.8	21.8	16.4
15.0	**38.0**	L7 V	Posterior Knee Crease to Sole	28.5	25.3	19.0	12.7	9.5
14.6	**37.0**	L8 V	Tibia at Knee to Sole	27.8	24.7	18.5	12.3	9.3
2.3	**5.8**	L9 V	Med. Malleolus of Ankle to Sole	4.4	3.9	2.9	1.9	1.5
2.0	**5.0**	L10 V	Lat. Malleolus of Ankle to Sole	3.8	3.3	2.5	1.7	1.3
9.8	**25.0**	L11 T	Breadth of Hips, Sitting	18.8	16.7	12.5	8.3	6.3
4.7	**12.0**	L12 T	Breadth Mid Thigh	9.0	8.0	6.0	4.0	3.0
3.7	**9.5**	L13 T	Breadth of Knee	7.1	6.3	4.8	3.2	2.4
3.3	**8.5**	L14 T	Breadth of Calf	6.4	5.7	4.3	2.8	2.1
1.9	**4.8**	L15 T	Breadth Ankle	3.6	3.2	2.4	1.6	1.2
4.7	**12.0**	L16 D	Mid Thigh, Anterior-Posterior	9.0	8.0	6.0	4.0	3.0
3.8	**9.7**	L17 D	Knee, Anterior-Posterior	7.3	6.5	4.9	3.2	2.4
3.4	**8.7**	L18 D	Mid Calf, Anterior-Posterior	6.5	5.8	4.4	2.9	2.2
2.8	**7.0**	L19 D	Ankle, Anterior-Posterior	5.3	4.7	3.5	2.3	1.8
8.5	**21.6**	L20 D	Length of Foot	16.2	14.4	10.8	7.2	5.4
6.3	**16.0**	L21 D	Heel to 5th Toe	12.0	10.7	8.0	5.3	4.0
3.3	**8.3**	L22 T	Breadth Foot	6.2	5.5	4.2	2.8	2.1
1.8	**4.6**	L23 T	Width of Heel	3.5	3.1	2.3	1.5	1.2
		** These are also listed with Vertical Measurements.*						

ELEVEN-YEAR-OLD GIRLS

There is less similarity between boys and girls at age eleven years. The practical lists of body proportions are shown separately in the tables of measurements. The range of twenty-fifth to fiftieth and to seventy-fifth percentile is listed here for comparison:

Percentile	Twenty-fifth		Fiftieth		Seventy-fifth	
Girls:	Short		Medium		Tall	
Stature	140 cm,	55 in.	145 cm,	57 in.	149 cm,	56.3 in.
Weight	32.5 kg,	71.8 lb.	40 kg,	81.5 lb.	42.8 kg,	94.5 lb.[1]

At this age, girls have begun to grow faster than boys. Girls begin to mature earlier and may gain weight and height more rapidly than boys of the same age. Many have experienced menarche and an early stage of sexual maturity. Breast development is about phase II with some development under the nipples. Many of the girls have early feminine contours in face shoulders and hips. Other noticeable difference is hair styling, dress, and interests in athletics. School level is sixth grade.

BODY MEASUREMENTS COMPARED TO HEIGHT

	Feature Measure in cm	Feature Compared to Stature	Approx. Ratio of Feature	Compared to Median Adult
Stature: Crown to sole	**145**			**86%**
Head Height: Crown to chin	**19.8**	**14%**	**1/7**	**97%**
Sitting Height:	**75.5**	**52%**	**1/2**	**85%**
Trunk Measure: Sternal notch to symphysis pubis	**44.5**	**31%**	**1/3–**	**87%**
Leg Length: Greater trochanter of the femur to sole	**75**	**52%**	**1/2**	**87%**
Thigh Length: Greater trochanter of the femur to tibia at knee	**37**	**26%**	**1/4**	**86%**
Tibia at Knee to Sole:	**38**	**26%**	**1/4**	**88%**
Arm Length: Acromion proccss at shoulder to dependent fingertip	**61**	**42%**	**2/5**	**85%**
Hand Length:	**16**	**11%**	**1/9**	**89%**
Foot Length:	**23**	**15%**	**1/6**	**96%**

The mask of the face, brow to chin, continues to grow as dentition increases. The first permanent upper and lower premolars may erupt at ten to twelve years of age, followed soon by the second upper and lower premolars.[2]

1. Nelson, W.E., p. 39.
2. Lockhart, R.D., p. 57.

ELEVEN-YEAR-OLD BOYS

The list of body proportions, for each sex is shown separately in the tables of measurements. The range of twenty-fifth to fiftieth and to seventy-fifth percentile is listed here for comparison:

Percentile	Twenty-fifth		Fiftieth		Seventy-fifth	
Boys:	Short		Medium		Tall	
Stature	138.7 cm,	54.5 in.	143.3 cm,	56.5 in.	147.8 cm,	58.3 in.
Weight	31.3 kg,	69 lb	35.3 kg,	77.8 lb.	40.4 kg,	89 lb.[1]

In younger ages, boys were on average, slightly taller and slightly heavier than girls, but girls begin to have more rapid growth as they begin to mature. They are usually taller and weigh more than boys. At this age, boys demonstrate more interest and participation in adult type of athletics like Little League. Cub Scout activities and out door adventures are popular. Muscular development begins to increases with activity. The school level is sixth grade.

BODY MEASUREMENTS COMPARED TO HEIGHT

	Feature Measure in cm	Feature Compared to Stature	Approx. Ratio of Feature	Compared to Median Adult
Stature: Crown to sole	143			78%
Head Height: Crown to chin	20	14%	1/7	87%
Sitting Height:	74	52%	1/2	77%
Trunk Measure: Sternal notch to symphysis pubis	42	29%	1/3	79%
Leg Length: Greater trochanter of the femur to sole	74	52%	1/2	76%
Thigh Length: Greater trochanter of the femur to tibia at knee	36	25%	1/4	77%
Tibia at Knee to Sole:	38	27%	1/4	74%
Arm Length: Acromion process at shoulder to dependent fingertip	62.5	44%	2/5+	78%
Hand Length:	15.6	11%	1/9	78%
Foot Length:	22.5	16%	1/6	79%

The mask of the face, brow to chin, grows more as dentition increases. The first permanent upper and lower premolar teeth may erupt at ten to twelve years of age, followed soon by the second upper and lower premolars.[2]

1. Nelson, W.E., p. 38.
2. Lockhart, R.D., p. 57.

Eleven-Year-Old Girl, anterior and lateral head

all measurements in centimeters

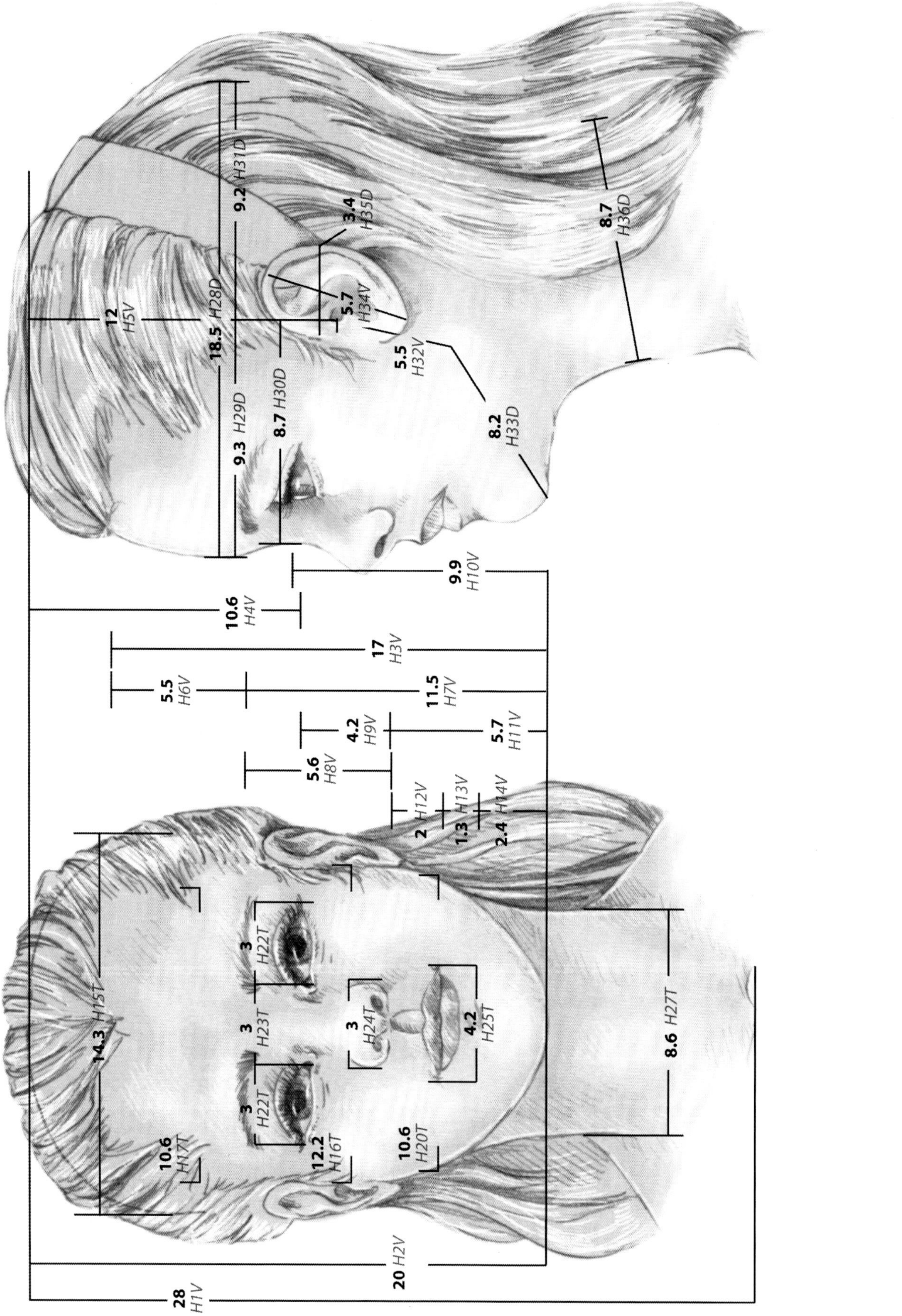

Eleven-Year-Old Girl, anterior and lateral figure

all measurements in centimeters

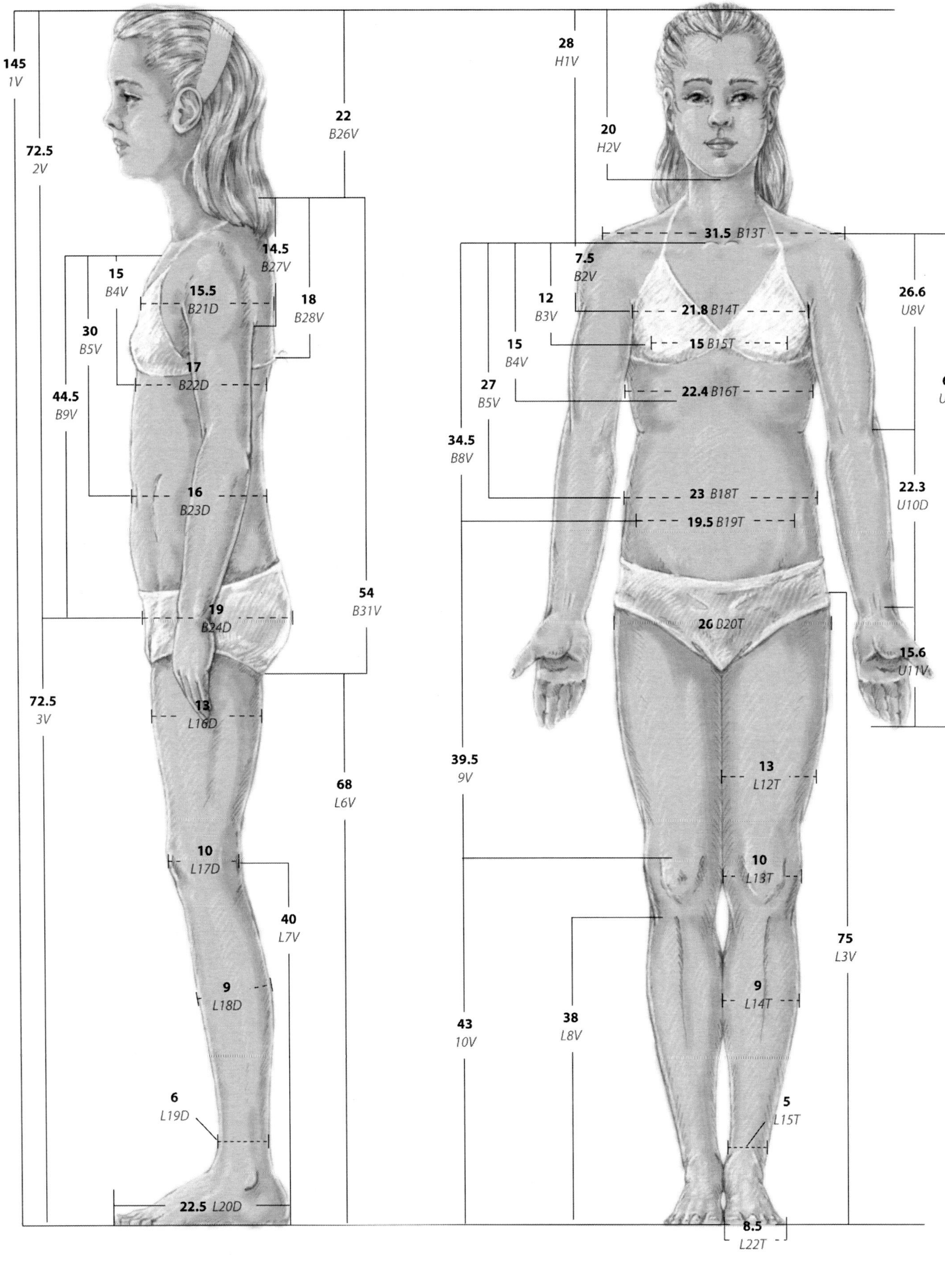

Life size			ELEVEN-YEAR-OLD GIRL	Fractions of life size				
				3/4	2/3	1/2	1/3	1/4
Inches	cm	Code		cm	cm	cm	cm	cm
			VERTICAL MEASUREMENTS					
57.1	**145.0**	1 V	Stature, Height, Vertex to Sole	108.8	96.7	72.5	48.3	36.3
28.5	**72.5**	2 V	Vertex to Symphysis Pubis	54.4	48.3	36.3	24.2	18.1
28.5	**72.5**	* 3 V	Symphysis Pubis to Sole	54.4	48.3	36.3	24.2	18.1
46.1	**117.0**	4 V	Sternal Notch to Sole	87.8	78.0	58.5	39.0	29.3
11.0	**28.0**	5 V	Vertex to Acromion Process	21.0	18.7	14.0	9.3	7.0
46.1	**117.0**	6 V	Acromion Process to Sole	87.8	78.0	58.5	39.0	29.3
32.5	**82.5**	* 7 V	Ant.Sup.Spine of Ilium to Sole	61.9	55.0	41.3	27.5	20.6
13.6	**34.5**	8 V	Acromion pr. to Ant. Sup. Spine	25.9	23.0	17.3	11.5	8.6
15.6	**39.5**	9 V	Ant. Sup. Spine of Ilium to Knee	29.6	26.3	19.8	13.2	9.9
16.9	**43.0**	10 V	Knee at Patella to Sole	32.3	28.7	21.5	14.3	10.8
34.3	**87.0**	11 V	Umbilicus to Sole	65.3	58.0	43.5	29.0	21.8
25.8	**65.5**	* 12 V	Crotch to Sole	49.1	43.7	32.8	21.8	16.4
42.9	**109.0**	13 V	Kneeling Height	81.8	72.7	54.5	36.3	27.3
29.7	**75.5**	14 V	Sitting Height	56.6	50.3	37.8	25.2	18.9
32.7	**83.0**	15 V	Buttock, Straight Leg to Sole	62.3	55.3	41.5	27.7	20.8
19.1	**48.5**	16 V	Buttock to Flexed Knee	36.4	32.3	24.3	16.2	12.1
18.1	**46.0**	17 V	Flexed Knee to Sole	34.5	30.7	23.0	15.3	11.5
		** These are also listed with Lower Extremity Measurements.*						
			VERTICAL HEAD MEASUREMENTS					
11.0	**28.0**	H1 V	Vertex to Sternal Notch	21.0	18.7	14.0	9.3	7.0
7.9	**20.0**	H2 V	Vertex to Chin, Head Height	15.0	13.3	10.0	6.7	5.0
6.7	**17.0**	H3 V	Hairline to Chin, Face Height	12.8	11.3	8.5	5.7	4.3
4.2	**10.6**	H4 V	Vertex to Lateral Corner of Eye	8.0	7.1	5.3	3.5	2.7
4.7	**12.0**	H5 V	Vertex to Tragion of Ear	9.0	8.0	6.0	4.0	3.0
2.2	**5.5**	H6 V	Hairline to Brow, Glabella	4.1	3.7	2.8	1.8	1.4
4.5	**11.5**	H7 V	Brow, Glabella, to Chin, Menton	8.6	7.7	5.8	3.8	2.9
2.2	**5.6**	H8 V	Brow to Subnasal Point	4.2	3.7	2.8	1.9	1.4
1.7	**4.2**	H9 V	Sellion to Subnasal, Nose Length	3.2	2.8	2.1	1.4	1.1
3.9	**9.9**	H10 V	Sellion, Nose Saddle, to Chin	7.4	6.6	5.0	3.3	2.5
2.2	**5.7**	H11 V	Subnasal point to Chin	4.3	3.8	2.9	1.9	1.4
0.8	**2.0**	H12 V	Subnasal to Lip Aperature	1.5	1.3	1.0	0.7	0.5
0.5	**1.3**	H13 V	Lip Aperat. to Sublabial Crease	1.0	0.9	0.7	0.4	0.3
0.9	**2.4**	H14 T	Sublabial Crease to Chin	1.8	1.6	1.2	0.8	0.6
			TRANSVERSE HEAD MEASUREMENTS					
5.6	**14.3**	H15 T	Bi-Parietal, Head Breadth	10.7	9.5	7.2	4.8	3.6
4.8	**12.2**	H16 T	Bi-Zygomatic, Cheekbones	9.2	8.1	6.1	4.1	3.1
4.2	**10.6**	H17 T	Bi-Fronto Temporal, Forehead	8.0	7.1	5.3	3.5	2.7

Life size			ELEVEN-YEAR-OLD GIRL	Fractions of life size				
				3/4	2/3	1/2	1/3	1/4
Inches	cm	Code		cm	cm	cm	cm	cm
			TRANSVERSE HEAD MEASUREMENTS					
3.8	**9.7**	H18 T	Inter Brow Ridge	7.3	6.5	4.9	3.2	2.4
4.6	**11.7**	H19 T	Inter Tragion of Ears	8.8	7.8	5.9	3.9	2.9
4.2	**10.6**	H20 T	Inter Angle of Mandible	8.0	7.1	5.3	3.5	2.7
3.5	**9.0**	H21 T	Inter Lateral Corners of Eyes	6.8	6.0	4.5	3.0	2.3
1.2	**3.0**	H22 T	Width of Eye	2.3	2.0	1.5	1.0	0.8
1.2	**3.0**	H23 T	Inter Inner Corner of eyes	2.3	2.0	1.5	1.0	0.8
1.2	**3.0**	H24 T	Breadth of Nose	2.3	2.0	1.5	1.0	0.8
1.7	**4.2**	H25 T	Breadth of Mouth	3.2	2.8	2.1	1.4	1.1
1.3	**3.2**	H26 T	Breadth of Chin	2.4	2.1	1.6	1.1	0.8
3.4	**8.6**	H27 T	Breadth of Neck	6.5	5.7	4.3	2.9	2.2
			ANTERIOR-POSTERIOR HEAD MEASUREMENTS					
7.3	**18.5**	H28 D	Brow to Occiput	13.9	12.3	9.3	6.2	4.6
3.7	**9.3**	H29 D	Brow to Tragus	7.0	6.2	4.7	3.1	2.3
3.4	**8.7**	H30 D	Sellion to Tragus	6.5	5.8	4.4	2.9	2.2
3.6	**9.2**	H31 D	Tragus to Occiput	6.9	6.1	4.6	3.1	2.3
2.2	**5.5**	H32 V	Tragus to Angle Mandible	4.1	3.7	2.8	1.8	1.4
3.2	**8.2**	H33 D	Angle of Mandible to Chin	6.2	5.5	4.1	2.7	2.1
2.2	**5.7**	H34 V	Length of Ear, Vertical	4.3	3.8	2.9	1.9	1.4
1.3	**3.4**	H35 D	Width of Ear, Anterior-Posterior	2.6	2.3	1.7	1.1	0.9
3.4	**8.7**	H36 D	Anterior to Posterior of Neck	6.5	5.8	4.4	2.9	2.2
			VERTICAL BODY MEASUREMENTS					
3.1	**8.0**	B1 V	Chin to Sternal Notch	6.0	5.3	4.0	2.7	2.0
3.0	**7.5**	B2 V	Sternal Notch to Axilla	5.6	5.0	3.8	2.5	1.9
4.7	**12.0**	B3 V	Sternal Notch to Nipples	9.0	8.0	6.0	4.0	3.0
5.9	**15.0**	B4 V	Sternal Notch to Xyphoid	11.3	10.0	7.5	5.0	3.8
11.8	**30.0**	B5 V	Sternal Notch to Umbilicus	22.5	20.0	15.0	10.0	7.5
5.9	**15.0**	B6 V	Xyphoid Process to Umbilicus	11.3	10.0	7.5	5.0	3.8
11.0	**28.0**	B7 V	Sternal Notch to Crest Ilium	21.0	18.7	14.0	9.3	7.0
13.6	**34.5**	B8 V	Sternal Notch to Ant.S.Sp.Ilium	25.9	23.0	17.3	11.5	8.6
17.5	**44.5**	B9 V	Sternal Notch to Pubis	33.4	29.7	22.3	14.8	11.1
5.5	**14.0**	B10 V	Umbilicus to Pubis	10.5	9.3	7.0	4.7	3.5
			TRANSVERSE BODY MEASUREMENTS					
13.4	**34.0**	B11 T	Shoulder Breadth	25.5	22.7	17.0	11.3	8.5
6.1	**15.5**	B12 T	Clavicle to Acromion Process	11.6	10.3	7.8	5.2	3.9
12.4	**31.5**	B13 T	Biacromial Process Breadth	23.6	21.0	15.8	10.5	7.9
8.6	**21.8**	B14 T	Inter Anterior Axillary Fold	16.4	14.5	10.9	7.3	5.5
5.9	**15.0**	B15 T	Inter Nipples	11.3	10.0	7.5	5.0	3.8

Life size			ELEVEN-YEAR-OLD GIRL	Fractions of life size				
				3/4	2/3	1/2	1/3	1/4
Inches	cm	Code		cm	cm	cm	cm	cm
			TRANSVERSE BODY MEASUREMENTS					
8.8	**22.4**	B16 T	Breadth of Thorax	16.8	14.9	11.2	7.5	5.6
8.7	**22.0**	B17 T	Inter Costal Margin	16.5	14.7	11.0	7.3	5.5
9.1	**23.0**	B18 T	Inter Crest of Ilium	17.3	15.3	11.5	7.7	5.8
7.7	**19.5**	B19 T	Inter Anterior Superior Spine	14.6	13.0	9.8	6.5	4.9
10.2	**26.0**	B20 T	Inter Greater Trochanter Femur	19.5	17.3	13.0	8.7	6.5
			ANTERIOR-POSTERIOR BODY MEASUREMENTS					
6.1	**15.5**	B21 D	Manubrium to Dorsal Spine	11.6	10.3	7.8	5.2	3.9
6.7	**17.0**	B22 D	Xyphoid Process to Dorsum	12.8	11.3	8.5	5.7	4.3
6.3	**16.0**	B23 D	Umbilicus to Dorsum	12.0	10.7	8.0	5.3	4.0
7.5	**19.0**	B24 D	Pubis of Pelvis to Buttock	14.3	12.7	9.5	6.3	4.8
5.9	**15.0**	B25 D	Thigh to Gluteal Furrow	11.3	10.0	7.5	5.0	3.8
			VERTICAL BACK MEASUREMENTS					
8.7	**22.0**	B26 V	Vertex to 7th Cervical Spine	16.5	14.7	11.0	7.3	5.5
			7th Cervical Vertebra Spine to:					
5.7	**15.0**	B27 V	Posterior Axillary Fold	10.9	9.7	7.3	4.8	3.6
7.1	**18.0**	B28 V	Inferior Tip of Scapula	13.5	12.0	9.0	6.0	4.5
14.6	**37.0**	B29 V	Posterior Superior Spine Ilium	27.8	24.7	18.5	12.3	9.3
18.1	**46.0**	B30 V	Midline Upper Buttock Crease	34.5	30.7	23.0	15.3	11.5
21.5	**54.5**	B31 V	to Gluteal Furrow	40.9	36.3	27.3	18.2	13.6
33.1	**84.0**	B32 V	to Popliteal Crease of Knee	63.0	56.0	42.0	28.0	21.0
			TRANSVERSE BACK MEASUREMENTS					
10.6	**27.0**	B33 T	Inter Posterior Axillary Fold	20.3	18.0	13.5	9.0	6.8
4.9	**12.4**	B34 T	Inter Scapulae Medial Border	9.3	8.3	6.2	4.1	3.1
2.8	**7.0**	B35 T	Inter Post. Sup. Spines Sacrum	5.3	4.7	3.5	2.3	1.8
			UPPER EXTREMITY MEASUREMENTS					
57.1	**145.0**	U1 V	Span of Arms	108.8	96.7	72.5	48.3	36.3
6.1	**15.5**	U2 V	Sternum to Acromion Process	11.6	10.3	7.8	5.2	3.9
24.0	**61.0**	U3 V	Acromion to Fingertip	45.8	40.7	30.5	20.3	15.3
72.8	**185.0**	U4 V	Vertical Reach to Fingertip	138.8	123.3	92.5	61.7	46.3
34.6	**88.0**	U5 T	Lateral Reach to Fingertip	66.0	58.7	44.0	29.3	22.0
27.2	**69.0**	U6 D	Forward Reach to Fingertip	51.8	46.0	34.5	23.0	17.3
11.8	**30.0**	U7 V	Shoulder to Elbow Flexed	22.5	20.0	15.0	10.0	7.5
10.5	**26.6**	U8 V	Acromion to Lat. Condyle Elbow	20.0	17.7	13.3	8.9	6.7
14.9	**37.9**	U9 D	Flexed Elbow to Fingertip	28.4	25.3	19.0	12.6	9.5
8.8	**22.3**	U10 D	Flexed Elbow Pivot to Wrist Pivot	16.7	14.9	11.2	7.4	5.6

Life size			ELEVEN-YEAR-OLD GIRL	Fractions of life size				
				3/4	2/3	1/2	1/3	1/4
Inches	cm	Code		cm	cm	cm	cm	cm
			UPPER EXTREMITY MEASUREMENTS					
6.1	**15.6**	U11 V	Wrist to Fingertip	11.7	10.4	7.8	5.2	3.9
4.1	**10.5**	U12 V	Wrist to Thumb	7.9	7.0	5.3	3.5	2.6
3.5	**9.0**	U13 V	Length of Palm	6.8	6.0	4.5	3.0	2.3
2.6	**6.6**	U14 V	Length 3rd Finger	5.0	4.4	3.3	2.2	1.7
4.2	**10.6**	U15 V	Thumb Crotch to 3rd Finger	8.0	7.1	5.3	3.5	2.7
2.2	**5.5**	U16 V	Thumb Crotch to Thumb Tip	4.1	3.7	2.8	1.8	1.4
2.7	**6.8**	U17 D	Anterior-Post. Arm at Biceps	5.1	4.5	3.4	2.3	1.7
2.4	**6.1**	U18 D	Anterior-Posterior at Forearm	4.6	4.1	3.1	2.0	1.5
1.4	**3.6**	U19 D	Depth of Wrist	2.7	2.4	1.8	1.2	0.9
2.0	**5.0**	U20 T	Breadth of Wrist	3.8	3.3	2.5	1.7	1.3
3.3	**8.4**	U21 T	Breadth of Hand	6.3	5.6	4.2	2.8	2.1
2.7	**6.9**	U22 T	Breadth of Palm	5.2	4.6	3.5	2.3	1.7
			LOWER EXTREMITY MEASUREMENTS					
35.4	**90.0**	L1 V	Crest of Ilium to Sole	67.5	60.0	45.0	30.0	22.5
32.5	**82.5**	* L2 V	Ant.Sup.Spine of Ilium to Sole	61.9	55.0	41.3	27.5	20.6
29.5	**75.0**	L3 V	Greater Trochanter Femur-Sole	56.3	50.0	37.5	25.0	18.8
28.5	**72.5**	* L4 V	Symphysis Pubis to Sole	54.4	48.3	36.3	24.2	18.1
25.8	**65.5**	* L5 V	Crotch to Sole	49.1	43.7	32.8	21.8	16.4
26.8	**68.0**	L6 V	Gluteal Furrow to Sole	51.0	45.3	34.0	22.7	17.0
15.7	**40.0**	L7 V	Posterior Knee Crease to Sole	30.0	26.7	20.0	13.3	10.0
15.0	**38.0**	L8 V	Tibia to Sole	28.5	25.3	19.0	12.7	9.5
2.4	**6.0**	L9 V	Med. Malleolus to Sole	4.5	4.0	3.0	2.0	1.5
2.0	**5.0**	L10 V	Lat. Malleolus to Sole	3.8	3.3	2.5	1.7	1.3
10.6	**27.0**	L11 V	Breadth of Hips, Sitting	20.3	18.0	13.5	9.0	6.8
5.1	**13.0**	L12 V	Inter Mid Thigh	9.8	8.7	6.5	4.3	3.3
3.9	**10.0**	L13 T	Breadth of Knee	7.5	6.7	5.0	3.3	2.5
3.5	**9.0**	L14 T	Breadth of Calf	6.8	6.0	4.5	3.0	2.3
2.0	**5.0**	L15 T	Breadth One Ankle	3.8	3.3	2.5	1.7	1.3
5.1	**13.0**	L16 T	Mid Thigh, Anterior-Posterior	9.8	8.7	6.5	4.3	3.3
3.9	**10.0**	L17 D	Knee, Anterior-Posterior	7.5	6.7	5.0	3.3	2.5
3.5	**9.0**	L18 D	Mid Calf, Anterior-Posterior	6.8	6.0	4.5	3.0	2.3
2.4	**6.0**	L19 D	Ankle, Anterior-Posterior	4.5	4.0	3.0	2.0	1.5
8.9	**22.5**	L20 D	Length of Foot	16.9	15.0	11.3	7.5	5.6
6.7	**17.0**	L21 D	Heel to 5th Toe	12.8	11.3	8.5	5.7	4.3
3.3	**8.5**	L22 T	Breadth Foot	6.4	5.7	4.3	2.8	2.1
2.0	**5.0**	L23 T	Width of Heel	3.8	3.3	2.5	1.7	1.3
		** These are also listed with Vertical Measurements.*						

Eleven-Year-Old Boy, anterior and lateral head

all measurements in centimeters

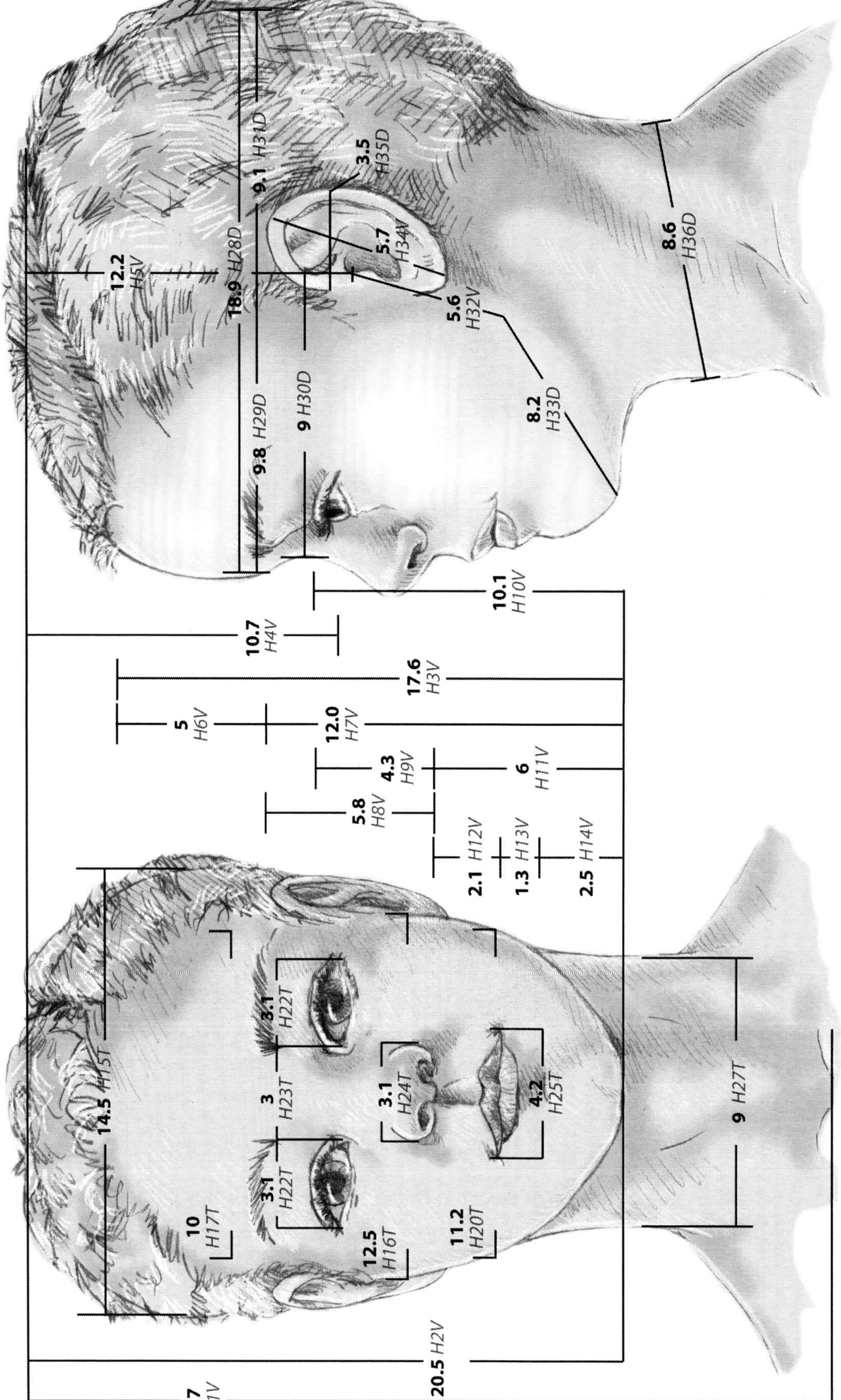

Eleven-Year-Old Boy, anterior and lateral figure

all measurements in centimeters

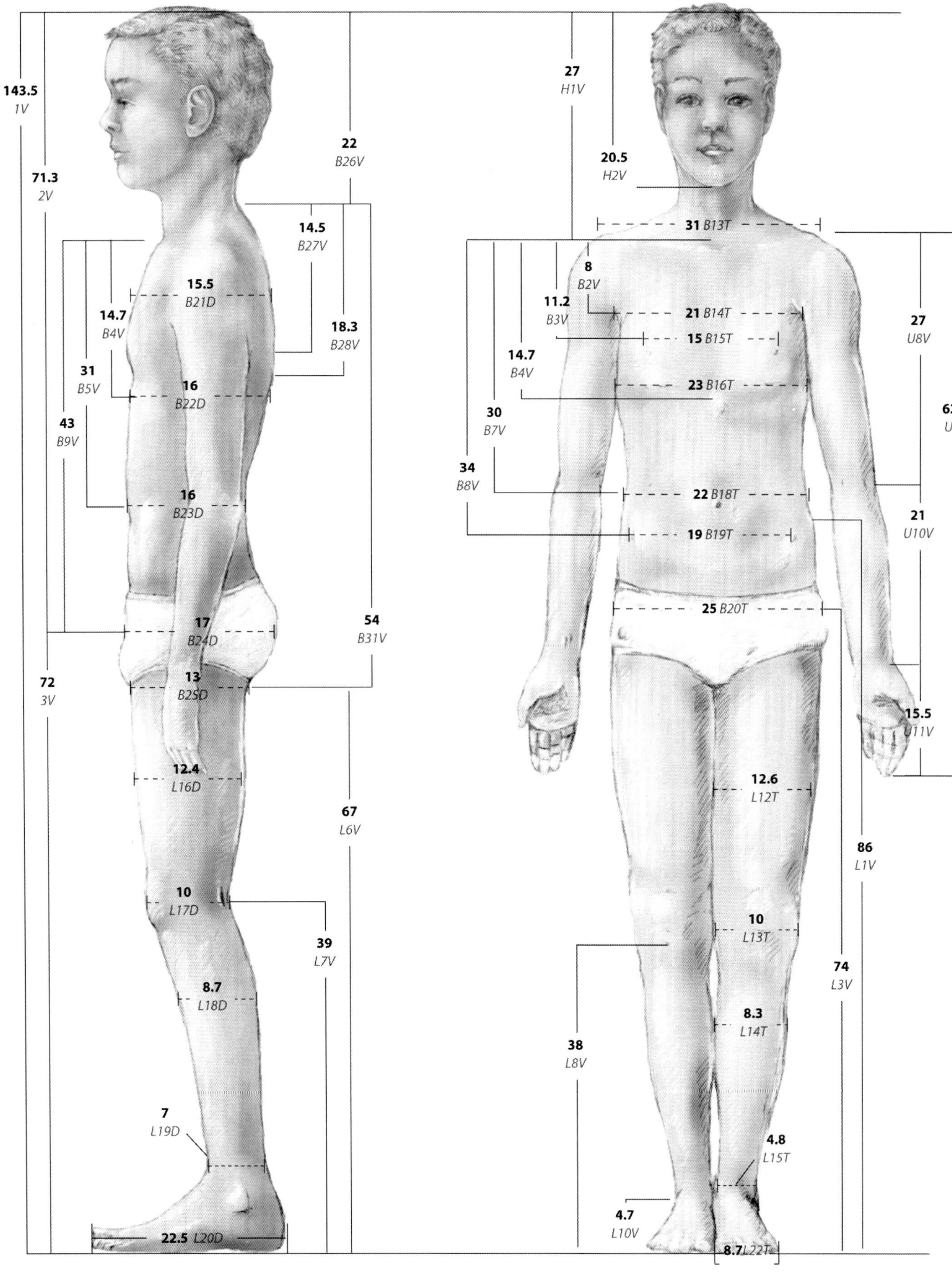

Life size			ELEVEN-YEAR-OLD BOY	Fractions of life size				
				3/4	2/3	1/2	1/3	1/4
Inches	cm	Code		cm	cm	cm	cm	cm
			VERTICAL MEASUREMENTS					
56.5	**143.5**	1 V	Stature, Height, Vertex to Sole	107.6	95.7	71.8	47.8	35.9
28.1	**71.3**	2 V	Vertex to Symphysis Pubis	53.5	47.5	35.7	23.8	17.8
28.3	**72.0**	* 3 V	Symphysis Pubis to Sole	54.0	48.0	36.0	24.0	18.0
46.1	**117.0**	4 V	Sternal Notch to Sole	87.8	78.0	58.5	39.0	29.3
10.8	**27.5**	5 V	Vertex to Acromion Process	20.6	18.3	13.8	9.2	6.9
45.7	**116.0**	6 V	Acromion Process to Sole	87.0	77.3	58.0	38.7	29.0
32.1	**81.5**	* 7 V	Ant.Sup.Spine of Ilium to Sole	61.1	54.3	40.8	27.2	20.4
13.6	**34.5**	8 V	Acromion pr. to Ant. Sup. Spine	25.9	23.0	17.3	11.5	8.6
15.6	**39.5**	9 V	Ant. Sup. Spine of Ilium to Knee	29.6	26.3	19.8	13.2	9.9
16.5	**42.0**	10 V	Knee at Patella to Sole	31.5	28.0	21.0	14.0	10.5
33.5	**85.0**	11 V	Umbilicus to Sole	63.8	56.7	42.5	28.3	21.3
25.2	**64.0**	* 12 V	Crotch to Sole	48.0	42.7	32.0	21.3	16.0
42.5	**108.0**	13 V	Kneeling Height	81.0	72.0	54.0	36.0	27.0
29.5	**75.0**	14 V	Sitting Height	56.3	50.0	37.5	25.0	18.8
29.5	**75.0**	15 V	Buttock, Straight Leg to Sole	56.3	50.0	37.5	25.0	18.8
18.7	**47.5**	16 V	Buttock to Flexed Knee	35.6	31.7	23.8	15.8	11.9
17.9	**45.5**	17 V	Flexed Knee to Sole	34.1	30.3	22.8	15.2	11.4
		** These are also listed with Lower Extremity Measurements.*						
			VERTICAL HEAD MEASUREMENTS					
10.6	**27.0**	H1 V	Vertex to Sternal Notch	20.3	18.0	13.5	9.0	6.8
8.1	**20.5**	H2 V	Vertex to Chin, Head Height	15.4	13.7	10.3	6.8	5.1
6.8	**17.2**	H3 V	Hairline to Chin, Face Height	12.9	11.5	8.6	5.7	4.3
4.2	**10.7**	H4 V	Vertex to Lateral Corner of Eye	8.0	7.1	5.4	3.6	2.7
4.8	**12.2**	H5 V	Vertex to Tragion of Ear	9.2	8.1	6.1	4.1	3.1
2.0	**5.0**	H6 V	Hairline to Brow, Glabella	3.8	3.3	2.5	1.7	1.3
4.6	**11.8**	H7 V	Brow, Glabella, to Chin, Menton	8.9	7.9	5.9	3.9	3.0
2.3	**5.8**	H8 V	Brow to Subnasal Point	4.4	3.9	2.9	1.9	1.5
1.7	**4.3**	H9 V	Sellion to Subnasal, Nose Length	3.2	2.9	2.2	1.4	1.1
4.0	**10.1**	H10 V	Sellion, Nose Saddle, to Chin	7.6	6.7	5.1	3.4	2.5
2.4	**6.0**	H11 V	Subnasal point to Chin	4.5	4.0	3.0	2.0	1.5
0.8	**2.1**	H12 V	Subnasal to Lip Aperature	1.6	1.4	1.1	0.7	0.5
0.5	**1.3**	H13 V	Lip Aperat. to Sublabial Crease	1.0	0.9	0.7	0.4	0.3
1.0	**2.5**	H14 V	Sublabial Crease to Chin	1.9	1.7	1.3	0.8	0.6
			TRANSVERSE HEAD MEASUREMENTS					
5.7	**14.5**	H15 T	Bi-Parietal, Head Breadth	10.9	9.7	7.3	4.8	3.6
4.9	**12.5**	H16 T	Bi-Zygomatic, Cheekbones	9.4	8.3	6.3	4.2	3.1
4.4	**11.2**	H17 T	Bi-Fronto Temporal, Forehead	8.4	7.5	5.6	3.7	2.8

Life size			ELEVEN-YEAR-OLD BOY	Fractions of life size				
				3/4	2/3	1/2	1/3	1/4
Inches	cm	Code		cm	cm	cm	cm	cm
			TRANSVERSE HEAD MEASUREMENTS					
3.7	9.4	H18 T	Inter Brow Ridge	7.1	6.3	4.7	3.1	2.4
4.7	12.0	H19 T	Inter Tragion of Ears	9.0	8.0	6.0	4.0	3.0
4.4	11.2	H20 T	Inter Angle of Mandible	8.4	7.5	5.6	3.7	2.8
3.6	9.2	H21 T	Inter Lateral Corners of Eyes	6.9	6.1	4.6	3.1	2.3
1.2	3.1	H22 T	Width of Eye	2.3	2.1	1.6	1.0	0.8
1.2	3.0	H23 T	Inter Inner Corner of eyes	2.3	2.0	1.5	1.0	0.8
1.2	3.1	H24 T	Breadth of Nose	2.3	2.1	1.6	1.0	0.8
1.7	4.2	H25 T	Breadth of Mouth	3.2	2.8	2.1	1.4	1.1
1.3	3.2	H26 T	Breadth of Chin	2.4	2.1	1.6	1.1	0.8
3.5	9.0	H27 T	Breadth of Neck	6.8	6.0	4.5	3.0	2.3
			ANTERIOR-POSTERIOR HEAD MEASUREMENTS					
7.4	18.9	H28 D	Brow to Occiput	14.2	12.6	9.5	6.3	4.7
3.9	9.8	H29 D	Brow to Tragus	7.4	6.5	4.9	3.3	2.5
3.5	9.0	H30 D	Sellion to Tragus	6.8	6.0	4.5	3.0	2.3
3.6	9.1	H31 D	Tragus to Occiput	6.8	6.1	4.6	3.0	2.3
2.2	5.6	H32 V	Tragus to Angle Mandible	4.2	3.7	2.8	1.9	1.4
3.2	8.2	H33 D	Angle of Mandible to Chin	6.2	5.5	4.1	2.7	2.1
2.2	5.7	H34 V	Length of Ear, Vertical	4.3	3.8	2.9	1.9	1.4
1.4	3.5	H35 D	Width of Ear, Anterior-Posterior	2.6	2.3	1.8	1.2	0.9
3.4	8.6	H36 D	Anterior to Posterior of Neck	6.5	5.7	4.3	2.9	2.2
			VERTICAL BODY MEASUREMENTS					
3.1	8.0	B1 V	Chin to Sternal Notch, neck pit	6.0	5.3	4.0	2.7	2.0
3.1	8.0	B2 V	Sternal Notch to Axilla, arm pit	6.0	5.3	4.0	2.7	2.0
4.4	11.2	B3 V	Sternal Notch to Nipples	8.4	7.5	5.6	3.7	2.8
5.8	14.7	B4 V	Sternal Notch to Xyphoid Proc.	11.0	9.8	7.4	4.9	3.7
12.2	31.0	B5 V	Sternal Notch to Umbilicus	23.3	20.7	15.5	10.3	7.8
6.3	16.0	B6 V	Xyphoid Process to Umbilicus	12.0	10.7	8.0	5.3	4.0
11.8	30.0	B7 V	Sternal Notch to Crest Ilium	22.5	20.0	15.0	10.0	7.5
13.4	34.0	B8 V	Sternal Notch to Ant.S.Sp.Ilium	25.5	22.7	17.0	11.3	8.5
16.9	43.0	B9 V	Sternal Notch to Pubis	32.3	28.7	21.5	14.3	10.8
5.1	13.0	B10 V	Umbilicus to Pubis	9.8	8.7	6.5	4.3	3.3
			TRANSVERSE BODY MEASUREMENTS					
13.4	34.0	B11 T	Shoulder Breadth	25.5	22.7	17.0	11.3	8.5
6.0	15.2	B12 T	Clavicle to Acromion Process	11.4	10.1	7.6	5.1	3.8
12.2	31.0	B13 T	Biacromial Process Breadth	23.3	20.7	15.5	10.3	7.8
8.3	21.0	B14 T	Inter Anterior Axillary Fold	15.8	14.0	10.5	7.0	5.3
5.9	15.0	B15 T	Inter Nipples	11.3	10.0	7.5	5.0	3.8

Life size			ELEVEN-YEAR-OLD BOY	Fractions of life size				
				3/4	2/3	1/2	1/3	1/4
Inches	cm	Code		cm	cm	cm	cm	cm
			TRANSVERSE BODY MEASUREMENTS					
9.1	**23.0**	B16 T	Breadth of Thorax	17.3	15.3	11.5	7.7	5.8
8.5	**21.5**	B17 T	Inter Costal Margin	16.1	14.3	10.8	7.2	5.4
8.7	**22.0**	B18 T	Inter Crest of Ilium	16.5	14.7	11.0	7.3	5.5
7.5	**19.0**	B19 T	Inter Anterior Superior Spine	14.3	12.7	9.5	6.3	4.8
9.8	**25.0**	B20 T	Inter Greater Trochanter Femur	18.8	16.7	12.5	8.3	6.3
			ANTERIOR-POSTERIOR BODY MEASUREMENTS					
6.1	**15.5**	B21 D	Manubrium to Dorsal Spine	11.6	10.3	7.8	5.2	3.9
6.3	**16.0**	B22 D	Xyphoid Process to Dorsum	12.0	10.7	8.0	5.3	4.0
6.3	**16.0**	B23 D	Umbilicus to Dorsum	12.0	10.7	8.0	5.3	4.0
6.7	**17.0**	B24 D	Pubis to Buttock	12.8	11.3	8.5	5.7	4.3
5.1	**13.0**	B25 D	Thigh to Gluteal Furrow	9.8	8.7	6.5	4.3	3.3
			VERTICAL BACK MEASUREMENTS					
8.7	**22.0**	B26 V	Vertex to 7th Cervical Spine	16.5	14.7	11.0	7.3	5.5
			7th Cervical Vertebra Spine to:					
5.7	**14.5**	B27 V	Posterior Axillary Fold	10.9	9.7	7.3	4.8	3.6
7.2	**18.3**	B28 V	Inferior Tip of Scapula	13.7	12.2	9.2	6.1	4.6
15.0	**38.0**	B29 V	Posterior Superior Spine Ilium	28.5	25.3	19.0	12.7	9.5
16.9	**43.0**	B30 V	to Midline Upper Buttock Crease	32.3	28.7	21.5	14.3	10.8
21.3	**54.0**	B31 V	to Gluteal Furrow	40.5	36.0	27.0	18.0	13.5
32.7	**83.0**	B32 V	to Popliteal Crease of Knee	62.3	55.3	41.5	27.7	20.8
			TRANSVERSE BACK MEASUREMENTS					
11.0	**28.0**	B33 T	Inter Posterior Axillary Fold	21.0	18.7	14.0	9.3	7.0
5.1	**13.0**	B34 T	Inter Scapulae Medial Border	9.8	8.7	6.5	4.3	3.3
2.8	**7.0**	B35 T	Inter Post. Sup. Spines Sacrum	5.3	4.7	3.5	2.3	1.8
			UPPER EXTREMITY MEASUREMENTS					
56.3	**143.0**	U1 V	Span of Arms	107.3	95.3	71.5	47.7	35.8
5.9	**15.0**	U2 V	Sternum to Acromion Process	11.3	10.0	7.5	5.0	3.8
24.6	**62.5**	U3 V	Acromion to Fingertip	46.9	41.7	31.3	20.8	15.6
70.5	**179.0**	U4 V	Vertical Reach to Fingertip	134.3	119.3	89.5	59.7	44.8
34.8	**88.5**	U5 T	Lateral Reach to Fingertip	66.4	59.0	44.3	29.5	22.1
28.0	**71.0**	U6 D	Forward Reach to Fingertip	53.3	47.3	35.5	23.7	17.8
11.7	**29.6**	U7 V	Shoulder to Elbow Flexed	22.2	19.7	14.8	9.9	7.4
10.6	**27.0**	U8 V	Acromion to Lat. Condyle Elbow	20.3	18.0	13.5	9.0	6.8
15.0	**38.0**	U9 D	Flexed Elbow to Fingertip	28.5	25.3	19.0	12.7	9.5
8.3	**21.0**	U10 D	Flexed Elbow to Wrist Pivot	15.8	14.0	10.5	7.0	5.3

Life size			ELEVEN-YEAR-OLD BOY	Fractions of life size				
				3/4	2/3	1/2	1/3	1/4
Inches	cm	Code		cm	cm	cm	cm	cm
			UPPER EXTREMITY MEASUREMENTS					
6.1	**15.5**	U11 V	Wrist to Fingertip	11.6	10.3	7.8	5.2	3.9
4.1	**10.5**	U12 V	Wrist to Thumb	7.9	7.0	5.3	3.5	2.6
3.5	**9.0**	U13 V	Length of Palm	6.8	6.0	4.5	3.0	2.3
2.6	**6.7**	U14 V	Length 3rd Finger	5.0	4.5	3.4	2.2	1.7
4.3	**11.0**	U15 V	Thumb Crotch to 3rd Finger	8.3	7.3	5.5	3.7	2.8
2.2	**5.5**	U16 V	Thumb Crotch to Thumb Tip	4.1	3.7	2.8	1.8	1.4
2.6	**6.5**	U17 T	Anterior-Post. Arm at Biceps	4.9	4.3	3.3	2.2	1.6
2.4	**6.0**	U18 T	Anterior-Posterior at Forearm	4.5	4.0	3.0	2.0	1.5
1.3	**3.3**	U19 T	Depth of Wrist	2.5	2.2	1.7	1.1	0.8
2.0	**5.0**	U20 T	Breadth of Wrist	3.8	3.3	2.5	1.7	1.3
3.5	**8.8**	U21 T	Breadth of Hand	6.6	5.9	4.4	2.9	2.2
2.8	**7.1**	U22 T	Breadth of Palm	5.3	4.7	3.6	2.4	1.8
			LOWER EXTREMITY MEASUREMENTS					
33.9	**86.0**	L1 V	Crest of Ilium to Sole	64.5	57.3	43.0	28.7	21.5
32.1	**81.5**	* L2 V	Ant.Sup.Spine of Ilium to Sole	61.1	54.3	40.8	27.2	20.4
29.1	**74.0**	L3 V	Greater Trochanter Femur-Sole	55.5	49.3	37.0	24.7	18.5
29.1	**74.0**	* L4 V	Symphysis Pubis to Sole	55.5	49.3	37.0	24.7	18.5
25.2	**64.0**	* L5 V	Crotch to Sole	48.0	42.7	32.0	21.3	16.0
26.4	**67.0**	L6 V	Gluteal Furrow to Sole	50.3	44.7	33.5	22.3	16.8
15.4	**39.0**	L7 V	Posterior Knee Crease to Sole	29.3	26.0	19.5	13.0	9.8
15.0	**38.0**	L8 V	Tibia to Sole	28.5	25.3	19.0	12.7	9.5
2.4	**6.0**	L9 V	Med. Malleolus to Sole	4.5	4.0	3.0	2.0	1.5
2.0	**5.0**	L10 V	Lat. Malleolus to Sole	3.8	3.3	2.5	1.7	1.3
9.8	**25.0**	L11 T	Breadth of Hips, Sitting	18.8	16.7	12.5	8.3	6.3
5.0	**12.6**	L12 T	Breadth Mid Thigh	9.5	8.4	6.3	4.2	3.2
3.9	**10.0**	L13 T	Breadth of Knee	7.5	6.7	5.0	3.3	2.5
3.5	**8.8**	L14 T	Breadth of Calf	6.6	5.9	4.4	2.9	2.2
1.9	**4.8**	L15 T	Breadth Ankle	3.6	3.2	2.4	1.6	1.2
4.9	**12.4**	L16 D	Mid Thigh, Anterior-Posterior	9.3	8.3	6.2	4.1	3.1
3.9	**10.0**	L17 D	Knee, Anterior-Posterior	7.5	6.7	5.0	3.3	2.5
3.4	**8.7**	L18 D	Mid Calf, Anterior-Posterior	6.5	5.8	4.4	2.9	2.2
2.8	**7.0**	L19 D	Ankle, Anterior-Posterior	5.3	4.7	3.5	2.3	1.8
8.9	**22.5**	L20 D	Length of Foot	16.9	15.0	11.3	7.5	5.6
6.7	**17.0**	L21 D	Heel to 5th Toe	12.8	11.3	8.5	5.7	4.3
3.4	**8.7**	L22 T	Breadth Foot	6.5	5.8	4.4	2.9	2.2
2.0	**5.0**	L23 T	Width of Heel	3.8	3.3	2.5	1.7	1.3
		**These are also listed with Vertical Measurements.*						

TWELVE-YEAR-OLD GIRLS

The practical lists of body proportions are shown in the tables. The range of twenty-fifth to fiftieth and to seventy-fifth percentile is listed here for comparison:

Percentile	Twenty-fifth		Fiftieth		Seventy-fifth	
Girls:	Short		Medium		Tall	
Stature	147 cm,	57.8 in.	151.5 cm,	59.4 in.	156 cm,	61.3 in.
Weight	36.5 kg,	80.5 lb.	41.5 kg,	91.5 lb.	48 kg,	106 lb.[1]

At this age, girls are on average, slightly taller and weigh slightly more than boys. Girls begin to mature earlier and may gain weight and height more rapidly. Some may be a half head taller than the boys in the same school class. Many girls may already have feminine contours in face shoulders and hips. Many have experienced onset of menstruation. Mammary development varies with individuals, but most are at Phase II with an elevation of the nipples. There may also be some increase of pectoral, or anterior chest, fat.[2] The school level is seventh grade.

BODY MEASUREMENTS COMPARED TO HEIGHT

	Feature Measure in cm	Feature Compared to Stature	Approx. Ratio of Feature	Compared to Median Adult
Stature: Crown to sole	151.5			90%
Head Height: Crown to chin	20	13%	1/8	98%
Sitting Height:	79	52%	1/2	88%
Trunk Measure: Sternal notch to symphysis pubis	46.5	31%	1/3–	95%
Leg Length: Greater trochanter of the femur to sole	77.5	51%	1/2	90%
Thigh Length: Greater trochanter of the femur to tibia at knee	36.5	29%	1/4	90%
Tibia at Knee to Sole:	41	28%	1/4	90%
Arm Length: Acromion process at shoulder to dependent fingertip	64	42%	2/5	85%
Hand Length:	16	11%	1/9	89%
Foot Length:	23	15%	1/7	96%

The stature approaches adult female measurements. Cranium growth is nearly complete. The mask of the face, from brow to chin, increases as teeth grow and erupt. The first permanent molars erupted by seven years. The lower canines, (cuspids) erupted at about nine years and the upper permanent canines erupt at about twelve years. The upper and lower first and second permanent premolars will have erupted by twelve years of age. The maxilla and mandible bones have grown to accommodate the permanent dentition. Other molars are growing.[3] The face assumes adolescent features.

1. Nelson, W.E., p. 39.
2. Lowery, G., p. 351.
3. Lockhart, R.D., p.57.

TWELVE-YEAR-OLD BOYS

The practical lists of body proportions are shown in the tables. The range of twenty-fifth to fiftieth and to seventy-fifth percentile is listed here for comparison:

Percentile	Twenty-fifth		Fiftieth		Seventy-fifth	
Boys:	Short		Medium		Tall	
Stature	144 cm,	56.8 in.	150 cm,	59 in.	155 cm,	60.8 in.
Weight	35 kg,	77.3 lb.	40 kg,	87.8 lb.	45.8 kg,	101 lb.[1]

Girls begin to mature earlier than boys and are usually taller than their male classmates at this age. Boys' interests continue to follow adult male sports and outdoor activities. The bicycle is important. The school level is seventh grade.

BODY MEASUREMENTS COMPARED TO HEIGHT

	Feature Measure in cm	Feature Compared to Stature	Approx. Ratio of Feature	Compared to Median Adult
Stature: Crown to sole	150			80%
Head Height: Crown to chin	20	13%	1/7–	87%
Sitting Height:	76	51%	1/2	79%
Trunk Measure: Sternal notch to symphysis pubis	46	31%	1/3–	87%
Leg Length: Greater trochanter of the femur to sole	77	51%	1/2	79%
Thigh Length: Greater trochanter of the femure to tibia at knee	36.5	24%	1/4	79%
Tibia at Knee to Sole:	40.5	27%	1/4	78%
Arm Length: Acromion process at shoulder to dependent fingertip	64	43%	2/5+	78%
Hand Length:	16.5	11%	1/9	82%
Foot Length:	23.5	16%	1/6	85%

Cranium growth is nearly complete. The mask of the face, from brow to the chin, grows as teeth increase and erupt. The first permanent molars erupted by seven years. The lower canines, (cuspids) erupted at about nine years and the upper permanent canines erupt at about twelve years. The upper and lower first and second permanent premolars will have erupted by twelve years of age. The maxilla and mandible bones have grown to accommodate the permanent dentition. Other molars are growing.[2] The face assumes adolescent features.

1. Nelson, W.E., p. 38.
2. Lockhart, R.D., p. 57.

Twelve-Year-Old Girl, anterior and lateral head

all measurements in centimeters

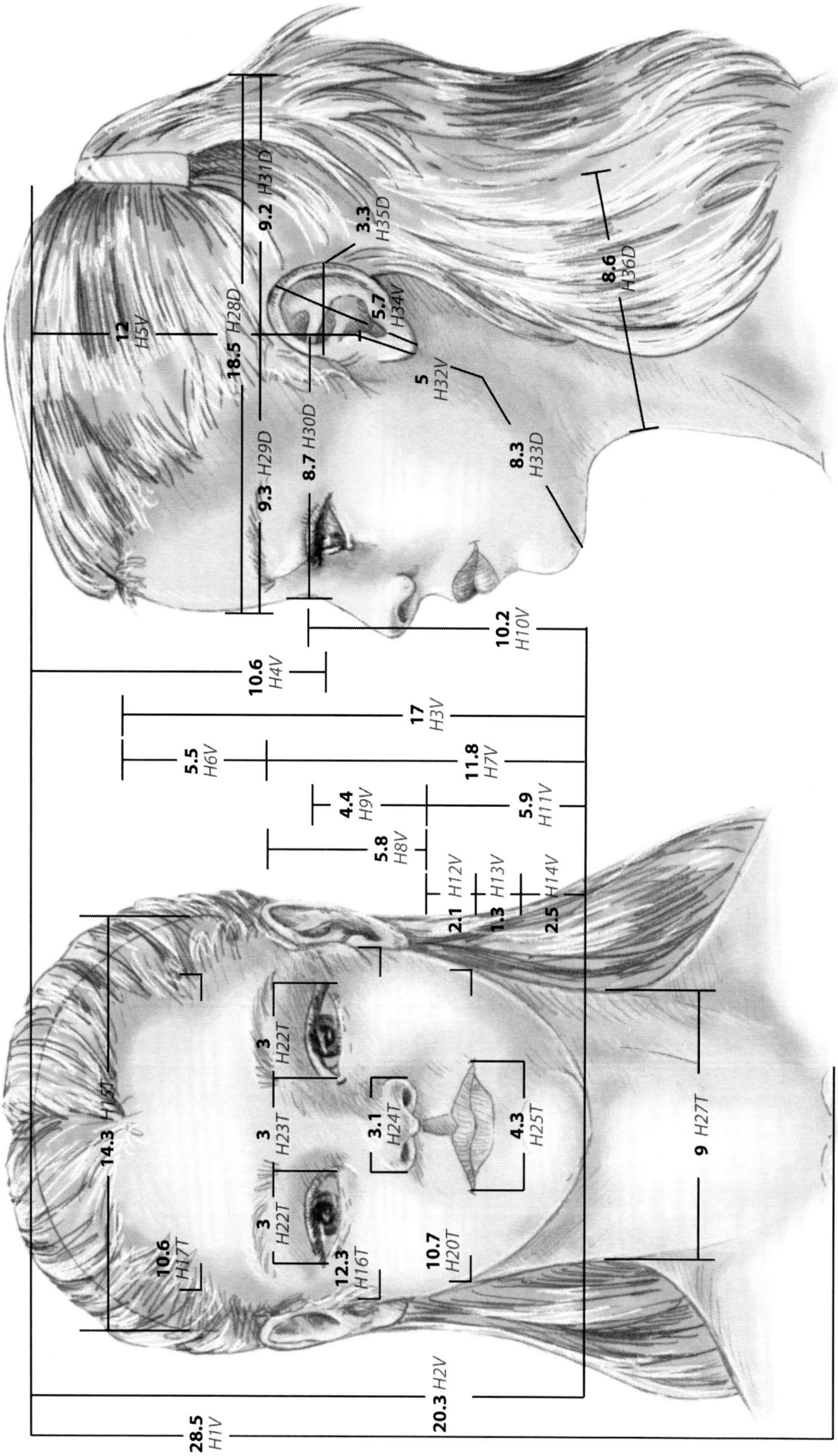

Twelve-Year-Old Girl, anterior and lateral figure

all measurements in centimeters

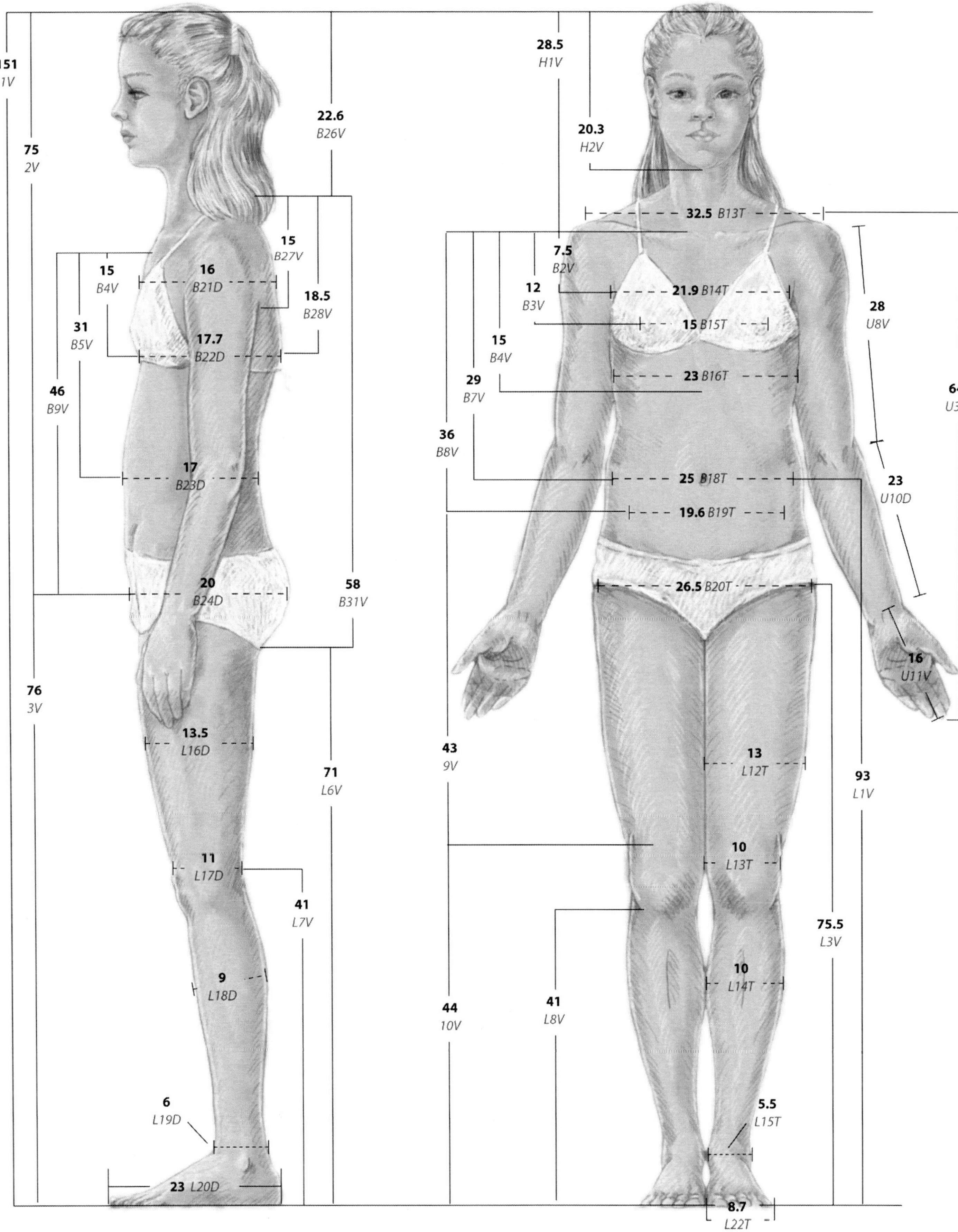

Life size			TWELVE-YEAR-OLD GIRL	Fractions of life size				
				3/4	2/3	1/2	1/3	1/4
Inches	cm	Code		cm	cm	cm	cm	cm
			VERTICAL MEASUREMENTS					
59.4	**151.0**	1 V	Stature, Height, Vertex to Sole	113.3	100.7	75.5	50.3	37.8
29.5	**75.0**	2 V	Vertex to Symphysis Pubis	56.3	50.0	37.5	25.0	18.8
29.9	**76.0**	* 3 V	Symphysis Pubis to Sole	57.0	50.7	38.0	25.3	19.0
47.6	**121.0**	4 V	Sternal Notch to Sole	90.8	80.7	60.5	40.3	30.3
11.0	**28.0**	5 V	Vertex to Acromion Process	21.0	18.7	14.0	9.3	7.0
48.2	**122.5**	6 V	Acromion Process to Sole	91.9	81.7	61.3	40.8	30.6
34.3	**87.0**	* 7 V	Ant.Sup.Spine of Ilium to Sole	65.3	58.0	43.5	29.0	21.8
14.0	**35.5**	8 V	Acrom. proc. to Ant. Sup. Spine	26.6	23.7	17.8	11.8	8.9
16.9	**43.0**	9 V	Ant.Sup.Spine of Ilium to Knee	32.3	28.7	21.5	14.3	10.8
17.3	**44.0**	10 V	Knee at Patella to Sole	33.0	29.3	22.0	14.7	11.0
35.8	**91.0**	11 V	Umbilicus to Sole	68.3	60.7	45.5	30.3	22.8
28.0	**71.0**	* 12 V	Crotch to Sole	53.3	47.3	35.5	23.7	17.8
44.7	**113.5**	13 V	Kneeling Height	85.1	75.7	56.8	37.8	28.4
31.1	**79.0**	14 V	Sitting Height	59.3	52.7	39.5	26.3	19.8
33.1	**84.0**	15 V	Buttock, Straight Leg to Sole	63.0	56.0	42.0	28.0	21.0
20.5	**52.0**	16 V	Buttock to Flexed Knee	39.0	34.7	26.0	17.3	13.0
18.9	**48.0**	17 V	Flexed Knee to Sole	36.0	32.0	24.0	16.0	12.0
		** These are also listed with Lower Extremity Measurements.*						
			VERTICAL HEAD MEASUREMENTS					
11.2	**28.5**	H1 V	Vertex to Sternal Notch	21.4	19.0	14.3	9.5	7.1
8.0	**20.3**	H2 V	Vertex to Chin, Head Height	15.2	13.5	10.2	6.8	5.1
6.7	**17.0**	H3 V	Hairline to Chin, Face Height	12.8	11.3	8.5	5.7	4.3
4.2	**10.6**	H4 V	Vertex to Lateral Corner of Eye	8.0	7.1	5.3	3.5	2.7
4.7	**12.0**	H5 V	Vertex to Tragion of Ear	9.0	8.0	6.0	4.0	3.0
2.2	**5.5**	H6 V	Hairline to Brow, Glabella	4.1	3.7	2.8	1.8	1.4
4.6	**11.6**	H7 V	Brow, Glabella, to Chin, Menton	8.7	7.7	5.8	3.9	2.9
2.2	**5.7**	H8 V	Brow to Subnasal Point	4.3	3.8	2.9	1.9	1.4
1.7	**4.4**	H9 V	Sellion to Subnasal, Nose Length	3.3	2.9	2.2	1.5	1.1
4.0	**10.2**	H10 V	Sellion, Nose Saddle, to Chin	7.7	6.8	5.1	3.4	2.6
2.3	**5.9**	H11 V	Subnasal point to Chin	4.4	3.9	3.0	2.0	1.5
0.8	**2.1**	H12 V	Subnasal to Lip Aperature	1.6	1.4	1.1	0.7	0.5
0.5	**1.3**	H13 V	Lip Aperat. to Sublabial Crease	1.0	0.9	0.7	0.4	0.3
1.0	**2.5**	H14 V	Sublabial Crease to Chin	1.9	1.7	1.3	0.8	0.6
			TRANSVERSE HEAD MEASUREMENTS					
5.6	**14.3**	H15 T	Bi-Parietal, Head Breadth	10.7	9.5	7.2	4.8	3.6
4.8	**12.3**	H16 T	Bi-Zygomatic, Cheekbones	9.2	8.2	6.2	4.1	3.1
4.2	**10.6**	H17 T	Bi-Fronto Temporal, Forehead	8.0	7.1	5.3	3.5	2.7

Life size			TWELVE-YEAR-OLD GIRL	Fractions of life size				
				3/4	2/3	1/2	1/3	1/4
Inches	cm	Code		cm	cm	cm	cm	cm
			TRANSVERSE HEAD MEASUREMENTS					
3.9	**9.8**	H18 T	Inter Brow Ridge	7.4	6.5	4.9	3.3	2.5
4.7	**11.9**	H19 T	Inter Tragion of Ears	8.9	7.9	6.0	4.0	3.0
4.2	**10.7**	H20 T	Inter Angle of Mandible	8.0	7.1	5.4	3.6	2.7
3.5	**9.0**	H21 T	Inter Lateral Corners of Eyes	6.8	6.0	4.5	3.0	2.3
1.2	**3.0**	H22 T	Width of Eye	2.3	2.0	1.5	1.0	0.8
1.2	**3.0**	H23 T	Inter Inner Corner of eyes	2.3	2.0	1.5	1.0	0.8
1.2	**3.1**	H24 T	Breadth of Nose	2.3	2.1	1.6	1.0	0.8
1.7	**4.3**	H25 T	Breadth of Mouth	3.2	2.9	2.2	1.4	1.1
1.3	**3.2**	H26 T	Breadth of Chin	2.4	2.1	1.6	1.1	0.8
3.5	**9.0**	H27 T	Breadth of Neck	6.8	6.0	4.5	3.0	2.3
			ANTERIOR-POSTERIOR HEAD MEASUREMENTS					
7.3	**18.5**	H28 D	Brow to Occiput	13.9	12.3	9.3	6.2	4.6
3.7	**9.3**	H29 D	Brow to Tragus	7.0	6.2	4.7	3.1	2.3
3.4	**8.7**	H30 D	Sellion to Tragus	6.5	5.8	4.4	2.9	2.2
3.6	**9.2**	H31 D	Tragus to Occiput	6.9	6.1	4.6	3.1	2.3
2.2	**5.5**	H32 V	Tragus to Angle Mandible	4.1	3.7	2.8	1.8	1.4
3.3	**8.3**	H33 D	Angle of Mandible to Chin	6.2	5.5	4.2	2.8	2.1
2.3	**5.8**	H34 V	Length of Ear, Vertical	4.4	3.9	2.9	1.9	1.5
1.3	**3.4**	H35 D	Width of Ear, Anterior-Posterior	2.6	2.3	1.7	1.1	0.9
3.3	**8.8**	H36 D	Anterior to Posterior of Neck	6.2	5.5	4.2	2.8	2.1
			VERTICAL BODY MEASUREMENTS					
3.9	**10.0**	B1 V	Chin to Sternal Notch	7.5	6.7	5.0	3.3	2.5
3.0	**7.5**	B2 V	Sternal Notch to Axilla	5.6	5.0	3.8	2.5	1.9
4.7	**12.0**	B3 V	Sternal Notch to Nipples	9.0	8.0	6.0	4.0	3.0
5.9	**15.0**	B4 V	Sternal Notch to Xyphoid Proc.	11.3	10.0	7.5	5.0	3.8
12.2	**31.0**	B5 V	Sternal Notch to Umbilicus	23.3	20.7	15.5	10.3	7.8
5.9	**15.0**	B6 V	Xyphoid Process to Umbilicus	11.3	10	7.5	5.0	3.8
11.4	**29.0**	B7 V	Sternal Notch to Crest Ilium	21.8	19.3	14.5	9.7	7.3
14.2	**36.0**	B8 V	Sternal Notch to Ant.S.Sp.Ilium	27.0	24.0	18.0	12.0	9.0
18.1	**46.0**	B9 V	Sternal Notch to Pubis	34.5	30.7	23.0	15.3	11.5
5.5	**14.0**	B10 V	Umbilicus to Pubis	10.5	9.3	7.0	4.7	3.5
			TRANSVERSE BODY MEASUREMENTS					
14.2	**36.0**	B11 T	Shoulder Breadth	27.0	24.0	18.0	12.0	9.0
6.2	**15.8**	B12 T	Clavicle to Acromion Process	11.9	10.5	7.9	5.3	4.0
12.8	**32.5**	B13 T	Biacromial Process Breadth	24.4	21.7	16.3	10.8	8.1
8.6	**21.9**	B14 T	Inter Anterior Axillary Fold	16.4	14.6	11.0	7.3	5.5
5.9	**15.0**	B15 T	Inter Nipples	11.3	10.0	7.5	5.0	3.8

Life size			TWELVE-YEAR-OLD GIRL	Fractions of life size				
				3/4	2/3	1/2	1/3	1/4
Inches	cm	Code		cm	cm	cm	cm	cm
			TRANSVERSE BODY MEASUREMENTS					
9.1	**23.0**	B16 T	Breadth of Thorax	17.3	15.3	11.5	7.7	5.8
8.8	**22.4**	B17 T	Inter Costal Margin Breadth	16.8	14.9	11.2	7.5	5.6
9.8	**25.0**	B18 T	Inter Crest of Ilium Breadth	18.8	16.7	12.5	8.3	6.3
7.7	**19.6**	B19 T	Inter Anterior Superior Spine	14.7	13.1	9.8	6.5	4.9
10.4	**26.5**	B20 T	Inter Greater Trochanter Femur	19.9	17.7	13.3	8.8	6.6
			ANTERIOR-POSTERIOR BODY MEASUREMENTS					
6.3	**16.0**	B21 D	Manubrium to Dorsal Spine	12.0	10.7	8.0	5.3	4.0
7.0	**17.7**	B22 D	Xyphoid Process to Dorsum	13.3	11.8	8.9	5.9	4.4
6.7	**17.0**	B23 D	Umbilicus to Dorsum	12.8	11.3	8.5	5.7	4.3
7.9	**20.0**	B24 D	Pubis to Buttock	15.0	13.3	10.0	6.7	5.0
6.1	**15.6**	B25 D	Thigh to Gluteal Furrow	11.7	10.4	7.8	5.2	3.9
			VERTICAL BACK MEASUREMENTS					
8.9	**22.6**	B26 V	Vertex to 7th Cervical Spine	17.0	15.1	11.3	7.5	5.7
			7th Cervical Vertebra Spine to:					
5.9	**15.0**	B27 V	To Posterior Axillary Fold	11.3	10.0	7.5	5.0	3.8
7.3	**18.5**	B28 V	To Inferior Tip of Scapula	13.9	12.3	9.3	6.2	4.6
15.0	**38.0**	B29 V	To Posterior Superior Spine Ilium	28.5	25.3	19.0	12.7	9.5
18.5	**47.0**	B30 V	To Midline Upper Buttock Crease	35.3	31.3	23.5	15.7	11.8
22.8	**58.0**	B31 V	To Gluteal Furrow	43.5	38.7	29.0	19.3	14.5
33.9	**86.0**	B32 V	To Popliteal Crease of Knee	64.5	57.3	43.0	28.7	21.5
			TRANSVERSE BACK MEASUREMENTS					
11.2	**28.5**	B33 T	Inter Posterior Axillary Folds	21.4	19	14.3	9.5	7.1
4.9	**12.5**	B34 T	Inter Scapulae Medial Borders	9.4	8.3	6.3	4.2	3.1
2.8	**7.0**	B35 T	Inter Post. Sup. Spines Sacrum	5.3	4.7	3.5	2.3	1.8
			UPPER EXTREMITY MEASUREMENTS					
59.4	**151.0**	U1 V	Span of Arms	113.3	100.7	75.5	50.3	37.8
6.3	**16.0**	U2 V	Sternum to Acromion Process	12.0	10.7	8.0	5.3	4.0
25.2	**64.0**	U3 V	Acromion Process to Fingertip	48.0	42.7	32.0	21.3	16.0
75.2	**191.0**	U4 V	Vertical Reach to Fingertip	143.3	127.3	95.5	63.7	47.8
35.4	**90.0**	U5 T	Lateral Reach to Fingertip	67.5	60.0	45.0	30.0	22.5
27.6	**70.0**	U6 D	Forward Reach to Fingertip	52.5	46.7	35.0	23.3	17.5
12.2	**31.0**	U7 V	Shoulder to Elbow Flexed	23.3	20.7	15.5	10.3	7.8
11.0	**28.0**	U8 V	Acromion to Lat. Condyle Elbow	21.0	18.7	14.0	9.3	7.0
15.7	**40.0**	U9 D	Flexed Elbow to Fingertip	30.0	26.7	20.0	13.3	10.0
9.1	**23.0**	U10 D	Flexed Elbow Pivot to Wrist Pivot	17.3	15.3	11.5	7.7	5.8

Life size			TWELVE-YEAR-OLD GIRL	Fractions of life size				
				3/4	2/3	1/2	1/3	1/4
Inches	cm	Code		cm	cm	cm	cm	cm
			UPPER EXTREMITY MEASUREMENTS					
6.3	**16.0**	U11 V	Wrist to Fingertip	12.0	10.7	8.0	5.3	4.0
4.1	**10.5**	U12 V	Wrist to Thumb Tip	7.9	7.0	5.3	3.5	2.6
3.6	**9.2**	U13 V	Length of Palm	6.9	6.1	4.6	3.1	2.3
2.8	**7.0**	U14 V	Length 3rd Finger	5.3	4.7	3.5	2.3	1.8
4.4	**11.2**	U15 V	Thumb Crotch to 3rd Finger	8.4	7.5	5.6	3.7	2.8
2.2	**5.6**	U16 V	Thumb Crotch to Thumb Tip	4.2	3.7	2.8	1.9	1.4
2.8	**7.0**	U17 D	Anterior-Post. Arm at Biceps	5.3	4.7	3.5	2.3	1.8
2.6	**6.5**	U18 D	Anterior-Posterior at Forearm	4.9	4.3	3.3	2.2	1.6
1.4	**3.6**	U19 D	Depth of Wrist	2.7	2.4	1.8	1.2	0.9
2.0	**5.0**	U20 T	Breadth of Wrist	3.8	3.3	2.5	1.7	1.3
3.5	**9.0**	U21 T	Breadth of Hand	6.8	6.0	4.5	3.0	2.3
2.9	**7.3**	U22 T	Breadth of Palm	5.5	4.9	3.7	2.4	1.8
			LOWER EXTREMITY MEASUREMENTS					
36.6	**93.0**	L1 V	Crest of Ilium to Sole	69.8	62.0	46.5	31.0	23.3
34.3	**87.0**	* L2 V	Ant.Sup.Spine of Ilium to Sole	65.3	58.0	43.5	29.0	21.8
30.5	**77.5**	L3 V	Greater Trochanter Femur-Sole	58.1	51.7	38.8	25.8	19.4
29.9	**76.0**	* L4 V	Symphysis Pubis to Sole	57.0	50.7	38.0	25.3	19.0
28.0	**71.0**	* L5 V	Crotch to Sole	53.3	47.3	35.5	23.7	17.8
28.0	**71.0**	L6 V	Gluteal Furrow to Sole	53.3	47.3	35.5	23.7	17.8
16.1	**41.0**	L7 V	Posterior Knee Crease to Sole	30.8	27.3	20.5	13.7	10.3
16.1	**41.0**	L8 V	Tibia at Knee to Sole	30.8	27.3	20.5	13.7	10.3
2.4	**6.0**	L9 V	Med. Malleolus to Sole	4.5	4.0	3.0	2.0	1.5
2.0	**5.0**	L10 V	Lat. Malleolus to Sole	3.8	3.3	2.5	1.7	1.3
11.4	**29.0**	L11 T	Breadth of Hips, Sitting	21.8	19.3	14.5	9.7	7.3
5.1	**13.0**	L12 T	Breadth Mid Thigh	9.8	8.7	6.5	4.3	3.3
3.9	**10.0**	L13 T	Breadth of Knee	7.5	6.7	5.0	3.3	2.5
3.9	**10.0**	L14 T	Breadth of Calf	7.5	6.7	5.0	3.3	2.5
2.2	**5.5**	L15 T	Breadth Ankle	4.1	3.7	2.8	1.8	1.4
5.3	**13.5**	L16 D	Mid Thigh, Anterior-Posterior	10.1	9.0	6.8	4.5	3.4
4.3	**11.0**	L17 D	Knee, Anterior-Posterior	8.3	7.3	5.5	3.7	2.8
3.5	**9.0**	L18 D	Mid Calf, Anterior-Posterior	6.8	6.0	4.5	3.0	2.3
2.4	**6.0**	L19 D	Ankle, Anterior-Posterior	4.5	4.0	3.0	2.0	1.5
9.1	**23.0**	L20 D	Length of Foot	17.3	15.3	11.5	7.7	5.8
6.9	**17.5**	L21 D	Heel to 5th Toe	13.1	11.7	8.8	5.8	4.4
3.4	**8.7**	L22 T	Breadth Foot	6.5	5.8	4.4	2.9	2.2
2.0	**5.0**	L23 T	Width of Heel	3.8	3.3	2.5	1.7	1.3
		** These are also listed with Vertical Measurements.*						

Twelve-Year-Old Boy, anterior and lateral head

all measurements in centimeters

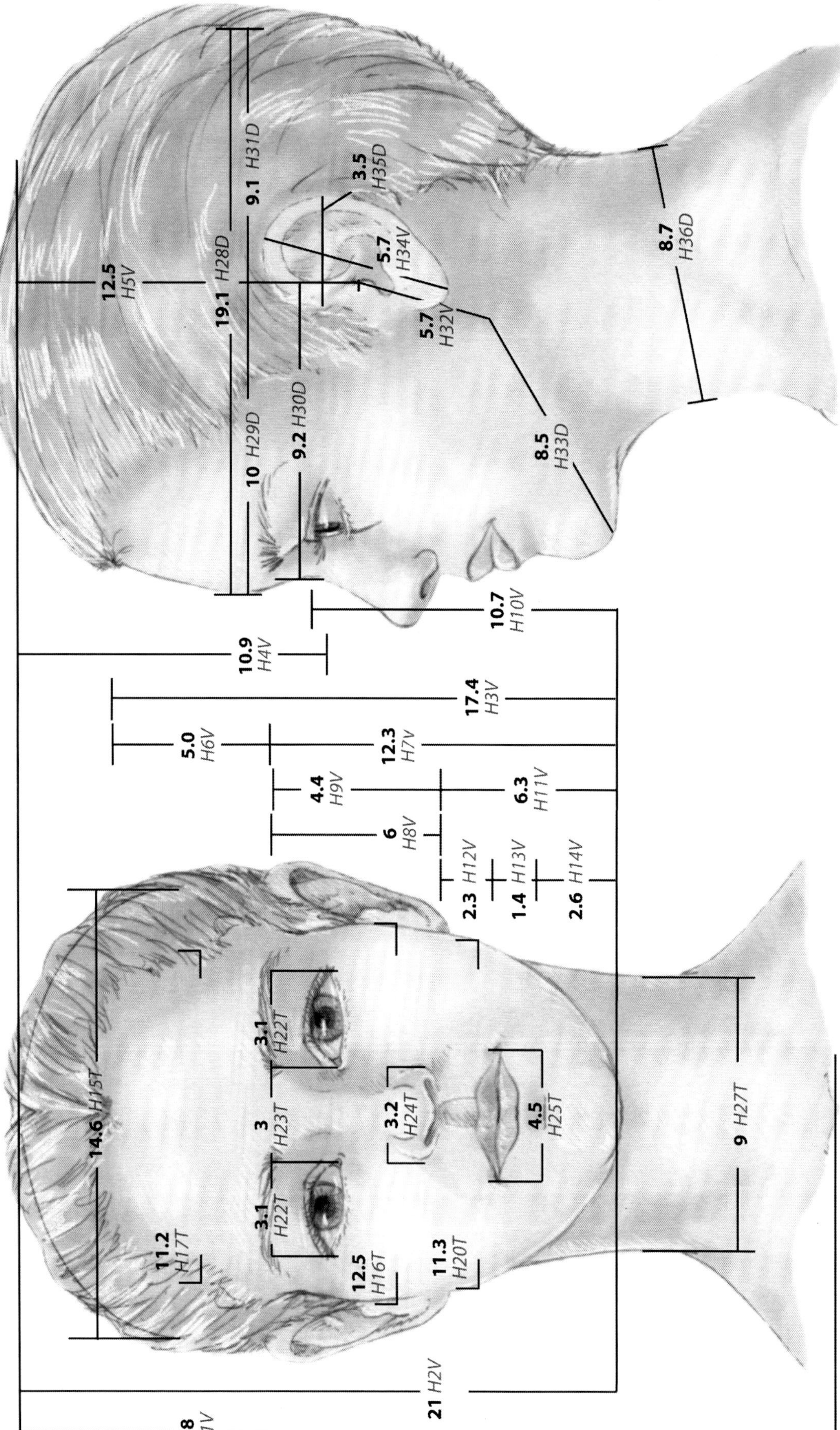

Twelve-Year-Old Boy, anterior and lateral figure

all measurements in centimeters

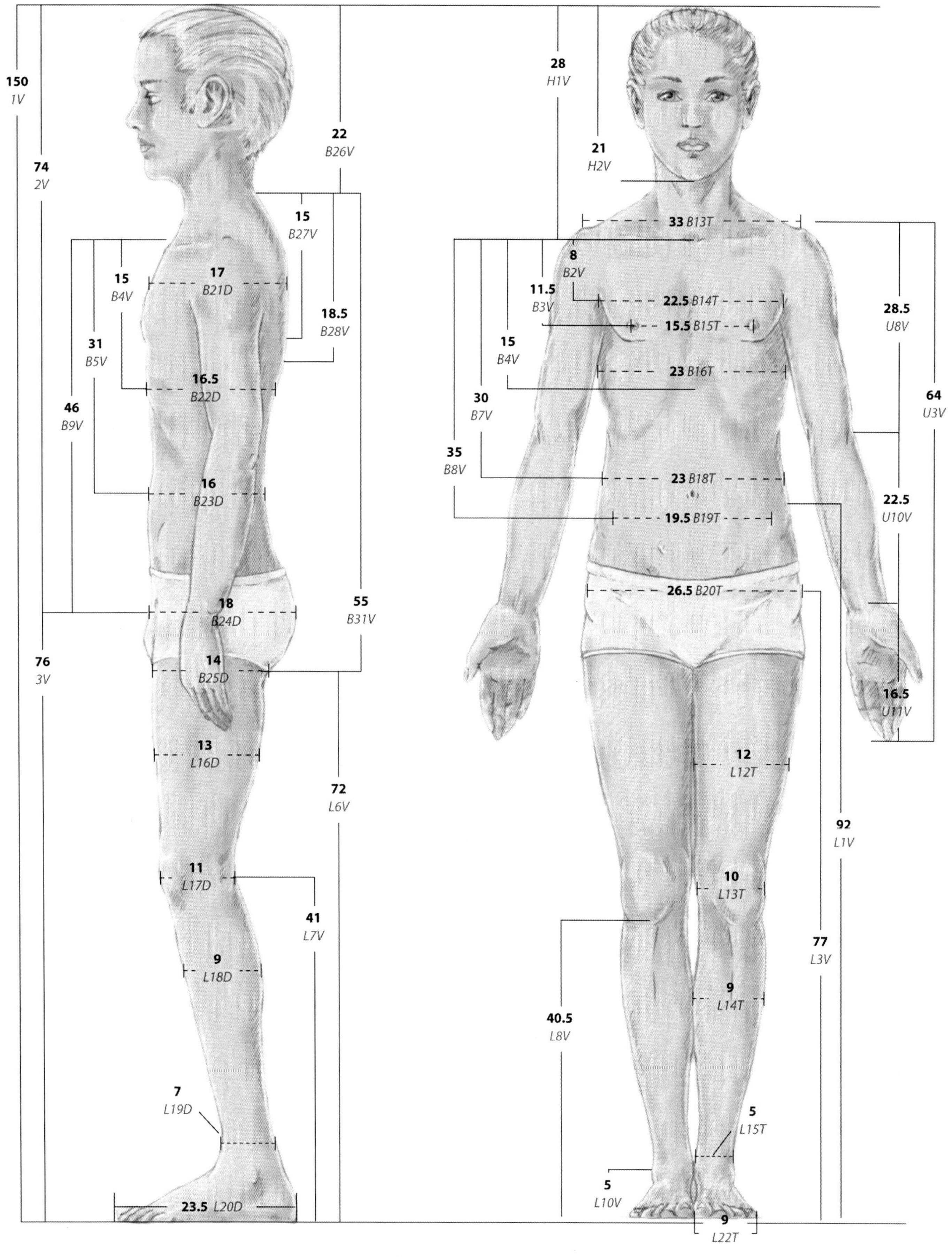

Life size			TWELVE-YEAR-OLD BOY	Fractions of life size				
				3/4	2/3	1/2	1/3	1/4
Inches	cm	Code		cm	cm	cm	cm	cm
			VERTICAL MEASUREMENTS					
59.1	**150.0**	1 V	Stature, Height, Vertex to Sole	112.5	100.0	75.0	50.0	37.5
29.1	**74.0**	2 V	Vertex to Symphysis Pubis	55.5	49.3	37.0	24.7	18.5
29.9	**76.0**	* 3 V	Symphysis Pubis to Sole	57.0	50.7	38.0	25.3	19.0
47.8	**121.5**	4 V	Sternal Notch to Sole	91.1	81.0	60.8	40.5	30.4
10.8	**27.5**	5 V	Vertex to Acromion Process	20.6	18.3	13.8	9.2	6.9
48.2	**122.5**	6 V	Acromion Process to Sole	91.9	81.7	61.3	40.8	30.6
33.9	**86.0**	* 7 V	Ant.Sup.Spine of Ilium to Sole	64.5	57.3	43.0	28.7	21.5
14.2	**36.0**	8 V	Acrom. proc. to Ant. Sup. Spine	27.0	24.0	18.0	12.0	9.0
16.9	**43.0**	9 V	Ant.Sup.Spine of Ilium to Knee	32.3	28.7	21.5	14.3	10.8
17.5	**44.5**	10 V	Knee at Patella to Sole	33.4	29.7	22.3	14.8	11.1
35.4	**90.0**	11 V	Umbilicus to Sole	67.5	60.0	45.0	30.0	22.5
27.2	**69.0**	* 12 V	Crotch to Sole	51.8	46.0	34.5	23.0	17.3
44.1	**112.0**	13 V	Kneeling Height	84.0	74.7	56.0	37.3	28.0
29.9	**76.0**	14 V	Sitting Height	57.0	50.7	38.0	25.3	19.0
29.9	**76.0**	15 V	Buttock, Straight Leg to Sole	57.0	50.7	38.0	25.3	19.0
19.7	**50.0**	16 V	Buttock to Flexed Knee	37.5	33.3	25.0	16.7	12.5
18.9	**48.0**	17 V	Flexed Knee to Sole	36.0	32.0	24.0	16.0	12.0
		These are also listed with Lower Extremity Measurements.						
			VERTICAL HEAD MEASUREMENTS					
11.0	**28.0**	H1 V	Vertex to Sternal Notch	21.0	18.7	14.0	9.3	7.0
8.3	**21.0**	H2 V	Vertex to Chin, Head Height	15.8	14.0	10.5	7.0	5.3
6.9	**17.4**	H3 V	Hairline to Chin, Face Height	13.1	11.6	8.7	5.8	4.4
4.3	**10.9**	H4 V	Vertex to Lateral Corner of Eye	8.2	7.3	5.5	3.6	2.7
4.9	**12.5**	H5 V	Vertex to Tragion of Ear	9.4	8.3	6.3	4.2	3.1
2.0	**5.0**	H6 V	Hairline to Brow, Glabella	3.8	3.3	2.5	1.7	1.3
4.8	**12.3**	H7 V	Brow, Glabella, to Chin, Menton	9.2	8.2	6.2	4.1	3.1
2.4	**6.0**	H8 V	Brow to Subnasal Point	4.5	4.0	3.0	2.0	1.5
1.7	**4.4**	H9 V	Sellion to Subnasal, Nose Length	3.3	2.9	2.2	1.5	1.1
4.2	**10.7**	H10 V	Sellion, Nose Saddle, to Chin	8.0	7.1	5.4	3.6	2.7
2.5	**6.3**	H11 V	Subnasal Point to Chin	4.7	4.2	3.2	2.1	1.6
0.9	**2.3**	H12 V	Subnasal to Lip Aperature	1.7	1.5	1.2	0.8	0.6
0.5	**1.3**	H13 V	Lip Aperat. to Sublabial Crease	1.0	0.9	0.7	0.4	0.3
1.0	**2.6**	H14 V	Sublabial Crease to Chin	2.0	1.7	1.3	0.9	0.7
			TRANSVERSE HEAD MEASUREMENTS					
5.7	**14.6**	H15 T	Bi-Parietal, Head Breadth	11.0	9.7	7.3	4.9	3.7
4.9	**12.5**	H16 T	Bi-Zygomatic, Cheekbones	9.4	8.3	6.3	4.2	3.1
4.4	**11.2**	H17 T	Bi-Fronto Temporal, Forehead	8.4	7.5	5.6	3.7	2.8

Life size			TWELVE-YEAR-OLD BOY	Fractions of life size				
				3/4	2/3	1/2	1/3	1/4
Inches	cm	Code		cm	cm	cm	cm	cm
			TRANSVERSE HEAD MEASUREMENTS					
3.7	**9.4**	H18 T	Inter Brow Ridge	7.1	6.3	4.7	3.1	2.4
4.8	**12.2**	H19 T	Inter Tragion of Ears	9.2	8.1	6.1	4.1	3.1
4.4	**11.3**	H20 T	Inter Angle of Mandible	8.5	7.5	5.7	3.8	2.8
3.6	**9.2**	H21 T	Inter Lateral Corners of Eyes	6.9	6.1	4.6	3.1	2.3
1.2	**3.1**	H22 T	Width of Eye	2.3	2.1	1.6	1.0	0.8
1.2	**3.0**	H23 T	Inter Inner Corner of eyes	2.3	2.0	1.5	1.0	0.8
1.3	**3.2**	H24 T	Breadth of Nose	2.4	2.1	1.6	1.1	0.8
1.8	**4.5**	H25 T	Breadth of Mouth	3.4	3.0	2.3	1.5	1.1
1.3	**3.4**	H26 T	Breadth of Chin	2.6	2.3	1.7	1.1	0.9
3.5	**9.0**	H27 T	Breadth of Neck	6.8	6.0	4.5	3.0	2.3
			ANTERIOR-POSTERIOR HEAD MEASUREMENTS					
7.5	**19.1**	H28 D	Brow to Occiput	14.3	12.7	9.6	6.4	4.8
3.9	**10.0**	H29 D	Brow to Tragus	7.5	6.7	5.0	3.3	2.5
3.6	**9.2**	H30 D	Sellion to Tragus	6.9	6.1	4.6	3.1	2.3
3.6	**9.1**	H31 D	Tragus to Occiput	6.8	6.1	4.6	3.0	2.3
2.2	**5.7**	H32 V	Tragus to Angle Mandible	4.3	3.8	2.9	1.9	1.4
3.3	**8.5**	H33 D	Angle of Mandible to Chin	6.4	5.7	4.3	2.8	2.1
2.2	**5.7**	H34 V	Length of Ear, Vertical	4.3	3.8	2.9	1.9	1.4
1.4	**3.5**	H35 D	Width of Ear, Anterior-Posterior	2.6	2.3	1.8	1.2	0.9
3.4	**8.7**	H36 D	Anterior to Posterior of Neck	6.5	5.8	4.4	2.9	2.2
			VERTICAL BODY MEASUREMENTS					
3.1	**8.0**	B1 V	Chin to Sternal Notch, Neck Pit	6.0	5.3	4.0	2.7	2.0
3.1	**8.0**	B2 V	Sternal Notch to Axilla, Arm Pit	6.0	5.3	4.0	2.7	2.0
4.5	**11.5**	B3 V	Sternal Notch to Nipples	8.6	7.7	5.8	3.8	2.9
5.9	**15.0**	B4 V	Sternal Notch to Xyphoid Proc.	11.3	10.0	7.5	5.0	3.8
12.2	**31.0**	B5 V	Sternal Notch to Umbilicus	23.3	20.7	15.5	10.3	7.8
6.3	**16.0**	B6 V	Xyphoid Process to Umbilicus	12.0	10.7	8.0	5.3	4.0
11.8	**30.0**	B7 V	Sternal Notch to Crest Ilium	22.5	20.0	15.0	10.0	7.5
13.8	**35.0**	B8 V	Sternal Notch to Ant.S.Sp.Ilium	26.3	23.3	17.5	11.7	8.8
18.1	**46.0**	B9 V	Sternal Notch to Pubis	34.5	30.7	23.0	15.3	11.5
5.9	**15.0**	B10 V	Umbilicus to Pubis	11.3	10.0	7.5	5.0	3.8
			TRANSVERSE BODY MEASUREMENTS					
14.2	**36.0**	B11 T	Shoulder Breadth	27.0	24.0	18.0	12.0	9.0
6.3	**16.0**	B12 T	Clavicle to Acromion Process	12.0	10.7	8.0	5.3	4.0
13.0	**33.0**	B13 T	Biacromial Process Breadth	24.8	22.0	16.5	11.0	8.3
8.9	**22.5**	B14 T	Inter Anterior Axillary Fold	16.9	15.0	11.3	7.5	5.6
6.1	**15.5**	B15 T	Inter Nipples	11.6	10.3	7.8	5.2	3.9

Life size			TWELVE-YEAR-OLD BOY	Fractions of life size				
				3/4	2/3	1/2	1/3	1/4
Inches	cm	Code		cm	cm	cm	cm	cm
			TRANSVERSE BODY MEASUREMENTS					
9.1	**23.0**	B16 T	Breadth of Thorax	17.3	15.3	11.5	7.7	5.8
8.7	**22.0**	B17 T	Inter Costal Margin Breadth	16.5	14.7	11.0	7.3	5.5
9.1	**23.0**	B18 T	Inter Crest of Ilium Breadth	17.3	15.3	11.5	7.7	5.8
7.7	**19.5**	B19 T	Inter Anterior Superior Spine	14.6	13.0	9.8	6.5	4.9
10.4	**26.5**	B20 T	Inter Greater Trochanter Femur	19.9	17.7	13.3	8.8	6.6
			ANTERIOR-POSTERIOR BODY MEASUREMENTS					
6.7	**17.0**	B21 D	Manubrium to Dorsal Spine	12.8	11.3	8.5	5.7	4.3
6.5	**16.5**	B22 D	Xyphoid Process to Dorsum	12.4	11.0	8.3	5.5	4.1
6.3	**16.0**	B23 D	Umbilicus to Dorsum	12.0	10.7	8.0	5.3	4.0
7.1	**18.0**	B24 D	Pubis to Buttock	13.5	12.0	9.0	6.0	4.5
5.5	**14.0**	B25 D	Thigh to Gluteal Furrow	10.5	9.3	7.0	4.7	3.5
			VERTICAL BACK MEASUREMENTS					
8.7	**22.0**	B26 V	Vertex to 7th Cervical Spine	16.5	14.7	11.0	7.3	5.5
			7th Cervical Vertebra Spine to:					
5.9	**15.0**	B27 V	To Posterior Axillary Fold	11.3	10.0	7.5	5.0	3.8
7.3	**18.5**	B28 V	To Inferior Tip of Scapula	13.9	12.3	9.3	6.2	4.6
15.4	**39.0**	B29 V	To Posterior Superior Spine Ilium	29.3	26.0	19.5	13.0	9.8
18.1	**46.0**	B30 V	To Midline Upper Buttock Crease	34.5	30.7	23.0	15.3	11.5
21.7	**55.0**	B31 V	To Gluteal Furrow	41.3	36.7	27.5	18.3	13.8
33.9	**86.0**	B32 V	To Popliteal Crease of Knee	64.5	57.3	43.0	28.7	21.5
			TRANSVERSE BACK MEASUREMENTS					
11.4	**29.0**	B33 T	Inter Posterior Axillary Folds	21.8	19.3	14.5	9.7	7.3
5.1	**13.0**	B34 T	Inter Scapulae Medial Borders	9.8	8.7	6.5	4.3	3.3
2.8	**7.0**	B35 T	Inter Post. Sup. Spines Sacrum	5.3	4.7	3.5	2.3	1.8
			UPPER EXTREMITY MEASUREMENTS					
58.7	**149.0**	U1 V	Span of Arms	111.8	99.3	74.5	49.7	37.3
6.3	**16.0**	U2 V	Sternum to Acromion Process	12.0	10.7	8.0	5.3	4.0
25.2	**64.0**	U3 V	Acromion Process to Fingertip	48.0	42.7	32.0	21.3	16.0
74.8	**190.0**	U4 V	Vertical Reach to Fingertip	142.5	126.7	95.0	63.3	47.5
35.8	**91.0**	U5 T	Lateral Reach to Fingertip	68.3	60.7	45.5	30.3	22.8
29.1	**74.0**	U6 D	Forward Reach to Fingertip	55.5	49.3	37.0	24.7	18.5
12.2	**31.0**	U7 V	Shoulder to Elbow Flexed	23.3	20.7	15.5	10.3	7.8
11.2	**28.5**	U8 V	Acromion to Lat. Condyle Elbow	21.4	19.0	14.3	9.5	7.1
15.7	**40.0**	U9 D	Flexed Elbow to Fingertip	30.0	26.7	20.0	13.3	10.0
8.9	**22.5**	U10 D	Flexed Elbow Pivot to Wrist Pivot	16.9	15.0	11.3	7.5	5.6

Life size			TWELVE-YEAR-OLD BOY	Fractions of life size				
				3/4	2/3	1/2	1/3	1/4
Inches	cm	Code		cm	cm	cm	cm	cm
			UPPER EXTREMITY MEASUREMENTS					
6.5	**16.5**	U11 V	Wrist to Fingertip	12.4	11.0	8.3	5.5	4.1
4.3	**11.0**	U12 V	Wrist to Thumb Tip	8.3	7.3	5.5	3.7	2.8
3.7	**9.3**	U13 V	Length of Palm	7.0	6.2	4.7	3.1	2.3
2.8	**7.0**	U14 V	Length 3rd Finger	5.3	4.7	3.5	2.3	1.8
4.5	**11.4**	U15 V	Thumb Crotch to 3rd Finger	8.6	7.6	5.7	3.8	2.9
2.2	**5.7**	U16 V	Thumb Crotch to Thumb Tip	4.3	3.8	2.9	1.9	1.4
2.6	**6.5**	U17 D	Anterior-Post. Arm at Biceps	4.9	4.3	3.3	2.2	1.6
2.6	**6.5**	U18 D	Anterior-Posterior at Forearm	4.9	4.3	3.3	2.2	1.6
1.4	**3.5**	U19 D	Depth of Wrist	2.6	2.3	1.8	1.2	0.9
2.0	**5.0**	U20 T	Breadth of Wrist	3.8	3.3	2.5	1.7	1.3
3.6	**9.1**	U21 T	Breadth of Hand	6.8	6.1	4.6	3.0	2.3
2.9	**7.4**	U22 T	Breadth of Palm	5.6	4.9	3.7	2.5	1.9
			LOWER EXTREMITY MEASUREMENTS					
36.2	**92.0**	L1 V	Crest of Ilium to Sole	69.0	61.3	46.0	30.7	23.0
33.9	**86.0**	* L2 V	Ant.Sup.Spine of Ilium to Sole	64.5	57.3	43.0	28.7	21.5
30.3	**77.0**	L3 V	Greater Trochanter Femur-Sole	57.8	51.3	38.5	25.7	19.3
29.9	**76.0**	* L4 V	Symphysis Pubis to Sole	57.0	50.7	38.0	25.3	19.0
27.2	**69.0**	* L5 V	Crotch to Sole	51.8	46.0	34.5	23.0	17.3
28.3	**72.0**	L6 V	Gluteal Furrow to Sole	54.0	48.0	36.0	24.0	18.0
16.1	**41.0**	L7 V	Posterior Knee Crease to Sole	30.8	27.3	20.5	13.7	10.3
15.9	**40.5**	L8 V	Tibia at Knee to Sole	30.4	27.0	20.3	13.5	10.1
2.4	**6.0**	L9 V	Medial Maleolus to Sole	4.5	4.0	3.0	2.0	1.5
2.0	**5.0**	L10 V	Latreal Maleolus to Sole	3.8	3.3	2.5	1.7	1.3
10.2	**26.0**	L11 T	Breadt of Hips, Sitting	19.5	17.3	13.0	8.7	6.5
5.1	**13.0**	L12 T	Breadth Mid Thigh	9.8	8.7	6.5	4.3	3.3
3.9	**10.0**	L13 T	Breadth of Knee	7.5	6.7	5.0	3.3	2.5
3.5	**9.0**	L14 T	Breadth of Calf	6.8	6.0	4.5	3.0	2.3
2.0	**5.0**	L15 T	Breadth Ankle	3.8	3.3	2.5	1.7	1.3
5.1	**13.0**	L16 D	Mid Thigh, Anterior-Posterior	9.8	8.7	6.5	4.3	3.3
4.3	**11.0**	L17 D	Knee, Anerior- Posterior	8.3	7.3	5.5	3.7	2.8
3.5	**9.0**	L18 D	Mid Calf, Anterior-Posterior	6.8	6.0	4.5	3.0	2.3
2.8	**7.0**	L19 D	Ankle, Anterior-Posterior	5.3	4.7	3.5	2.3	1.8
9.3	**23.5**	L20 D	Length of Foot	17.6	15.7	11.8	7.8	5.9
7.1	**18.0**	L21 D	Heel to 5th Toe	13.5	12.0	9.0	6.0	4.5
3.5	**9.0**	L22 T	Breadth Foot	6.8	6.0	4.5	3.0	2.3
2.0	**5.0**	L23 T	Width of Heel	3.8	3.3	2.5	1.7	1.3
		** These are also listed with Vertical Measurements.*						

Phases of Female Breast Development

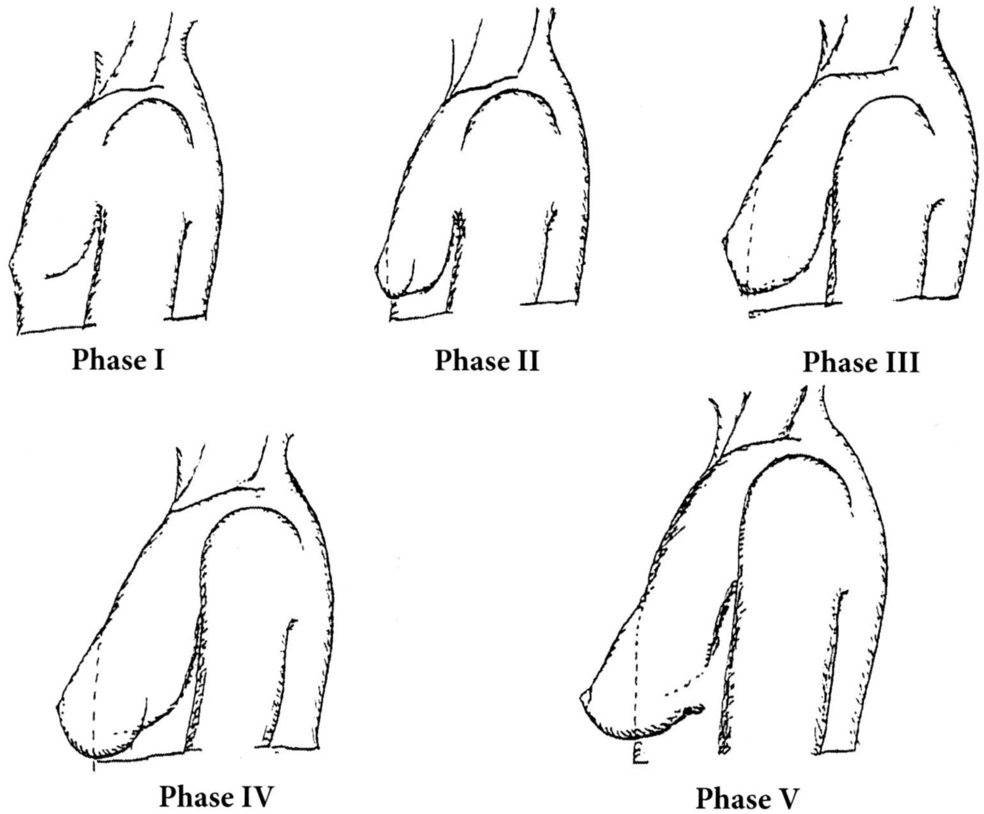

Phase I is preadolescent. It may be up to twelve years but is usually prior to age eleven years. The breasts are flat and the areola is unpigmented and immature.

Phase II is early adolescent. Age is variable and the gradual onset may be between age eight and thirteen years. It is usually after age eleven. The breasts demonstrate an elevation of the nipple with a small mound underneath the areola that begins to have some pigmentation. A small growth of pubertal hair develops on the mons. A rapid growth of stature usually begins.

Phase III is mid-adolescent. The age is from ten to fourteen. The peak of rapid growth of stature has passed, but growth continues. Menarche is most often at this stage. Breasts demonstrate further development with pigmentation of the nipples and areolas. Glandular development and contours increase.

Phase IV is post mid-adolescence. The age is variable between twelve and fifteen, but usually after thirteen years. The breasts have increased development and contour. The areola projects above the surrounding skin of the breast.

Phase V demonstrates a mature breast configuration and hemispherical contour. The age is from thirteen to seventeen, but is usually after fourteen. The nipple is erect and the areola is mature. (Lowrey, G.H., pg, 351)

Male Growth and Development by L.A.J. Quetelet

Lambert Adolphe Jaques Quetelet, a Belgian statistician and astronomer was also interested in anthropology. He applied mathematical methods in the study of human proportion and development. His studies were published in *Sur l'homme et le developpement de ses faculties* (1869) and in *L' anthropometrie* (1871).

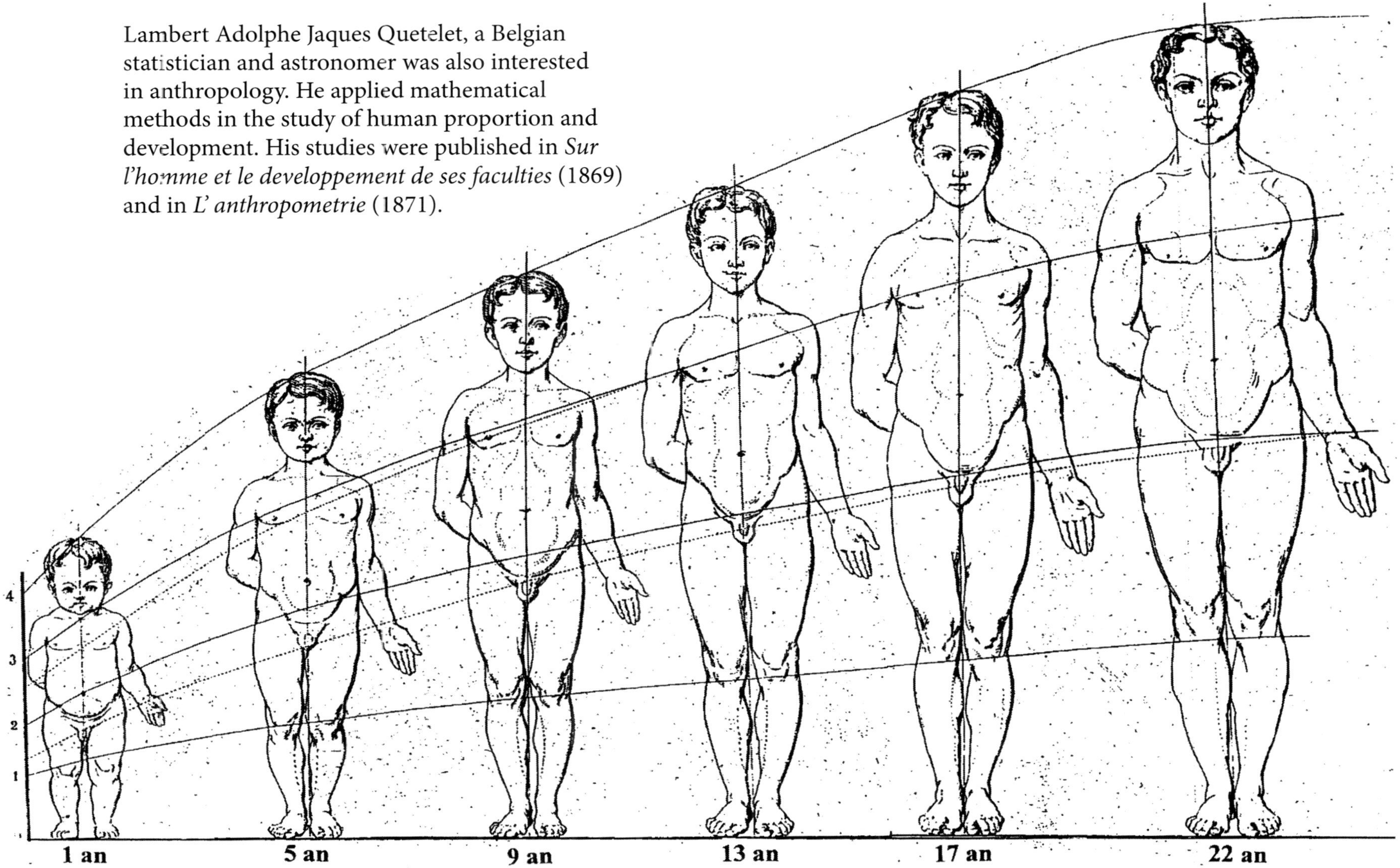

Female Profile Comparisons During Growth

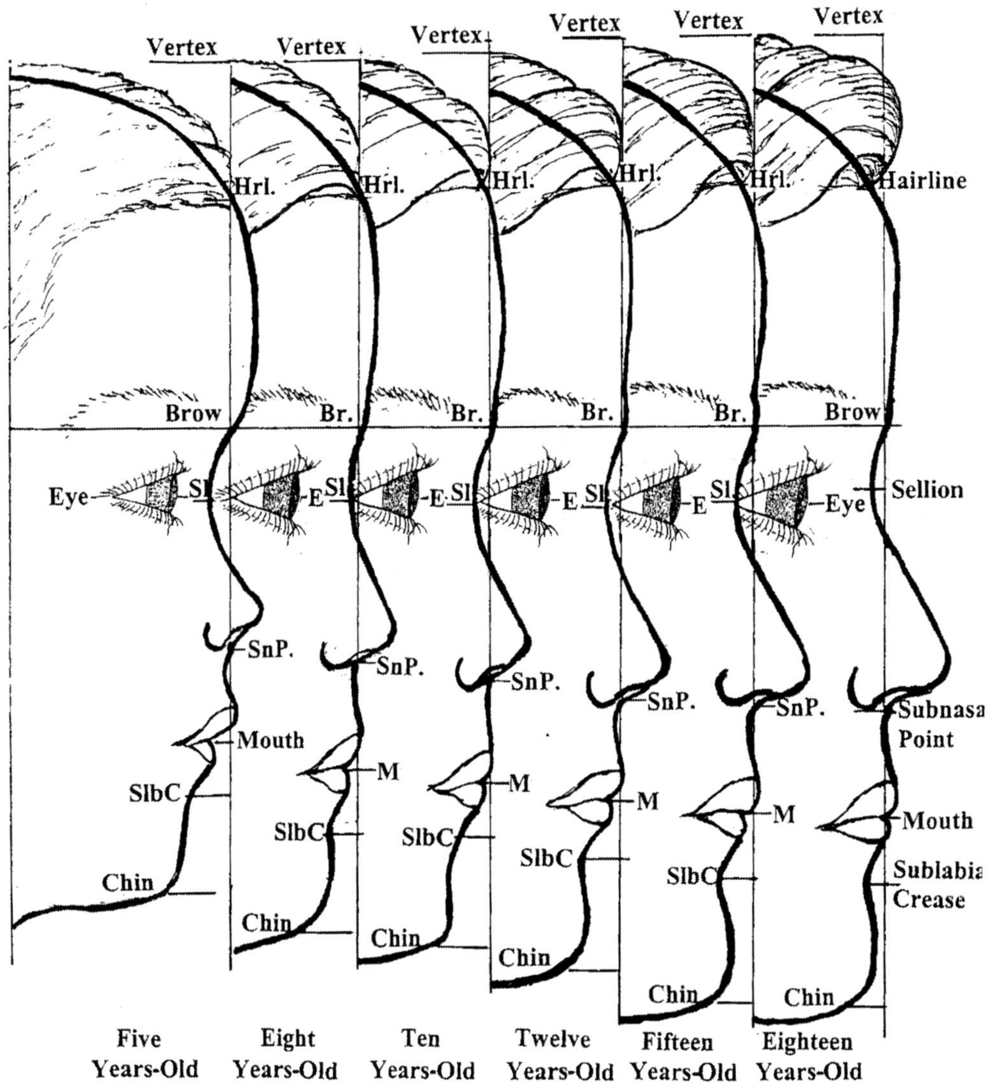

PROFILES OF FEMALE FACES

Superimposed female face profiles are illustrated to demonstrate progressive changes in age five through eighteen years of age. These are intended to be representative, using measurements in the tables. There are considerable individual variations, particularly in soft tissue measurements, and a live model is recommended for more accurate details.

For comparison, horizontal alignments are at each image's brow. Vertical alignments are from the brow to the subnasal point. From age five years to adult, there is little increase in the brow to vertex measurement. In early teens, the growth of the mid-face features, from the brow to the subnasal point, may be more rapid than the lower face growth, from the subnasal point to the chin. Most of the female face growth occurs by fifteen years of age. By eighteen years of age, the face is nearly mature. The brow, nose, and chin are less prominent than similar features in male subjects.

Male Profile Comparisons During Growth

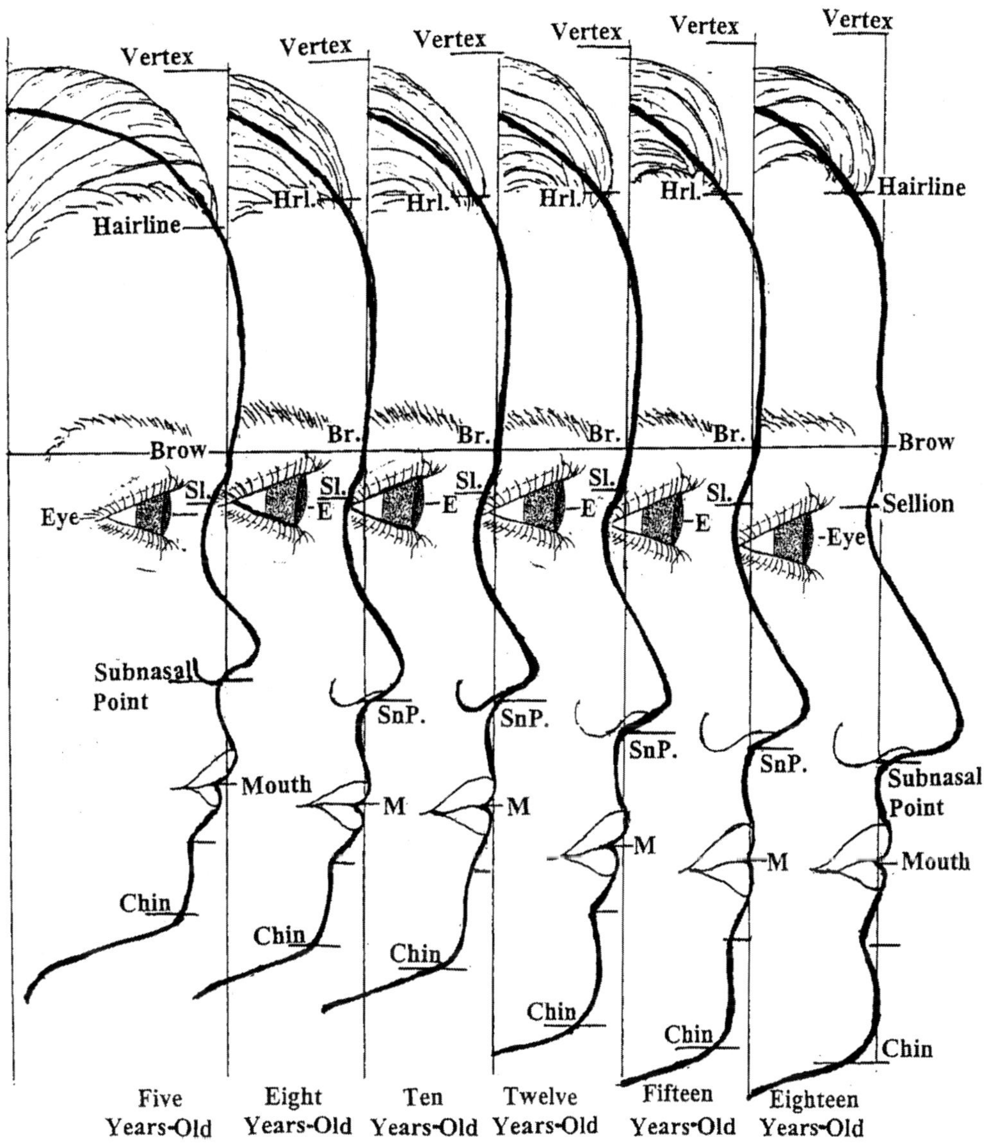

PROFILES OF MALE FACES

Representative face profiles are illustrated of males ages five through eighteen years of age. Since there are considerable individual variations, this is intended to demonstrate general trends of facial growth. Average measurements are used as listed in charts.

The faces are aligned horizontally to individual brows. Vertical alignments are with the brows and subnasal points. The comparative illustrations demonstrate that there is a slight increase growth of the brow to vertex from childhood to nearly adulthood. More growth of the mid face is significant from the brow to the subnasal point. Even greater growth of the lower face occurs, from the subnasal point to the chin, when permanent dentition matures in the upper and lower jaws. Male features include a prominent brow, a longer nose, a more prominent chin, and more angular features than a female face.

THIRTEEN-YEAR-OLD GIRLS

At this age, most girls have begun to mature. It is earlier than boys and they gain weight and height more rapidly. Many at this age are plump. They develop an interest in dancing and more attention to clothing. Hair styling becomes more important and some may begin using cosmetics. Breast development is Phase III with more developed nipples and areola. There is an increase in glandular and pectoral fatty tissue.[1] The range of stature is listed here for comparison. School level is eighth grade.

Percentile	Twenty-fifth		Fiftieth		Seventy-fifth	
Girls:	Short		Medium		Tall	
Stature	153 cm,	60.3 in.	157 cm,	61.8 in.	161 cm,	63.5 in.
Weight	40.5 kg,	89.5 lb.	46 kg,	101.8 lb.	53 kg,	116.8 lb.[2]

BODY MEASUREMENTS COMPARED TO HEIGHT

	Feature Measure in cm	Feature Compared to Stature	Approx. Ratio of Feature	Compared to Median Adult
Stature: Crown to sole	**157**			**93%**
Head Height: Crown to chin	**20.2**	**13%**	**1/8**	**99%**
Sitting Height:	**82**	**52%**	**1/2**	**92%**
Trunk Measure: Sternal notch to symphysis pubis	**48**	**31%**	**1/3–**	**90%**
Leg Length: Greater trochanter of the femur to sole	**80.5**	**51%**	**1/2**	**94%**
Thigh Length: Greater trochanter of the femur to tibia at knee	**39.5**	**29%**	**1/3–**	**93%**
Tibia at Knee to Sole:	**41**	**29%**	**1/4**	**94%**
Arm Length: Acromion process to dependent fingertip	**69**	**44%**	**2/5+**	**92%**
Hand Length:	**17**	**11%**	**1/9**	**94%**
Foot Length:	**23.4**	**15%**	**1/7**	**97%**

By this age, girls achieve nearly adult female measurements for most features with about the same ratios, except thigh length. Many facial features are similar to boys, since the growth of bones and dentition is similar. The upper and lower second molars erupt between twelve and thirteen years of age.[3] The face of most teenage girls has an oval outline with graceful contours. At this age, the relative proportions are nearly the same for the adult female, as noted in the third and fourth column.

1. Lowery, G., p. 351.
2. Nelson, W.E., p. 39.
3. Lockhart, G., p. 57.

THIRTEEN-YEAR-OLD BOYS

Before age ten, boys on average, are slightly taller and slightly heavier than female children. However, in early youth, about ten years, girls begin to mature earlier and grow more rapidly for several years. At age thirteen, girls may be a half head taller than boys of the same school class. Participation in vigorous and outdoor sports is popular for boys. The range of height and weights are listed here for comparison. The school level is eighth grade.

Percentile	Twenty-fifth		Fiftieth		Seventy-fifth	
Boys:	Short		Medium		Tall	
Stature	150.5,	59.3 in.	156.5 cm,	61.5 in.	161.8 cm,	63.8 in.
Weight	39.7 kg,	87.5 lb.	45 kg,	99 lb.	51.8 kg,	114.3 lb.[1]

BODY MEASUREMENTS COMPARED TO HEIGHT

	Feature Measure in cm	Feature Compared to Stature	Approx. Ratio of Feature	Compared to Median Adult
Stature: Crown to sole	**157**			**86%**
Head Height: Crown to chin	**20.5**	**13%**	**1/8**	**93%**
Sitting Height:	**78**	**50%**	**1/2**	**85%**
Trunk Measure: Sternal notch to symphysis pubis	**45**	**29%**	**1/3–**	**85%**
Leg Length: Greater trochanter of the femur to sole	**80**	**51%**	**1/2**	**81%**
Thigh Length: Greater trochanter of the femur to tibia at knee	**38**	**24%**	**1/4**	**81%**
Tibia at Knee to Sole:	**42**	**27%**	**1/4+**	**82%**
Arm Length: Acromion process at shoulder to dependent fingertip	**70**	**45%**	**1/2–**	**88%**
Hand Length:	**17**	**11%**	**1/9**	**89%**
Foot Length:	**24**	**16%**	**1/6**	**90%**

By this age, boys have achieved between eighty and ninety percent of adult male measurements of most features, except leg and thigh length. Muscular development has improved. Many facial features are similar to girls, since the growth of bones and dentition is similar. The face of most teenage youth has an oval outline with graceful contours. The upper and lower second molars erupt between twelve and thirteen years of age.[2]

1. Nelson, W.E., p. 38.
2. Lockhart, R.D., p. 57.

Thirteen-Year-Old Girl, anterior and lateral head

all measurements in centimeters

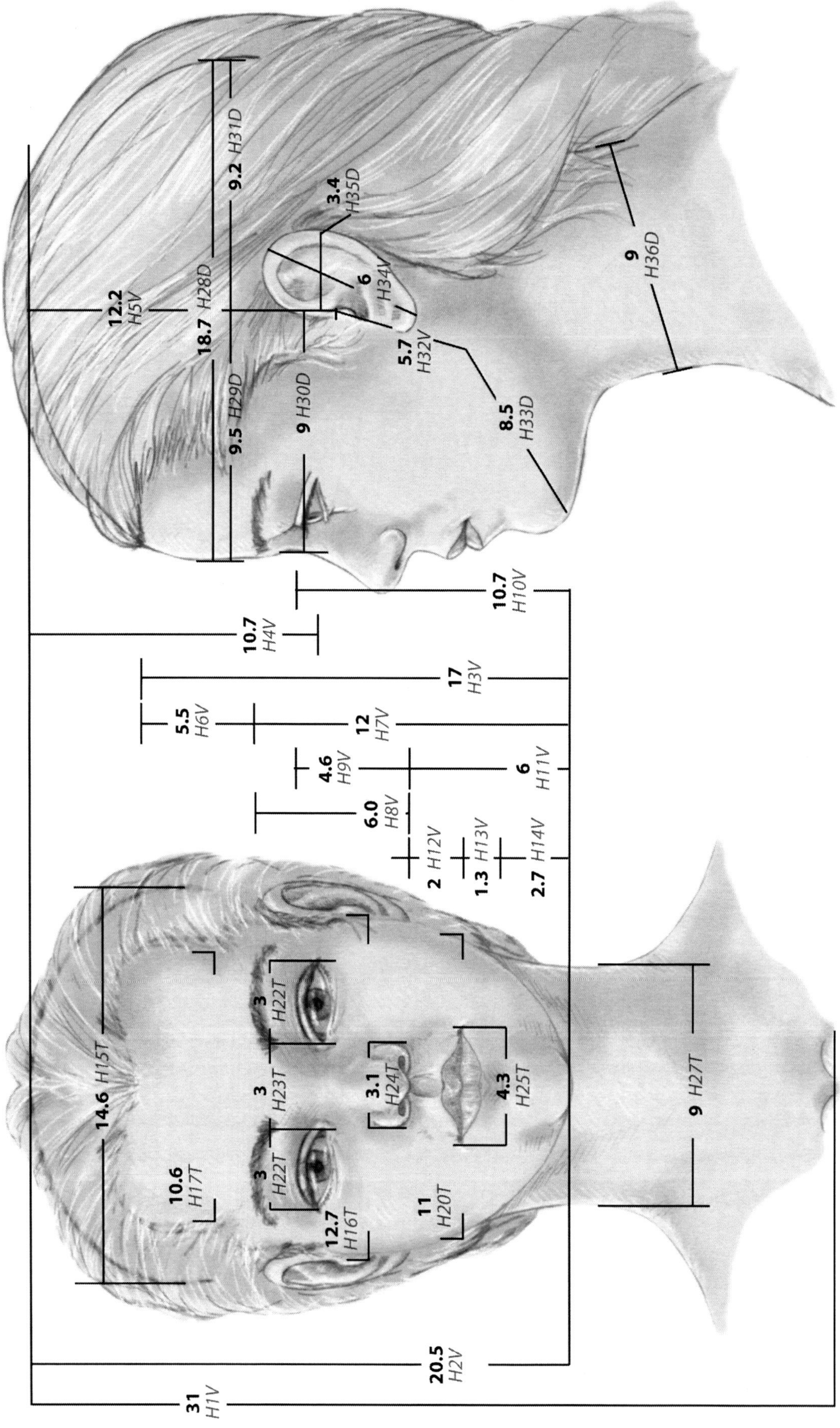

Thirteen-Year-Old Girl, anterior and lateral figure

all measurements in centimeters

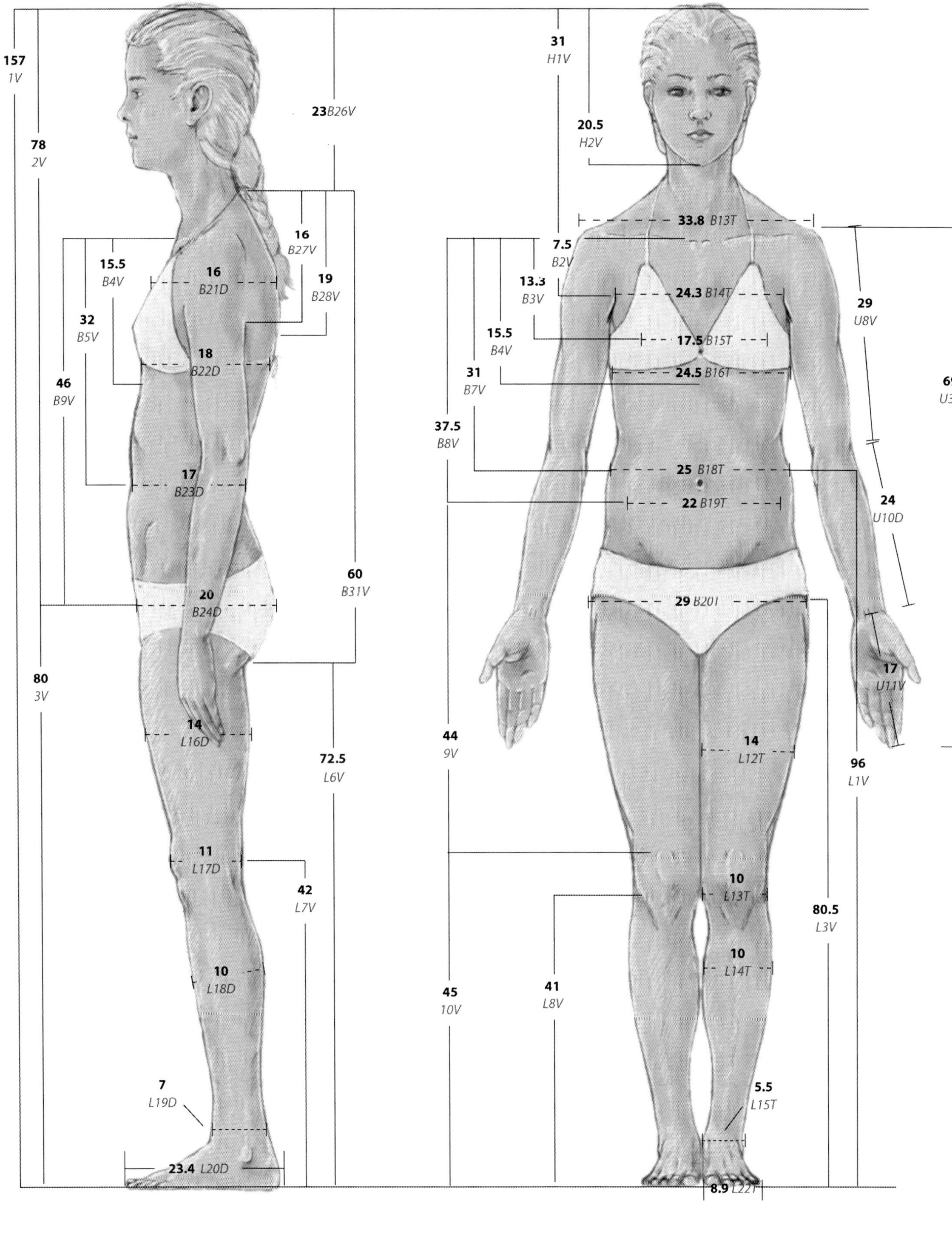

Life size			THIRTEEN-YEAR-OLD GIRL	Fractions of life size				
				3/4	2/3	1/2	1/3	1/4
Inches	cm	Code		cm	cm	cm	cm	cm
			VERTICAL MEASUREMENTS					
61.8	**157.0**	1 V	Stature, Height, Vertex to Sole	117.8	104.7	78.5	52.3	39.3
30.7	**78.0**	2 V	Vertex to Symphysis Pubis	58.5	52.0	39.0	26.0	19.5
31.5	**80.0**	* 3 V	Symphysis Pubis to Sole	60.0	53.3	40.0	26.7	20.0
49.6	**126.0**	4 V	Sternal Notch to Sole	94.5	84.0	63.0	42.0	31.5
11.8	**30.0**	5 V	Vertex to Acromion Process	22.5	20.0	15.0	10.0	7.5
50.0	**127.0**	6 V	Acromion Process to Sole	95.3	84.7	63.5	42.3	31.8
35.0	**89.0**	* 7 V	Ant.Sup.Spine of Ilium to Sole	66.8	59.3	44.5	29.7	22.3
15.0	**38.0**	8 V	Acrom. proc. to Ant. Sup. Spine	28.5	25.3	19.0	12.7	9.5
17.3	**44.0**	9 V	Ant.Sup.Spine of Ilium to Knee	33.0	29.3	22.0	14.7	11.0
17.7	**45.0**	10 V	Knee at Patella to Sole	33.8	30.0	22.5	15.0	11.3
37.0	**94.0**	11 V	Umbilicus to Sole	70.5	62.7	47.0	31.3	23.5
28.0	**71.0**	* 12 V	Crotch to Sole	53.3	47.3	35.5	23.7	17.8
45.3	**115.0**	13 V	Kneeling Height	86.3	76.7	57.5	38.3	28.8
32.3	**82.0**	14 V	Sitting Height	61.5	54.7	41.0	27.3	20.5
33.1	**84.0**	15 V	Buttock, Straight Leg to Sole	63.0	56.0	42.0	28.0	21.0
20.9	**53.0**	16 V	Buttock to Flexed Knee	39.8	35.3	26.5	17.7	13.3
19.1	**48.5**	17 V	Flexed Knee to Sole	36.4	32.3	24.3	16.2	12.1
		** These are also listed with Lower Extremity Measurements.*						
			VERTICAL HEAD MEASUREMENTS					
12.2	**31.0**	H1 V	Vertex to Sternal Notch	23.3	20.7	15.5	10.3	7.8
8.1	**20.5**	H2 V	Vertex to Chin, Head Height	15.4	13.7	10.3	6.8	5.1
6.7	**17.0**	H3 V	Hairline to Chin, Face Height	12.8	11.3	8.5	5.7	4.3
4.2	**10.7**	H4 V	Vertex to Lateral Corner of Eye	8.0	7.1	5.4	3.6	2.7
4.8	**12.2**	H5 V	Vertex to Tragion of Ear	9.2	8.1	6.1	4.1	3.1
2.2	**5.5**	H6 V	Hairline to Brow, Glabella	4.1	3.7	2.8	1.8	1.4
4.6	**11.7**	H7 V	Brow, Glabella, to Chin, Menton	8.8	7.8	5.9	3.9	2.9
2.2	**5.7**	H8 V	Brow to Subnasal Point	4.3	3.8	2.9	1.9	1.4
1.8	**4.6**	H9 V	Sellion to Subnasal, Nose Length	3.5	3.1	2.3	1.5	1.2
4.2	**10.7**	H10 V	Sellion, Nose Saddle, to Chin	8.0	7.1	5.4	3.6	2.7
2.4	**6.0**	H11 V	Subnasal Point to Chin	4.5	4.0	3.0	2.0	1.5
0.8	**2.1**	H12 V	Subnasal to Lip Aperature	1.6	1.4	1.1	0.7	0.5
0.5	**1.3**	H13 V	Lip Aperat. to Sublabial Crease	1.0	0.9	0.7	0.4	0.3
1.1	**2.7**	H14 V	Sublabial Crease to Chin	2.0	1.8	1.4	0.9	0.7
			TRANSVERSE HEAD MEASUREMENTS					
5.7	**14.6**	H15 T	Bi-Parietal, Head Breadth	11.0	9.7	7.3	4.9	3.7
5.0	**12.7**	H16 T	Bi-Zygomatic, Cheekbones	9.5	8.5	6.4	4.2	3.2
4.2	**10.6**	H17 T	Bi-Fronto Temporal, Forehead	8.0	7.1	5.3	3.5	2.7

Life size			THIRTEEN-YEAR-OLD GIRL	Fractions of life size				
				3/4	2/3	1/2	1/3	1/4
Inches	cm	Code		cm	cm	cm	cm	cm
			TRANSVERSE HEAD MEASUREMENTS					
3.9	**9.8**	H18 T	Inter Brow Ridge	7.4	6.5	4.9	3.3	2.5
4.8	**12.3**	H19 T	Inter Tragion of Ears	9.2	8.2	6.2	4.1	3.1
4.3	**11.0**	H20 T	Inter Angle of Mandible	8.3	7.3	5.5	3.7	2.8
3.5	**9.0**	H21 T	Inter Lateral Corners of Eyes	6.8	6.0	4.5	3.0	2.3
1.2	**3.0**	H22 T	Width of Eye	2.3	2.0	1.5	1.0	0.8
1.2	**3.0**	H23 T	Inter Inner Corner of eyes	2.3	2.0	1.5	1.0	0.8
1.2	**3.1**	H24 T	Breadth of Nose	2.3	2.1	1.6	1.0	0.8
1.7	**4.3**	H25 T	Breadth of Mouth	3.2	2.9	2.2	1.4	1.1
1.2	**3.0**	H26 T	Breadth of Chin	2.3	2.0	1.5	1.0	0.8
3.5	**9.0**	H27 T	Breadth of Neck	6.8	6.0	4.5	3.0	2.3
			ANTERIOR-POSTERIOR HEAD MEASUREMENTS					
7.4	**18.7**	H28 D	Brow to Occiput	14.0	12.5	9.4	6.2	4.7
3.7	**9.5**	H29 D	Brow to Tragus	7.1	6.3	4.8	3.2	2.4
3.5	**9.0**	H30 D	Sellion to Tragus	6.8	6.0	4.5	3.0	2.3
3.6	**9.2**	H31 D	Tragus to Occiput	6.9	6.1	4.6	3.1	2.3
2.2	**5.7**	H32 V	Tragus to Angle Mandible	4.3	3.8	2.9	1.9	1.4
3.3	**8.5**	H33 D	Angle of Mandible to Chin	6.4	5.7	4.3	2.8	2.1
2.4	**6.0**	H34 V	Length of Ear, Vertical	4.5	4.0	3.0	2.0	1.5
1.3	**3.4**	H35 D	Width of Ear, Anterior-Posterior	2.6	2.3	1.7	1.1	0.9
3.5	**9.0**	H36 D	Anterior to Posterior of Neck	6.8	6.0	4.5	3.0	2.3
			VERTICAL BODY MEASUREMENTS					
3.9	**10.0**	B1 V	Chin to Sternal Notch	7.5	6.7	5.0	3.3	2.5
3.0	**7.5**	B2 V	Sternal Notch to Axilla	5.6	5.0	3.8	2.5	1.9
5.2	**13.2**	B3 V	Sternal Notch to Nipples	9.9	8.8	6.6	4.4	3.3
6.1	**15.5**	B4 V	Sternal Notch to Xyphoid Proc.	11.6	10.3	7.8	5.2	3.9
12.6	**32.0**	B5 V	Sternal Notch to Umbilicus	24.0	21.3	16.0	10.7	8.0
5.9	**15.0**	B6 V	Xyphoid Process to Umbilicus	11.3	10.0	7.5	5.0	3.8
12.2	**31.0**	B7 V	Sternal Notch to Crest Ilium	23.3	20.7	15.5	10.3	7.8
14.8	**37.5**	B8 V	Sternal Notch to Ant.S.Sp.Ilium	28.1	25.0	18.8	12.5	9.4
18.1	**46.0**	B9 V	Sternal Notch to Pubis	34.5	30.7	23.0	15.3	11.5
5.5	**14.0**	B10 V	Umbilicus to Pubis	10.5	9.3	7.0	4.7	3.5
			TRANSVERSE BODY MEASUREMENTS					
14.6	**37.0**	B11 T	Shoulder Breadth	27.8	24.7	18.5	12.3	9.3
6.5	**16.5**	B12 T	Clavicle to Acromion Process	12.4	11.0	8.3	5.5	4.1
13.3	**33.8**	B13 T	Biacromial Process Breadth	25.4	22.5	16.9	11.3	8.5
9.6	**24.3**	B14 T	Inter Anterior Axillary Fold	18.2	16.2	12.2	8.1	6.1
6.9	**17.5**	B15 T	Inter Nipples	13.1	11.7	8.8	5.8	4.4

Life size			THIRTEEN-YEAR-OLD GIRL	Fractions of life size				
				3/4	2/3	1/2	1/3	1/4
Inches	cm	Code		cm	cm	cm	cm	cm
			TRANSVERSE BODY MEASUREMENTS					
9.6	**24.5**	B16 T	Breadth of Thorax	18.4	16.3	12.3	8.2	6.1
9.4	**24.0**	B17 T	Inter Costal Margin Breadth	18.0	16.0	12.0	8.0	6.0
9.8	**25.0**	B18 T	Inter Crest of Ilium Breadth	18.8	16.7	12.5	8.3	6.3
8.7	**22.0**	B19 T	Inter Anterior Superior Spine	16.5	14.7	11.0	7.3	5.5
11.4	**29.0**	B20 T	Inter Greater Trochanter Femur	21.8	19.3	14.5	9.7	7.3
			ANTERIOR-POSTERIOR BODY MEASUREMENTS					
6.3	**16.0**	B21 D	Manubrium to Dorsal Spine	12.0	10.7	8.0	5.3	4.0
7.1	**18.0**	B22 D	Xyphoid Process to Dorsum	13.5	12.0	9.0	6.0	4.5
6.7	**17.0**	B23 D	Umbilicus to Dorsum	12.8	11.3	8.5	5.7	4.3
7.9	**20.0**	B24 D	Pubis to Buttock	15.0	13.3	10.0	6.7	5.0
6.3	**16.0**	B25 D	Thigh to Gluteal Furrow	12.0	10.7	8.0	5.3	4.0
			VERTICAL BACK MEASUREMENTS					
9.1	**23.0**	B26 V	Vertex to 7th Cervical Spine	17.3	15.3	11.5	7.7	5.8
			7th Cervical Vertebra Spine to:					
6.3	**16.0**	B27 V	To Posterior Axillary Fold	12.0	10.7	8.0	5.3	4.0
7.5	**19.0**	B28 V	To Inferior Tip of Scapula	14.3	12.7	9.5	6.3	4.8
15.4	**39.0**	B29 V	To Posterior Superior Spine Ilium	29.3	26.0	19.5	13.0	9.8
18.9	**48.0**	B30 V	To Midline Upper Buttock Crease	36.0	32.0	24.0	16.0	12.0
23.6	**60.0**	B31 V	To Gluteal Furrow	45.0	40.0	30.0	20.0	15.0
35.4	**90.0**	B32 V	To Popliteal Crease of Knee	67.5	60.0	45.0	30.0	22.5
			TRANSVERSE BACK MEASUREMENTS					
11.4	**29.0**	B33 T	Inter Posterior Axillary Folds	21.8	19.3	14.5	9.7	7.3
5.1	**13.0**	B34 T	Inter Scapulae Medial Borders	9.8	8.7	6.5	4.3	3.3
3.1	**8.0**	B35 T	Inter Post. Sup. Spines Sacrum	6.0	5.3	4.0	2.7	2.0
			UPPER EXTREMITY MEASUREMENTS					
61.4	**156.0**	U1 V	Span of Arms	117.0	104.0	78.0	52.0	39.0
6.5	**16.5**	U2 V	Sternum to Acromion Process	12.4	11.0	8.3	5.5	4.1
27.2	**69.0**	U3 V	Acromion Process to Fingertip	51.8	46.0	34.5	23.0	17.3
76.4	**194.0**	U4 V	Vertical Reach to Fingertip	145.5	129.3	97.0	64.7	48.5
37.8	**96.0**	U5 T	Lateral Reach to Fingertip	72.0	64.0	48.0	32.0	24.0
29.5	**75.0**	U6 D	Forward Reach to Fingertip	56.3	50.0	37.5	25.0	18.8
12.6	**32.0**	U7 V	Shoulder to Elbow Flexed	24.0	21.3	16.0	10.7	8.0
11.4	**29.0**	U8 V	Acromion to Lat. Condyle Elbow	21.8	19.3	14.5	9.7	7.3
16.3	**41.5**	U9 D	Flexed Elbow to Fingertip	31.1	27.7	20.8	13.8	10.4
9.3	**23.0**	U10 D	Flexed Elbow Pivot to Wrist Pivot	17.6	15.7	11.8	7.8	5.9

Life size			THIRTEEN-YEAR-OLD GIRL	Fractions of life size				
				3/4	2/3	1/2	1/3	1/4
Inches	cm	Code		cm	cm	cm	cm	cm
			UPPER EXTREMITY MEASUREMENTS					
6.7	**17.0**	U11 V	Wrist to Fingertip	12.8	11.3	8.5	5.7	4.3
4.1	**10.5**	U12 V	Wrist to Thumb Tip	7.9	7.0	5.3	3.5	2.6
3.7	**9.4**	U13 V	Length of Palm	7.1	6.3	4.7	3.1	2.4
2.8	**7.2**	U14 V	Length 3rd Finger	5.4	4.8	3.6	2.4	1.8
4.5	**11.4**	U15 V	Thumb Crotch to 3rd Finger	8.6	7.6	5.7	3.8	2.9
2.3	**5.8**	U16 V	Thumb Crotch to Thumb Tip	4.4	3.9	2.9	1.9	1.5
2.8	**7.0**	U17 D	Anterior-Post. Arm at Biceps	5.3	4.7	3.5	2.3	1.8
2.8	**7.0**	U18 D	Anterior-Posterior at Forearm	5.3	4.7	3.5	2.3	1.8
1.5	**3.7**	U19 D	Depth of Wrist	2.8	2.5	1.9	1.2	0.9
2	**5.0**	U20 T	Breadth of Wrist	3.8	3.3	2.5	1.7	1.3
3.6	**9.2**	U21 T	Breadth of Hand	6.9	6.1	4.6	3.1	2.3
3.0	**7.5**	U22 T	Breadth of Palm	5.6	5.0	3.8	2.5	1.9
			LOWER EXTREMITY MEASUREMENTS					
37.8	**96.0**	L1 V	Crest of Ilium to Sole	72.0	64.0	48.0	32.0	24.0
35.2	**89.5**	* L2 V	Ant.Sup.Spine of Ilium to Sole	67.1	59.7	44.8	29.8	22.4
31.7	**80.5**	L3 V	Greater Trochanter Femur-Sole	60.4	53.7	40.3	26.8	20.1
31.5	**80.0**	* L4 V	Symphysis Pubis to Sole	60.0	53.3	40.0	26.7	20.0
28.0	**71.0**	* L5 V	Crotch to Sole	53.3	47.3	35.5	23.7	17.8
28.5	**72.5**	L6 V	Gluteal Furrow to Sole	54.4	48.3	36.3	24.2	18.1
16.5	**42.0**	L7 V	Posterior Knee Crease to Sole	31.5	28.0	21.0	14.0	10.5
16.1	**41.0**	L8 V	Tibia at Knee to Sole	30.8	27.3	20.5	13.7	10.3
2.4	**6.0**	L9 V	Med. Malleolus to Sole	4.5	4.0	3.0	2.0	1.5
2.0	**5.0**	L10 V	Lat. Malleolus to Sole	3.8	3.3	2.5	1.7	1.3
12.2	**31.0**	L11 T	Breadth of Hips, Sitting	23.3	20.7	15.5	10.3	7.8
5.5	**14.0**	L12 T	Breadth Mid Thigh	10.5	9.3	7.0	4.7	3.5
3.9	**10.0**	L13 T	Breadth of Knee	7.5	6.7	5.0	3.3	2.5
3.9	**10.0**	L14 T	Breadth of Calf	7.5	6.7	5.0	3.3	2.5
2.2	**5.5**	L15 T	Breadth Ankle	4.1	3.7	2.8	1.8	1.4
5.5	**14.0**	L16 D	Mid Thigh, Anterior-Posterior	10.5	9.3	7.0	4.7	3.5
4.3	**11.0**	L17 D	Knee, Anterior-Posterior	8.3	7.3	5.5	3.7	2.8
3.9	**10.0**	L18 D	Mid Calf, Anterior-Posterior	7.5	6.7	5.0	3.3	2.5
2.8	**7.0**	L19 D	Ankle, Anterior-Posterior	5.3	4.7	3.5	2.3	1.8
9.2	**23.4**	L20 D	Length of Foot	17.6	15.6	11.7	7.8	5.9
7.5	**19.0**	L21 D	Heel to 5th Toe	14.3	12.7	9.5	6.3	4.8
3.5	**8.9**	L22 T	Breadth Foot	6.7	5.9	4.5	3.0	2.2
2.0	**5.0**	L23 T	Width of Heel	3.8	3.3	2.5	1.7	1.3
		** These are also listed with Vertical Measurements.*						

Thirteen-Year-Old Boy, anterior and lateral head

all measurements in centimeters

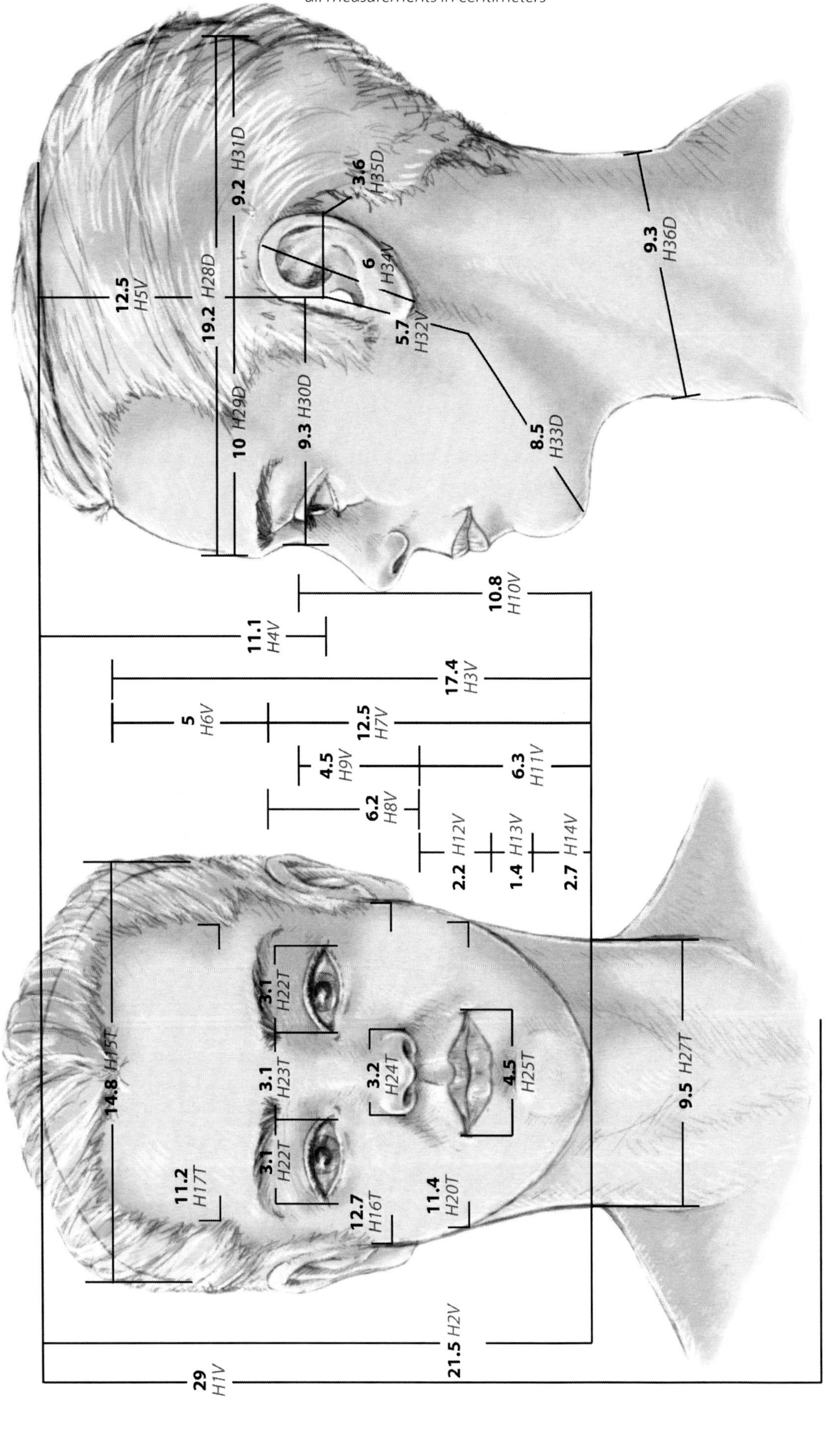

Thirteen-Year-Old Boy, anterior and lateral figure

all measurements in centimeters

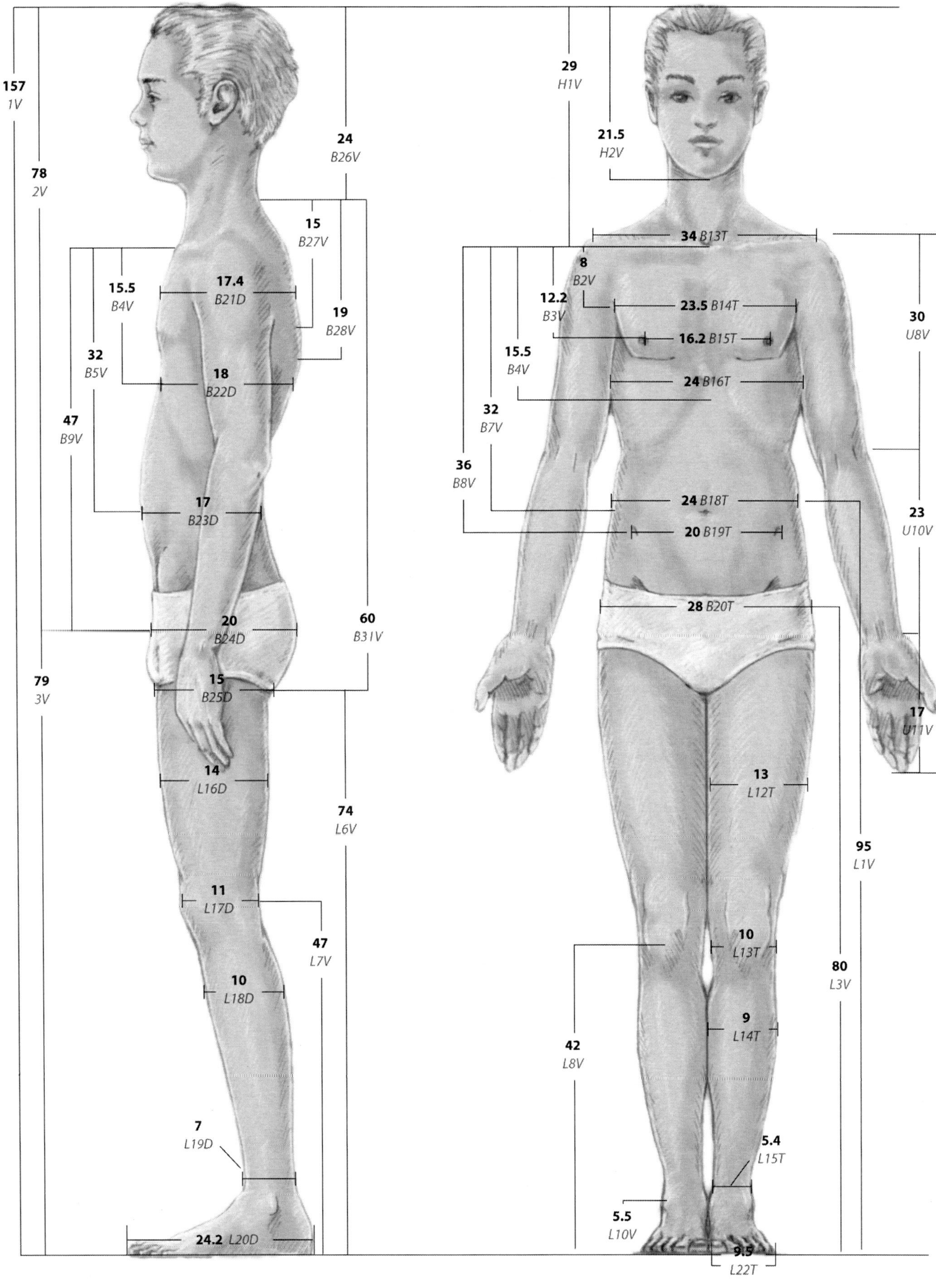

Life size			THIRTEEN-YEAR-OLD BOY	Fractions of life size				
				3/4	2/3	1/2	1/3	1/4
Inches	cm	Code		cm	cm	cm	cm	cm
			VERTICAL MEASUREMENTS					
61.8	**157.0**	1 V	Stature, Height, Vertex to Sole	117.8	104.7	78.5	52.3	39.3
30.4	**77.3**	2 V	Vertex to Symphysis Pubis	58.0	51.5	38.7	25.8	19.3
31.1	**79.0**	* 3 V	Symphysis Pubis to Sole	59.3	52.7	39.5	26.3	19.8
50.0	**127.0**	4 V	Sternal Notch to Sole	95.3	84.7	63.5	42.3	31.8
11.8	**30.0**	5 V	Vertex to Acromion Process	22.5	20.0	15.0	10.0	7.5
50.2	**127.5**	6 V	Acromion Process to Sole	95.6	85.0	63.8	42.5	31.9
35.4	**90.0**	* 7 V	Ant.Sup.Spine of Ilium to Sole	67.5	60.0	45.0	30.0	22.5
14.8	**37.5**	8 V	Acromion pr. to Ant. Sup. Spine	28.1	25.0	18.8	12.5	9.4
16.9	**43.0**	9 V	Ant. Sup. Spine of Ilium to Knee	32.3	28.7	21.5	14.3	10.8
18.5	**47.0**	10 V	Knee at Patella to Sole	35.3	31.3	23.5	15.7	11.8
37.0	**94.0**	11 V	Umbilicus to Sole	70.5	62.7	47.0	31.3	23.5
28.3	**72.0**	* 12 V	Crotch to Sole	54.0	48.0	36.0	24.0	18.0
45.7	**116.0**	13 V	Kneeling Height	87.0	77.3	58.0	38.7	29.0
31.1	**79.0**	14 V	Sitting Height	59.3	52.7	39.5	26.3	19.8
31.9	**81.0**	15 V	Buttock, Straight Leg to Sole	60.8	54.0	40.5	27.0	20.3
20.5	**52.0**	16 V	Buttock to Flexed Knee	39.0	34.7	26.0	17.3	13.0
19.7	**50.0**	17 V	Flexed Knee to Sole	37.5	33.3	25.0	16.7	12.5
		** These are also listed with Lower Extremity Measurements.*						
			VERTICAL HEAD MEASUREMENTS					
11.4	**29.0**	H1 V	Vertex to Sternal Notch	21.8	19.3	14.5	9.7	7.3
8.5	**21.5**	H2 V	Vertex to Chin, Head Height	16.1	14.3	10.8	7.2	5.4
6.9	**17.4**	H3 V	Hairline to Chin, Face Height	13.1	11.6	8.7	5.8	4.4
4.4	**11.1**	H4 V	Vertex to Lateral Corner of Eye	9.0	7.4	5.6	3.7	2.8
4.9	**12.5**	H5 V	Vertex to Tragion of Ear	9.4	8.3	6.3	4.2	3.1
2.0	**5.0**	H6 V	Hairline to Brow, Glabella	3.8	3.3	2.5	1.7	1.3
4.9	**12.5**	H7 V	Brow, Glabella, to Chin, Menton	9.5	8.3	6.3	4.2	3.1
2.4	**6.2**	H8 V	Brow to Subnasal Point	4.6	4.1	3.1	2.1	1.6
1.8	**4.5**	H9 V	Sellion to Subnasal, Nose Length	3.4	3.0	2.3	1.5	1.1
4.3	**10.8**	H10 V	Sellion, Nose Saddle, to Chin	8.1	7.2	5.4	3.6	2.7
2.5	**6.3**	H11 V	Subnasal point to Chin	4.5	4.2	3.2	2.1	1.6
0.9	**2.2**	H12 V	Subnasal to Lip Aperature	2.2	1.5	1.1	0.7	0.6
0.6	**1.4**	H13 V	Lip Aperat. to Sublabial Crease	1.6	0.9	0.7	0.5	0.4
1.1	**2.7**	H14 V	Sublabial Crease to Chin	2.7	1.8	1.4	0.9	0.7
			TRANSVERSE HEAD MEASUREMENTS					
5.8	**14.8**	H15 T	Bi-Parietal, Head Breadth	11.1	9.9	7.4	4.9	3.7
5.0	**12.7**	H16 T	Bi-Zygomatic, Cheekbones	9.5	8.5	6.4	4.2	3.2
4.4	**11.2**	H17 T	Bi-Fronto Temporal, Forehead	8.4	7.5	5.6	3.7	2.8

Life size			THIRTEEN-YEAR-OLD BOY	Fractions of life size				
				3/4	2/3	1/2	1/3	1/4
Inches	cm	Code		cm	cm	cm	cm	cm
			TRANSVERSE HEAD MEASUREMENTS					
3.7	**9.5**	H18 T	Inter Brow Ridge	7.1	6.3	4.8	3.2	2.4
4.9	**12.4**	H19 T	Inter Tragion of Ears	9.3	8.3	6.2	4.1	3.1
4.5	**11.4**	H20 T	Inter Angle of Mandible	8.6	7.6	5.7	3.8	2.9
3.7	**9.3**	H21 T	Inter Lateral Corners of Eyes	6.9	6.2	4.7	3.1	2.3
1.2	**3.1**	H22 T	Width of Eye	2.3	2.1	1.6	1.0	0.8
1.2	**3.1**	H23 T	Inter Inner Corner of eyes	2.3	2.1	1.6	1.0	0.8
1.3	**3.2**	H24 T	Breadth of Nose	2.4	2.1	1.6	1.1	0.8
1.8	**4.5**	H25 T	Breadth of Mouth	3.4	3.0	2.3	1.5	1.1
1.3	**3.4**	H26 T	Breadth of Chin	2.6	2.3	1.7	1.1	0.9
3.7	**9.5**	H27 T	Breadth of Neck	7.1	6.3	4.8	3.2	2.4
			ANTERIOR-POSTERIOR HEAD MEASUREMENTS					
7.6	**19.2**	H28 D	Brow to Occiput, Posterior Head	14.4	12.8	9.6	6.4	4.8
3.9	**10.0**	H29 D	Brow to Tragus of Ear	7.5	6.7	5.0	3.3	2.5
3.7	**9.3**	H30 D	Sellion to Tragus	7.0	6.2	4.7	3.1	2.3
3.6	**9.2**	H31 D	Tragus to Occiput	6.9	6.1	4.6	3.1	2.3
2.2	**5.7**	H32 V	Tragus to Angle Mandible	4.3	3.8	2.9	1.9	1.4
3.3	**8.5**	H33 D	Angle of Mandible to Chin	6.4	5.7	4.3	2.8	2.1
2.4	**6.0**	H34 V	Length of Ear, Vertical	4.5	4.0	3.0	2.0	1.5
1.4	**3.6**	H35 D	Width of Ear, Anterior-Posterior	2.7	2.4	1.8	1.2	0.9
3.7	**9.3**	H36 D	Anterior to Posterior of Neck	7.0	6.2	4.7	3.1	2.3
			VERTICAL BODY MEASUREMENTS					
3.5	**9.0**	B1 V	Chin to Sternal Notch	6.8	6.0	4.5	3.0	2.3
3.1	**8.0**	B2 V	Sternal Notch to Axilla, Arm Pit	6.0	5.3	4.0	2.7	2.0
4.8	**12.2**	B3 V	Sternal Notch to Nipples	9.2	8.1	6.1	4.1	3.1
6.1	**15.5**	B4 V	Sternal Notch to Xyphoid	11.6	10.3	7.8	5.2	3.9
12.6	**32.0**	B5 V	Sternal Notch to Umbilicus	24.0	21.3	16.0	10.7	8.0
6.5	**16.5**	B6 V	Xyphoid to Umbilicus	12.4	11.0	8.3	5.5	4.1
12.6	**32.0**	B7 V	Sternal Notch to Crest Ilium	24.0	21.3	16.0	10.7	8.0
14.2	**36.0**	B8 V	Sternal Notch to Ant.S.Sp.Ilium	27.0	24.0	18.0	12.0	9.0
18.5	**47.0**	B9 V	Sternal Notch to Pubis	35.3	31.3	23.5	15.7	11.8
5.9	**15.0**	B10 V	Umbilicus to Pubis	12.0	10.0	7.5	5.0	3.8
			TRANSVERSE BODY MEASUREMENTS					
14.6	**37.0**	B11 T	Shoulder Breadth	27.8	24.7	18.5	12.3	9.3
6.4	**16.3**	B12 T	Clavicle to Acromion Process	12.2	10.9	8.2	5.4	4.1
13.4	**34.0**	B13 T	Biacromial Process Breadth	25.5	22.7	17.0	11.3	8.5
9.3	**23.5**	B14 T	Inter Anterior Axillary Fold	17.6	15.7	11.8	7.8	5.9
6.4	**16.2**	B15 T	Inter Nipples	12.2	10.8	8.1	5.4	4.1

Life size			THIRTEEN-YEAR-OLD BOY	Fractions of life size				
				3/4	2/3	1/2	1/3	1/4
Inches	cm	Code		cm	cm	cm	cm	cm
			TRANSVERSE BODY MEASUREMENTS					
9.4	**24.0**	B16 T	Breadth of Thorax	18.0	16.0	12.0	8.0	6.0
8.9	**22.5**	B17 T	Inter Costal Margin	16.9	15.0	11.3	7.5	5.6
9.4	**24.0**	B18 T	Inter Crest of Ilium	18.0	16.0	12.0	8.0	6.0
7.9	**20.0**	B19 T	Inter Anterior Superior Spine	15.0	13.3	10.0	6.7	5.0
11.0	**28.0**	B20 T	Inter Greater Trochanter Femur	21.0	18.7	14.0	9.3	7.0
			ANTERIOR-POSTERIOR BODY MEASUREMENTS					
6.9	**17.4**	B21 D	Manubrium to Dorsal Spine	13.1	11.6	8.7	5.8	4.4
7.1	**18.0**	B22 D	Xyphoid Process to Dorsum	13.5	12.0	9.0	6.0	4.5
6.7	**17.0**	B23 D	Umbilicus to Dorsum	12.8	11.3	8.5	5.7	4.3
7.9	**20.0**	B24 D	Pubis to Buttock	15.0	13.3	10.0	6.7	5.0
5.9	**15.0**	B25 D	Thigh to Gluteal Furrow	11.3	10.0	7.5	5.0	3.8
			VERTICAL BACK MEASUREMENTS					
9.4	**24.0**	B26 V	Vertex to 7th Cervical Spine	18.0	16.0	12.0	8.0	6.0
			7th Cervical Vertebra Spine to:					
5.9	**15.0**	B27 V	Posterior Axillary Fold	11.3	10.0	7.5	5.0	3.8
7.5	**19.0**	B28 V	Inferior Tip of Scapula	13.5	12.7	9.5	6.3	4.8
15.4	**39.0**	B29 V	Posterior Superior Spine Ilium	29.3	26.0	19.5	13.0	9.8
18.1	**46.0**	B30 V	Midline Upper Buttock Crease	34.5	30.7	23.0	15.3	11.5
23.6	**60.0**	B31 V	to Gluteal Furrow	45.0	40.0	30.0	20.0	15.0
34.3	**87.0**	B32 V	to Popliteal Crease of Knee	65.3	58.0	43.5	29.0	21.8
			TRANSVERSE BACK MEASUREMENTS					
11.8	**30.0**	B33 T	Inter Posterior Axillary Fold	22.5	20.0	15.0	10.0	7.5
5.3	**13.5**	B34 T	Inter Scapulae Medial Border	10.1	9.0	6.8	4.5	3.4
3.2	**8.2**	B35 T	Inter Post. Sup. Spines Sacrum	6.2	5.5	4.1	2.7	2.1
			UPPER EXTREMITY MEASUREMENTS					
61.4	**156.0**	U1 V	Span of Arms	117.0	104.0	78.0	52.0	39.0
6.5	**16.5**	U2 V	Sternum to Acromion Process	12.4	11.0	8.3	5.5	4.1
26.8	**68.0**	U3 V	Acromion to Fingertip	51.0	45.3	34.0	22.7	17.0
77.6	**197.0**	U4 V	Vertical Reach to Fingertip	147.8	131.3	98.5	65.7	49.3
37.0	**94.0**	U5 T	Lateral Reach to Fingertip	70.5	62.7	47.0	31.3	23.5
29.1	**74.0**	U6 D	Forward Reach to Fingertip	55.5	49.3	37.0	24.7	18.5
13.0	**33.0**	U7 V	Shoulder to Elbow Flexed	24.8	22.0	16.5	11.0	8.3
11.8	**30.0**	U8 V	Acromion to Lat. Condyle Elbow	22.5	20.0	15.0	10.0	7.5
16.1	**41.0**	U9 D	Flexed Elbow to Fingertip	30.8	27.3	20.5	13.7	10.3
9.1	**23.0**	U10 D	Flexed Elbow Pivot to Wrist Pivot	17.3	15.3	11.5	7.7	5.8

Life size			THIRTEEN-YEAR-OLD BOY	Fractions of life size				
				3/4	2/3	1/2	1/3	1/4
Inches	cm	Code		cm	cm	cm	cm	cm
			UPPER EXTREMITY MEASUREMENTS					
6.7	**17.0**	U11 V	Wrist to Fingertip	12.8	11.3	8.5	5.7	4.3
4.3	**11.0**	U12 V	Wrist to Thumb Tip	8.3	7.3	5.5	3.7	2.8
3.7	**9.4**	U13 V	Length of Palm	7.1	6.3	4.7	3.1	2.4
2.9	**7.3**	U14 V	Length 3rd Finger	5.5	4.9	3.7	2.4	1.8
4.7	**12.0**	U15 T	Thumb Crotch to 3rd Finger	9.0	8.0	6.0	4.0	3.0
2.3	**5.8**	U16 T	Thumb Crotch to Thumb Tip	4.4	3.9	2.9	1.9	1.5
2.7	**6.8**	U17 D	Anterior-Post. Arm at Biceps	5.1	4.5	3.4	2.3	1.7
2.7	**6.9**	U18 D	Anterior-Posterior at Forearm	5.2	4.6	3.5	2.3	1.7
1.5	**3.7**	U19 D	Depth of Wrist	2.8	2.5	1.9	1.2	0.9
2.0	**5.0**	U20 T	Breadth of Wrist	3.8	3.3	2.5	1.7	1.3
3.7	**9.3**	U21 T	Breadth of Hand	7.0	6.2	4.7	3.1	2.3
3.0	**7.6**	U22 T	Breadth of Palm	5.7	5.1	3.8	2.5	1.9
			LOWER EXTREMITY MEASUREMENTS					
37.4	**95.0**	L1 V	Crest of Ilium to Sole	71.3	63.3	47.5	31.7	23.8
35.4	**90.0**	* L2 V	Ant.Sup.Spine of Ilium to Sole	67.5	60.0	45.0	30.0	22.5
31.5	**80.0**	L3 V	Greater Trochanter Femur-Sole	60.0	53.3	40.0	26.7	20.0
31.1	**79.0**	* L4 V	Symphysis Pubis to Sole	59.3	52.7	39.5	26.3	19.8
28.3	**72.0**	* L5 V	Crotch to Sole	54.0	48.0	36.0	24.0	18.0
29.1	**74.0**	L6 V	Gluteal Furrow to Sole	55.5	49.3	37.0	24.7	18.5
18.5	**47.0**	L7 V	Posterior Knee Crease to Sole	35.3	31.3	23.5	15.7	11.8
16.5	**42.0**	L8 V	Tibia at knee to Sole	31.5	28.0	21.0	14.0	10.5
2.6	**6.5**	L9 V	Med. Malleolus to Sole	4.9	4.3	3.3	2.2	1.6
2.1	**5.4**	L10 V	Lat. Malleolus to Sole	3.4	3.6	2.7	1.8	1.4
10.6	**27.0**	L11 T	Breadth of Hips, Sitting	20.3	18.0	13.5	9.0	6.8
5.1	**13.0**	L12 T	Breadth Mid Thigh	9.8	8.7	6.5	4.3	3.3
3.9	**10.0**	L13 T	Breadth of Knee	7.5	6.7	5.0	3.3	2.5
3.5	**9.0**	L14 T	Breadth of Calf	6.8	6.0	4.5	3.0	2.3
2.1	**5.4**	L15 T	Breadth Ankle	4.1	3.6	2.7	1.8	1.4
5.5	**14.0**	L16 D	Mid Thigh, Anterior-Posterior	10.5	9.3	7.0	4.7	3.5
4.3	**11.0**	L17 D	Knee, Anterior-Posterior	8.3	7.3	5.5	3.7	2.8
3.9	**10.0**	L18 D	Mid Calf, Anterior-Posterior	7.5	6.7	5.0	3.3	2.5
2.8	**7.0**	L19 D	Ankle, Anterior-Posterior	5.3	4.7	3.5	2.3	1.8
9.5	**24.2**	L20 D	Length of Foot	18.2	16.1	12.1	8.1	6.1
7.5	**19.0**	L21 D	Heel to 5th Toe	14.3	12.7	9.5	6.3	4.8
3.7	**9.5**	L22 T	Breadth Foot	7.1	6.3	4.8	3.2	2.4
2.2	**5.5**	L23 T	Width of Heel	4.1	3.7	2.8	1.8	1.4
		**These are also listed with Vertical Measurements.*						

FOURTEEN-YEAR-OLD GIRLS

By this age, girls are more mature than boys. Most girls are taller, but soon the boys begin to mature and become taller than girls. The girls may have more rounded contours and are developing feminine physique. Breast development is at Phase IV, with more nipple and glandular development.[1] The range of height and weight is listed here for comparison. School level is ninth grade.

Percentile	Twenty-fifth		Fiftieth		Seventy-fifth	
Girls:	Short		Medium		Tall	
Stature	156 cm,	61.5 in.	160 cm,	63.3 in.	165 cm,	64.8 in.
Weight	44.5 kg,	98 lb.	50 kg,	110 lb.	57 kg,	125 lb.[2]

BODY MEASUREMENTS COMPARED TO HEIGHT

	Feature Measure in cm	Feature Compared to Stature	Approx. Ratio of Feature	Compared to Median Adult
Stature: Crown to sole	160.4			95%
Head Height: Crown to chin	20.4	13%	1/8	100%
Sitting Height:	82.4	51%	1/2	93%
Trunk Measure: Sternal notch to symphysis pubis	45	28%	1/4+	92%
Leg Length: Greater trochanter of the femur to sole	81	50%	1/2	94%
Thigh Length: Greater trochanter of the femur to tibia at knee	40	25%	1/4	93%
Tibia at Knee to Sole:	41	26%	1/4	93%
Arm Length: Acromion process at shoulder to dependent fingertip	69	43%	2/5+	92%
Hand Length:	17	11%	1/9	95%
Foot Length:	23.4	15%	1/7	98%

By this age, most girls achieve stature and extremity length for nearly expected adult measurements. There will be some additional growth and maturation up to the age of twenty, but the majority of bone development is nearly complete. Permanent dentition is nearly complete. Third molars continue to grow for two to five more years before eruption.[3] The comparison with adult females reveals almost adult proportions. There will be additional growth, but of subtle changes that are not apparent in the usual measurements.

1. Lowery, G., p. 351.
2. Nelson, W.E., p. 39.
3. Lockhart, R.D., p. 57.

FOURTEEN-YEAR-OLD BOYS

Before ten years of age, boys are, on average, slightly taller and heavier than girls. However, in early youth, girls mature earlier, starting at about twelve and grow more rapidly. Boys begin to mature at twelve, but girls are on average slightly taller until after thirteen. By fourteen years of age, boys are growing more rapidly and soon become taller than girls. Boys continue to be interested in competitive athletics, outdoor activities, including camping, bicycling, fishing, hiking, and ball games. The school level is ninth grade.

Percentile	Twenty-fifth		Fiftieth		Seventy-fifth	
Boys:	Short		Medium		Tall	
Stature	157 cm,	61.8. in.	163 cm,	64.3 in.	168.5 cm,	66.3 in.
Weight	45 kg,	100 lb.	51 kg,	112 lb.	58.3 kg,	128.5 lb.[1]

BODY MEASUREMENTS COMPARED TO HEIGHT

	Feature Measure in cm	Feature Compared to Stature	Approx. Ratio of Feature	Compared to Median Adult
Stature: Crown to sole	163			89%
Head Height: Crown to chin	21	13%	1/8+	95%
Sitting Height:	82.5	51%	1/2	89%
Trunk Measure: Sternal notch to symphysis pubis	48.4	30%	1/3–	92%
Leg Length: Greater trochanter of the femur to sole	85	52%	1/2	94%
Thigh Length: Greater trochanter of the femur to tibia at knee	42	28%	1/4+	98%
Tibia at Knee to Sole:	43	29%	1/4+	89%
Arm Length: Acromion process at shoulder to dependent fingertip	70	43%	2/5	88%
Hand Length:	18	11%	1/9	95%
Foot Length:	25.5	16%	1/6	95%

By this age, boys achieve about ninety percent of adult male measurements of most features. Muscular development has improved. Many facial features are similar to girls, since the growth of bones and dentition is similar, yet there are subtle differences. The upper and lower second molars have erupted. The third molars are growing to erupt in two to five more years.[2] The face of most teenage youth has an oval outline with graceful contours.

1. Nelson, W.E., p. 38.
2. Locjhart, R.D., p. 57.

Fourteen-Year-Old Girl, anterior and lateral head

all measurements in centimeters

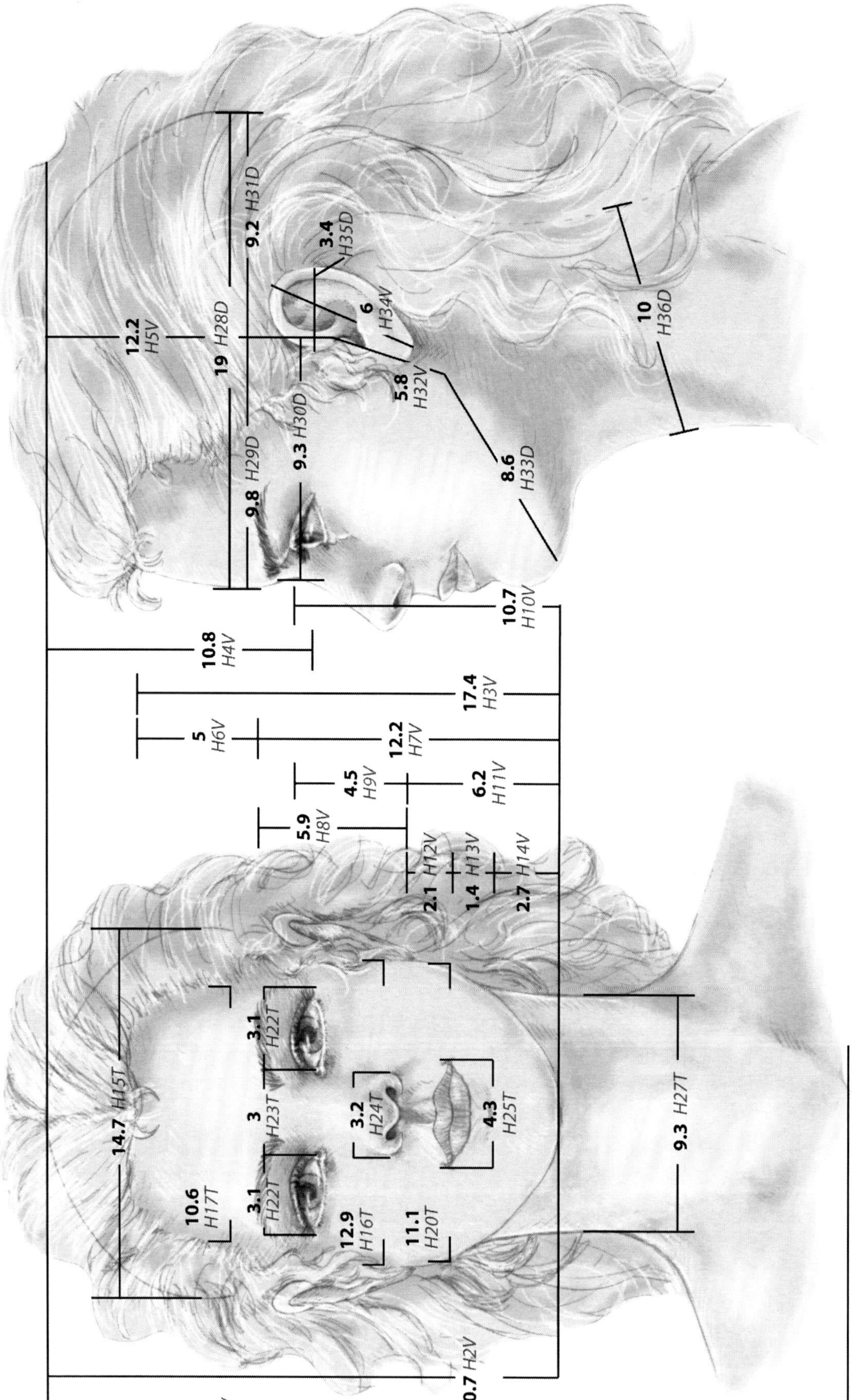

Fourteen-Year-Old Girl, anterior and lateral figure

all measurements in centimeters

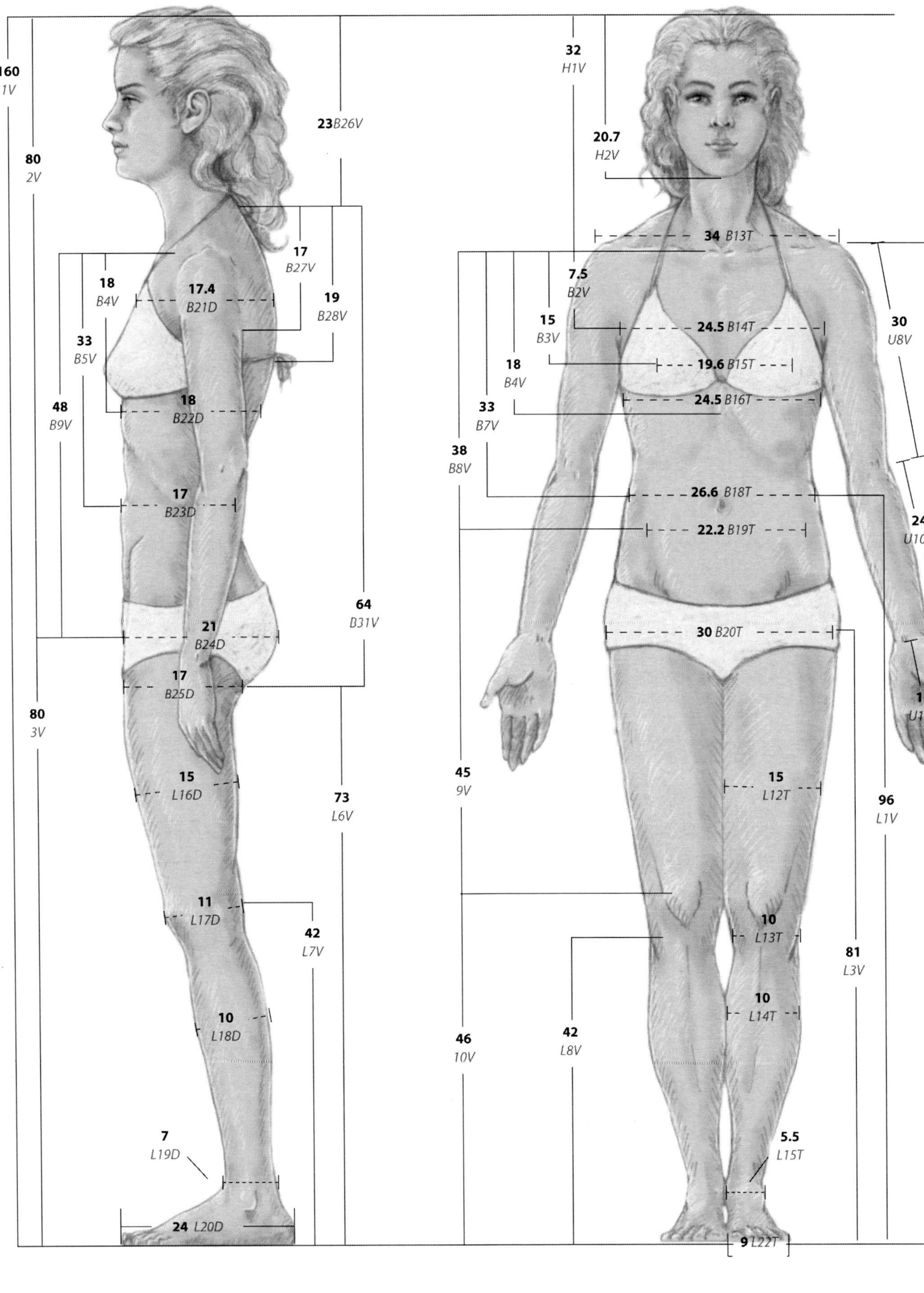

Life size			FOURTEEN-YEAR-OLD GIRL	Fractions of life size				
				3/4	2/3	1/2	1/3	1/4
Inches	cm	Code		cm	cm	cm	cm	cm
			VERTICAL MEASUREMENTS					
63.0	**160.0**	1 V	Stature, Height, Vertex to Sole	120.0	106.7	80.0	586.7	40.0
31.5	**80.0**	2 V	Vertex to Symphysis Pubis	60.0	53.3	40.0	293.3	20.0
31.5	**80.0**	* 3 V	Symphysis Pubis to Sole	60.0	53.3	40.0	293.3	20.0
50.4	**128.0**	4 V	Sternal Notch to Sole	96.0	85.3	64.0	469.3	32.0
12.2	**31.0**	5 V	Vertex to Acromion Process	23.3	20.7	15.5	113.7	7.8
50.8	**129.0**	6 V	Acromion Process to Sole	96.8	86.0	64.5	473.0	32.3
35.8	**91.0**	* 7 V	Ant.Sup.Spine of Ilium to Sole	68.3	60.7	45.5	333.7	22.8
15.0	**38.0**	8 V	Acrom. proc. to Ant. Sup. Spine	28.5	25.3	19.0	139.3	9.5
17.7	**45.0**	9 V	Ant.Sup.Spine of Ilium to Knee	33.8	30.0	22.5	165.0	11.3
18.1	**46.0**	10 V	Knee at Patella to Sole	34.5	30.7	23.0	168.7	11.5
37.8	**96.0**	11 V	Umbilicus to Sole	72.0	64.0	48.0	352.0	24.0
28.3	**72.0**	* 12 V	Crotch to Sole	54.0	48.0	36.0	264.0	18.0
46.5	**118.0**	13 V	Kneeling Height	88.5	78.7	59.0	432.7	29.5
32.7	**83.0**	14 V	Sitting Height	62.3	55.3	41.5	304.3	20.8
33.9	**86.0**	15 V	Buttock, Straight Leg to Sole	64.5	57.3	43.0	315.3	21.5
21.3	**54.0**	16 V	Buttock to Flexed Knee	40.5	36.0	27.0	198.0	13.5
19.7	**50.0**	17 V	Flexed Knee to Sole	37.5	33.3	25.0	183.3	12.5
		** These are also listed with Lower Extremity Measurements.*						
			VERTICAL HEAD MEASUREMENTS					
12.6	**32.0**	H1 V	Vertex to Sternal Notch	24.0	21.3	16.0	117.3	8.0
8.1	**20.7**	H2 V	Vertex to Chin, Head Height	15.5	13.8	10.4	75.9	5.2
6.9	**17.4**	H3 V	Hairline to Chin, Face Height	13.1	11.6	8.7	63.8	4.4
4.3	**10.8**	H4 V	Vertex to Lateral Corner of Eye	8.1	7.2	5.4	39.6	2.7
4.8	**12.2**	H5 V	Vertex to Tragion of Ear	9.2	8.1	6.1	44.7	3.1
2.1	**5.3**	H6 V	Hairline to Brow, Glabella	4.0	3.5	2.7	19.4	1.3
4.8	**12.2**	H7 V	Brow, Glabella, to Chin, Mcnton	9.2	8.1	6.1	44.7	3.1
2.3	**5.9**	H8 V	Brow to Subnasal Point	4.4	3.9	3.0	21.6	1.5
1.8	**4.6**	H9 V	Sellion to Subnasal, Nose Length	3.5	3.1	2.3	16.9	1.2
4.2	**10.7**	H10 V	Sellion, Nose Saddle, to Chin	8.0	7.1	5.4	39.2	2.7
2.4	**6.2**	H11 V	Subnasal point to Chin	4.7	4.1	3.1	22.7	1.6
0.8	**2.1**	H12 V	Subnasal to Lip Aperature	1.6	1.4	1.1	7.7	0.5
0.6	**1.4**	H13 V	Lip Aperat. to Sublabial Crease	1.1	0.9	0.7	5.1	0.4
1.1	**2.7**	H14 T	Sublabial Crease to Chin	2.0	1.8	1.4	9.9	0.7
			TRANSVERSE HEAD MEASUREMENTS					
5.8	**14.7**	H15 T	Bi-Parietal, Head Breadth	11.0	9.8	7.4	53.9	3.7
5.1	**12.9**	H16 T	Bi-Zygomatic, Cheekbones	9.7	8.6	6.5	47.3	3.2
4.2	**10.6**	H17 T	Bi-Fronto Temporal, Forehead	8.0	7.1	5.3	38.9	2.7

Life size			FOURTEEN-YEAR-OLD GIRL	Fractions of life size				
				3/4	2/3	1/2	1/3	1/4
Inches	cm	Code		cm	cm	cm	cm	cm
			TRANSVERSE HEAD MEASUREMENTS					
3.9	10.0	H18 T	Inter Brow Ridge	7.5	6.7	5.0	36.7	2.5
4.8	12.3	H19 T	Inter Tragion of Ears	9.2	8.2	6.2	45.1	3.1
4.4	11.1	H20 T	Inter Angle of Mandible	8.3	7.4	5.6	40.7	2.8
3.6	9.2	H21 T	Inter Lateral Corners of Eyes	6.9	6.1	4.6	33.7	2.3
1.2	3.1	H22 T	Width of Eye	2.3	2.1	1.6	11.4	0.8
1.2	3.0	H23 T	Inter Inner Corner of eyes	2.3	2.0	1.5	11.0	0.8
1.3	3.2	H24 T	Breadth of Nose	2.4	2.1	1.6	11.7	0.8
1.7	4.3	H25 T	Breadth of Mouth	3.2	2.9	2.2	15.8	1.1
1.3	3.2	H26 T	Breadth of Chin	2.4	2.1	1.6	11.7	0.8
3.7	9.3	H27 T	Breadth of Neck	7.0	6.2	4.7	34.1	2.3
			ANTERIOR-POSTERIOR HEAD MEASUREMENTS					
7.5	19.0	H28 D	Brow to Occiput	14.3	12.7	9.5	69.7	4.8
3.9	9.8	H29 D	Brow to Tragus	7.4	6.5	4.9	35.9	2.5
3.7	9.3	H30 D	Sellion to Tragus	7.0	6.2	4.7	34.1	2.3
3.6	9.2	H31 D	Tragus to Occiput	6.9	6.1	4.6	33.7	2.3
2.3	5.8	H32 V	Tragus to Angle Mandible	4.4	3.9	2.9	21.3	1.5
3.4	8.6	H33 D	Angle of Mandible to Chin	6.5	5.7	4.3	31.5	2.2
2.4	6.0	H34 V	Length of Ear, Vertical	4.5	4.0	3.0	22.0	1.5
1.3	3.4	H35 D	Width of Ear, Anterior-Posterior	2.6	2.3	1.7	12.5	0.9
3.9	10.0	H36 D	Anterior to Posterior of Neck	7.5	6.7	5.0	36.7	2.5
			VERTICAL BODY MEASUREMENTS					
3.9	10.0	B1 V	Chin to Sternal Notch	7.5	6.7	5.0	36.7	2.5
3.0	7.5	B2 V	Sternal Notch to Axilla	5.6	5.0	3.8	27.5	1.9
5.9	15.0	B3 V	Sternal Notch to Nipples	11.3	10.0	7.5	55.0	3.8
7.1	18.0	B4 V	Sternal Notch to Xyphoid Proc.	13.5	12.0	9.0	66.0	4.5
13.0	33.0	B5 V	Sternal Notch to Umbilicus	24.8	22.0	16.5	121.0	8.3
5.9	15.0	B6 V	Xyphoid Process to Umbilicus	11.3	10.0	7.5	55.0	3.8
12.6	32.0	B7 V	Sternal Notch to Crest Ilium	24	21.3	16	117.3	8
15.0	38.0	B8 V	Sternal Notch to Ant.S.Sp.Ilium	28.5	25.3	19.0	139.3	9.5
18.9	48.0	B9 V	Sternal Notch to Pubis	36.0	32.0	24.0	176.0	12.0
5.9	15.0	B10 V	Umbilicus to Pubis	11.3	10.0	7.5	55.0	3.8
			TRANSVERSE BODY MEASUREMENTS					
15.0	38.0	B11 T	Shoulder Breadth	28.5	25.3	19.0	139.3	9.5
6.6	16.7	B12 T	Clavicle to Acromion Process	12.5	11.1	8.4	61.2	4.2
13.4	34.0	B13 T	Biacromial Process Breadth	25.5	22.7	17.0	124.7	8.5
9.6	24.5	B14 T	Inter Anterior Axillary Fold	18.4	16.3	12.3	89.8	6.1
7.7	19.6	B15 T	Inter Nipples	14.7	13.1	9.8	71.9	4.9

Life size			FOURTEEN-YEAR-OLD GIRL	Fractions of life size				
				3/4	2/3	1/2	1/3	1/4
Inches	cm	Code		cm	cm	cm	cm	cm
			TRANSVERSE BODY MEASUREMENTS					
9.6	24.5	B16 T	Breadth of Thorax	18.4	16.3	12.3	89.8	6.1
9.4	24.0	B17 T	Inter Costal Margin Breadth	18.0	16.0	12.0	88.0	6.0
10.5	26.6	B18 T	Inter Crest of Ilium Breadth	20.0	17.7	13.3	97.5	6.7
8.7	22.2	B19 T	Inter Anterior Superior Spine	16.7	14.8	11.1	81.4	5.6
11.8	30.0	B20 T	Inter Greater Trochanter Femur	22.5	20.0	15.0	110.0	7.5
			ANTERIOR-POSTERIOR BODY MEASUREMENTS					
6.9	17.4	B21 D	Manubrium to Dorsal Spine	13.1	11.6	8.7	63.8	4.4
7.1	18.0	B22 D	Xyphoid Process to Dorsum	13.5	12.0	9.0	66.0	4.5
6.7	17.0	B23 D	Umbilicus to Dorsum	12.8	11.3	8.5	62.3	4.3
8.3	21.0	B24 D	Pubis to Buttock	15.8	14.0	10.5	77.0	5.3
6.7	17.0	B25 D	Thigh to Gluteal Furrow	12.8	11.3	8.5	62.3	4.3
			VERTICAL BACK MEASUREMENTS					
9.1	23.0	B26 V	Vertex to 7th Cervical Spine	17.3	15.3	11.5	84.3	5.8
			7th Cervical Vertebra Spine to:					
7.1	18.0	B27 V	To Posterior Axillary Fold	13.5	12.0	9.0	66.0	4.5
7.5	19.0	B28 V	To Inferior Tip of Scapula	14.3	12.7	9.5	69.7	4.8
16.1	41.0	B29 V	To Posterior Superior Spine Ilium	30.8	27.3	20.5	150.3	10.3
19.3	49.0	B30 V	To Midline Upper Buttock Crease	36.8	32.7	24.5	179.7	12.3
25.2	64.0	B31 V	To Gluteal Furrow	48.0	42.7	32.0	234.7	16.0
36.2	92.0	B32 V	To Popliteal Crease of Knee	69.0	61.3	46.0	337.3	23.0
			TRANSVERSE BACK MEASUREMENTS					
12.2	31.0	B33 T	Inter Posterior Axillary Folds	23.3	20.7	15.5	113.7	7.8
5.3	13.5	B34 T	Inter Scapulae Medial Borders	10.1	9.0	6.8	49.5	3.4
3.3	8.5	B35 T	Inter Post. Sup. Spines Sacrum	6.4	5.7	4.3	31.2	2.1
			UPPER EXTREMITY MEASUREMENTS					
62.6	159.0	U1 V	Span of Arms	119.3	106.0	79.5	583.0	39.8
6.6	16.8	U2 V	Sternum to Acromion Process	12.6	11.2	8.4	61.6	4.2
27.2	69.0	U3 V	Acromion Process to Fingertip	51.8	46.0	34.5	253.0	17.3
78.0	198.0	U4 V	Vertical Reach to Fingertip	153.0	136.0	102.0	748.0	51.0
38.2	97.0	U5 T	Lateral Reach to Fingertip	75.4	67.0	50.3	368.5	25.1
30.3	77.0	U6 D	Forward Reach to Fingertip	57.8	51.3	38.5	282.3	19.3
13.0	33.0	U7 V	Shoulder to Elbow Flexed	24.8	22.0	16.5	121.0	8.3
11.8	30.0	U8 V	Acromion to Lat. Condyle Elbow	22.5	20.0	15.0	110.0	7.5
16.5	42.0	U9 D	Flexed Elbow to Fingertip	31.5	28.0	21.0	154.0	10.5
9.4	24.0	U10 D	Flexed Elbow Pivot to Wrist Pivot	18.0	16.0	12.0	88.0	6.0

Life size			FOURTEEN-YEAR-OLD GIRL	Fractions of life size				
				3/4	2/3	1/2	1/3	1/4
Inches	cm	Code		cm	cm	cm	cm	cm
			UPPER EXTREMITY MEASUREMENTS					
6.7	**17.0**	U11 V	Wrist to Fingertip	12.8	11.3	8.5	62.3	4.3
4.1	**10.5**	U12 V	Wrist to Thumb Tip	7.9	7.0	5.3	38.5	2.6
3.8	**9.6**	U13 V	Length of Palm	7.2	6.4	4.8	35.2	2.4
2.9	**7.3**	U14 V	Length 3rd Finger	5.5	4.9	3.7	26.8	1.8
4.5	**11.4**	U15 V	Thumb Crotch to 3rd Finger	8.6	7.6	5.7	41.8	2.9
2.3	**5.8**	U16 V	Thumb Crotch to Thumb Tip	4.4	3.9	2.9	21.3	1.5
2.8	**7.0**	U17 D	Anterior-Post. Arm at Biceps	5.3	4.7	3.5	25.7	1.8
2.8	**7.0**	U18 D	Anterior-Posterior at Forearm	5.3	4.7	3.5	25.7	1.8
1.5	**3.7**	U19 D	Depth of Wrist	2.8	2.5	1.9	13.6	0.9
2.0	**5.0**	U20 T	Breadth of Wrist	3.8	3.3	2.5	18.3	1.3
3.7	**9.3**	U21 T	Breadth of Hand	7.0	6.2	4.7	34.1	2.3
3.0	**7.6**	U22 T	Breadth of Palm	5.7	5.1	3.8	27.9	1.9
			LOWER EXTREMITY MEASUREMENTS					
37.8	**96.0**	L1 V	Crest of Ilium to Sole	72.0	64.0	48.0	352.0	24.0
35.8	**91.0**	* L2 V	Ant.Sup.Spine of Ilium to Sole	68.3	60.7	45.5	333.7	22.8
31.9	**81.0**	L3 V	Greater Trochanter Femur-Sole	60.8	54.0	40.5	297.0	20.3
31.5	**80.0**	* L4 V	Symphysis Pubis to Sole	60.0	53.3	40.0	293.3	20.0
28.3	**72.0**	* L5 V	Crotch to Sole	54.0	48.0	36.0	264.0	18.0
28.7	**73.0**	L6 V	Gluteal Furrow to Sole	54.8	48.7	36.5	267.7	18.3
16.5	**42.0**	L7 V	Posterior Knee Crease to Sole	31.5	28.0	21.0	154.0	10.5
16.5	**42.0**	L8 V	Tibia at Knee to Sole	31.5	28.0	21.0	154.0	10.5
2.4	**6.0**	L9 V	Med. Malleolus to Sole	4.5	4.0	3.0	22.0	1.5
2.0	**5.0**	L10 V	Lat. Malleolus to Sole	3.8	3.3	2.5	18.3	1.3
12.8	**32.4**	L11 T	Breadth of Hips, Sitting	24.3	21.6	16.2	118.8	8.1
5.9	**15.0**	L12 T	Breadth Mid Thigh	11.3	10.0	7.5	55.0	3.8
4.3	**11.0**	L13 T	Breadth of Knee	8.3	7.3	5.5	40.3	2.8
3.9	**10.0**	L14 T	Breadth of Calf	7.5	6.7	5.0	36.7	2.5
2.2	**5.5**	L15 T	Breadth Ankle	4.1	3.7	2.8	20.2	1.4
5.9	**15.0**	L16 D	Mid Thigh, Anterior-Posterior	11.3	10.0	7.5	55.0	3.8
4.3	**11.0**	L17 D	Knee, Anterior-Posterior	8.3	7.3	5.5	40.3	2.8
3.9	**10.0**	L18 D	Mid Calf, Anterior-Posterior	7.5	6.7	5.0	36.7	2.5
2.8	**7.0**	L19 D	Ankle, Anterior-Posterior	5.3	4.7	3.5	25.7	1.8
9.4	**24.0**	L20 D	Length of Foot	18.0	16.0	12.0	88.0	6.0
7.5	**19.0**	L21 D	Heel to 5th Toe	14.3	12.7	9.5	69.7	4.8
3.5	**9.0**	L22 T	Breadth Foot	6.8	6.0	4.5	33.0	2.3
2.0	**5.0**	L23 T	Width of Heel	3.8	3.3	2.5	18.3	1.3
		** These are also listed with Vertical Measurements.*						

Fourteen-Year-Old Boy, anterior and lateral head

all measurements in centimeters

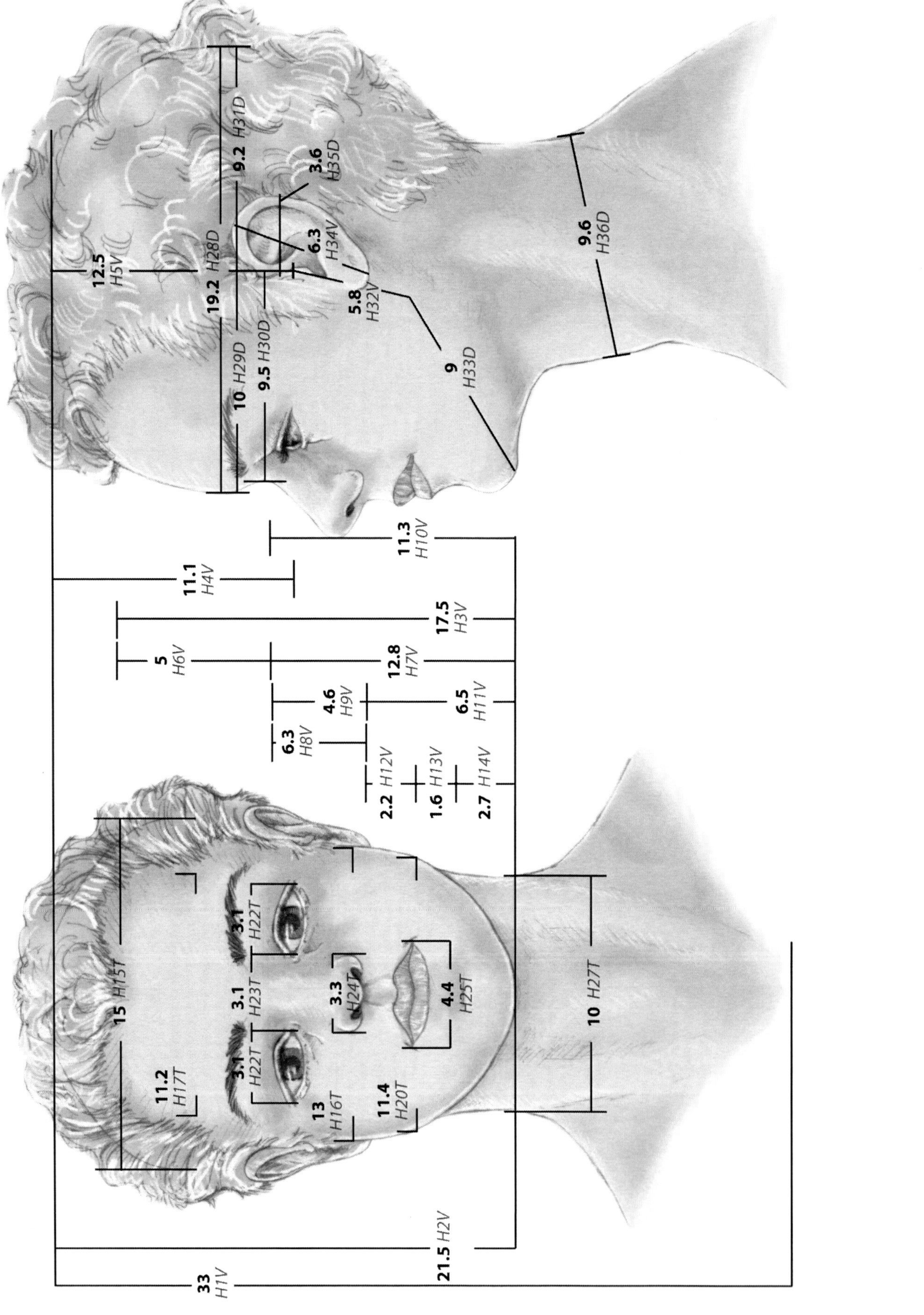

Fourteen-Year-Old Boy, anterior and lateral figure

all measurements in centimeters

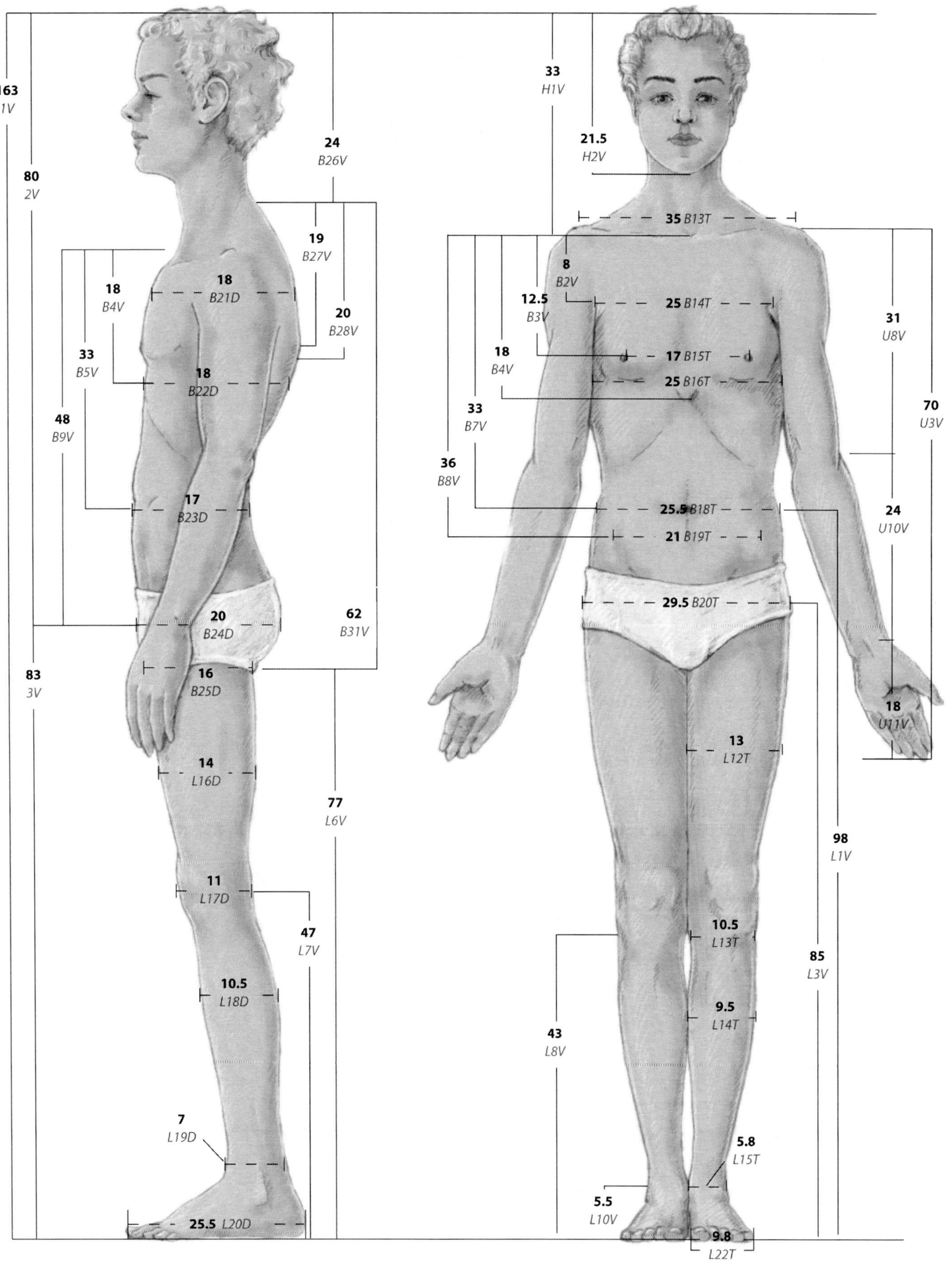

Life size			FOURTEEN-YEAR-OLD BOY	Fractions of life size				
				3/4	2/3	1/2	1/3	1/4
Inches	cm	Code		cm	cm	cm	cm	cm
			VERTICAL MEASUREMENTS					
64.2	**163.0**	1 V	Stature, Height, Vertex to Sole	122.3	108.7	81.5	54.3	40.8
31.5	**80.0**	2 V	Vertex to Symphysis Pubis	60.0	53.3	40.0	26.7	20.0
32.7	**83.0**	* 3 V	Symphysis Pubis to Sole	62.3	55.3	41.5	27.7	20.8
50.4	**128.0**	4 V	Sternal Notch to Sole	96.0	85.3	64.0	42.7	32.0
12.2	**31.0**	5 V	Vertex to Acromion Process	23.3	20.7	15.5	10.3	7.8
51.2	**130.0**	6 V	Acromion Process to Sole	97.5	86.7	65.0	43.3	32.5
36.6	**93.0**	* 7 V	Ant.Sup.Spine of Ilium to Sole	69.8	62.0	46.5	31.0	23.3
15.0	**38.0**	8 V	Acromion pr. to Ant. Sup. Spine	28.5	25.3	19.0	12.7	9.5
18.1	**46.0**	9 V	Ant. Sup. Spine of Ilium to Knee	34.5	30.7	23.0	15.3	11.5
18.1	**46.0**	10 V	Knee at Patella to Sole	34.5	30.7	23.0	15.3	11.5
39.8	**101.0**	11 V	Umbilicus to Sole	75.8	67.3	50.5	33.7	25.3
29.9	**76.0**	* 12 V	Crotch to Sole	57.0	50.7	38.0	25.3	19.0
48.0	**122.0**	13 V	Kneeling Height	91.5	81.3	61.0	40.7	30.5
32.5	**82.5**	14 V	Sitting Height	61.9	55.0	41.3	27.5	20.6
34.3	**87.0**	15 V	Buttock, Straight Leg to Sole	65.3	58.0	43.5	29.0	21.8
21.7	**55.0**	16 V	Buttock to Flexed Knee	41.3	36.7	27.5	18.3	13.8
20.5	**52.0**	17 V	Flexed Knee to Sole	39.0	34.7	26.0	17.3	13.0
		** These are also listed with Lower Extremity Measurements.*						
			VERTICAL HEAD MEASUREMENTS					
13.0	**33.0**	H1 V	Vertex to Sternal Notch	24.8	22.0	16.5	11.0	8.3
8.5	**21.5**	H2 V	Vertex to Chin, Head Height	16.1	14.3	10.8	7.2	5.4
6.9	**17.5**	H3 V	Hairline to Chin, Face Height	13.1	11.7	8.8	5.8	4.4
4.4	**11.1**	H4 V	Vertex to Lateral Corner of Eye	8.3	7.4	5.6	3.7	2.8
4.9	**12.5**	H5 V	Vertex to Tragion of Ear	9.4	8.3	6.3	4.2	3.1
2.0	**5.0**	H6 V	Hairline to Brow, Glabella	3.8	3.3	2.5	1.7	1.3
5.0	**12.8**	H7 V	Brow, Glabella, to Chin, Menton	9.6	8.5	6.4	4.3	3.2
2.5	**6.3**	H8 V	Brow to Subnasal Point	4.7	4.2	3.2	2.1	1.6
1.8	**4.6**	H9 V	Sellion to Subnasal, Nose Length	3.5	3.1	2.3	1.5	1.2
4.3	**11.0**	H10 V	Sellion, Nose Saddle, to Chin	8.3	7.3	5.5	3.7	2.8
2.6	**6.5**	H11 V	Subnasal point to Chin	4.9	4.3	3.3	2.2	1.6
0.9	**2.2**	H12 V	Subnasal to Lip Aperature	1.7	1.5	1.1	0.7	0.6
0.6	**1.6**	H13 V	Lip Aperat. to Sublabial Crease	1.2	1.1	0.8	0.5	0.4
1.1	**2.7**	H14 V	Sublabial Crease to Chin	2.0	1.8	1.4	0.9	0.7
			TRANSVERSE HEAD MEASUREMENTS					
5.9	**15.0**	H15 T	Bi-Parietal, Head Breadth	11.3	10.0	7.5	5.0	3.8
5.1	**13.0**	H16 T	Bi-Zygomatic, Cheekbones	9.8	8.7	6.5	4.3	3.3
4.4	**11.2**	H17 T	Bi-Fronto Temporal, Forehead	8.4	7.5	5.6	3.7	2.8

Life size			FOURTEEN-YEAR-OLD BOY	Fractions of life size				
				3/4	2/3	1/2	1/3	1/4
Inches	cm	Code		cm	cm	cm	cm	cm
			TRANSVERSE HEAD MEASUREMENTS					
3.8	**9.7**	H18 T	Inter Brow Ridge	7.3	6.5	4.9	3.2	2.4
4.9	**12.5**	H19 T	Inter Tragion of Ears	9.4	8.3	6.3	4.2	3.1
4.5	**11.4**	H20 T	Inter Angle of Mandible	8.6	7.6	5.7	3.8	2.9
3.7	**9.3**	H21 T	Inter Lateral Corners of Eyes	7.0	6.2	4.7	3.1	2.3
1.2	**3.1**	H22 T	Width of Eye	2.3	2.1	1.6	1.0	0.8
1.2	**3.1**	H23 T	Inter Inner Corner of eyes	2.3	2.1	1.6	1.0	0.8
1.3	**3.3**	H24 T	Breadth of Nose	2.5	2.2	1.7	1.1	0.8
1.8	**4.6**	H25 T	Breadth of Mouth	3.5	3.1	2.3	1.5	1.2
1.3	**3.4**	H26 T	Breadth of Chin	2.6	2.3	1.7	1.1	0.9
3.9	**10.0**	H27 T	Breadth of Neck	7.5	6.7	5.0	3.3	2.5
			ANTERIOR-POSTERIOR HEAD MEASUREMENTS					
7.6	**19.2**	H28 D	Brow to Occiput, Posterior Head	14.4	12.8	9.6	6.4	4.8
3.9	**10.0**	H29 D	Brow to Tragus of Ear	7.5	6.7	5.0	3.3	2.5
3.7	**9.5**	H30 D	Sellion to Tragus	7.1	6.3	4.8	3.2	2.4
3.6	**9.2**	H31 D	Tragus to Occiput	6.9	6.1	4.6	3.1	2.3
2.3	**5.8**	H32 V	Tragus to Angle Mandible	4.4	3.9	2.9	1.9	1.5
3.5	**9.0**	H33 D	Angle of Mandible to Chin	6.8	6.0	4.5	3.0	2.3
2.4	**6.2**	H34 V	Length of Ear, Vertical	4.7	4.1	3.1	2.1	1.6
1.4	**3.6**	H35 D	Width of Ear, Anterior-Posterior	2.7	2.4	1.8	1.2	0.9
3.8	**9.6**	H36 D	Anterior to Posterior of Neck	7.2	6.4	4.8	3.2	2.4
			VERTICAL BODY MEASUREMENTS					
3.9	**10.0**	B1 V	Chin to Sternal Notch, Neck Pit	7.5	6.7	5.0	3.3	2.5
3.1	**8.0**	B2 V	Sternal Notch to Axilla, Arm Pit	6.0	5.3	4.0	2.7	2.0
4.9	**12.5**	B3 V	Sternal Notch to Nipples	9.4	8.3	6.3	4.2	3.1
7.1	**18.0**	B4 V	Sternal Notch to Xyphoid	13.5	12.0	9.0	6.0	4.5
13.0	**33.0**	B5 V	Sternal Notch to Umbilicus	24.8	22.0	16.5	11.0	8.3
6.7	**17.0**	B6 V	Xyphoid Process to Umbilicus	12.8	11.3	8.5	5.7	4.3
13.0	**33.0**	B7 V	Sternal Notch to Crest Ilium	24.8	22.0	16.5	11.0	8.3
14.2	**36.0**	B8 V	Sternal Notch to Ant.S.Sp.Ilium	27.0	24.0	18.0	12.0	9.0
18.9	**48.0**	B9 V	Sternal Notch to Pubis	36.0	32.0	24.0	16.0	12.0
5.9	**15.0**	B10 V	Umbilicus to Pubis of Pelvis	11.3	10.0	7.5	5.0	3.8
			TRANSVERSE BODY MEASUREMENTS					
15.4	**39.0**	B11 T	Shoulder Breadth	29.3	26.0	19.5	13.0	9.8
6.7	**17.0**	B12 T	Clavicle to Acromion Process	12.8	11.3	8.5	5.7	4.3
13.8	**35.0**	B13 T	Biacromial Process Breadth	26.3	23.3	17.5	11.7	8.8
9.8	**25.0**	B14 T	Inter Anterior Axillary Fold	18.8	16.7	12.5	8.3	6.3
6.7	**17.0**	B15 T	Inter Nipples	12.8	11.3	8.5	5.7	4.3

Life size			FOURTEEN-YEAR-OLD BOY	Fractions of life size				
				3/4	2/3	1/2	1/3	1/4
Inches	cm	Code		cm	cm	cm	cm	cm
			TRANSVERSE BODY MEASUREMENTS					
9.8	**25.0**	B16 T	Breadth of Thorax	18.8	16.7	12.5	8.3	6.3
9.4	**24.0**	B17 T	Inter Costal Margin	18.0	16.0	12.0	8.0	6.0
10.0	**25.5**	B18 T	Inter Crest of Ilium	19.1	17.0	12.8	8.5	6.4
8.3	**21.0**	B19 T	Inter Anterior Superior Spine	15.8	14.0	10.5	7.0	5.3
11.6	**29.5**	B20 T	Inter Greater Trochanter Femur	22.1	19.7	14.8	9.8	7.4
			ANTERIOR-POSTERIOR BODY MEASUREMENTS					
7.1	**18.0**	B21 D	Manubrium to Dorsal Spine	13.5	12.0	9.0	6.0	4.5
7.1	**18.0**	B22 D	Xyphoid Process to Dorsum	13.5	12.0	9.0	6.0	4.5
6.7	**17.0**	B23 D	Umbilicus to Dorsum	12.8	11.3	8.5	5.7	4.3
7.9	**20.0**	B24 D	Pubis to Buttock	15.0	13.3	10.0	6.7	5.0
6.3	**16.0**	B25 D	Thigh to Gluteal Furrow	12.0	10.7	8.0	5.3	4.0
			VERTICAL BACK MEASUREMENTS					
9.4	**24.0**	B26 V	Vertex to 7th Cervical Spine	18.0	16.0	12.0	8.0	6.0
			7th Cervical Vertebra Spine to:					
7.5	**19.0**	B27 V	Posterior Axillary Fold	14.3	12.7	9.5	6.3	4.8
7.9	**20.0**	B28 V	Inferior Tip of Scapula	15.0	13.3	10.0	6.7	5.0
16.1	**41.0**	B29 V	Posterior Superior Spine Ilium	30.8	27.3	20.5	13.7	10.3
19.3	**49.0**	B30 V	Midline Upper Buttock Crease	36.8	32.7	24.5	16.3	12.3
24.4	**62.0**	B31 V	to Gluteal Furrow	46.5	41.3	31.0	20.7	15.5
36.6	**93.0**	B32 V	to Popliteal Crease of Knee	69.8	62.0	46.5	31.0	23.3
			TRANSVERSE BACK MEASUREMENTS					
12.2	**31.0**	B33 T	Inter Posterior Axillary Fold	23.3	20.7	15.5	10.3	7.8
5.5	**14.0**	B34 T	Inter Scapulae Medial Border	10.5	9.3	7.0	4.7	3.5
3.2	**8.2**	B35 T	Inter Post. Sup. Spines Sacrum	6.2	5.5	4.1	2.7	2.1
			UPPER EXTREMITY MEASUREMENTS					
63.8	**162.0**	U1 V	Span of Arms	121.5	108.0	81.0	54.0	40.5
6.7	**17.0**	U2 V	Sternum to Acromion Process	12.8	11.3	8.5	5.7	4.3
27.6	**70.0**	U3 V	Acromion to Fingertip	52.5	46.7	35.0	23.3	17.5
80.7	**205.0**	U4 V	Vertical Reach to Fingertip	153.8	136.7	102.5	68.3	51.3
38.6	**98.0**	U5 T	Lateral Reach to Fingertip	73.5	65.3	49.0	32.7	24.5
30.7	**78.0**	U6 D	Forward Reach to Fingertip	58.5	52.0	39.0	26.0	19.5
13.4	**34.0**	U7 V	Shoulder to Elbow Flexed	25.5	22.7	17.0	11.3	8.5
12.2	**31.0**	U8 V	Acromion to Lat. Condyle Elbow	23.3	20.7	15.5	10.3	7.8
16.9	**43.0**	U9 D	Flexed Elbow to Fingertip	32.3	28.7	21.5	14.3	10.8
9.4	**24.0**	U10 D	Flexed Elbow to Wrist Pivot	18.0	16.0	12.0	8.0	6.0

Life size			FOURTEEN-YEAR-OLD BOY	Fractions of life size				
				3/4	2/3	1/2	1/3	1/4
Inches	cm	Code		cm	cm	cm	cm	cm
			UPPER EXTREMITY MEASUREMENTS					
7.1	**18.0**	U11 V	Wrist to Fingertip, hand length	13.5	12.0	9.0	6.0	4.5
4.7	**12.0**	U12 V	Wrist to Thumb	9.0	8.0	6.0	4.0	3.0
4.1	**10.5**	U13 V	Length of Palm	7.9	7.0	5.3	3.5	2.6
3.1	**8.0**	U14 V	Length of Third Finger	6.0	5.3	4.0	2.7	2.0
4.7	**12.0**	U15 V	Thumb Crotch to Third Finger	9.0	8.0	6.0	4.0	3.0
2.4	**6.1**	U16 V	Thumb Crotch to Thumb Tip	4.6	4.1	3.1	2.0	1.5
3.1	**8.0**	U17 D	Anterior-Post. Arm at Biceps	6.0	5.3	4.0	2.7	2.0
3.1	**7.8**	U18 D	Anterior-Posterior at Forearm	5.9	5.2	3.9	2.6	2.0
1.6	**4.0**	U19 D	Depth of Wrist	3.0	2.7	2.0	1.3	1.0
2.2	**5.5**	U20 T	Breadth of Wrist	4.1	3.7	2.8	1.8	1.4
3.9	**9.8**	U21 T	Breadth of Hand	7.4	6.5	4.9	3.3	2.5
3.1	**8.0**	U22 T	Breadth of Palm	6.0	5.3	4.0	2.7	2.0
			LOWER EXTREMITY MEASUREMENTS					
38.6	**98.0**	L1 V	Crest of Ilium to Sole	73.5	65.3	49.0	32.7	24.5
36.6	**93.0**	* L2 V	Ant.Sup.Spine of Ilium to Sole	69.8	62.0	46.5	31.0	23.3
33.5	**85.0**	L3 V	Greater Trochanter Femur-Sole	63.8	56.7	42.5	28.3	21.3
32.7	**83.0**	* L4 V	Symphysis Pubis to Sole	62.3	55.3	41.5	27.7	20.8
29.9	**76.0**	* L5 V	Crotch to Sole	57.0	50.7	38.0	25.3	19.0
30.3	**77.0**	L6 V	Gluteal Furrow to Sole	57.8	51.3	38.5	25.7	19.3
18.5	**47.0**	L7 V	Posterior Knee Crease to Sole	35.3	31.3	23.5	15.7	11.8
16.9	**43.0**	L8 V	Tibia at knee to Sole	32.3	28.7	21.5	14.3	10.8
2.8	**7.0**	L9 V	Med. Malleolus to Sole	5.3	4.7	3.5	2.3	1.8
2.2	**5.5**	L10 V	Lat. Malleolus to Sole	4.1	3.7	2.8	1.8	1.4
11.4	**29.0**	L11 T	Breadth of Hips, Sitting	21.8	19.3	14.5	9.7	7.3
5.1	**13.0**	L12 T	Breadth Mid Thigh	9.8	8.7	6.5	4.3	3.3
4.1	**10.5**	L13 T	Breadth of Knee	7.9	7.0	5.3	3.5	2.6
3.7	**9.5**	L14 T	Breadth of Calf	7.1	6.3	4.8	3.2	2.4
2.3	**5.8**	L15 T	Breadth Ankle	4.4	3.9	2.9	1.9	1.5
5.5	**14.0**	L16 D	Mid Thigh, Anterior-Posterior	10.5	9.3	7.0	4.7	3.5
4.3	**11.0**	L17 D	Knee, Anterior-Posterior	8.3	7.3	5.5	3.7	2.8
4.1	**10.5**	L18 D	Mid Calf, Anterior-Posterior	7.9	7.0	5.3	3.5	2.6
2.8	**7.0**	L19 D	Ankle, Anterior-Posterior	5.3	4.7	3.5	2.3	1.8
10.0	**25.5**	L20 D	Length of Foot	19.1	17.0	12.8	8.5	6.4
8.3	**21.0**	L21 D	Heel to 5th Toe	15.8	14.0	10.5	7.0	5.3
3.9	**9.8**	L22 T	Breadth Foot	7.4	6.5	4.9	3.3	2.5
2.2	**5.5**	L23 T	Width of Heel	4.1	3.7	2.8	1.8	1.4
		** These are also listed with Vertical Measurements.*						

FIFTEEN-YEAR-OLD GIRLS

The lists of body measurements are shown separately in the tables. The range of twenty-fifth, to fiftieth, and to seventy-fifth percentile is listed here for comparison:

Percentile	Twenty-fifth		Fiftieth		Seventy-fifth	
Girlss:	Short		Medium		Tall	
Stature	158 cm,	52.3 in.	162.5 cm,	64 in.	167 cm,	65.3 in.
Weight	50 kg,	110.5 lb.	56 kg,	137.3 lb.	62.3 kg,	137.3 lb.[1]

By this age, girls achieve considerable maturity with feminine attributes with rounded contours in face shoulders and hips. Breast development is at phase V with nearly mature female breast contours.[2] Hair styling is important, and cosmetics are used regularly. They do share interest in athletics and other outdoor activities with boys, but dancing and social events are important. They are beginning to express desire to have drivers' training and drive automobiles. School level is tenth grade.

BODY MEASUREMENTS COMPARED TO HEIGHT

	Feature Measure in cm	Feature Compared to Stature	Approx. Ratio of Feature	Compared to Median Adult
Stature: Crown to sole	152.5			97%
Head Height: Crown to chin	20.5	13%	1/8	100%
Sitting Height:	83.6	52%	1/2	94%
Trunk Measure: Sternal notch to symphysis pubis	50	31%	1/3	96%
Leg Length: Greater trochanter of the femur to sole	81	51%	1/2	95%
Thigh Length: Greater trochanter of the femur to tibia at knee	39	24%	1/4	92%
Tibia at Knee to Sole:	42	26%	1/4	96%
Arm Length: Acromion process at shoulder to dependent fingertip	70.4	55%	1/2	94%
Hand Length:	17.3	11%	1/9	96%
Foot Length:	23.6	15%	1/7	98%

After this age there will be some growth and maturation up to twenty years of age with some subtle changes, but the majority of bone development is nearly complete. Permanent dentition is nearly complete. Third molars continue to grow for two to four more years before eruption. The comparison to adult female ratios reveals very near adult proportions.

1. Nelson, W.E., p. 39.
2. Lowery, G., p. 351.

FIFTEEN-YEAR-OLD BOYS

The lists of body measurements are shown separately in the tables. The range of twenty-fifth to fiftieth and to seventy-fifth percentile is listed here for comparison:

Percentile	Twenty-fifth		Fiftieth		Seventy-fifth	
Boys:	Short		Medium		Tall	
Stature	158 cm,	64.3 in.	169 cm,	66.5 in	174 cm,	68.5 in.
Weight	51 kg,	112.3 lb.	56.7 kg,	125 lb.	64.7 kg,	142.3 lb.[1]

By this age most boys have achieved some maturity and surpass girls in height and weight. Hair style and care has become important. Clothing has received more attention. Increase muscular and bone development facilitate competitive sports that consumes much of their free time. Drivers' training becomes important as they have increased interest in automobiles. School level is tenth grade.

BODY MEASUREMENTS COMPARED TO HEIGHT

	Feature Measure in cm	Feature Compared to Stature	Approx. Ratio of Feature	Compared to Median Adult
Stature: Crown to sole	169			93%
Head Height: Crown to chin	21.5	13%	1/8	94%
Sitting Height:	86	50%	1/2	90%
Trunk Measure: Sternal notch to symphysis pubis	50	30%	1/3–	94%
Leg Length: Greater trochanter of femur to sole	88	50%	1/2	93%
Thigh Length: Greater trochanter of the femur to tibia at knee	44	25%	1/4	94%
Tibia at Knee to Sole:	44	25%	1/4	94%
Arm Length: Acromion process at shoulder to dependent fingertip	71.5	42%	2/5	88%
Hand Length:	18.5	11%	1/9	93%
Foot Length:	26	15%	1/7	93%

By this age, boys achieve about ninety-five percent of adult male measurements for most features. Many facial features begin to resemble those of adult males. The bones of the mid and lower face, the maxilla and mandible have grown to accommodate the permanent teeth. The upper and lower second molars have erupted. The third molars are growing and will erupt in one to four more years.[2] The face is still youthful.

1. Nelson, W.E., p. 38.
2. Lockhart, R.D., p. 57.

Fifteen-Year-Old Girl, anterior and lateral head

all measurements in centimeters

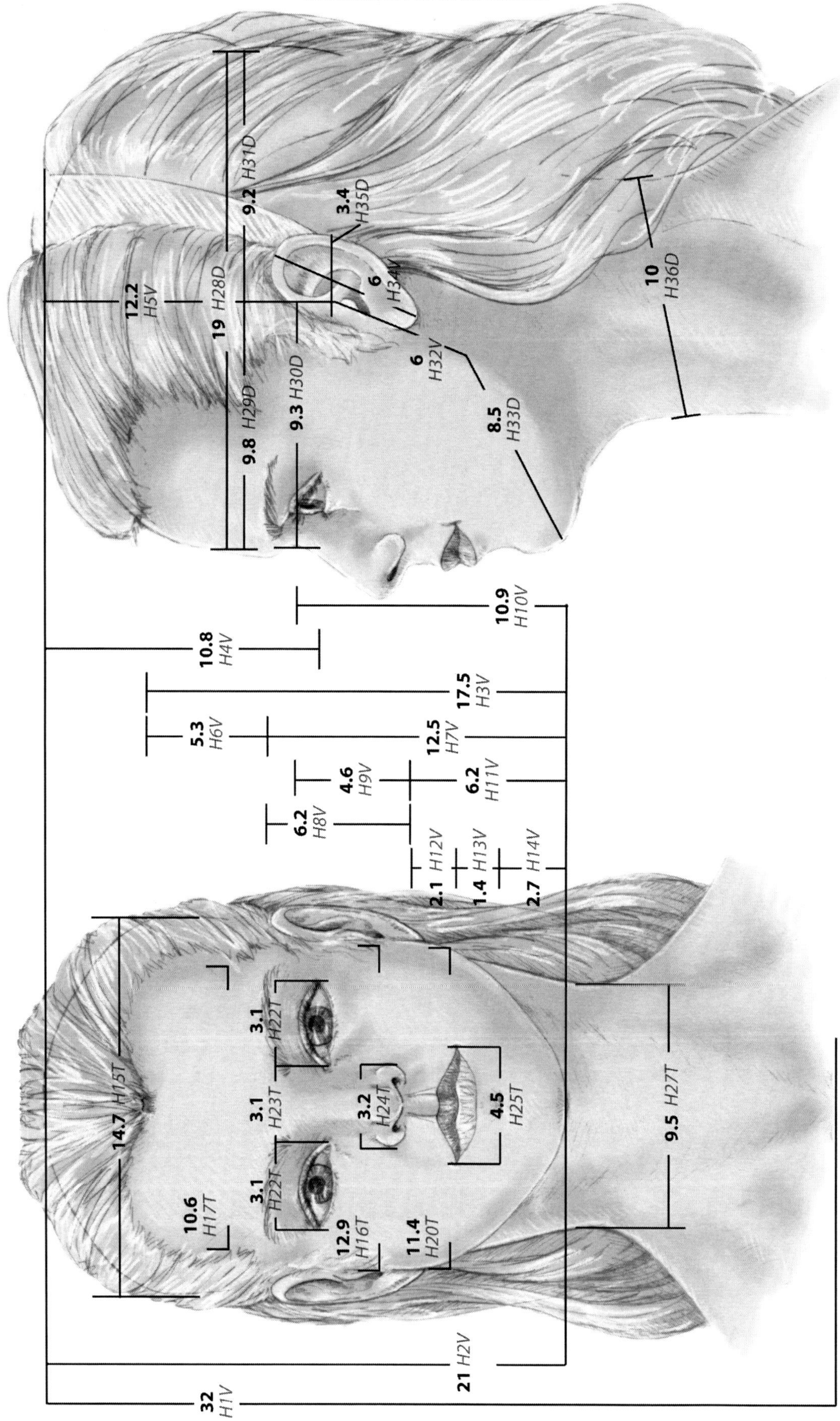

Fifteen-Year-Old Girl, anterior and lateral figure

all measurements in centimeters

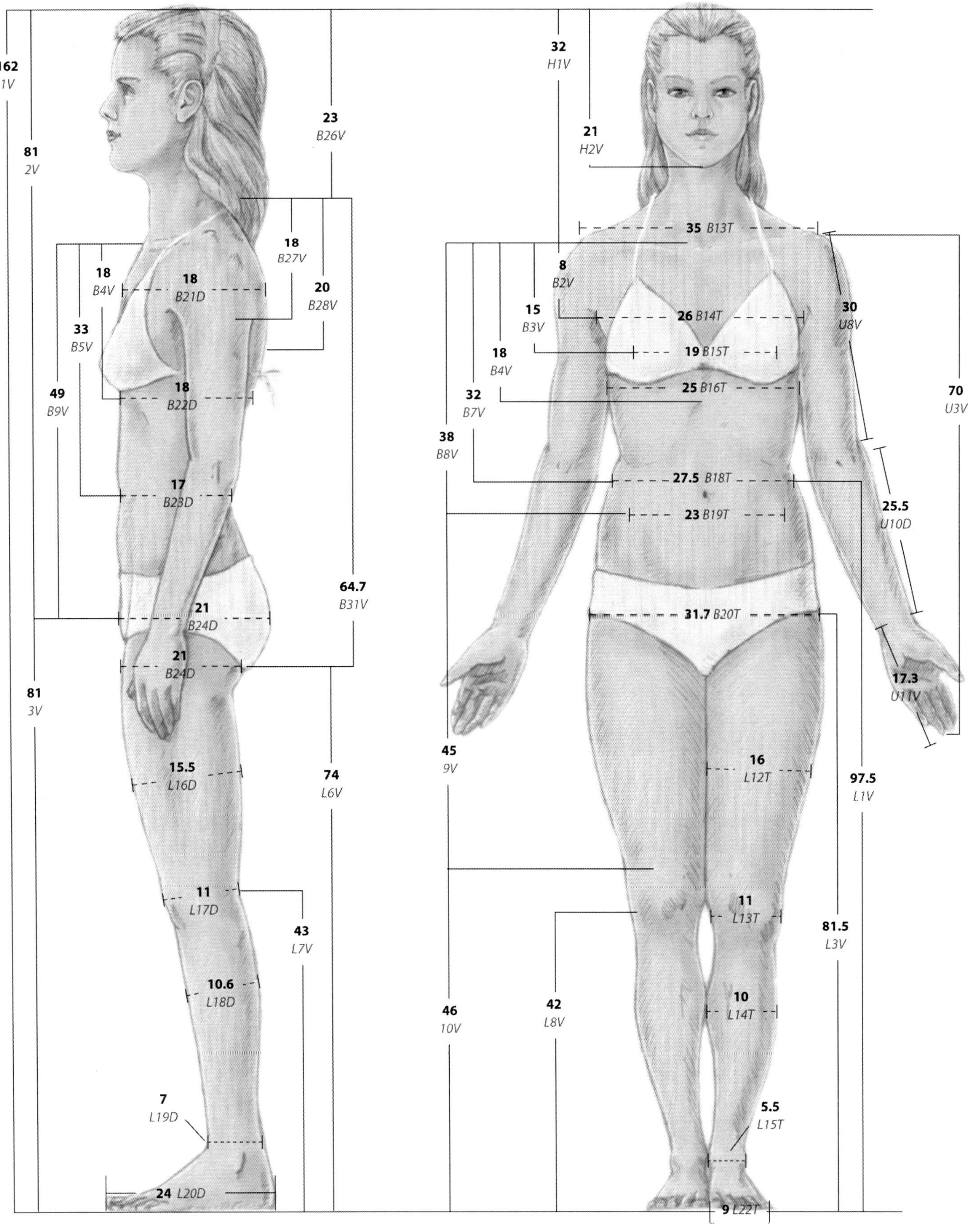

Life size			FIFTEEN-YEAR-OLD GIRL	Fractions of life size				
				3/4	2/3	1/2	1/3	1/4
Inches	cm	Code		cm	cm	cm	cm	cm
			VERTICAL MEASUREMENTS					
63.8	**162.0**	1 V	Stature, Height, Vertex to Sole	121.5	108.0	81.0	54.0	40.5
31.9	**81.0**	2 V	Vertex to Symphysis Pubis	60.8	54.0	40.5	27.0	20.3
31.9	**81.0**	* 3 V	Symphysis Pubis to Sole	60.8	54.0	40.5	27.0	20.3
51.0	**129.5**	4 V	Sternal Notch to Sole	97.1	86.3	64.8	43.2	32.4
12.2	**31.0**	5 V	Vertex to Acromion Process	23.3	20.7	15.5	10.3	7.8
51.2	**130.0**	6 V	Acromion Process to Sole	97.5	86.7	65.0	43.3	32.5
36.0	**91.5**	* 7 V	Ant.Sup.Spine of Ilium to Sole	68.6	61.0	45.8	30.5	22.9
15.2	**38.5**	8 V	Acrom. proc. to Ant. Sup. Spine	28.9	25.7	19.3	12.8	9.6
17.7	**45.0**	9 V	Ant.Sup.Spine of Ilium to Knee	33.8	30.0	22.5	15.0	11.3
18.1	**46.0**	10 V	Knee at Patella to Sole	34.5	30.7	23.0	15.3	11.5
37.8	**96.0**	11 V	Umbilicus to Sole	72.0	64.0	48.0	32.0	24.0
28.3	**72.0**	* 12 V	Crotch to Sole	54.0	48.0	36.0	24.0	18.0
48.0	**122.0**	13 V	Kneeling Height	91.5	81.3	61.0	40.7	30.5
33.1	**84.0**	14 V	Sitting Height	63.0	56.0	42.0	28.0	21.0
33.9	**86.0**	15 V	Buttock, Straight Leg to Sole	64.5	57.3	43.0	28.7	21.5
21.7	**55.0**	16 V	Buttock to Flexed Knee	41.3	36.7	27.5	18.3	13.8
19.7	**50.0**	17 V	Flexed Knee to Sole	37.5	33.3	25.0	16.7	12.5
		** These are also listed with Lower Extremity Measurements.*						
			VERTICAL HEAD MEASUREMENTS					
12.6	**32.0**	H1 V	Vertex to Sternal Notch	24.0	21.3	16.0	10.7	8.0
8.3	**21.0**	H2 V	Vertex to Chin, Head Height	15.8	14.0	10.5	7.0	5.3
6.9	**17.5**	H3 V	Hairline to Chin, Face Height	13.1	11.7	8.8	5.8	4.4
4.3	**10.8**	H4 V	Vertex to Lateral Corner of Eye	8.1	7.2	5.4	3.6	2.7
4.8	**12.2**	H5 V	Vertex to Tragion of Ear	9.2	8.1	6.1	4.1	3.1
2.1	**5.3**	H6 V	Hairline to Brow, Glabella	4.0	3.5	2.7	1.8	1.3
4.9	**12.5**	H7 V	Brow, Glabella, to Chin, Menton	9.4	8.3	6.3	4.2	3.1
2.4	**6.2**	H8 V	Brow to Subnasal Point	4.7	4.1	3.1	2.1	1.6
1.8	**4.6**	H9 V	Sellion to Subnasal, Nose Length	3.5	3.1	2.3	1.5	1.2
4.3	**10.9**	H10 V	Sellion, Nose Saddle, to Chin	8.2	7.3	5.5	3.6	2.7
2.4	**6.2**	H11 V	Subnasal point to Chin	4.7	4.1	3.1	2.1	1.6
0.8	**2.1**	H12 V	Subnasal to Lip Aperature	1.6	1.4	1.1	0.7	0.5
0.6	**1.4**	H13 V	Lip Aperat. to Sublabial Crease	1.1	0.9	0.7	0.5	0.4
1.1	**2.7**	H14 V	Sublabial Crease to Chin	2.0	1.8	1.4	0.9	0.7
			TRANSVERSE HEAD MEASUREMENTS					
5.8	**14.7**	H15 T	Bi-Parietal, Head Breadth	11.0	9.8	7.4	4.9	3.7
5.1	**12.9**	H16 T	Bi-Zygomatic, Cheekbones	9.7	8.6	6.5	4.3	3.2
4.2	**10.6**	H17 T	Bi-Fronto Temporal, Forehead	8.0	7.1	5.3	3.5	2.7

Life size			FIFTEEN-YEAR-OLD GIRL	Fractions of life size				
				3/4	2/3	1/2	1/3	1/4
Inches	cm	Code		cm	cm	cm	cm	cm
			TRANSVERSE HEAD MEASUREMENTS					
4.0	**10.1**	H18 T	Inter Brow Ridge	7.6	6.7	5.1	3.4	2.5
4.9	**12.4**	H19 T	Inter Tragion of Ears	9.3	8.3	6.2	4.1	3.1
4.5	**11.4**	H20 T	Inter Angle of Mandible	8.6	7.6	5.7	3.8	2.9
3.6	**9.2**	H21 T	Inter Lateral Corners of Eyes	6.9	6.1	4.6	3.1	2.3
1.2	**3.0**	H22 T	Width of Eye	2.3	2.0	1.5	1.0	0.8
1.2	**3.1**	H23 T	Inter Inner Corner of eyes	2.3	2.1	1.6	1.0	0.8
1.3	**3.2**	H24 T	Breadth of Nose	2.4	2.1	1.6	1.1	0.8
1.8	**4.5**	H25 T	Breadth of Mouth	3.4	3.0	2.3	1.5	1.1
1.3	**3.3**	H26 T	Breadth of Chin	2.5	2.2	1.7	1.1	0.8
3.7	**9.5**	H27 T	Breadth of Neck	7.1	6.3	4.8	3.2	2.4
			ANTERIOR-POSTERIOR HEAD MEASUREMENTS					
7.5	**19.0**	H28 D	Brow to Occiput	14.3	12.7	9.5	6.3	4.8
3.9	**9.8**	H29 D	Brow to Tragus	7.4	6.5	4.9	3.3	2.5
3.7	**9.3**	H30 D	Sellion to Tragus	7.0	6.2	4.7	3.1	2.3
3.6	**9.2**	H31 D	Tragus to Occiput	6.9	6.1	4.6	3.1	2.3
2.4	**6.0**	H32 V	Tragus to Angle Mandible	4.5	4.0	3.0	2.0	1.5
3.4	**8.7**	H33 D	Angle of Mandible to Chin	6.5	5.8	4.4	2.9	2.2
2.4	**6.0**	H34 V	Length of Ear, Vertical	4.5	4.0	3.0	2.0	1.5
1.3	**3.4**	H35 D	Width of Ear, Anterior-Posterior	2.6	2.3	1.7	1.1	0.9
3.9	**10.0**	H36 D	Anterior to Posterior of Neck	7.5	6.7	5.0	3.3	2.5
			VERTICAL BODY MEASUREMENTS					
3.9	**10.0**	B1 V	Chin to Sternal Notch	7.5	6.7	5.0	3.3	2.5
3.1	**8.0**	B2 V	Sternal Notch to Axilla	6.0	5.3	4.0	2.7	2.0
5.9	**15.0**	B3 V	Sternal Notch to Nipples	11.3	10.0	7.5	5.0	3.8
7.1	**18.0**	B4 V	Sternal Notch to Xyphoid Proc.	13.5	12.0	9.0	6.0	4.5
13.0	**33.0**	B5 V	Sternal Notch to Umbilicus	24.8	22.0	16.5	11.0	8.3
5.9	**15.0**	B6 V	Xyphoid Process to Umbilicus	11.3	10.0	7.5	5.0	3.8
12.6	**32.0**	B7 V	Sternal Notch to Crest Ilium	24.0	21.3	16.0	10.7	8.0
15.0	**38.0**	B8 V	Sternal Notch to Ant.S.Sp.Ilium	28.5	25.3	19.0	12.7	9.5
19.3	**49.0**	B9 V	Sternal Notch to Pubis	36.8	32.7	24.5	16.3	12.3
5.9	**15.0**	B10 V	Umbilicus to Pubis	11.3	10.0	7.5	5.0	3.8
			TRANSVERSE BODY MEASUREMENTS					
15.4	**39.0**	B11 T	Shoulder Breadth	29.3	26.0	19.5	13.0	9.8
6.9	**17.4**	B12 T	Clavicle to Acromion Process	13.1	11.6	8.7	5.8	4.4
13.8	**35.0**	B13 T	Biacromial Process Breadth	26.3	23.3	17.5	11.7	8.8
10.2	**26.0**	B14 T	Inter Anterior Axillary Fold	19.5	17.3	13.0	8.7	6.5
7.5	**19.0**	B15 T	Inter Nipples	14.3	12.7	9.5	6.3	4.8

Life size			FIFTEEN-YEAR-OLD GIRL	Fractions of life size				
				3/4	2/3	1/2	1/3	1/4
Inches	cm	Code		cm	cm	cm	cm	cm
			TRANSVERSE BODY MEASUREMENTS					
9.8	**25.0**	B16 T	Breadth of Thorax	18.8	16.7	12.5	8.3	6.3
9.4	**24.0**	B17 T	Inter Costal Margin Breadth	18.0	16.0	12.0	8.0	6.0
10.8	**27.5**	B18 T	Inter Crest of Ilium Breadth	20.6	18.3	13.8	9.2	6.9
9.1	**23.0**	B19 T	Inter Anterior Superior Spine	17.3	15.3	11.5	7.7	5.8
12.5	**31.7**	B20 T	Inter Greater Trochanter Femur	23.8	21.1	15.9	10.6	7.9
			ANTERIOR-POSTERIOR BODY MEASUREMENTS					
7.1	**18.0**	B21 D	Manubrium to Dorsal Spine	13.5	12.0	9.0	6.0	4.5
7.1	**18.0**	B22 D	Xyphoid Process to Dorsum	13.5	12.0	9.0	6.0	4.5
6.7	**17.0**	B23 D	Umbilicus to Dorsum	12.8	11.3	8.5	5.7	4.3
8.3	**21.0**	B24 D	Pubis to Buttock	15.8	14.0	10.5	7.0	5.3
6.7	**17.0**	B25 D	Thigh to Gluteal Furrow	12.8	11.3	8.5	5.7	4.3
			VERTICAL BACK MEASUREMENTS					
9.1	**23.0**	B26 V	Vertex to 7th Cervical Spine	17.3	15.3	11.5	7.7	5.8
			7th Cervical Vertebra Spine to:					
7.1	**18.0**	B27 V	To Posterior Axillary Fold	13.5	12.0	9.0	6.0	4.5
7.9	**20.0**	B28 V	To Inferior Tip of Scapula	15.0	13.3	10.0	6.7	5.0
16.5	**42.0**	B29 V	To Posterior Superior Spine Ilium	31.5	28.0	21.0	14.0	10.5
19.7	**50.0**	B30 V	To Midline Upper Buttock Crease	37.5	33.3	25.0	16.7	12.5
25.5	**64.7**	B31 V	To Gluteal Furrow	48.5	43.1	32.4	21.6	16.2
37.4	**95.0**	B32 V	To Popliteal Crease of Knee	71.3	63.3	47.5	31.7	23.8
			TRANSVERSE BACK MEASUREMENTS					
12.2	**31.0**	B33 T	Inter Posterior Axillary Folds	23.3	20.7	15.5	10.3	7.8
5.5	**14.0**	B34 T	Inter Scapulae Medial Borders	10.5	9.3	7.0	4.7	3.5
3.1	**8.0**	B35 T	Inter Post. Sup. Spines Sacrum	6.0	5.3	4.0	2.7	2.0
			UPPER EXTREMITY MEASUREMENTS					
63.0	**160.0**	U1 V	Span of Arms	120.0	106.7	80.0	53.3	40.0
6.9	**17.4**	U2 V	Sternum to Acromion Process	13.1	11.6	8.7	5.8	4.4
27.6	**70.0**	U3 V	Acromion Process to Fingertip	52.5	46.7	35.0	23.3	17.5
79.9	**203.0**	U4 V	Vertical Reach to Fingertip	154.5	137.3	103.0	68.7	51.5
38.6	**98.0**	U5 T	Lateral Reach to Fingertip	76.5	68.0	51.0	34.0	25.5
30.7	**78.0**	U6 D	Forward Reach to Fingertip	58.5	52.0	39.0	26.0	19.5
13.2	**33.5**	U7 V	Shoulder to Elbow Flexed	25.1	22.3	16.8	11.2	8.4
11.8	**30.0**	U8 V	Acromion to Lat. Condyle Elbow	22.5	20.0	15.0	10.0	7.5
16.9	**43.0**	U9 D	Flexed Elbow to Fingertip	32.3	28.7	21.5	14.3	10.8
10.0	**25.5**	U10 D	Flexed Elbow Pivot to Wrist Pivot	19.1	17.0	12.8	8.5	6.4

Life size			FIFTEEN-YEAR-OLD GIRL	Fractions of life size				
				3/4	2/3	1/2	1/3	1/4
Inches	cm	Code		cm	cm	cm	cm	cm
			UPPER EXTREMITY MEASUREMENTS					
6.8	**17.3**	U11 V	Wrist to Fingertip	13.0	11.5	8.7	5.8	4.3
4.3	**11.0**	U12 V	Wrist to Thumb Tip	8.3	7.3	5.5	3.7	2.8
3.8	**9.7**	U13 V	Length of Palm	7.3	6.5	4.9	3.2	2.4
3.0	**7.5**	U14 V	Length 3rd Finger	5.6	5.0	3.8	2.5	1.9
4.5	**11.5**	U15 V	Thumb Crotch to 3rd Finger	8.6	7.7	5.8	3.8	2.9
2.3	**5.9**	U16 V	Thumb Crotch to Thumb Tip	4.4	3.9	3.0	2.0	1.5
3.0	**7.5**	U17 D	Anterior-Post. Arm at Biceps	5.6	5.0	3.8	2.5	1.9
2.8	**7.0**	U18 D	Anterior-Posterior at Forearm	5.3	4.7	3.5	2.3	1.8
1.5	**3.7**	U19 D	Depth of Wrist	2.8	2.5	1.9	1.2	0.9
2.0	**5.0**	U20 T	Breadth of Wrist	3.8	3.3	2.5	1.7	1.3
3.7	**9.3**	U21 T	Breadth of Hand	7.0	6.2	4.7	3.1	2.3
3.0	**7.6**	U22 T	Breadth of Palm	5.7	5.1	3.8	2.5	1.9
			LOWER EXTREMITY MEASUREMENTS					
38.4	**97.5**	L1 V	Crest of Ilium to Sole	73.1	65.0	48.8	32.5	24.4
36.0	**91.5**	* L2 V	Ant.Sup.Spine of Ilium to Sole	68.6	61.0	45.8	30.5	22.9
32.1	**81.5**	L3 V	Greater Trochanter Femur-Sole	61.1	54.3	40.8	27.2	20.4
31.9	**81.0**	* L4 V	Symphysis Pubis to Sole	60.8	54.0	40.5	27.0	20.3
28.3	**72.0**	* L5 V	Crotch to Sole	54.0	48.0	36.0	24.0	18.0
29.1	**74.0**	L6 V	Gluteal Furrow to Sole	55.5	49.3	37.0	24.7	18.5
16.9	**43.0**	L7 V	Posterior Knee Crease to Sole	32.3	28.7	21.5	14.3	10.8
16.5	**42.0**	L8 V	Tibia at Knee to Sole	31.5	28.0	21.0	14.0	10.5
2.6	**6.5**	L9 V	Med. Malleolus to Sole	4.9	4.3	3.3	2.2	1.6
2.0	**5.0**	L10 V	Lat. Malleolus to Sole	3.8	3.3	2.5	1.7	1.3
13.2	**33.5**	L11 T	Breadth of Hips, Sitting	25.1	22.3	16.8	11.2	8.4
6.3	**16.0**	L12 T	Breadth Mid Thigh	12.0	10.7	8.0	5.3	4.0
4.3	**11.0**	L13 T	Breadth of Knee	8.3	7.3	5.5	3.7	2.8
3.9	**10.0**	L14 T	Breadth of Calf	7.5	6.7	5.0	3.3	2.5
2.2	**5.5**	L15 T	Breadth Ankle	4.1	3.7	2.8	1.8	1.4
6.1	**15.5**	L16 D	Mid Thigh, Anterior-Posterior	11.6	10.3	7.8	5.2	3.9
4.3	**11.0**	L17 D	Knee, Anterior-Posterior	8.3	7.3	5.5	3.7	2.8
4.1	**10.5**	L18 D	Mid Calf, Anterior-Posterior	7.9	7.0	5.3	3.5	2.6
2.8	**7.0**	L19 D	Ankle, Anterior-Posterior	5.3	4.7	3.5	2.3	1.8
9.4	**24.0**	L20 D	Length of Foot	18.0	16.0	12.0	8.0	6.0
7.5	**19.0**	L21 D	Heel to 5th Toe	14.3	12.7	9.5	6.3	4.8
3.5	**9.0**	L22 T	Breadth Foot	6.8	6.0	4.5	3.0	2.3
2.4	**6.0**	L23 T	Width of Heel	4.5	4.0	3.0	2.0	1.5
		** These are also listed with Vertical Measurements.*						

Fifteen-Year-Old Boy, anterior and lateral head

all measurements in centimeters

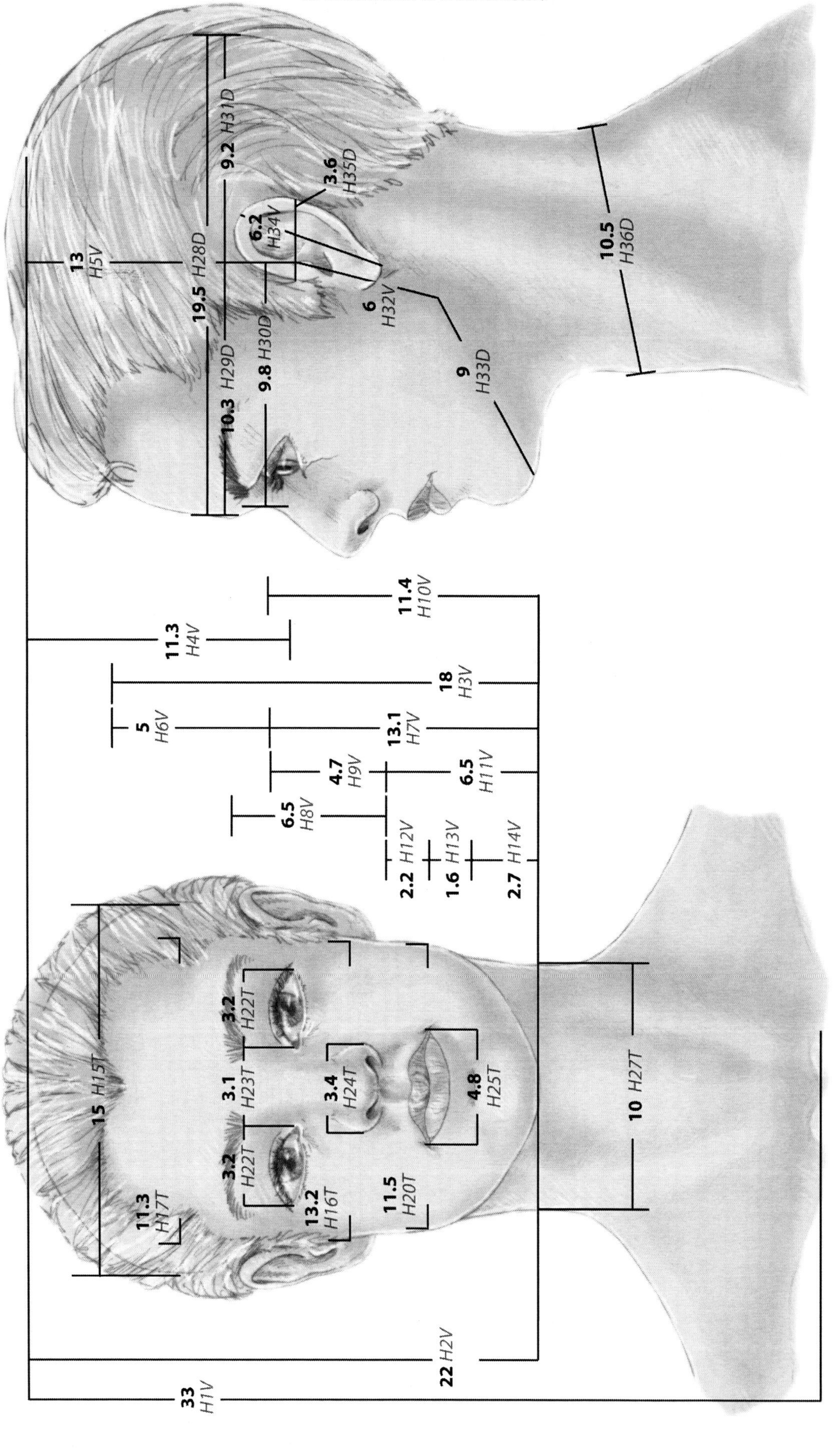

Fifteen-Year-Old Boy, anterior and lateral figure

all measurements in centimeters

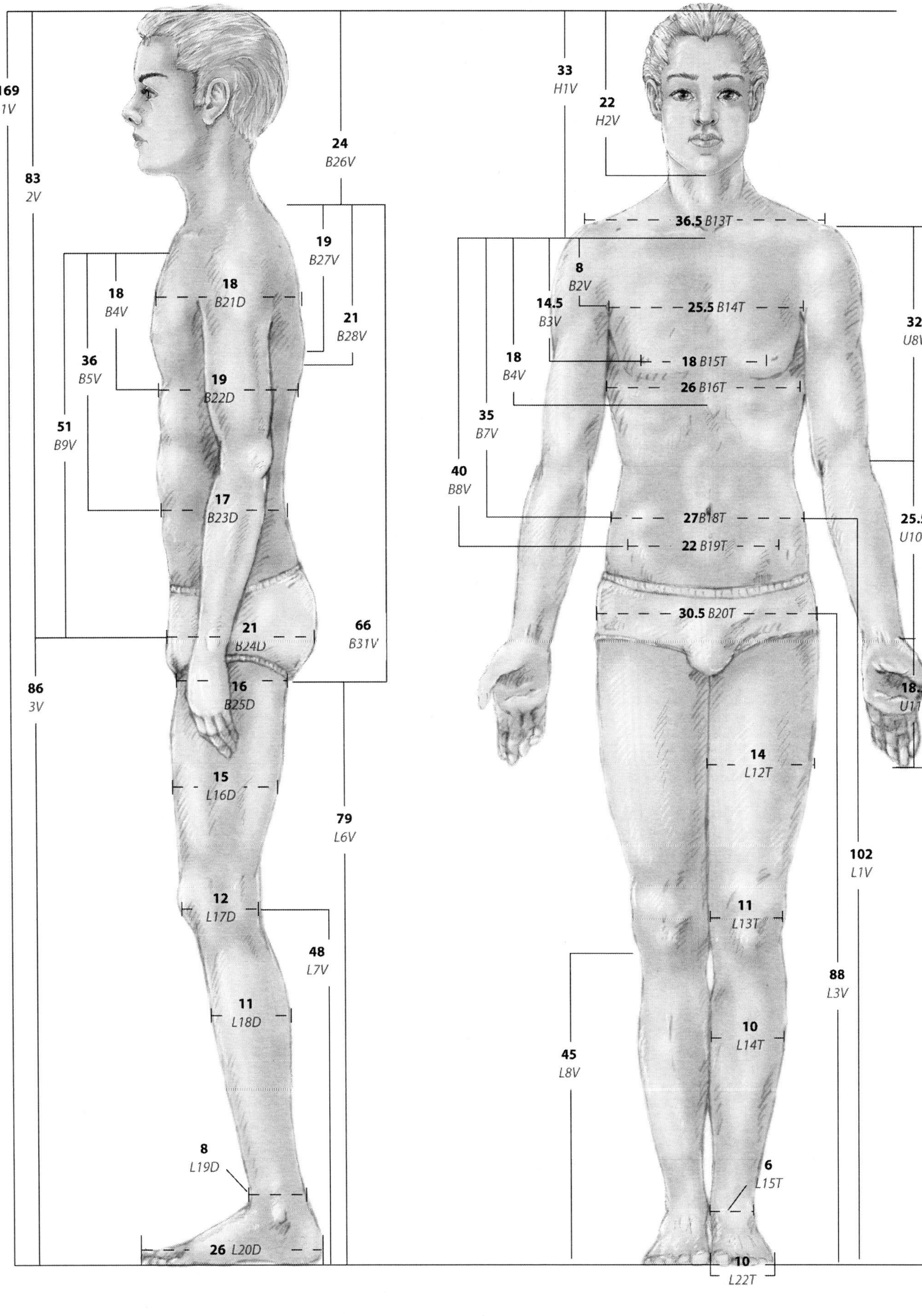

Life size			FIFTEEN-YEAR-OLD BOY	Fractions of life size				
				3/4	2/3	1/2	1/3	1/4
Inches	cm	Code		cm	cm	cm	cm	cm
			VERTICAL MEASUREMENTS					
66.5	**169.0**	1 V	Stature, Height, Vertex to Sole	126.8	112.7	84.5	56.3	42.3
32.7	**83.0**	2 V	Vertex to Symphysis Pubis	62.3	55.3	41.5	27.7	20.8
33.9	**86.0**	* 3 V	Symphysis Pubis to Sole	64.5	57.3	43.0	28.7	21.5
53.9	**137.0**	4 V	Sternal Notch to Sole	102.8	91.3	68.5	45.7	34.3
12.4	**31.5**	5 V	Vertex to Acromion Process	23.6	21.0	15.8	10.5	7.9
54.3	**138.0**	6 V	Acromion Process to Sole	103.5	92.0	69.0	46.0	34.5
37.4	**95.0**	* 7 V	Ant.Sup.Spine of Ilium to Sole	71.3	63.3	47.5	31.7	23.8
16.9	**43.0**	8 V	Acrom. proc. to Ant. Sup. Spine	32.3	28.7	21.5	14.3	10.8
18.1	**46.0**	9 V	Ant.Sup.Spine of Ilium to Knee	34.5	30.7	23.0	15.3	11.5
19.3	**49.0**	10 V	Knee at Patella to Sole	36.8	32.7	24.5	16.3	12.3
40.6	**103.0**	11 V	Umbilicus to Sole	77.3	68.7	51.5	34.3	25.8
31.5	**80.0**	* 12 V	Crotch to Sole	60.0	53.3	40.0	26.7	20.0
49.6	**126.0**	13 V	Kneeling Height	94.5	84.0	63.0	42.0	31.5
33.9	**86.0**	14 V	Sitting Height	64.5	57.3	43.0	28.7	21.5
35.4	**90.0**	15 V	Buttock, Straight Leg to Sole	67.5	60.0	45.0	30.0	22.5
22.4	**57.0**	16 V	Buttock to Flexed Knee	42.8	38.0	28.5	19.0	14.3
21.1	**53.5**	17 V	Flexed Knee to Sole	40.1	35.7	26.8	17.8	13.4
		**These are also listed with Lower Extremity Measurements.*						
			VERTICAL HEAD MEASUREMENTS					
13.0	**33.0**	H1 V	Vertex to Sternal Notch	24.8	22.0	16.5	11.0	8.3
8.7	**22.0**	H2 V	Vertex to Chin, Head Height	16.5	14.7	11.0	7.3	5.5
7.1	**18.0**	H3 V	Hairline to Chin, Face Height	13.5	12.0	9.0	6.0	4.5
4.4	**11.3**	H4 V	Vertex to Lateral Corner of Eye	8.5	7.5	5.7	3.8	2.8
5.1	**13.0**	H5 V	Vertex to Tragion of Ear	9.8	8.7	6.5	4.3	3.3
2.0	**5.0**	H6 V	Hairline to Brow, Glabella	3.8	3.3	2.5	1.7	1.3
5.2	**13.1**	H7 V	Brow, Glabella, to Chin, Menton	9.8	8.7	6.6	4.4	3.3
2.6	**6.5**	H8 V	Brow to Subnasal Point	4.9	4.3	3.3	2.2	1.6
1.9	**4.7**	H9 V	Sellion to Subnasal, Nose Length	3.5	3.1	2.4	1.6	1.2
4.5	**11.4**	H10 V	Sellion, Nose Saddle, to Chin	8.6	7.6	5.7	3.8	2.9
2.6	**6.5**	H11 V	Subnasal point to Chin	4.9	4.3	3.3	2.2	1.6
0.9	**2.2**	H12 V	Subnasal to Lip Aperature	1.7	1.5	1.1	0.7	0.6
0.6	**1.6**	H13 V	Lip Aperat. to Sublabial Crease	1.2	1.1	0.8	0.5	0.4
1.1	**2.7**	H14 V	Sublabial Crease to Chin	2.0	1.8	1.4	0.9	0.7
			TRANSVERSE HEAD MEASUREMENTS					
5.9	**15.0**	H15 T	Bi-Parietal, Head Breadth	11.3	10.0	7.5	5.0	3.8
5.2	**13.2**	H16 T	Bi-Zygomatic, Cheekbones	9.9	8.8	6.6	4.4	3.3
4.4	**11.3**	H17 T	Bi-Fronto Temporal, Forehead	8.5	7.5	5.7	3.8	2.8

Life size			FIFTEEN-YEAR-OLD BOY	Fractions of life size				
				3/4	2/3	1/2	1/3	1/4
Inches	cm	Code		cm	cm	cm	cm	cm
			TRANSVERSE HEAD MEASUREMENTS					
3.9	**10.0**	H18 T	Inter Brow Ridge	7.5	6.7	5.0	3.3	2.5
5.0	**12.6**	H19 T	Inter Tragion of Ears	9.5	8.4	6.3	4.2	3.2
4.5	**11.5**	H20 T	Inter Angle of Mandible	8.6	7.7	5.8	3.8	2.9
3.7	**9.5**	H21 T	Inter Lateral Corners of Eyes	7.1	6.3	4.8	3.2	2.4
1.3	**3.2**	H22 T	Width of Eye	2.4	2.1	1.6	1.1	0.8
1.2	**3.1**	H23 T	Inter Inner Corner of eyes	2.3	2.1	1.6	1.0	0.8
1.3	**3.4**	H24 T	Breadth of Nose	2.6	2.3	1.7	1.1	0.9
1.9	**4.8**	H25 T	Breadth of Mouth	3.6	3.2	2.4	1.6	1.2
1.3	**3.4**	H26 T	Breadth of Chin	2.6	2.3	1.7	1.1	0.9
3.9	**10.0**	H27 T	Breadth of Neck	7.5	6.7	5.0	3.3	2.5
			ANTERIOR-POSTERIOR HEAD MEASUREMENTS					
7.7	**19.5**	H28 D	Brow to Occiput	14.6	13.0	9.8	6.5	4.9
4.1	**10.3**	H29 D	Brow to Tragus	7.7	6.9	5.2	3.4	2.6
3.9	**9.8**	H30 D	Sellion to Tragus	7.4	6.5	4.9	3.3	2.5
3.6	**9.2**	H31 D	Tragus to Occiput	6.9	6.1	4.6	3.1	2.3
2.4	**6.0**	H32 V	Tragus to Angle Mandible	4.5	4.0	3.0	2.0	1.5
3.5	**9.0**	H33 D	Angle of Mandible to Chin	6.8	6.0	4.5	3.0	2.3
2.4	**6.2**	H34 V	Length of Ear, Vertical	4.7	4.1	3.1	2.1	1.6
1.4	**3.6**	H35 D	Width of Ear, Anterior-Posterior	2.7	2.4	1.8	1.2	0.9
4.1	**10.5**	H36 D	Anterior to Posterior of Neck	7.9	7.0	5.3	3.5	2.6
			VERTICAL BODY MEASUREMENTS					
4.1	**10.5**	B1 V	Chin to Sternal Notch	7.9	7.0	5.3	3.5	2.6
3.1	**8.0**	B2 V	Sternal Notch to Axilla	6.0	5.3	4.0	2.7	2.0
5.7	**14.5**	B3 V	Sternal Notch to Nipples	10.9	9.7	7.3	4.8	3.6
7.1	**18.0**	B4 V	Sternal Notch to Xyphoid Proc.	13.5	12.0	9.0	6.0	4.5
14.2	**36.0**	B5 V	Sternal Notch to Umbilicus	27.0	24.0	18.0	12.0	9.0
7.1	**18.0**	B6 V	Xyphoid Process to Umbilicus	13.5	12.0	9.0	6.0	4.5
13.8	**35.0**	B7 V	Sternal Notch to Crest Ilium	26.3	23.3	17.5	11.7	8.8
15.7	**40.0**	B8 V	Sternal Notch to Ant.S.Sp.Ilium	30.0	26.7	20.0	13.3	10.0
20.1	**51.0**	B9 V	Sternal Notch to Pubis	38.3	34.0	25.5	17.0	12.8
5.9	**15.0**	B10 V	Umbilicus to Pubis	11.3	10.0	7.5	5.0	3.8
			TRANSVERSE BODY MEASUREMENTS					
16.1	**41.0**	B11 T	Shoulder Breadth	30.8	27.3	20.5	13.7	10.3
7.1	**18.0**	B12 T	Clavicle to Acromion Process	13.5	12.0	9.0	6.0	4.5
14.4	**36.5**	B13 T	Biacromial Process Breadth	27.4	24.3	18.3	12.2	9.1
10.0	**25.5**	B14 T	Inter Anterior Axillary Fold	19.1	17.0	12.8	8.5	6.4
7.1	**18.0**	B15 T	Inter Nipples	13.5	12.0	9.0	6.0	4.5

Life size			FIFTEEN-YEAR-OLD BOY	Fractions of life size				
				3/4	2/3	1/2	1/3	1/4
Inches	cm	Code		cm	cm	cm	cm	cm
			TRANSVERSE BODY MEASUREMENTS					
10.2	**26.0**	B16 T	Breadth of Thorax	19.5	17.3	13.0	8.7	6.5
9.8	**25.0**	B17 T	Inter Costal Margin Breadth	18.8	16.7	12.5	8.3	6.3
10.6	**27.0**	B18 T	Inter Crest of Ilium Breadth	20.3	18.0	13.5	9.0	6.8
8.7	**22.0**	B19 T	Inter Anterior Superior Spine	16.5	14.7	11.0	7.3	5.5
12.0	**30.5**	B20 T	Inter Greater Trochanter Femur	22.9	20.3	15.3	10.2	7.6
			ANTERIOR-POSTERIOR BODY MEASUREMENTS					
7.1	**18.0**	B21 D	Manubrium to Dorsal Spine	13.5	12.0	9.0	6.0	4.5
7.5	**19.0**	B22 D	Xyphoid Process to Dorsum	14.3	12.7	9.5	6.3	4.8
6.7	**17.0**	B23 D	Umbilicus to Dorsum	12.8	11.3	8.5	5.7	4.3
8.3	**21.0**	B24 D	Pubis to Buttock	15.8	14.0	10.5	7.0	5.3
6.3	**16.0**	B25 D	Thigh to Gluteal Furrow	12.0	10.7	8.0	5.3	4.0
			VERTICAL BACK MEASUREMENTS					
9.4	**24.0**	B26 V	Vertex to 7th Cervical Spine	18.0	16.0	12.0	8.0	6.0
			7th Cervical Vertebra Spine to:					
7.5	**19.0**	B27 V	To Posterior Axillary Fold	14.3	12.7	9.5	6.3	4.8
8.3	**21.0**	B28 V	To Inferior Tip of Scapula	15.8	14.0	10.5	7.0	5.3
16.5	**42.0**	B29 V	To Posterior Superior Spine Ilium	31.5	28.0	21.0	14.0	10.5
20.9	**53.0**	B30 V	To Midline Upper Buttock Crease	39.8	35.3	26.5	17.7	13.3
26.0	**66.0**	B31 V	To Gluteal Furrow	49.5	44.0	33.0	22.0	16.5
38.6	**98.0**	B32 V	To Popliteal Crease of Knee	73.5	65.3	49.0	32.7	24.5
			TRANSVERSE BACK MEASUREMENTS					
13.4	**34.0**	B33 T	Inter Posterior Axillary Folds	25.5	22.7	17.0	11.3	8.5
5.9	**15.0**	B34 T	Inter Scapulae Medial Borders	11.3	10.0	7.5	5.0	3.8
3.3	**8.5**	B35 T	Inter Post. Sup. Spines Sacrum	6.4	5.7	4.3	2.8	2.1
			UPPER EXTREMITY MEASUREMENTS					
66.1	**168.0**	U1 V	Span of Arms	126.0	112.0	84.0	56.0	42.0
7.1	**18.0**	U2 V	Sternum to Acromion Process	13.5	12.0	9.0	6.0	4.5
28.3	**72.0**	U3 V	Acromion Process to Fingertip	54.0	48.0	36.0	24.0	18.0
82.7	**210.0**	U4 V	Vertical Reach to Fingertip	155.3	138.0	103.5	69.0	51.8
39.4	**100.0**	U5 T	Lateral Reach to Fingertip	73.5	65.3	49.0	32.7	24.5
31.1	**79.0**	U6 D	Forward Reach to Fingertip	59.3	52.7	39.5	26.3	19.8
13.8	**35.0**	U7 V	Shoulder to Elbow Flexed	26.3	23.3	17.5	11.7	8.8
12.6	**32.0**	U8 V	Acromion to Lat. Condyle Elbow	24.0	21.3	16.0	10.7	8.0
17.7	**45.0**	U9 D	Flexed Elbow to Fingertip	33.8	30.0	22.5	15.0	11.3
10.0	**25.5**	U10 D	Flexed Elbow Pivot to Wrist Pivot	19.1	17.0	12.8	8.5	6.4

Life size			FIFTEEN-YEAR-OLD BOY	Fractions of life size				
				3/4	2/3	1/2	1/3	1/4
Inches	cm	Code		cm	cm	cm	cm	cm
			UPPER EXTREMITY MEASUREMENTS					
7.3	**18.5**	U11 V	Wrist to Fingertip	13.9	12.3	9.3	6.2	4.6
4.7	**12.0**	U12 V	Wrist to Thumb Tip	9.0	8.0	6.0	4.0	3.0
4.1	**10.5**	U13 V	Length of Palm	7.9	7.0	5.3	3.5	2.6
3.1	**8.0**	U14 V	Length 3rd Finger	6.0	5.3	4.0	2.7	2.0
4.9	**12.4**	U15 V	Thumb Crotch to 3rd Finger	9.3	8.3	6.2	4.1	3.1
2.6	**6.5**	U16 V	Thumb Crotch to Thumb Tip	4.9	4.3	3.3	2.2	1.6
3.1	**8.0**	U17 D	Anterior-Post. Arm at Biceps	6.0	5.3	4.0	2.7	2.0
3.1	**7.8**	U18 D	Anterior-Posterior at Forearm	5.9	5.2	3.9	2.6	2.0
1.6	**4.0**	U19 D	Depth of Wrist	3.0	2.7	2.0	1.3	1.0
2.2	**5.5**	U20 T	Breadth of Wrist	4.1	3.7	2.8	1.8	1.4
3.9	**10.0**	U21 T	Breadth of Hand	7.5	6.7	5.0	3.3	2.5
3.2	**8.2**	U22 T	Breadth of Palm	6.2	5.5	4.1	2.7	2.1
			LOWER EXTREMITY MEASUREMENTS					
40.2	**102.0**	L1 V	Crest of Ilium to Sole	76.5	68.0	51.0	34.0	25.5
37.8	**96.0**	* L2 V	Ant.Sup.Spine of Ilium to Sole	72.0	64.0	48.0	32.0	24.0
34.6	**88.0**	L3 V	Greater Trochanter Femur-Sole	66.0	58.7	44.0	29.3	22.0
33.9	**86.0**	* L4 V	Symphysis Pubis to Sole	64.5	57.3	43.0	28.7	21.5
31.5	**80.0**	* L5 V	Crotch to Sole	60.0	53.3	40.0	26.7	20.0
31.1	**79.0**	L6 V	Gluteal Furrow to Sole	59.3	52.7	39.5	26.3	19.8
18.9	**48.0**	L7 V	Posterior Knee Crease to Sole	36.0	32.0	24.0	16.0	12.0
17.7	**45.0**	L8 V	Tibia at Knee to Sole	33.8	30.0	22.5	15.0	11.3
2.8	**7.0**	L9 V	Med. Malleolus to Sole	5.3	4.7	3.5	2.3	1.8
2.2	**5.5**	L10 V	Lat. Malleolus to Sole	4.1	3.7	2.8	1.8	1.4
11.8	**30.0**	L11 T	Breadth of Hips, Sitting	22.5	20.0	15.0	10.0	7.5
5.5	**14.0**	L12 T	Breadth Mid Thigh	10.5	9.3	7.0	4.7	3.5
4.3	**11.0**	L13 T	Breadth of Knee	8.3	7.3	5.5	3.7	2.8
3.9	**10.0**	L14 T	Breadth of Calf	7.5	6.7	5.0	3.3	2.5
2.4	**6.0**	L15 T	Breadth Ankle	4.5	4.0	3.0	2.0	1.5
5.9	**15.0**	L16 D	Mid Thigh, Anterior-Posterior	11.3	10.0	7.5	5.0	3.8
4.7	**12.0**	L17 D	Knee, Anterior-Posterior	9.0	8.0	6.0	4.0	3.0
4.3	**11.0**	L18 D	Mid Calf, Anterior-Posterior	8.3	7.3	5.5	3.7	2.8
3.1	**8.0**	L19 D	Ankle, Anterior-Posterior	6.0	5.3	4.0	2.7	2.0
10.2	**26.0**	L20 D	Length of Foot	19.5	17.3	13.0	8.7	6.5
8.3	**21.0**	L21 D	Heel to 5th Toe	15.8	14.0	10.5	7.0	5.3
3.9	**10.0**	L22 T	Breadth Foot	7.5	6.7	5.0	3.3	2.5
2.3	**5.8**	L23 T	Width of Heel	4.4	3.9	2.9	1.9	1.5
		These are also listed with Vertical Measurements.						

SIXTEEN-YEAR-OLD GIRLS

The lists of body proportions are shown in the tables of measurements. A range of weight and stature is listed here for comparison:

Percentile	Twenty-fifth		Fiftieth		Seventy-fifth	
Girls:	Short		Medium		Tall	
Stature	158 cm,	62.3 in.	167 cm,	65.8 in.	171 cm,	67.3 in.
Weight	50 kg,	110.5 lb.	56 kg,	123 lb.	72 kg,	158 lb.[1]

After age fifteen, girls gain stature less rapidly, and the average height of boys is taller. By age sixteen, girls have more mature feminine features. Although they participate in outdoor sports, hair and dress style, dancing and socializing have increased importance.

Breast development is mature female. The variation in sizes of breasts depends primarily on the amount of fatty tissue in the pectoral area. At this age, many of them have earned a driver's license, and automobiles become an important part of school and social life. The school level is eleventh grade.

BODY MEASUREMENTS COMPARED TO HEIGHT

	Feature Measure in cm	Feature Compared to Stature	Approx. Ratio of Feature	Compared to Median Adult
Stature: Crown to sole	162.4			97%
Head Height: Crown to chin	20.5	13%	1/8	100%
Sitting Height:	88	54%	1/2	97%
Trunk Measure: Sternal notch to symphysis pubis	51	31%	1/3–	99%
Leg Length: Greater trochanter of the femur to sole	82	50%	1/2	95%
Thigh Length: Greater trochanter of the femur to tibia at knee	40	25%	1/4	98%
Tibia at Knee to Sole:	42	26%	1/4 98%	
Arm Length: Acromion process at shoulder to dependent fingertip	70	43%	1/2–	98%
Hand Length:	17.3	11%	1/9	96%
Foot Length:	24	15%	1/7	99%

Most of the growth, particularly of stature, is complete by this age. There will be some growth and additional maturation of the face up to twenty years of age. The permanent dentition is nearly complete. Third molars continue to grow from one to four more years before eruption.[2] The comparison of major measurements and ratios of this age to adult female ratios reveals the numbers are very near adult proportions.

1. Nelson, W.E., p. 39.

2. Lockhart, R.D., p. 57.

Sixteen-Year-Old Girl, anterior and lateral head

all measurements in centimeters

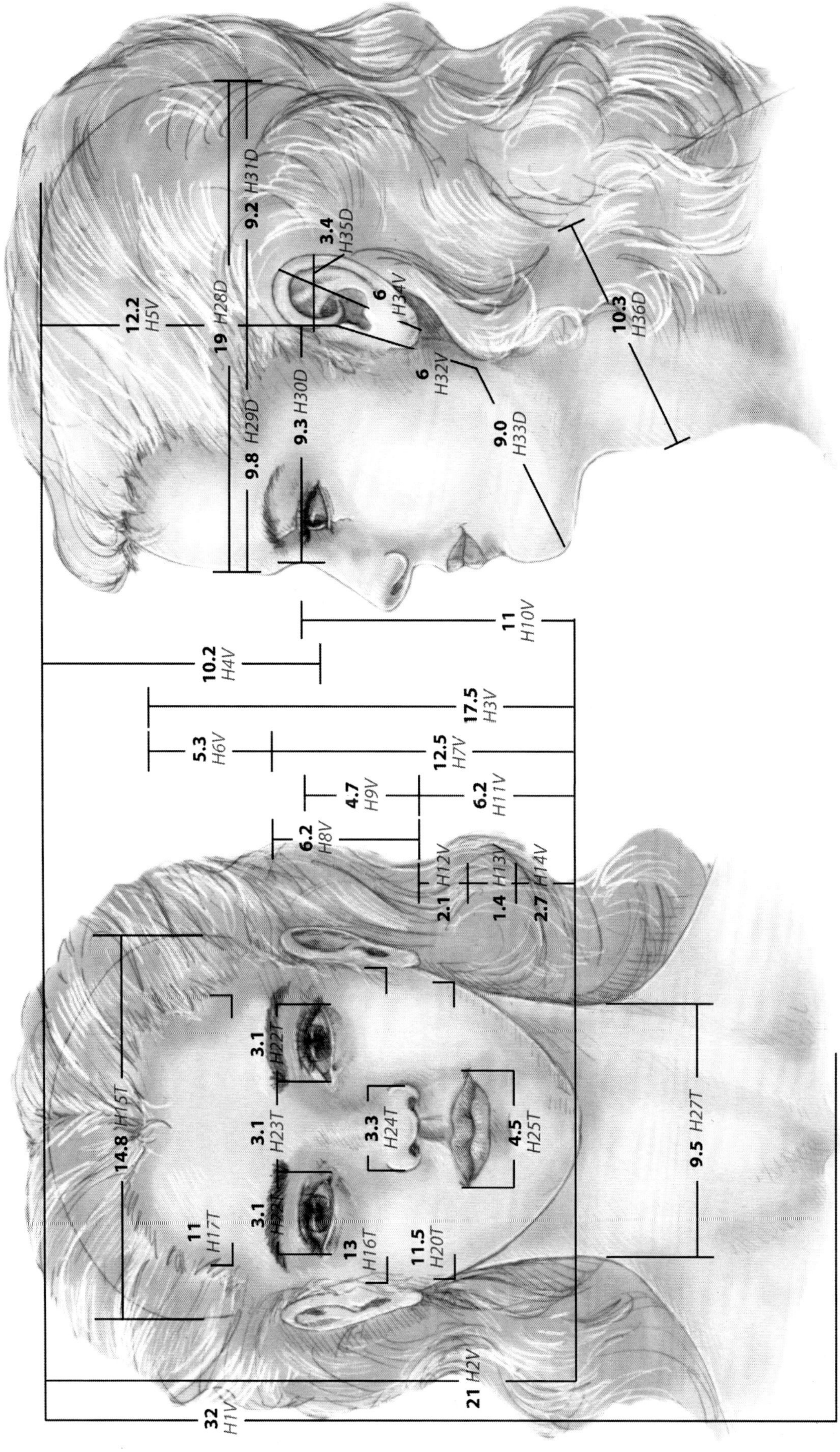

Life size			SIXTEEN-YEAR-OLD GIRL	Fractions of life size				
				3/4	2/3	1/2	1/3	1/4
Inches	cm	Code		cm	cm	cm	cm	cm
			VERTICAL MEASUREMENTS					
64.2	**163.0**	1 V	Stature, Height, Vertex to Sole	122.3	108.7	81.5	54.3	40.8
32.1	**81.5**	2 V	Vertex to Symphysis Pubis	61.1	54.3	40.8	27.2	20.4
32.1	**81.5**	3 V	Symphysis Pubis to Sole	61.1	54.3	40.8	27.2	20.4
51.6	**131.0**	4 V	Sternal Notch to Sole	98.3	87.3	65.5	43.7	32.8
12.2	**31.0**	5 V	Vertex to Acromion Process	23.3	20.7	15.5	10.3	7.8
52.0	**132.0**	6 V	Acromion Process to Sole	99.0	88.0	66.0	44.0	33.0
36.2	**92.0**	7 V	Ant.Sup.Spine of Ilium to Sole	69.0	61.3	46.0	30.7	23.0
15.7	**40.0**	8 V	Acrom. proc. to Ant. Sup. Spine	30.0	26.7	20.0	13.3	10.0
17.7	**45.0**	9 V	Ant.Sup.Spine of Ilium to Knee	33.8	30.0	22.5	15.0	11.3
18.5	**47.0**	10 V	Knee at Patella to Sole	35.3	31.3	23.5	15.7	11.8
38.2	**97.0**	11 V	Umbilicus to Sole	72.8	64.7	48.5	32.3	24.3
28.3	**72.0**	12 V	Crotch to Sole	54.0	48.0	36.0	24.0	18.0
48.0	**122.0**	13 V	Kneeling Height	91.5	81.3	61.0	40.7	30.5
33.9	**86.0**	14 V	Sitting Height	64.5	57.3	43.0	28.7	21.5
34.1	**86.7**	15 V	Buttock, Straight Leg to Sole	65.0	57.8	43.4	28.9	21.7
21.7	**55.0**	16 V	Buttock to Flexed Knee	41.3	36.7	27.5	18.3	13.8
19.7	**50.0**	17 V	Flexed Knee to Sole	37.5	33.3	25.0	16.7	12.5
		** These are also listed with Lower Extremity Measurements.*						
			VERTICAL HEAD MEASUREMENTS					
12.6	**32.0**	H1 V	Vertex to Sternal Notch	24.0	21.3	16.0	10.7	8.0
8.3	**21.0**	H2 V	Vertex to Chin, Head Height	15.8	14.0	10.5	7.0	5.3
6.9	**17.5**	H3 V	Hairline to Chin, Face Height	13.1	11.7	8.8	5.8	4.4
4.3	**10.8**	H4 V	Vertex to Lateral Corner of Eye	8.1	7.2	5.4	3.6	2.7
4.8	**12.2**	H5 V	Vertex to Tragion of Ear	9.2	8.1	6.1	4.1	3.1
2.1	**5.3**	H6 V	Hairline to Brow, Glabella	4.0	3.5	2.7	1.8	1.3
4.9	**12.5**	H7 V	Brow, Glabella, to Chin, Menton	9.4	8.3	6.3	4.2	3.1
2.4	**6.2**	H8 V	Brow to Subnasal Point	4.7	4.1	3.1	2.1	1.6
1.9	**4.7**	H9 V	Sellion to Subnasal, Nose Length	3.5	3.1	2.4	1.6	1.2
4.3	**11.0**	H10 V	Sellion, Nose Saddle, to Chin	8.3	7.3	5.5	3.7	2.8
2.4	**6.2**	H11 V	Subnasal point to Chin	4.7	4.1	3.1	2.1	1.6
0.8	**2.1**	H12 V	Subnasal to Lip Aperature	1.6	1.4	1.1	0.7	0.5
0.6	**1.4**	H13 V	Lip Aperat. to Sublabial Crease	1.1	0.9	0.7	0.5	0.4
1.1	**2.7**	H14 V	Sublabial Crease to Chin	2.0	1.8	1.4	0.9	0.7
			TRANSVERSE HEAD MEASURE					
5.8	**14.8**	H15 T	Bi-Parietal, Head Breadth	11.1	9.9	7.4	4.9	3.7
5.1	**13.0**	H16 T	Bi-Zygomatic, Cheekbones	9.8	8.7	6.5	4.3	3.3
4.3	**11.0**	H17 T	Bi-Fronto Temporal, Forehead	8.3	7.3	5.5	3.7	2.8

Life size			SIXTEEN-YEAR-OLD GIRL	Fractions of life size				
				3/4	2/3	1/2	1/3	1/4
Inches	cm	Code		cm	cm	cm	cm	cm
			TRANSVERSE HEAD MEASUREMENTS					
4.0	**10.2**	H18 T	Inter Brow Ridge	7.7	6.8	5.1	3.4	2.6
5.0	**12.7**	H19 T	Inter Tragion of Ears	9.5	8.5	6.4	4.2	3.2
4.5	**11.5**	H20 T	Inter Angle of Mandible	8.6	7.7	5.8	3.8	2.9
3.7	**9.3**	H21 T	Inter Lateral Corners of Eyes	7.0	6.2	4.7	3.1	2.3
1.2	**3.1**	H22 T	Width of Eye	2.3	2.1	1.6	1.0	0.8
1.2	**3.1**	H23 T	Inter Inner Corner of eyes	2.3	2.1	1.6	1.0	0.8
1.3	**3.3**	H24 T	Breadth of Nose	2.5	2.2	1.7	1.1	0.8
1.8	**4.5**	H25 T	Breadth of Mouth	3.4	3.0	2.3	1.5	1.1
1.4	**3.5**	H26 T	Breadth of Chin	2.6	2.3	1.8	1.2	0.9
3.7	**9.5**	H27 T	Breadth of Neck	7.1	6.3	4.8	3.2	2.4
			ANTERIOR-POSTERIOR HEAD MEASUREMENTS					
7.5	**19.0**	H28 D	Brow to Occiput	14.3	12.7	9.5	6.3	4.8
3.9	**9.8**	H29 D	Brow to Tragus	7.4	6.5	4.9	3.3	2.5
3.7	**9.3**	H30 D	Sellion to Tragus	7.0	6.2	4.7	3.1	2.3
3.6	**9.2**	H31 D	Tragus to Occiput, Back of Head	6.9	6.1	4.6	3.1	2.3
2.4	**6.0**	H32 V	Tragus to Angle Mandible	4.5	4.0	3.0	2.0	1.5
3.5	**9.0**	H33 D	Angle of Mandible to Chin	6.8	6.0	4.5	3.0	2.3
2.4	**6.0**	H34 V	Length of Ear, Vertical	4.5	4.0	3.0	2.0	1.5
1.3	**3.4**	H35 D	Width of Ear, Anterior-Posterior	2.6	2.3	1.7	1.1	0.9
4.1	**10.3**	H36 D	Anterior to Posterior of Neck	7.7	6.9	5.2	3.4	2.6
			VERTICAL BODY MEASUREMENTS					
4.1	**10.5**	B1 V	Chin to Sternal Notch	7.9	7.0	5.3	3.5	2.6
3.1	**8.0**	B2 V	Sternal Notch to Axilla	6.0	5.3	4.0	2.7	2.0
6.3	**16.0**	B3 V	Sternal Notch to Nipples	12.0	10.7	8.0	5.3	4.0
7.1	**18.0**	B4 V	Sternal Notch to Xyphoid Proc.	13.5	12.0	9.0	6.0	4.5
13.4	**34.0**	B5 V	Sternal Notch to Umbilicus	25.5	22.7	17.0	11.3	8.5
6.3	**16.0**	B6 V	Xyphoid Process to Umbilicus	12.0	10.7	8.0	5.3	4.0
12.6	**32.0**	B7 V	Sternal Notch to Crest Ilium	24.0	21.3	16.0	10.7	8.0
15.4	**39.0**	B8 V	Sternal Notch to Ant.S.Sp.Ilium	29.3	26.0	19.5	13.0	9.8
19.7	**50.0**	B9 V	Sternal Notch to Pubis	37.5	33.3	25.0	16.7	12.5
5.9	**15.0**	B10 V	Umbilicus to Pubis of Pelvis	11.3	10.0	7.5	5.0	3.8
			TRANSVERSE BODY MEASUREMENTS					
15.4	**39.0**	B11 T	Shoulder Breadth	29.3	26.0	19.5	13.0	9.8
7.1	**18.0**	B12 T	Clavicle to Acromion Process	13.5	12.0	9.0	6.0	4.5
14.0	**35.5**	B13 T	Biacromial Process Breadth	26.6	23.7	17.8	11.8	8.9
10.2	**26.0**	B14 T	Inter Anterior Axillary Fold	19.5	17.3	13.0	8.7	6.5
7.5	**19.0**	B15 T	Inter Nipples	14.3	12.7	9.5	6.3	4.8

Life size			SIXTEEN-YEAR-OLD GIRL	Fractions of life size				
				3/4	2/3	1/2	1/3	1/4
Inches	cm	Code		cm	cm	cm	cm	cm
			TRANSVERSE BODY MEASUREMENTS					
10.0	**25.5**	B16 T	Breadth of Thorax	19.1	17.0	12.8	8.5	6.4
9.4	**24.0**	B17 T	Inter Costal Margin Breadth	18.0	16.0	12.0	8.0	6.0
11.0	**28.0**	B18 T	Inter Crest of Ilium Breadth	21.0	18.7	14.0	9.3	7.0
9.3	**23.5**	B19 T	Inter Anterior Superior Spine	17.6	15.7	11.8	7.8	5.9
12.6	**32.0**	B20 T	Inter Greater Trochanter Femur	24.0	21.3	16.0	10.7	8.0
			ANTERIOR-POSTERIOR BODY MEASUREMENTS					
7.1	**18.0**	B21 D	Manubrium to Dorsal Spine	13.5	12.0	9.0	6.0	4.5
7.5	**19.0**	B22 D	Xyphoid Process to Dorsum	14.3	12.7	9.5	6.3	4.8
6.7	**17.0**	B23 D	Umbilicus to Dorsum	12.8	11.3	8.5	5.7	4.3
8.7	**22.0**	B24 D	Pubis to Buttock	16.5	14.7	11.0	7.3	5.5
6.7	**17.0**	B25 D	Thigh to Gluteal Furrow	12.8	11.3	8.5	5.7	4.3
			VERTICAL BACK MEASUREMENTS					
9.4	**24.0**	B26 V	Vertex to 7th Cervical Spine	18.0	16.0	12.0	8.0	6.0
			7th Cervical Vertebra Spine to:					
7.1	**18.0**	B27 V	To Posterior Axillary Fold, Armpit	13.5	12.0	9.0	6.0	4.5
7.9	**20.0**	B28 V	To Inferior Tip of Scapula	15.0	13.3	10.0	6.7	5.0
16.9	**43.0**	B29 V	To Posterior Superior Spine Ilium	32.3	28.7	21.5	14.3	10.8
20.1	**51.0**	B30 V	To Midline Upper Buttock Crease	38.3	34.0	25.5	17.0	12.8
25.6	**65.0**	B31 V	To Gluteal Furrow	48.8	43.3	32.5	21.7	16.3
37.8	**96.0**	B32 V	To Popliteal Crease of Knee	72.0	64.0	48.0	32.0	24.0
			TRANSVERSE BACK MEASUREMENTS					
12.2	**31.0**	B33 T	Inter Posterior Axillary Folds	23.3	20.7	15.5	10.3	7.8
5.5	**14.0**	B34 T	Inter Scapulae Medial Borders	10.5	9.3	7.0	4.7	3.5
3.1	**8.0**	B35 T	Inter Post. Sup. Spines Sacrum	6.0	5.3	4.0	2.7	2.0
			UPPER EXTREMITY MEASUREMENTS					
63.0	**160.0**	U1 V	Span of Arms	120.0	106.7	80.0	53.3	40.0
7.0	**17.8**	U2 V	Sternum to Acromion Process	13.4	11.9	8.9	5.9	4.5
27.6	**70.0**	U3 V	Acromion Process to Fingertip	52.5	46.7	35.0	23.3	17.5
80.3	**204.0**	U4 V	Vertical Reach to Fingertip	155.3	138.0	103.5	69.0	51.8
39.0	**99.0**	U5 T	Lateral Reach to Fingertip	77.3	68.7	51.5	34.3	25.8
31.1	**79.0**	U6 D	Forward Reach to Fingertip	59.3	52.7	39.5	26.3	19.8
13.2	**33.5**	U7 V	Shoulder to Elbow Flexed	25.1	22.3	16.8	11.2	8.4
11.8	**30.0**	U8 V	Acromion to Lat. Condyle Elbow	22.5	20.0	15.0	10.0	7.5
16.9	**43.0**	U9 D	Flexed Elbow to Fingertip	32.3	28.7	21.5	14.3	10.8
10.0	**25.5**	U10 D	Flexed Elbow Pivot to Wrist Pivot	19.1	17.0	12.8	8.5	6.4

Life size			SIXTEEN-YEAR-OLD GIRL	Fractions of life size				
				3/4	2/3	1/2	1/3	1/4
Inches	cm	Code		cm	cm	cm	cm	cm
			UPPER EXTREMITY MEASUREMENTS					
6.8	**17.3**	U11 V	Wrist to Fingertip	13.0	11.5	8.7	5.8	4.3
4.3	**11.0**	U12 V	Wrist to Thumb Tip	8.3	7.3	5.5	3.7	2.8
3.8	**9.6**	U13 V	Length of Palm	7.2	6.4	4.8	3.2	2.4
3.0	**7.5**	U14 V	Length 3rd Finger	5.6	5.0	3.8	2.5	1.9
4.5	**11.5**	U15 V	Thumb Crotch to 3rd Finger	8.6	7.7	5.8	3.8	2.9
2.4	**6.0**	U16 V	Thumb Crotch to Thumb Tip	4.5	4.0	3.0	2.0	1.5
3.1	**8.0**	U17 D	Anterior-Post. Arm at Biceps	6.0	5.3	4.0	2.7	2.0
3.0	**7.5**	U18 D	Anterior-Posterior at Forearm	5.6	5.0	3.8	2.5	1.9
1.5	**3.7**	U19 D	Depth of Wrist	2.8	2.5	1.9	1.2	0.9
2.2	**5.5**	U20 T	Breadth of Wrist	4.1	3.7	2.8	1.8	1.4
3.7	**9.4**	U21 T	Breadth of Hand	7.1	6.3	4.7	3.1	2.4
3.0	**7.6**	U22 T	Breadth of Palm	5.7	5.1	3.8	2.5	1.9
			LOWER EXTREMITY MEASUREMENTS					
38.6	**98.0**	L1 V	Crest of Ilium to Sole	73.5	65.3	49.0	32.7	24.5
36.2	**92.0**	* L2 V	Ant.Sup.Spine of Ilium to Sole	69.0	61.3	46.0	30.7	23.0
32.3	**82.0**	L3 V	Greater Trochanter Femur-Sole	61.5	54.7	41.0	27.3	20.5
32.1	**81.5**	* L4 V	Symphysis Pubis to Sole	61.1	54.3	40.8	27.2	20.4
28.3	**72.0**	* L5 V	Crotch to Sole	54.0	48.0	36.0	24.0	18.0
29.1	**74.0**	L6 V	Gluteal Furrow to Sole	55.5	49.3	37.0	24.7	18.5
16.9	**43.0**	L7 V	Posterior Knee Crease to Sole	32.3	28.7	21.5	14.3	10.8
16.5	**42.0**	L8 V	Tibia at Knee to Sole	31.5	28.0	21.0	14.0	10.5
2.6	**6.5**	L9 V	Med. Malleolus to Sole	4.9	4.3	3.3	2.2	1.6
2.0	**5.0**	L10 V	Lat. Malleolus to Sole	3.8	3.3	2.5	1.7	1.3
13.8	**35.0**	L11 T	Breadth of Hips, Sitting	26.3	23.3	17.5	11.7	8.8
6.7	**17.0**	L12 T	Breadth Mid Thigh	12.8	11.3	8.5	5.7	4.3
4.3	**11.0**	L13 T	Breadth of Knee	8.3	7.3	5.5	3.7	2.8
4.0	**10.2**	L14 T	Breadth of Calf	7.7	6.8	5.1	3.4	2.6
2.2	**5.5**	L15 T	Breadth Ankle	4.1	3.7	2.8	1.8	1.4
6.3	**16.0**	L16 D	Mid Thigh, Anterior-Posterior	12.0	10.7	8.0	5.3	4.0
4.5	**11.5**	L17 D	Knee, Anterior-Posterior	8.6	7.7	5.8	3.8	2.9
4.1	**10.5**	L18 D	Mid Calf, Anterior-Posterior	7.9	7.0	5.3	3.5	2.6
2.8	**7.0**	L19 D	Ankle, Anterior-Posterior	5.3	4.7	3.5	2.3	1.8
9.4	**24.0**	L20 D	Length of Foot	18.0	16.0	12.0	8.0	6.0
7.5	**19.0**	L21 D	Heel to 5th Toe	14.3	12.7	9.5	6.3	4.8
3.5	**9.0**	L22 T	Breadth Foot	6.8	6.0	4.5	3.0	2.3
2.4	**6.0**	L23 T	Width of Heel	4.5	4.0	3.0	2.0	1.5
		These are also listed with Vertical Measurements.						

SIXTEEN-YEAR-OLD BOYS

The lists of body proportions are recorded in the tables. A range of weight and stature is listed here for comparison:

Percentile	Twenty-fifth		Fiftieth		Seventy-fifth	
Boys:	Short		Medium		Tall	
Stature	169 cm,	66.5 in.	173.5 cm,	68.3 in.	178 cm,	70 in.
Weight	56 kg,	123.8 lb.	62 kg,	137 lb.	70 kg,	155 lb.[1]

By this age, boys are, on the average, taller than girls. Bone and muscular development has progressed. Interest continues in competitive and outdoor sports. More attention is paid to appearance. At sixteen years, most boys have earned a driver's license. They are expected to accept more responsibility. Some have part time jobs. The automobile becomes an important part of social and school life. Social events with girls and dances become more important. The school level is eleventh grade.

BODY MEASUREMENTS COMPARED TO HEIGHT

	Feature Measure in cm	Feature Compared to Stature	Approx. Ratio of Feature	Compared to Median Adult
Stature: Crown to sole	173.5			96%
Head Height: Crown to chin	22	13%	1/8	96%
Sitting Height:	90.5	52%	1/2	94%
Trunk Measure: Sternal notch to symphysis pubis	53	30%	1/3–	98%
Leg Length: Greater trochanter of the femur to sole	90	52%	1/2	92%
Thigh Length: Greater trochanter of the femur to tibia at knee	44	25%	1/4	94%
Tibia at Knee to Sole:	46	26%	1/4	96%
Arm Length: Acromion Process at shoulder to dependent fingertip	75	43%	1/2–	93%
Hand Length:	18	10%	1/10	90%
Foot Length:	26.5	15%	1/7	95%

By this age, many facial features resemble those of adult males. The bones of the mid and lower face, the maxilla and mandible or jaw, have grown to accommodate the permanent teeth. The third molars are growing and are expected to erupt in the next three years. The face continues to appear youthful, but some mature male features become apparent.

1. Nelson, W.E., p. 39.

Sixteen-Year-Old Boy, anterior and lateral head

all measurements in centimeters

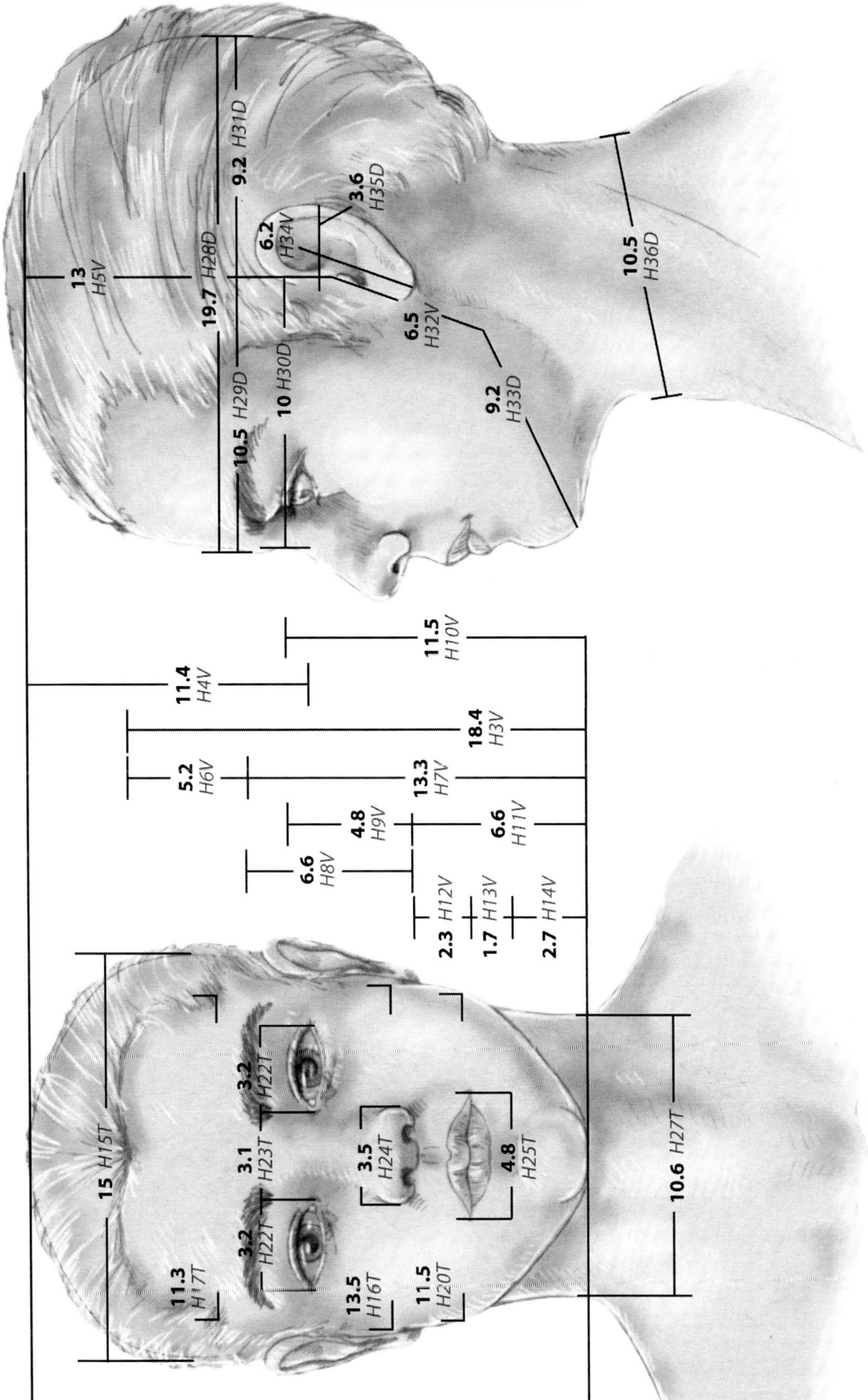

Life size			SIXTEEN-YEAR-OLD BOY	Fractions of life size				
				3/4	2/3	1/2	1/3	1/4
Inches	cm	Code		cm	cm	cm	cm	cm
			VERTICAL MEASUREMENTS					
68.5	**174.0**	1 V	Stature, Height, Vertex to Sole	130.5	116.0	87.0	58.0	43.5
33.9	**86.0**	2 V	Vertex to Symphysis Pubis	64.5	57.3	43.0	28.7	21.5
34.6	**88.0**	* 3 V	Symphysis Pubis to Sole	66.0	58.7	44.0	29.3	22.0
55.9	**142.0**	4 V	Sternal Notch to Sole	106.5	94.7	71.0	47.3	35.5
12.6	**32.0**	5 V	Vertex to Acromion Process	24.0	21.3	16.0	10.7	8.0
56.1	**142.5**	6 V	Acromion Process to Sole	106.9	95.0	71.3	47.5	35.6
38.6	**98.0**	* 7 V	Ant.Sup.Spine of Ilium to Sole	73.5	65.3	49.0	32.7	24.5
17.5	**44.5**	8 V	Acrom. proc. to Ant. Sup. Spine	33.4	29.7	22.3	14.8	11.1
18.9	**48.0**	9 V	Ant. Sup. Spine of Ilium to Knee	36.0	32.0	24.0	16.0	12.0
19.7	**50.0**	10 V	Knee at Patella to Sole	37.5	33.3	25.0	16.7	12.5
41.7	**106.0**	11 V	Umbilicus to Sole	79.5	70.7	53.0	35.3	26.5
32.5	**82.5**	* 12 V	Crotch to Sole	61.9	55.0	41.3	27.5	20.6
50.8	**129.0**	13 V	Kneeling Height	96.8	86.0	64.5	43.0	32.3
35.6	**90.5**	14 V	Sitting Height	67.9	60.3	45.3	30.2	22.6
36.2	**92.0**	15 V	Buttock, Straight Leg to Sole	69.0	61.3	46.0	30.7	23.0
23.2	**59.0**	16 V	Buttock to Flexed Knee	44.3	39.3	29.5	19.7	14.8
21.7	**55.0**	17 V	Flexed Knee to Sole	41.3	36.7	27.5	18.3	13.8
		** These are also listed with Lower Extremity Measurements.*						
			VERTICAL HEAD MEASUREMENTS					
13.0	**33.0**	H1 V	Vertex to Sternal Notch	24.8	22.0	16.5	11.0	8.3
8.8	**22.3**	H2 V	Vertex to Chin, Head Height	16.7	14.9	11.2	7.4	5.6
7.2	**18.4**	H3 V	Hairline to Chin, Face Height	13.8	12.3	9.2	6.1	4.6
4.5	**11.4**	H4 V	Vertex to Lateral Corner of Eye	8.6	7.6	5.7	3.8	2.9
5.1	**13.0**	H5 V	Vertex to Tragion of Ear	9.8	8.7	6.5	4.3	3.3
2.0	**5.2**	H6 V	Hairline to Brow, Glabella	3.9	3.5	2.6	1.7	1.3
5.2	**13.3**	H7 V	Brow, Glabella, to Chin, Menton	10.0	8.9	6.7	4.4	3.3
2.6	**6.6**	H8 V	Brow to Subnasal Point	5.0	4.4	3.3	2.2	1.7
1.9	**4.8**	H9 V	Sellion to Subnasal, Nose Length	3.6	3.2	2.4	1.6	1.2
4.5	**11.5**	H10 V	Sellion, Nose Saddle, to Chin	8.6	7.7	5.8	3.8	2.9
2.6	**6.6**	H11 V	Subnasal point to Chin	5.0	4.4	3.3	2.2	1.7
0.9	**2.3**	H12 V	Subnasal to Lip Aperature	1.7	1.5	1.2	0.8	0.6
0.7	**1.7**	H13 V	Lip Aperat. to Sublabial Crease	1.3	1.1	0.9	0.6	0.4
1.1	**2.7**	H14 V	Sublabial Crease to Chin	2.0	1.8	1.4	0.9	0.7
			TRANSVERSE HEAD MEASUREMENTS					
5.9	**15.0**	H15 T	Bi-Parietal, Head Breadth	11.3	10.0	7.5	5.0	3.8
5.3	**13.5**	H16 T	Bi-Zygomatic, Cheekbones	10.1	9.0	6.8	4.5	3.4
4.4	**11.3**	H17 T	Bi-Fronto Temporal, Forehead	8.5	7.5	5.7	3.8	2.8

Life size			SIXTEEN-YEAR-OLD BOY	Fractions of life size				
				3/4	2/3	1/2	1/3	1/4
Inches	cm	Code		cm	cm	cm	cm	cm
			TRANSVERSE HEAD MEASUREMENTS					
4.0	**10.2**	H18 T	Inter Brow Ridge	7.7	6.8	5.1	3.4	2.6
5.1	**12.9**	H19 T	Inter Tragion of Ears	9.7	8.6	6.5	4.3	3.2
4.5	**11.5**	H20 T	Inter Angle of Mandible	8.6	7.7	5.8	3.8	2.9
3.7	**9.5**	H21 T	Inter Lateral Corners of Eyes	7.1	6.3	4.8	3.2	2.4
1.3	**3.2**	H22 T	Width of Eye	2.4	2.1	1.6	1.1	0.8
1.2	**3.1**	H23 T	Inter Inner Corner of eyes	2.3	2.1	1.6	1.0	0.8
1.4	**3.5**	H24 T	Breadth of Nose	2.6	2.3	1.8	1.2	0.9
1.9	**4.8**	H25 T	Breadth of Mouth	3.6	3.2	2.4	1.6	1.2
1.3	**3.4**	H26 T	Breadth of Chin	2.6	2.3	1.7	1.1	0.9
4.2	**10.6**	H27 T	Breadth of Neck	8.0	7.1	5.3	3.5	2.7
			ANTERIOR-POSTERIOR HEAD MEASUREMENTS					
7.8	**19.7**	H28 D	Brow to Occiput, Posterior Head	14.8	13.1	9.9	6.6	4.9
4.1	**10.5**	H29 D	Brow to Tragus of Ear	7.9	7.0	5.3	3.5	2.6
3.9	**10.0**	H30 D	Sellion to Tragus	7.5	6.7	5.0	3.3	2.5
3.6	**9.2**	H31 D	Tragus to Occiput	6.9	6.1	4.6	3.1	2.3
2.6	**6.5**	H32 V	Tragus to Angle Mandible	4.9	4.3	3.3	2.2	1.6
3.6	**9.2**	H33 D	Angle of Mandible to Chin	6.9	6.1	4.6	3.1	2.3
2.4	**6.2**	H34 V	Length of Ear, Vertical	4.7	4.1	3.1	2.1	1.6
1.4	**3.6**	H35 D	Width of Ear, Anterior-Posterior	2.7	2.4	1.8	1.2	0.9
4.1	**10.5**	H36 D	Anterior to Posterior of Neck	7.9	7.0	5.3	3.5	2.6
			VERTICAL BODY MEASUREMENTS					
4.3	**11.0**	B1 V	Chin to Sternal Notch	8.3	7.3	5.5	3.7	2.8
3.1	**8.0**	B2 V	Sternal Notch to Axilla	6.0	5.3	4.0	2.7	2.0
5.5	**14.0**	B3 V	Sternal Notch to Nipples	10.5	9.3	7.0	4.7	3.5
7.1	**18.0**	B4 V	Sternal Notch to Xyphoid	13.5	12.0	9.0	6.0	4.5
14.2	**36.0**	B5 V	Sternal Notch to Umbilicus	27.0	24.0	18.0	12.0	9.0
7.1	**18.0**	B9 V	Xyphoid Process to Umbilicus	13.5	12.0	9.0	6.0	4.5
14.2	**36.0**	B6 V	Sternal Notch to Crest Ilium	27.0	24.0	18.0	12.0	9.0
16.9	**43.0**	B7 V	Sternal Notch to Ant.S.Sp.Ilium	32.3	28.7	21.5	14.3	10.8
20.9	**53.0**	B8 V	Sternal Notch to Pubis	39.8	35.3	26.5	17.7	13.3
7.1	**18.0**	B10 V	Umbilicus to Pubis of Pelvis	13.5	12.0	9.0	6.0	4.5
			TRANSVERSE BODY MEASUREMENTS					
17.1	**43.5**	B11 T	Shoulder Breadth	32.6	29.0	21.8	14.5	10.9
7.1	**18.0**	B12 T	Clavicle to Acromion Process	13.5	12.0	9.0	6.0	4.5
15.0	**38.0**	B13 T	Biacromial Process Breadth	28.5	25.3	19.0	12.7	9.5
10.6	**27.0**	B14 T	Inter Anterior Axillary Fold	20.3	18.0	13.5	9.0	6.8
7.1	**18.0**	B15 T	Inter Nipples	13.5	12.0	9.0	6.0	4.5

Life size			SIXTEEN-YEAR-OLD BOY	Fractions of life size				
				3/4	2/3	1/2	1/3	1/4
Inches	cm	Code		cm	cm	cm	cm	cm
			TRANSVERSE BODY MEASUREMENTS					
11.0	**28.0**	B16 T	Breadth of Thorax	21.0	18.7	14.0	9.3	7.0
10.2	**26.0**	B17 T	Inter Costal Margin	19.5	17.3	13.0	8.7	6.5
11.0	**28.0**	B18 T	Inter Crest of Ilium	21.0	18.7	14.0	9.3	7.0
9.3	**23.5**	B19 T	Inter Anterior Superior Spine	17.6	15.7	11.8	7.8	5.9
12.6	**32.0**	B20 T	Inter Greater Trochanter Femur	24.0	21.3	16.0	10.7	8.0
			ANTERIOR-POSTERIOR BODY MEASUREMENTS					
7.5	**19.0**	B21 D	Manubrium to Dorsal Spine	14.3	12.7	9.5	6.3	4.8
7.8	**19.7**	B22 D	Xyphoid Process to Dorsum	14.8	13.1	9.9	6.6	4.9
7.1	**18.0**	B23 D	Umbilicus to Dorsum	13.5	12.0	9.0	6.0	4.5
8.3	**21.0**	B24 D	Pubis to Buttock	15.8	14.0	10.5	7.0	5.3
6.3	**16.0**	B25 D	Thigh to Gluteal Furrow	12.0	10.7	8.0	5.3	4.0
			VERTICAL BACK MEASUREMENTS					
9.4	**24.0**	B26 V	Vertex to 7th Cervical Spine	18.0	16.0	12.0	8.0	6.0
			7th Cervical Vertebra Spine to:					
7.5	**19.0**	B27 V	Posterior Axillary Fold	14.3	12.7	9.5	6.3	4.8
8.3	**21.0**	B28 V	Inferior Tip of Scapula	15.8	14.0	10.5	7.0	5.3
17.7	**45.0**	B29 V	Posterior Superior Spine Ilium	33.8	30.0	22.5	15.0	11.3
21.7	**55.0**	B30 V	Midline Upper Buttock Crease	41.3	36.7	27.5	18.3	13.8
26.8	**68.0**	B31 V	to Gluteal Furrow	51.0	45.3	34.0	22.7	17.0
39.8	**101.0**	B32 V	to Popliteal Crease of Knee	75.8	67.3	50.5	33.7	25.3
			TRANSVERSE BACK MEASUREMENTS					
13.8	**35.0**	B33 T	Inter Posterior Axillary Fold	26.3	23.3	17.5	11.7	8.8
5.9	**15.0**	B34 T	Inter Scapulae Medial Border	11.3	10.0	7.5	5.0	3.8
3.5	**8.8**	B35 T	Inter Post. Sup. Spines Sacrum	6.6	5.9	4.4	2.9	2.2
			UPPER EXTREMITY MEASUREMENTS					
69.3	**176.0**	U1 V	Span of Arms	132.0	117.3	88.0	58.7	44.0
7.1	**18.0**	U2 V	Sternum to Acromion Process	13.5	12.0	9.0	6.0	4.5
29.5	**75.0**	U3 V	Acromion to Fingertip	56.3	50.0	37.5	25.0	18.8
87.4	**222.0**	U4 V	Vertical Reach to Fingertip	165.0	146.7	110.0	73.3	55.0
41.7	**106.0**	U5 T	Lateral Reach to Fingertip	76.5	68.0	51.0	34.0	25.5
31.5	**80.0**	U6 D	Forward Reach to Fingertip	60.0	53.3	40.0	26.7	20.0
13.8	**35.0**	U7 V	Shoulder to Elbow Flexed	26.3	23.3	17.5	11.7	8.8
12.6	**32.0**	U8 V	Acromion to Lat. Condyle Elbow	24.0	21.3	16.0	10.7	8.0
17.7	**45.0**	U9 D	Flexed Elbow to Fingertip	33.8	30.0	22.5	15.0	11.3
10.2	**26.0**	U10 D	Flexed Elbow Pivot to Wrist Pivot	19.5	17.3	13.0	8.7	6.5

Life size			SIXTEEN-YEAR-OLD BOY	Fractions of life size				
				3/4	2/3	1/2	1/3	1/4
Inches	cm	Code		cm	cm	cm	cm	cm
			UPPER EXTREMITY MEASUREMENTS					
7.1	**18.0**	U11 V	Wrist to Fingertip	13.5	12.0	9.0	6.0	4.5
4.7	**12.0**	U12 V	Wrist to Thumb	9.0	8.0	6.0	4.0	3.0
4.1	**10.5**	U13 V	Length of Palm	7.9	7.0	5.3	3.5	2.6
3.1	**8.0**	U14 V	Length 3rd Finger	6.0	5.3	4.0	2.7	2.0
5.0	**12.6**	U15 V	Thumb Crotch to 3rd Finger	9.5	8.4	6.3	4.2	3.2
2.6	**6.5**	U16 V	Thumb Crotch to Thumb Tip	4.9	4.3	3.3	2.2	1.6
3.1	**8.0**	U17 D	Anterior-Post. Arm at Biceps	6.0	5.3	4.0	2.7	2.0
3.1	**8.0**	U18 D	Anterior-Posterior at Forearm	6.0	5.3	4.0	2.7	2.0
1.6	**4.0**	U19 D	Depth of Wrist	3.0	2.7	2.0	1.3	1.0
2.2	**5.5**	U20 T	Breadth of Wrist	4.1	3.7	2.8	1.8	1.4
4.1	**10.5**	U21 T	Breadth of Hand	7.9	7.0	5.3	3.5	2.6
3.3	**8.5**	U22 T	Breadth of Palm	6.4	5.7	4.3	2.8	2.1
			LOWER EXTREMITY MEASUREMENTS					
41.3	**105.0**	L1 V	Crest of Ilium to Sole	78.8	70.0	52.5	35.0	26.3
38.6	**98.0**	* L2 V	Ant.Sup.Spine of Ilium to Sole	73.5	65.3	49.0	32.7	24.5
35.4	**90.0**	L3 V	Greater Trochanter Femur-Sole	67.5	60.0	45.0	30.0	22.5
34.6	**88.0**	* L4 V	Symphysis Pubis to Sole	66.0	58.7	44.0	29.3	22.0
32.5	**82.5**	* L5 V	Crotch to Sole	61.9	55.0	41.3	27.5	20.6
32.3	**82.0**	L6 V	Gluteal Furrow to Sole	61.5	54.7	41.0	27.3	20.5
19.3	**49.0**	L7 V	Posterior Knee Crease to Sole	36.8	32.7	24.5	16.3	12.3
18.1	**46.0**	L8 V	Tibia at Knee to Sole	34.5	30.7	23.0	15.3	11.5
2.8	**7.0**	L9 V	Med. Malleolus to Sole	5.3	4.7	3.5	2.3	1.8
2.2	**5.5**	L10 V	Lat. Malleolus to Sole	4.1	3.7	2.8	1.8	1.4
12.6	**32.0**	L11 T	Breadth of Hips, Sitting	24.0	21.3	16.0	10.7	8.0
5.9	**15.0**	L12 T	Breadth Mid Thigh	11.3	10.0	7.5	5.0	3.8
4.3	**11.0**	L13 T	Breadth of Knee	8.3	7.3	5.5	3.7	2.8
4.1	**10.5**	L14 T	Breadth of Calf	7.9	7.0	5.3	3.5	2.6
2.4	**6.0**	L15 T	Breadth Ankle	4.5	4.0	3.0	2.0	1.5
5.9	**15.0**	L16 D	Mid Thigh, Anterior-Posterior	11.3	10.0	7.5	5.0	3.8
4.7	**12.0**	L17 D	Knee, Anterior-Posterior	9.0	8.0	6.0	4.0	3.0
4.3	**11.0**	L18 D	Mid Calf, Anterior-Posterior	8.3	7.3	5.5	3.7	2.8
3.1	**8.0**	L19 D	Ankle, Anterior-Posterior	6.0	5.3	4.0	2.7	2.0
10.4	**26.5**	L20 D	Length of Foot	19.9	17.7	13.3	8.8	6.6
8.3	**21.0**	L21 D	Heel to 5th Toe	15.8	14.0	10.5	7.0	5.3
3.9	**10.0**	L22 T	Breadth Foot	7.5	6.7	5.0	3.3	2.5
2.4	**6.0**	L23 T	Width of Heel	4.5	4.0	3.0	2.0	1.5

* *These are also listed with Vertical Measurements.*

SEVENTEEN-YEAR-OLD GIRLS

The lists of body proportions are recorded separately in the tables of measurements. The range of twenty-fifth, fiftieth, and seventy-fifth percentile is shown here for comparison:

Percentile	Twenty-fifth		Fiftieth		Seventy-fifth	
Girls:	Short		Medium		Tall	
Stature	159 cm,	62.5 in.	163 cm,	65.8 in.	171 cm,	67.5 in.
Weight	51 kg,	113 lb.	57 kg,	125 lb.	72 kg,	159.5 lb.[1]

By this age girls have achieved nearly adult female features. They have rounded contours in face shoulders and hips. They are interested in style of hair and clothing. Outdoor athletics continues, but dancing and social activities receive more attention. School level is twelfth grade, and many are making plans for future employment, or further education including college.

BODY MEASUREMENTS COMPARED TO HEIGHT

	Feature Measure in cm	Feature Compared to Stature	Approx. Ratio of Feature	Compared to Median Adult
Stature: Crown to sole	163			99%
Head Height: Crown to chin	20.6	13%	1/8	100%
Sitting Height:	86	53%	1/2	97%
Trunk Measure: Sternal notch to symphysis pubis	50.5	31%	1/3-	100%
Leg Length: Greater trochanter of the femur to sole	82	50%	1/2	95%
Thigh Length: Greater trochanter of the femur to tibia at knee	40	25%	1/4	93%
Tibia at Knee to Sole:	42	26%	1/4	96%
Arm Length: Acromion process at shoulder to dependent fingertip	68	43%	1/2–	91%
Hand Length:	17.3	11%	1/9	96%
Foot Length:	24	15%	1/7	100%

Between fifteen and seventeen years of age there is a small amount of growth with some subtle changes, and some maturation that continues up to the age of twenty. Most of the bone development is nearly complete. The permanent dentition is nearly complete. Third molars continue to grow for one to two more years before eruption. The comparison of major measurements and ratios to that of adult females reveals very near adult proportions. Illustrations of proportions are omitted, since there are few changes after sixteen years of age.

1. Nelson, W.E., p. 39

Life size			SEVENTEEN-YEAR-OLD GIRL	Fractions of life size				
				3/4	2/3	1/2	1/3	1/4
Inches	cm	Code		cm	cm	cm	cm	cm
			VERTICAL MEASUREMENTS					
63.6	**163.0**	1 V	Stature, Height, Vertex to Sole	122.3	108.7	81.5	54.3	40.8
31.8	**81.5**	2 V	Vertex to Symphysis Pubis	61.1	54.3	40.8	27.2	20.4
31.8	**81.5**	* 3 V	Symphysis Pubis to Sole	61.1	54.3	40.8	27.2	20.4
51.1	**131.0**	4 V	Sternal Notch to Sole	98.3	87.3	65.5	43.7	32.8
12.1	**31.0**	5 V	Vertex to Acromion Process	23.3	20.7	15.5	10.3	7.8
51.7	**132.5**	6 V	Acromion Process to Sole	99.4	88.3	66.3	44.2	33.1
35.9	**92.0**	* 7 V	Ant.Sup.Spine of Ilium to Sole	69.0	61.3	46.0	30.7	23.0
15.8	**40.5**	8 V	Acrom. proc. to Ant. Sup. Spine	30.4	27.0	20.3	13.5	10.1
17.6	**45.0**	9 V	Ant.Sup.Spine of Ilium to Knee	33.8	30.0	22.5	15.0	11.3
18.3	**47.0**	10 V	Knee at Patella to Sole	35.3	31.3	23.5	15.7	11.8
38.2	**98.0**	11 V	Umbilicus to Sole	73.5	65.3	49.0	32.7	24.5
28.1	**72.0**	* 12 V	Crotch to Sole	54.0	48.0	36.0	24.0	18.0
48.0	**123.0**	13 V	Kneeling Height	92.3	82.0	61.5	41.0	30.8
33.5	**86.0**	14 V	Sitting Height	64.5	57.3	43.0	28.7	21.5
33.8	**86.7**	15 V	Buttock, Straight Leg to Sole	65.0	57.8	43.4	28.9	21.7
21.5	**55.0**	16 V	Buttock to Flexed Knee	41.3	36.7	27.5	18.3	13.8
19.5	**50.0**	17 V	Flexed Knee to Sole	37.5	33.3	25.0	16.7	12.5
		** These are also listed with Lower Extremity Measurements.*						
			VERTICAL HEAD MEASUREMENTS					
12.5	**32.0**	H1 V	Vertex to Sternal Notch	24.0	21.3	16.0	10.7	8.0
8.2	**21.0**	H2 V	Vertex to Chin, Head Height	15.8	14.0	10.5	7.0	5.3
7.0	**18.0**	H3 V	Hairline to Chin, Face Height	13.5	12.0	9.0	6.0	4.5
4.2	**10.8**	H4 V	Vertex to Lateral Corner of Eye	8.1	7.2	5.4	3.6	2.7
4.9	**12.5**	H5 V	Vertex to Tragion of Ear	9.4	8.3	6.3	4.2	3.1
2.0	**5.0**	H6 V	Hairline to Brow, Glabella	3.8	3.3	2.5	1.7	1.3
4.9	**12.5**	H7 V	Brow, Glabella, to Chin, Menton	9.4	8.3	6.3	4.2	3.1
2.4	**6.2**	H8 V	Brow to Subnasal Point	4.7	4.1	3.1	2.1	1.6
1.8	**4.7**	H9 V	Sellion to Subnasal, Nose Length	3.5	3.1	2.4	1.6	1.2
4.3	**11.1**	H10 V	Sellion, Nose Saddle, to Chin	8.3	7.4	5.6	3.7	2.8
2.4	**6.2**	H11 V	Subnasal point to Chin	4.7	4.1	3.1	2.1	1.6
0.9	**2.2**	H12 V	Subnasal to Lip Aperature	1.7	1.5	1.1	0.7	0.6
0.5	**1.4**	H13 V	Lip Aperat. to Sublabial Crease	1.1	0.9	0.7	0.5	0.4
1.1	**2.7**	H14 V	Sublabial Crease to Chin	2.0	1.8	1.4	0.9	0.7
			TRANSVERSE HEAD MEASUREMENTS					
5.8	**14.8**	H15 T	Bi-Parietal, Head Breadth	11.1	9.9	7.4	4.9	3.7
5.1	**13.0**	H16 T	Bi-Zygomatic, Cheekbones	9.8	8.7	6.5	4.3	3.3
4.3	**11.0**	H17 T	Bi-Fronto Temporal, Forehead	8.3	7.3	5.5	3.7	2.8

Life size			SEVENTEEN-YEAR-OLD GIRL	Fractions of life size				
				3/4	2/3	1/2	1/3	1/4
Inches	cm	Code		cm	cm	cm	cm	cm
			TRANSVERSE HEAD MEASUREMENTS					
4.0	**10.2**	H18 T	Inter Brow Ridge	7.7	6.8	5.1	3.4	2.6
5.0	**12.7**	H19 T	Inter Tragion of Ears	9.5	8.5	6.4	4.2	3.2
4.6	**11.7**	H20 T	Inter Angle of Mandible	8.8	7.8	5.9	3.9	2.9
3.6	**9.3**	H21 T	Inter Lateral Corners of Eyes	7.0	6.2	4.7	3.1	2.3
1.2	**3.1**	H22 T	Width of Eye	2.3	2.1	1.6	1.0	0.8
1.2	**3.1**	H23 T	Inter Inner Corner of eyes	2.3	2.1	1.6	1.0	0.8
1.3	**3.3**	H24 T	Breadth of Nose	2.5	2.2	1.7	1.1	0.8
1.8	**4.6**	H25 T	Breadth of Mouth	3.5	3.1	2.3	1.5	1.2
1.3	**3.4**	H26 T	Breadth of Chin	2.6	2.3	1.7	1.1	0.9
3.7	**9.5**	H27 T	Breadth of Neck	7.1	6.3	4.8	3.2	2.4
			ANTERIOR-POSTERIOR HEAD MEASUREMENTS					
7.4	**19.0**	H28 D	Brow to Occiput	14.3	12.7	9.5	6.3	4.8
3.8	**9.8**	H29 D	Brow to Tragus	7.4	6.5	4.9	3.3	2.5
3.6	**9.3**	H30 D	Sellion to Tragus	7.0	6.2	4.7	3.1	2.3
3.6	**9.2**	H31 D	Tragus to Occiput, Back of Head	6.9	6.1	4.6	3.1	2.3
2.3	**6.0**	H32 V	Tragus to Angle Mandible	4.5	4.0	3.0	2.0	1.5
3.5	**9.0**	H33 D	Angle of Mandible to Chin	6.8	6.0	4.5	3.0	2.3
2.3	**6.0**	H34 V	Length of Ear, Vertical	4.5	4.0	3.0	2.0	1.5
1.3	**3.4**	H35 D	Width of Ear, Anterior-Posterior	2.6	2.3	1.7	1.1	0.9
4.1	**10.4**	H36 D	Anterior to Posterior of Neck	7.8	6.9	5.2	3.5	2.6
			VERTICAL BODY MEASUREMENTS					
4.1	**10.5**	B1 V	Chin to Sternal Notch	7.9	7.0	5.3	3.5	2.6
3.1	**8.0**	B2 V	Sternal Notch to Axilla	6.0	5.3	4.0	2.7	2.0
6.2	**16.0**	B3 V	Sternal Notch to Nipples	12.0	10.7	8.0	5.3	4.0
7.0	**18.0**	B4 V	Sternal Notch to Xyphoid Proc.	13.5	12.0	9.0	6.0	4.5
13.7	**35.0**	B5 V	Sternal Notch to Umbilicus	26.3	23.3	17.5	11.7	8.8
6.2	**16.0**	B6 V	Xyphoid Process to Umbilicus	12.0	10.7	8.0	5.3	4.0
12.9	**33.0**	B7 V	Sternal Notch to Crest Ilium	24.8	22.0	16.5	11.0	8.3
15.6	**40.0**	B8 V	Sternal Notch to Ant.S.Sp.Ilium	30.0	26.7	20.0	13.3	10.0
19.9	**51.0**	B9 V	Sternal Notch to Pubis	38.3	34.0	25.5	17.0	12.8
6.2	**16.0**	B10 V	Umbilicus to Pubis	12.0	10.7	8.0	5.3	4.0
			TRANSVERSE BODY MEASUREMENTS					
15.6	**40.0**	B11 T	Shoulder Breadth	30.0	26.7	20.0	13.3	10.0
7.0	**18.0**	B12 T	Clavicle to Acromion Process	13.5	12.0	9.0	6.0	4.5
14.0	**36.0**	B13 T	Biacromial Process Breadth	27.0	24.0	18.0	12.0	9.0
10.3	**26.5**	B14 T	Inter Anterior Axillary Fold	19.9	17.7	13.3	8.8	6.6
7.8	**20.0**	B15 T	Inter Nipples	15.0	13.3	10.0	6.7	5.0

Life size			SEVENTEEN-YEAR-OLD GIRL	Fractions of life size				
				3/4	2/3	1/2	1/3	1/4
Inches	cm	Code		cm	cm	cm	cm	cm
			TRANSVERSE BODY MEASUREMENTS					
10.1	**26.0**	B16 T	Breadth of Thorax	19.5	17.3	13.0	8.7	6.5
9.8	**25.0**	B17 T	Inter Costal Margin Breadth	18.8	16.7	12.5	8.3	6.3
10.9	**28.0**	B18 T	Inter Crest of Ilium Breadth	21.0	18.7	14.0	9.3	7.0
9.8	**25.0**	B19 T	Inter Anterior Superior Spine	18.8	16.7	12.5	8.3	6.3
12.5	**32.0**	B20 T	Inter Greater Trochanter Femur	24.0	21.3	16.0	10.7	8.0
			ANTERIOR-POSTERIOR BODY MEASUREMENTS					
7.0	**18.0**	B21 D	Manubrium to Dorsal Spine	13.5	12.0	9.0	6.0	4.5
7.4	**19.0**	B22 D	Xyphoid Process to Dorsum	14.3	12.7	9.5	6.3	4.8
6.9	**17.7**	B23 D	Umbilicus to Dorsum	13.3	11.8	8.9	5.9	4.4
8.6	**22.0**	B24 D	Pubis to Buttock	16.5	14.7	11.0	7.3	5.5
6.8	**17.5**	B25 D	Thigh to Gluteal Furrow	13.1	11.7	8.8	5.8	4.4
			VERTICAL BACK MEASUREMENTS					
9.4	**24.0**	B26 V	Vertex to 7th Cervical Spine	18.0	16.0	12.0	8.0	6.0
			7th Cervical Vertebra Spine to:					
7.0	**18.0**	B27 V	To Posterior Axillary Fold	13.5	12.0	9.0	6.0	4.5
7.8	**20.0**	B28 V	To Inferior Tip of Scapula	15.0	13.3	10.0	6.7	5.0
16.8	**43.0**	B29 V	To Posterior Superior Spine Ilium	32.3	28.7	21.5	14.3	10.8
20.3	**52.0**	B30 V	To Midline Upper Buttock Crease	39.0	34.7	26.0	17.3	13.0
25.4	**65.0**	B31 V	To Gluteal Furrow	48.8	43.3	32.5	21.7	16.3
37.4	**96.0**	B32 V	To Popliteal Crease of Knee	72.0	64.0	48.0	32.0	24.0
			TRANSVERSE BACK MEASUREMENTS					
12.5	**32.0**	B33 T	Inter Posterior Axillary Folds	24.0	21.3	16.0	10.7	8.0
5.5	**14.0**	B34 T	Inter Scapulae Medial Borders	10.5	9.3	7.0	4.7	3.5
3.2	**8.3**	B35 T	Inter Post. Sup. Spines Sacrum	6.2	5.5	4.2	2.8	2.1
			UPPER EXTREMITY MEASUREMENTS					
63.2	**162.0**	U1 V	Span of Arms	121.5	108.0	81.0	54.0	40.5
6.9	**17.8**	U2 V	Sternum to Acromion Process	13.4	11.9	8.9	5.9	4.5
27.3	**70.0**	U3 V	Acromion Process to Fingertip	52.5	46.7	35.0	23.3	17.5
80.7	**207.0**	U4 V	Vertical Reach to Fingertip	156.8	139.3	104.5	69.7	52.3
39.4	**100.0**	U5 T	Lateral Reach to Fingertip	77.3	68.7	51.5	34.3	25.8
30.8	**79.0**	U6 D	Forward Reach to Fingertip	59.3	52.7	39.5	26.3	19.8
13.3	**34.0**	U7 V	Shoulder to Elbow Flexed	25.5	22.7	17.0	11.3	8.5
11.7	**30.0**	U8 V	Acromion to Lat. Condyle Elbow	22.5	20.0	15.0	10.0	7.5
16.8	**43.0**	U9 D	Flexed Elbow to Fingertip	32.3	28.7	21.5	14.3	10.8
9.9	**25.5**	U10 D	Flexed Elbow Pivot to Wrist Pivot	19.1	17.0	12.8	8.5	6.4

Life size			SEVENTEEN-YEAR-OLD GIRL	Fractions of life size				
				3/4	2/3	1/2	1/3	1/4
Inches	cm	Code		cm	cm	cm	cm	cm
			UPPER EXTREMITY MEASUREMENTS					
6.7	**17.3**	U11 V	Wrist to Fingertip	13.0	11.5	8.7	5.8	4.3
4.5	**11.5**	U12 V	Wrist to Thumb Tip	8.6	7.7	5.8	3.8	2.9
3.7	**9.6**	U13 V	Length of Palm	7.2	6.4	4.8	3.2	2.4
2.9	**7.5**	U14 V	Length 3rd Finger	5.6	5.0	3.8	2.5	1.9
4.5	**11.5**	U15 V	Thumb Crotch to 3rd Finger	8.6	7.7	5.8	3.8	2.9
2.3	**6.0**	U16 V	Thumb Crotch to Thumb Tip	4.5	4.0	3.0	2.0	1.5
3.1	**8.0**	U17 D	Anterior-Post. Arm at Biceps	6.0	5.3	4.0	2.7	2.0
2.9	**7.5**	U18 D	Anterior-Posterior at Forearm	5.6	5.0	3.8	2.5	1.9
1.4	**3.7**	U19 D	Depth of Wrist	2.8	2.5	1.9	1.2	0.9
2.1	**5.5**	U20 T	Breadth of Wrist	4.1	3.7	2.8	1.8	1.4
3.7	**9.5**	U21 T	Breadth of Hand	7.1	6.3	4.8	3.2	2.4
3.0	**7.6**	U22 T	Breadth of Palm	5.7	5.1	3.8	2.5	1.9
			LOWER EXTREMITY MEASUREMENTS					
38.6	**99.0**	L1 V	Crest of Ilium to Sole	74.3	66.0	49.5	33.0	24.8
35.9	**92.0**	* L2 V	Ant.Sup.Spine of Ilium to Sole	69.0	61.3	46.0	30.7	23.0
32.0	**82.0**	L3 V	Greater Trochanter Femur-Sole	61.5	54.7	41.0	27.3	20.5
31.8	**81.5**	* L4 V	Symphysis Pubis to Sole	61.1	54.3	40.8	27.2	20.4
28.5	**73.0**	* L5 V	Crotch to Sole	54.8	48.7	36.5	24.3	18.3
28.9	**74.0**	L6 V	Gluteal Furrow to Sole	55.5	49.3	37.0	24.7	18.5
16.8	**43.0**	L7 V	Posterior Knee Crease to Sole	32.3	28.7	21.5	14.3	10.8
16.6	**42.5**	L8 V	Tibia at Knee to Sole	31.9	28.3	21.3	14.2	10.6
2.6	**6.6**	L9 V	Med. Malleolus to Sole	5.0	4.4	3.3	2.2	1.7
2.0	**5.0**	L10 V	Lat. Malleolus to Sole	3.8	3.3	2.5	1.7	1.3
13.7	**35.0**	L11 T	Breadth of hips, Sitting	26.3	23.3	17.5	11.7	8.8
6.2	**16.0**	L12 T	Breadth Mid Thigh	12.0	10.7	8.0	5.3	4.0
4.5	**11.5**	L13 T	Breadth of Knee	8.6	7.7	5.8	3.8	2.9
4.1	**10.5**	L14 T	Breadth of Calf	7.9	7.0	5.3	3.5	2.6
2.3	**6.0**	L15 T	Breadth Ankle	4.5	4.0	3.0	2.0	1.5
6.2	**16.0**	L16 D	Mid Thigh, Anterior-Posterior	12.0	10.7	8.0	5.3	4.0
4.5	**11.5**	L17 D	Knee, Anterior-Posterior	8.6	7.7	5.8	3.8	2.9
4.1	**10.5**	L18 D	Mid Calf, Anterior-Posterior	7.9	7.0	5.3	3.5	2.6
3.1	**8.0**	L19 D	Ankle, Anterior-Posterior	6.0	5.3	4.0	2.7	2.0
9.4	**24.0**	L20 D	Length of Foot	18.0	16.0	12.0	8.0	6.0
7.8	**20.0**	L21 D	Heel to 5th Toe	15.0	13.3	10.0	6.7	5.0
3.5	**9.0**	L22 T	Breadth Foot	6.8	6.0	4.5	3.0	2.3
2.3	**6.0**	L23 T	Width of Heel	4.5	4.0	3.0	2.0	1.5
		** These are also listed with Vertical Measurements.*						

SEVENTEEN-YEAR-OLD BOYS

The lists of body proportions are recorded separately in the tables of measurements. The range of twenty-fifth, fiftieth, and seventy-fifth percentile is shown here for comparison:

Percentile	Twenty-fifth		Fiftieth		Seventy-fifth	
Boys:	Short		Medium		Tall	
Stature	171.9 cm,	67.8 in.	176.2 cm,	69.3 in.	180.5 cm,	71 in.
Weight	60 kg,	132.8 lb.	66 kg,	146.3 lb.	74 kg,	163.5 lb.[1]

By this age boys' average height is definitely taller than girls' average height. Bone and muscular development has progressed. Interest continues in competitive and outdoor sports. Social events with girls and dances continue to be very important. The automobile has become important to many activities. School level is twelfth grade. Many are making plans for vocation. For some this will include technical or college education.

BODY MEASUREMENTS COMPARED TO HEIGHT

	Feature Measure in cm	Feature Compared to Stature	Approx. Ratio of Feature	Compared to Median Adult
Stature: Crown to sole	176			96%
Head Height: Crown to chin	22	13%	1/8	96%
Sitting Height:	92	52%	1/2	96%
Trunk Measure: Sternal notch to symphysis pubis	53	30%	1/3–	100%
Leg Length: Greater trochanter of the femur to sole	90	51%	1/2	97%
Thigh Length: Greater trochanter of the femur to tibia at knee	43	25%	1/4	94%
Tibia at Knee to Sole:	47	26%	1/4	97%
Arm Length: Acromion process at shoulder to dependent fingertip	76	43%	1/2–	94%
Hand Length:	19	11%	1/9	95%
Foot Length:	26.8	15 %	1/7	95%

Most boys achieve about ninety-eight percent of adult male stature by age sixteen. There is small additional incremental growth to about age twenty-five years of age. Many facial features resemble those of adult males. The bones of the mid and lower face, the maxilla and mandible, have grown to accommodate the permanent teeth. In some the third molars have erupted; in others they will erupt soon. The face continues to appear youthful, but some mature male features become apparent as the mid-face and lower face increases vertical measurements. Illustrations of proportions are omitted, since there are few changes after age sixteen, except growth of arms and legs.

1. Nelson, W.E., p. 39.

Life size			SEVENTEEN-YEAR-OLD BOY	Fractions of life size				
				3/4	2/3	1/2	1/3	1/4
Inches	cm	Code		cm	cm	cm	cm	cm
			VERTICAL MEASUREMENTS					
69.3	**176.0**	1 V	Stature, Height, Vertex to Sole	132.0	117.3	88.0	58.7	44.0
33.9	**86.0**	2 V	Vertex to Symphysis Pubis	64.5	57.3	43.0	28.7	21.5
35.4	**90.0**	* 3 V	Symphysis Pubis to Sole	67.5	60.0	45.0	30.0	22.5
56.3	**143.0**	4 V	Sternal Notch to Sole	107.3	95.3	71.5	47.7	35.8
13.0	**33.0**	5 V	Vertex to Acromion Process	24.8	22.0	16.5	11.0	8.3
56.7	**144.0**	6 V	Acromion Process to Sole	108.0	96.0	72.0	48.0	36.0
39.0	**99.0**	* 7 V	Ant.Sup.Spine of Ilium to Sole	74.3	66.0	49.5	33.0	24.8
17.7	**45.0**	8 V	Acrom. proc. to Ant. Sup. Spine	33.8	30.0	22.5	15.0	11.3
19.3	**49.0**	9 V	Ant.Sup.Spine of Ilium to Knee	36.8	32.7	24.5	16.3	12.3
19.7	**50.0**	10 V	Knee at Patella to Sole	37.5	33.3	25.0	16.7	12.5
42.1	**107.0**	11 V	Umbilicus to Sole	80.3	71.3	53.5	35.7	26.8
33.1	**84.0**	* 12 V	Crotch to Sole	63.0	56.0	42.0	28.0	21.0
50.8	**129.0**	13 V	Kneeling Height	96.8	86.0	64.5	43.0	32.3
36.2	**92.0**	14 V	Sitting Height	69.0	61.3	46.0	30.7	23.0
36.2	**92.0**	15 V	Buttock, Straight Leg to Sole	69.0	61.3	46.0	30.7	23.0
23.2	**59.0**	16 V	Buttock to Flexed Knee	44.3	39.3	29.5	19.7	14.8
21.9	**55.5**	17 V	Flexed Knee to Sole	41.6	37.0	27.8	18.5	13.9
		** These are also listed with Lower Extremity Measurements.*						
			VERTICAL HEAD MEASUREMENTS					
13.0	**33.0**	H1 V	Vertex to Sternal Notch	24.8	22.0	16.5	11.0	8.3
8.9	**22.5**	H2 V	Vertex to Chin, Head Height	16.9	15.0	11.3	7.5	5.6
7.3	**18.5**	H3 V	Hairline to Chin, Face Height	13.9	12.3	9.3	6.2	4.6
4.5	**11.5**	H4 V	Vertex to Lateral Corner of Eye	8.6	7.7	5.8	3.8	2.9
5.1	**13.0**	H5 V	Vertex to Tragion of Ear	9.8	8.7	6.5	4.3	3.3
2.0	**5.1**	H6 V	Hairline to Brow, Glabella	3.8	3.4	2.6	1.7	1.3
5.3	**13.5**	H7 V	Brow, Glabella, to Chin, Menton	10.1	9.0	6.8	4.5	3.4
2.6	**6.7**	H8 V	Brow to Subnasal Point	5.0	4.5	3.4	2.2	1.7
2.0	**5.0**	H9 V	Sellion to Subnasal, Nose Length	3.8	3.3	2.5	1.7	1.3
4.6	**11.8**	H10 V	Sellion, Nose Saddle, to Chin	8.9	7.9	5.9	3.9	3.0
2.6	**6.6**	H11 V	Subnasal point to Chin	5.0	4.4	3.3	2.2	1.7
0.9	**2.3**	H12 V	Subnasal to Lip Aperature	1.7	1.5	1.2	0.8	0.6
0.7	**1.7**	H13 V	Lip Aperat. to Sublabial Crease	1.3	1.1	0.9	0.6	0.4
1.1	**2.7**	H14 V	Sublabial Crease to Chin	2.0	1.8	1.4	0.9	0.7
			TRANSVERSE HEAD MEASUREMENTS					
6.0	**15.3**	H15 T	Bi-Parietal, Head Breadth	11.5	10.2	7.7	5.1	3.8
5.4	**13.6**	H16 T	Bi-Zygomatic, Cheekbones	10.2	9.1	6.8	4.5	3.4
4.5	**11.4**	H17 T	Bi-Fronto Temporal, Forehead	8.6	7.6	5.7	3.8	2.9

Life size			SEVENTEEN-YEAR-OLD BOY	Fractions of life size				
				3/4	2/3	1/2	1/3	1/4
Inches	cm	Code		cm	cm	cm	cm	cm
			TRANSVERSE HEAD MEASUREMENTS					
4.1	**10.5**	H18 T	Inter Brow Ridge	7.9	7.0	5.3	3.5	2.6
5.1	**13.0**	H19 T	Inter Tragion of Ears	9.8	8.7	6.5	4.3	3.3
4.6	**11.7**	H20 T	Inter Angle of Mandible	8.8	7.8	5.9	3.9	2.9
3.9	**9.8**	H21 T	Inter Lateral Corners of Eyes	7.4	6.5	4.9	3.3	2.5
1.3	**3.3**	H22 T	Width of Eye	2.5	2.2	1.7	1.1	0.8
1.3	**3.2**	H23 T	Inter Inner Corner of eyes	2.4	2.1	1.6	1.1	0.8
1.4	**3.5**	H24 T	Breadth of Nose	2.6	2.3	1.8	1.2	0.9
2.0	**5.0**	H25 T	Breadth of Mouth	3.8	3.3	2.5	1.7	1.3
1.4	**3.6**	H26 T	Breadth of Chin	2.7	2.4	1.8	1.2	0.9
4.3	**11.0**	H27 T	Breadth of Neck	8.3	7.3	5.5	3.7	2.8
			ANTERIOR-POSTERIOR HEAD MEASUREMENTS					
7.8	**19.8**	H28 D	Brow to Occiput, Posterior Head	14.9	13.2	9.9	6.6	5.0
4.2	**10.6**	H29 D	Brow to Tragus of Ear	8.0	7.1	5.3	3.5	2.7
4.0	**10.1**	H30 D	Sellion to Tragus	7.6	6.7	5.1	3.4	2.5
3.6	**9.2**	H31 D	Tragus to Occiput	6.9	6.1	4.6	3.1	2.3
2.7	**6.8**	H32 V	Tragus to Angle Mandible	5.1	4.5	3.4	2.3	1.7
3.6	**9.2**	H33 D	Angle of Mandible to Chin	6.9	6.1	4.6	3.1	2.3
2.4	**6.2**	H34 V	Length of Ear, Vertical	4.7	4.1	3.1	2.1	1.6
1.4	**3.6**	H35 D	Width of Ear, Anterior-Posterior	2.7	2.4	1.8	1.2	0.9
4.2	**10.6**	H36 D	Anterior to Posterior of Neck	8.0	7.1	5.3	3.5	2.7
			VERTICAL BODY MEASUREMENTS					
4.3	**11.0**	B1 V	Chin to Sternal Notch	8.3	7.3	5.5	3.7	2.8
3.1	**8.0**	B2 V	Sternal Notch to Axilla	6.0	5.3	4.0	2.7	2.0
5.9	**15.0**	B3 V	Sternal Notch to Nipples	11.3	10.0	7.5	5.0	3.8
7.1	**18.0**	B4 V	Sternal Notch to Xyphoid Proc.	13.5	12.0	9.0	6.0	4.5
14.2	**36.0**	B5 V	Sternal Notch to Umbilicus	27.0	24.0	18.0	12.0	9.0
7.1	**18.0**	B6 V	Xyphoid Process to Umbilicus	13.5	12.0	9.0	6.0	4.5
14.2	**36.0**	B7 V	Sternal Notch to Crest Ilium	27.0	24.0	18.0	12.0	9.0
17.3	**44.0**	B8 V	Sternal Notch to Ant.S.Sp.Ilium	33.0	29.3	22.0	14.7	11.0
21.3	**54.0**	B9 V	Sternal Notch to Pubis	40.5	36.0	27.0	18.0	13.5
7.1	**18.0**	B10 V	Umbilicus to Pubis	13.5	12.0	9.0	6.0	4.5
			TRANSVERSE BODY MEASUREMENTS					
17.3	**44.0**	B11 T	Shoulder Breadth	33.0	29.3	22.0	14.7	11.0
7.5	**19.0**	B12 T	Clavicle to Acromion Process	14.3	12.7	9.5	6.3	4.8
15.4	**39.0**	B13 T	Biacromial Process Breadth	29.3	26.0	19.5	13.0	9.8
11.4	**29.0**	B14 T	Inter Anterior Axillary Fold	21.8	19.3	14.5	9.7	7.3
7.1	**18.0**	B15 T	Inter Nipples	13.5	12.0	9.0	6.0	4.5

Life size			SEVENTEEN-YEAR-OLD BOY	Fractions of life size				
				3/4	2/3	1/2	1/3	1/4
Inches	cm	Code		cm	cm	cm	cm	cm
			TRANSVERSE BODY MEASUREMENTS					
11.0	**28.0**	B16 T	Breadth of Thorax	21.0	18.7	14.0	9.3	7.0
10.6	**27.0**	B17 T	Inter Costal Margin Breadth	20.3	18.0	13.5	9.0	6.8
11.0	**28.0**	B18 T	Inter Crest of Ilium Breadth	21.0	18.7	14.0	9.3	7.0
9.4	**24.0**	B19 T	Inter Anterior Superior Spine	18.0	16.0	12.0	8.0	6.0
12.6	**32.0**	B20 T	Inter Greater Trochanter Femur	24.0	21.3	16.0	10.7	8.0
			ANTERIOR-POSTERIOR BODY MEASUREMENTS					
7.9	**20.0**	B21 D	Manubrium to Dorsal Spine	15.0	13.3	10.0	6.7	5.0
8.1	**20.5**	B22 D	Xyphoid Process to Dorsum	15.4	13.7	10.3	6.8	5.1
7.1	**18.0**	B23 D	Umbilicus to Dorsum	13.5	12.0	9.0	6.0	4.5
8.3	**21.0**	B24 D	Pubis to Buttock	15.8	14.0	10.5	7.0	5.3
6.3	**16.0**	B25 D	Thigh to Gluteal Furrow	12.0	10.7	8.0	5.3	4.0
			VERTICAL BACK MEASUREMENTS					
9.4	**24.0**	B26 V	Vertex to 7th Cervical Spine	18.0	16.0	12.0	8.0	6.0
			7th Cervical Vertebra Spine to:					
7.9	**20.0**	B27 V	To Posterior Axillary Fold	15.0	13.3	10.0	6.7	5.0
8.3	**21.0**	B28 V	To Inferior Tip of Scapula	15.8	14.0	10.5	7.0	5.3
18.5	**47.0**	B29 V	To Posterior Superior Spine Ilium	35.3	31.3	23.5	15.7	11.8
22.4	**57.0**	B30 V	To Midline Upper Buttock Crease	42.8	38.0	28.5	19.0	14.3
27.2	**69.0**	B31 V	To Gluteal Furrow	51.8	46.0	34.5	23.0	17.3
40.4	**102.5**	B32 V	To Popliteal Crease of Knee	76.9	68.3	51.3	34.2	25.6
			TRANSVERSE BACK MEASUREMENTS					
14.2	**36.0**	B33 T	Inter Posterior Axillary Folds	27.0	24.0	18.0	12.0	9.0
6.3	**16.0**	B34 T	Inter Scapulae Medial Borders	12.0	10.7	8.0	5.3	4.0
3.5	**9.0**	B35 T	Inter Post. Sup. Spines Sacrum	6.8	6.0	4.5	3.0	2.3
			UPPER EXTREMITY MEASUREMENTS					
69.3	**176.0**	U1 V	Span of Arms	132.0	117.3	88.0	58.7	44.0
7.5	**19.0**	U2 V	Sternum to Acromion Process	14.3	12.7	9.5	6.3	4.8
29.9	**76.0**	U3 V	Acromion Process to Fingertip	57.0	50.7	38.0	25.3	19.0
86.6	**222.0**	U4 V	Vertical Reach to Fingertip	165.0	146.7	110.0	73.3	55.0
42.5	**108.0**	U5 T	Lateral Reach to Fingertip	81.0	72.0	54.0	36.0	27.0
31.9	**81.0**	U6 D	Forward Reach to Fingertip	60.8	54.0	40.5	27.0	20.3
14.4	**36.5**	U7 V	Shoulder to Elbow Flexed	27.4	24.3	18.3	12.2	9.1
12.9	**32.7**	U8 V	Acromion to Lat. Condyle Elbow	24.5	21.8	16.4	10.9	8.2
18.7	**47.6**	U9 D	Flexed Elbow to Fingertip	35.7	31.7	23.8	15.9	11.9
10.2	**26.0**	U10 D	Flexed Elbow Pivot to Wrist Pivot	19.5	17.3	13.0	8.7	6.5

Life size			SEVENTEEN-YEAR-OLD BOY	Fractions of life size				
				3/4	2/3	1/2	1/3	1/4
Inches	cm	Code		cm	cm	cm	cm	cm
			UPPER EXTREMITY MEASUREMENTS					
7.5	**19.0**	U11 V	Wrist to Fingertip	14.3	12.7	9.5	6.3	4.8
4.9	**12.5**	U12 V	Wrist to Thumb Tip	9.4	8.3	6.3	4.2	3.1
4.3	**10.8**	U13 V	Length of Palm	8.1	7.2	5.4	3.6	2.7
3.2	**8.2**	U14 V	Length 3rd Finger	6.2	5.5	4.1	2.7	2.1
5.2	**13.2**	U15 V	Thumb Crotch to 3rd Finger	9.9	8.8	6.6	4.4	3.3
2.6	**6.7**	U16 V	Thumb Crotch to Thumb Tip	5.0	4.5	3.4	2.2	1.7
3.5	**9.0**	U17 D	Anterior-Post. Arm at Biceps	6.8	6.0	4.5	3.0	2.3
3.3	**8.5**	U18 D	Anterior-Posterior at Forearm	6.4	5.7	4.3	2.8	2.1
1.7	**4.3**	U19 D	Depth of Wrist	3.2	2.9	2.2	1.4	1.1
2.2	**5.5**	U20 T	Breadth of Wrist	4.1	3.7	2.8	1.8	1.4
4.3	**10.8**	U21 T	Breadth of Hand	8.1	7.2	5.4	3.6	2.7
3.5	**8.8**	U22 T	Breadth of Palm	6.6	5.9	4.4	2.9	2.2
			LOWER EXTREMITY MEASUREMENTS					
41.3	**105.0**	L1 V	Crest of Ilium to Sole	78.8	70.0	52.5	35.0	26.3
39.0	**99.0**	* L2 V	Ant.Sup.Spine of Ilium to Sole	74.3	66.0	49.5	33.0	24.8
35.4	**90.0**	L3 V	Greater Trochanter Femur-Sole	67.5	60.0	45.0	30.0	22.5
35.4	**90.0**	* L4 V	Symphysis Pubis to Sole	67.5	60.0	45.0	30.0	22.5
33.5	**85.0**	* L5 V	Crotch to Sole	63.8	56.7	42.5	28.3	21.3
32.7	**83.0**	L6 V	Gluteal Furrow to Sole	62.3	55.3	41.5	27.7	20.8
19.7	**50.0**	L7 V	Posterior Knee Crease to Sole	37.5	33.3	25.0	16.7	12.5
18.5	**47.0**	L8 V	Tibia at Knee to Sole	35.3	31.3	23.5	15.7	11.8
2.8	**7.0**	L9 V	Med. Malleolus to Sole	5.3	4.7	3.5	2.3	1.8
2.2	**5.5**	L10 V	Lat. Malleolus to Sole	4.1	3.7	2.8	1.8	1.4
13.0	**33.0**	L11 T	Breadth of Hips, Sitting	24.8	22.0	16.5	11.0	8.3
6.5	**16.6**	L12 T	Breadth Mid Thigh	12.5	11.1	8.3	5.5	4.2
4.3	**11.0**	L13 T	Breadth of Knee	8.3	7.3	5.5	3.7	2.8
4.3	**11.0**	L14 T	Breadth of Calf	8.3	7.3	5.5	3.7	2.8
2.4	**6.0**	L15 T	Breadth Ankle	4.5	4.0	3.0	2.0	1.5
6.3	**16.0**	L16 D	Mid Thigh, Anterior-Posterior	12.0	10.7	8.0	5.3	4.0
5.7	**14.5**	L17 D	Knee, Anterior-Posterior	10.9	9.7	7.3	4.8	3.6
4.3	**11.0**	L18 D	Mid Calf, Anterior-Posterior	8.3	7.3	5.5	3.7	2.8
3.1	**8.0**	L19 D	Ankle, Anterior-Posterior	6.0	5.3	4.0	2.7	2.0
10.6	**26.8**	L20 D	Length of Foot	20.1	17.9	13.4	8.9	6.7
8.3	**21.0**	L21 D	Heel to 5th Toe	15.8	14.0	10.5	7.0	5.3
4.1	**10.4**	L22 T	Breadth Foot	7.8	6.9	5.2	3.5	2.6
2.4	**6.0**	L23 T	Width of Heel	4.5	4.0	3.0	2.0	1.5
		** These are also listed with Vertical Measurements.*						

EIGHTEEN-YEAR-OLD GIRLS

The practical lists of body proportions are shown separately in the tables. The range of twenty-fifth to fiftieth and to seventy-fifth percentile is listed here for comparison:

Percentile	Twenty-fifth		Fiftieth		Seventy-fifth	
Girls:	Short		Medium		Tall	
Stature	160 cm,	62.8 in.	164 cm,	64.5 in.	168 cm,	66 in.
Weight	51.5 kg,	113.3 lb.	56.6 kg,	122.8 lb.	63 kg,	138.5 lb.[1]

By eighteen years of age, most girls have achieved their maximum stature and most have already developed typical feminine physique. There are yet some subtle changes of maturation that will develop between eighteen and twenty-two years of age. Most will have graduated from high school. Some will be gainfully employed, while others may be pursuing technical or college education.

BODY MEASUREMENTS COMPARED TO HEIGHT

	Feature Measure in cm	Feature Compared to Stature	Approx. Ratio of Feature	Compared to Median Adult
Stature: Crown to sole	164			98%
Head Height: Crown to chin	20.5	13%	1/8	100%
Sitting Height:	86	52%	1/2	97%
Trunk Measure: Sternal notch to symphysis pubis	51	31%	1/3–	96%
Leg Length: Greater trochanter of the femur to sole	80	49%	1/2	93%
Thigh Length: Greater trochanter of the femur to tibia at knee	49.7	30%	1/3–	100%
Tibia at Knee to Sole:	42	26%	1/4	98%
Arm Length: Acromion process at shoulder to dependent fingertip	71	43%	1/2–	95%
Hand Length:	17.3	11%	1/9	96%
Foot Length:	24	15%	1/7	100%

By this age most young women have completed their growth. There may be small increments of growth and some maturation that continue up to the age of twenty-two. The permanent dentition is nearly complete. Third molars may continue to grow until eruption. The comparison of major measurements and ratios to adult females reveals very near adult proportions. Illustrations of proportions are omitted, since there are few changes from age sixteen.

1. Nelson, W.E., p. 39.

Life size			EIGHTEEN-YEAR-OLD GIRL	Fractions of life size				
				3/4	2/3	1/2	1/3	1/4
Inches	cm	Code		cm	cm	cm	cm	cm
			VERTICAL MEASUREMENTS					
64.6	**164.0**	1 V	Stature, Height, Vertex to Sole	123.0	109.3	82.0	54.7	41.0
32.3	**82.0**	2 V	Vertex to Symphysis Pubis	61.5	54.7	41.0	27.3	20.5
32.3	**82.0**	* 3 V	Symphysis Pubis to Sole	61.5	54.7	41.0	27.3	20.5
51.8	**131.5**	4 V	Sternal Notch to Sole	98.6	87.7	65.8	43.8	32.9
12.2	**31.0**	5 V	Vertex to Acromion Process	23.3	20.7	15.5	10.3	7.8
52.4	**133.0**	6 V	Acromion Process to Sole	99.8	88.7	66.5	44.3	33.3
36.2	**92.0**	* 7 V	Ant.Sup.Spine of Ilium to Sole	69.0	61.3	46.0	30.7	23.0
16.1	**41.0**	8 V	Acrom. proc. to Ant. Sup. Spine	30.8	27.3	20.5	13.7	10.3
18.1	**46.0**	9 V	Ant.Sup.Spine of Ilium to Knee	34.5	30.7	23.0	15.3	11.5
18.5	**47.0**	10 V	Knee at Patella to Sole	35.3	31.3	23.5	15.7	11.8
39.0	**99.0**	11 V	Umbilicus to Sole	74.3	66.0	49.5	33.0	24.8
28.7	**73.0**	* 12 V	Crotch to Sole	54.8	48.7	36.5	24.3	18.3
48.4	**123.0**	13 V	Kneeling Height	92.3	82.0	61.5	41.0	30.8
33.9	**86.0**	14 V	Sitting Height	64.5	57.3	43.0	28.7	21.5
34.3	**87.0**	15 V	Buttock, Straight Leg to Sole	65.3	58.0	43.5	29.0	21.8
21.7	**55.0**	16 V	Buttock to Flexed Knee	41.3	36.7	27.5	18.3	13.8
19.7	**50.0**	17 V	Flexed Knee to Sole	37.5	33.3	25.0	16.7	12.5
		** These are also listed with Lower Extremity Measurements.*						
			VERTICAL HEAD MEASUREMENTS					
12.6	**32.0**	H1 V	Vertex to Sternal Notch	24.0	21.3	16.0	10.7	8.0
8.3	**21.0**	H2 V	Vertex to Chin, Head Height	15.8	14.0	10.5	7.0	5.3
7.1	**18.0**	H3 V	Hairline to Chin, Face Height	13.5	12.0	9.0	6.0	4.5
4.3	**10.8**	H4 V	Vertex to Lateral Corner of Eye	8.1	7.2	5.4	3.6	2.7
4.9	**12.5**	H5 V	Vertex to Tragion of Ear	9.4	8.3	6.3	4.2	3.1
2.0	**5.0**	H6 V	Hairline to Brow, Glabella	3.8	3.3	2.5	1.7	1.3
4.9	**12.5**	H7 V	Brow, Glabella, to Chin, Menton	9.4	8.3	6.3	4.2	3.1
2.4	**6.2**	H8 V	Brow to Subnasal Point	4.7	4.1	3.1	2.1	1.6
1.9	**4.7**	H9 V	Sellion to Subnasal, Nose Length	3.5	3.1	2.4	1.6	1.2
4.4	**11.1**	H10 V	Sellion, Nose Saddle, to Chin	8.3	7.4	5.6	3.7	2.8
2.5	**6.3**	H11 V	Subnasal point to Chin	4.7	4.2	3.2	2.1	1.6
0.9	**2.2**	H12 V	Subnasal to Lip Aperature	1.7	1.5	1.1	0.7	0.6
0.6	**1.4**	H13 V	Lip Aperat. to Sublabial Crease	1.1	0.9	0.7	0.5	0.4
1.1	**2.8**	H14 V	Sublabial Crease to Chin	2.1	1.9	1.4	0.9	0.7
			TRANSVERSE HEAD MEASUREMENTS					
5.8	**14.8**	H15 T	Bi-Parietal, Head Breadth	11.1	9.9	7.4	4.9	3.7
5.1	**13.0**	H16 T	Bi-Zygomatic, Cheekbones	9.8	8.7	6.5	4.3	3.3
4.5	**11.5**	H17 T	Bi-Fronto Temporal, Forehead	8.6	7.7	5.8	3.8	2.9

Life size			EIGHTEEN-YEAR-OLD GIRL	Fractions of life size				
				3/4	2/3	1/2	1/3	1/4
Inches	cm	Code		cm	cm	cm	cm	cm
			TRANSVERSE HEAD MEASUREMENTS					
4.1	**10.3**	H18 T	Inter Brow Ridge	7.7	6.9	5.2	3.4	2.6
5.0	**12.7**	H19 T	Inter Tragion of Ears	9.5	8.5	6.4	4.2	3.2
4.6	**11.8**	H20 T	Inter Angle of Mandible	8.9	7.9	5.9	3.9	3.0
3.7	**9.3**	H21 T	Inter Lateral Corners of Eyes	7.0	6.2	4.7	3.1	2.3
1.2	**3.1**	H22 T	Width of Eye	2.3	2.1	1.6	1.0	0.8
1.2	**3.1**	H23 T	Inter Inner Corner of eyes	2.3	2.1	1.6	1.0	0.8
1.3	**3.3**	H24 T	Breadth of Nose	2.5	2.2	1.7	1.1	0.8
1.8	**4.6**	H25 T	Breadth of Mouth	3.5	3.1	2.3	1.5	1.2
1.4	**3.5**	H26 T	Breadth of Chin	2.6	2.3	1.8	1.2	0.9
3.7	**9.5**	H27 T	Breadth of Neck	7.1	6.3	4.8	3.2	2.4
			ANTERIOR-POSTERIOR HEAD MEASUREMENTS					
7.5	**19.0**	H28 D	Brow to Occiput, Posterior Head	14.3	12.7	9.5	6.3	4.8
3.9	**9.8**	H29 D	Brow to Tragus of Ear	7.4	6.5	4.9	3.3	2.5
3.7	**9.3**	H30 D	Sellion, Nose Bridge, to Tragus	7.0	6.2	4.7	3.1	2.3
3.6	**9.2**	H31 D	Tragus to Occiput, Back of Head	6.9	6.1	4.6	3.1	2.3
2.6	**6.6**	H32 V	Tragus to Angle Mandible	5.0	4.4	3.3	2.2	1.7
3.5	**9.0**	H33 D	Angle of Mandible to Chin	6.8	6.0	4.5	3.0	2.3
2.4	**6.0**	H34 V	Length of Ear, Vertical	4.5	4.0	3.0	2.0	1.5
1.3	**3.3**	H35 D	Width of Ear, Anterior-Posterior	2.5	2.2	1.7	1.1	0.8
4.1	**10.5**	H36 D	Anterior to Posterior of Neck	7.9	7.0	5.3	3.5	2.6
			VERTICAL BODY MEASUREMENTS					
4.1	**10.5**	B1 V	Chin to Sternal Notch, Neck Pit	7.9	7.0	5.3	3.5	2.6
3.1	**8.0**	B2 V	Sternal Notch to Axilla, Arm Pit	6.0	5.3	4.0	2.7	2.0
6.7	**17.0**	B3 V	Sternal Notch to Nipples	12.8	11.3	8.5	5.7	4.3
7.1	**18.0**	B4 V	Sternal Notch to Xyphoid Proc.	13.5	12.0	9.0	6.0	4.5
13.8	**35.0**	B5 V	Sternal Notch to Umbilicus	26.3	23.3	17.5	11.7	8.8
6.3	**16.0**	B6 V	Xyphoid Process to Umbilicus	12.0	10.7	8.0	5.3	4.0
13.9	**35.3**	B7 V	Sternal Notch to Crest Ilium	26.5	23.5	17.7	11.8	8.8
16.1	**41.0**	B8 V	Sternal Notch to Ant.S.Sp.Ilium	30.8	27.3	20.5	13.7	10.3
20.1	**51.0**	B9 V	Sternal Notch to Pubis	38.3	34.0	25.5	17.0	12.8
6.7	**17.0**	B10 V	Umbilicus to Pubis	12.8	11.3	8.5	5.7	4.3
			TRANSVERSE BODY MEASUREMENTS					
15.7	**40.0**	B11 T	Shoulder Breadth	30.0	26.7	20.0	13.3	10.0
7.1	**18.0**	B12 T	Clavicle to Acromion Process	13.5	12.0	9.0	6.0	4.5
14.2	**36.0**	B13 T	Biacromial Process Breadth	27.0	24.0	18.0	12.0	9.0
10.2	**26.0**	B14 T	Inter Anterior Axillary Fold	19.5	17.3	13.0	8.7	6.5
7.9	**20.0**	B15 T	Inter Nipples	15.0	13.3	10.0	6.7	5.0

Life size			EIGHTEEN-YEAR-OLD GIRL	Fractions of life size				
				3/4	2/3	1/2	1/3	1/4
Inches	cm	Code		cm	cm	cm	cm	cm
			TRANSVERSE BODY MEASUREMENTS					
10.2	**26.0**	B16 T	Breadth of Thorax	19.5	17.3	13.0	8.7	6.5
9.8	**25.0**	B17 T	Inter Costal Margin Breadth	18.8	16.7	12.5	8.3	6.3
11.0	**28.0**	B18 T	Inter Crest of Ilium Breadth	21.0	18.7	14.0	9.3	7.0
9.8	**25.0**	B19 T	Inter Anterior Superior Spine	18.8	16.7	12.5	8.3	6.3
12.6	**32.0**	B20 T	Inter Greater Trochanter Femur	24.0	21.3	16.0	10.7	8.0
			ANTERIOR-POSTERIOR BODY MEASUREMENTS					
7.2	**18.4**	B21 D	Manubrium to Dorsal Spine	13.8	12.3	9.2	6.1	4.6
7.5	**19.1**	B22 D	Xyphoid Process to Dorsum	14.3	12.7	9.6	6.4	4.8
7.1	**18.0**	B23 D	Umbilicus to Dorsum	13.5	12.0	9.0	6.0	4.5
8.7	**22.0**	B24 D	Pubis to Buttock	16.5	14.7	11.0	7.3	5.5
6.9	**17.5**	B25 D	Thigh to Gluteal Furrow	13.1	11.7	8.8	5.8	4.4
			VERTICAL BACK MEASUREMENTS					
9.4	**24.0**	B26 V	Vertex to 7th Cervical Spine	18.0	16.0	12.0	8.0	6.0
			7th Cervical Vertebra Spine to:					
7.1	**18.0**	B27 V	To Posterior Axillary Fold	13.5	12.0	9.0	6.0	4.5
7.9	**20.0**	B28 V	To Inferior Tip of Scapula	15.0	13.3	10.0	6.7	5.0
17.1	**43.5**	B29 V	To Posterior Superior Spine Ilium	32.6	29.0	21.8	14.5	10.9
20.6	**52.2**	B30 V	To Midline Upper Buttock Crease	39.2	34.8	26.1	17.4	13.1
25.6	**65.0**	B31 V	To Gluteal Furrow	48.8	43.3	32.5	21.7	16.3
37.8	**96.0**	B32 V	To Popliteal Crease of Knee	72.0	64.0	48.0	32.0	24.0
			TRANSVERSE BACK MEASUREMENTS					
12.6	**32.0**	B33 T	Inter Posterior Axillary Folds	24.0	21.3	16.0	10.7	8.0
5.5	**14.0**	B34 T	Inter Scapulae Medial Borders	10.5	9.3	7.0	4.7	3.5
3.3	**8.3**	B35 T	Inter Post. Sup. Spines Sacrum	6.2	5.5	4.2	2.8	2.1
			UPPER EXTREMITY MEASUREMENTS					
64.2	**163.0**	U1 V	Span of Arms	122.3	108.7	81.5	54.3	40.8
7.0	**17.8**	U2 V	Sternum to Acromion Process	13.4	11.9	8.9	5.9	4.5
27.6	**70.0**	U3 V	Acromion Process to Fingertip	52.5	46.7	35.0	23.3	17.5
80.7	**207.0**	U4 V	Vertical Reach to Fingertip	156.8	139.3	104.5	69.7	52.3
39.4	**100.0**	U5 T	Lateral Reach to Fingertip	77.3	68.7	51.5	34.3	25.8
31.1	**79.0**	U6 D	Forward Reach to Fingertip	59.3	52.7	39.5	26.3	19.8
13.4	**34.0**	U7 V	Shoulder to Elbow Flexed	25.5	22.7	17.0	11.3	8.5
11.8	**30.0**	U8 V	Acromion to Lat. Condyle Elbow	22.5	20.0	15.0	10.0	7.5
16.9	**43.0**	U9 D	Flexed Elbow to Fingertip	32.3	28.7	21.5	14.3	10.8
10.0	**25.5**	U10 D	Flexed Elbow Pivot to Wrist Pivot	19.1	17.0	12.8	8.5	6.4

Life size			EIGHTEEN-YEAR-OLD GIRL	Fractions of life size				
				3/4	2/3	1/2	1/3	1/4
Inches	cm	Code		cm	cm	cm	cm	cm
			UPPER EXTREMITY MEASUREMENTS					
6.8	**17.3**	U11 V	Wrist to Fingertip	13.0	11.5	8.7	5.8	4.3
4.7	**12.0**	U12 V	Wrist to Thumb Tip	9.0	8.0	6.0	4.0	3.0
3.9	**10.0**	U13 V	Length of Palm	7.5	6.7	5.0	3.3	2.5
3.0	**7.5**	U14 V	Length 3rd Finger	5.6	5.0	3.8	2.5	1.9
4.5	**11.5**	U15 V	Thumb Crotch to 3rd Finger	8.6	7.7	5.8	3.8	2.9
2.4	**6.0**	U16 V	Thumb Crotch to Thumb Tip	4.5	4.0	3.0	2.0	1.5
3.1	**8.0**	U17 D	Anterior-Post. Arm at Biceps	6.0	5.3	4.0	2.7	2.0
3.0	**7.5**	U18 D	Anterior-Posterior at Forearm	5.6	5.0	3.8	2.5	1.9
1.5	**3.7**	U19 D	Depth of Wrist	2.8	2.5	1.9	1.2	0.9
2.2	**5.5**	U20 T	Breadth of Wrist	4.1	3.7	2.8	1.8	1.4
3.8	**9.7**	U21 T	Breadth of Hand	7.3	6.5	4.9	3.2	2.4
3.1	**7.8**	U22 T	Breadth of Palm	5.9	5.2	3.9	2.6	2.0
			LOWER EXTREMITY MEASUREMENTS					
39.0	**99.0**	L1 V	Crest of Ilium to Sole	74.3	66.0	49.5	33.0	24.8
36.2	**92.0**	* L2 V	Ant.Sup.Spine of Ilium to Sole	69.0	61.3	46.0	30.7	23.0
32.3	**82.0**	L3 V	Greater Trochanter Femur-Sole	61.5	54.7	41.0	27.3	20.5
32.3	**82.0**	* L4 V	Symphysis Pubis to Sole	61.5	54.7	41.0	27.3	20.5
28.7	**73.0**	* L5 V	Crotch to Sole	54.8	48.7	36.5	24.3	18.3
29.1	**74.0**	L6 V	Gluteal Furrow to Sole	55.5	49.3	37.0	24.7	18.5
16.9	**43.0**	L7 V	Posterior Knee Crease to Sole	32.3	28.7	21.5	14.3	10.8
16.7	**42.5**	L8 V	Tibia at Knee to Sole	31.9	28.3	21.3	14.2	10.6
2.6	**6.5**	L9 V	Med. Malleolus to Sole	4.9	4.3	3.3	2.2	1.6
2.0	**5.0**	L10 V	Lat. Malleolus to Sole	3.8	3.3	2.5	1.7	1.3
13.8	**35.0**	L11 T	Breadth of Hips, Sitting	26.3	23.3	17.5	11.7	8.8
6.3	**16.0**	L12 T	Breadth Mid Thigh	12.0	10.7	8.0	5.3	4.0
4.5	**11.5**	L13 T	Breadth of Knee	8.6	7.7	5.8	3.8	2.9
3.9	**10.0**	L14 T	Breadth of Calf	7.5	6.7	5.0	3.3	2.5
2.4	**6.0**	L15 T	Breadth Ankle	4.5	4.0	3.0	2.0	1.5
6.3	**16.0**	L16 D	Mid Thigh, Anterior-Posterior	12.0	10.7	8.0	5.3	4.0
4.5	**11.5**	L17 D	Knee, Anterior-Posterior	8.6	7.7	5.8	3.8	2.9
4.1	**10.5**	L18 D	Mid Calf, Anterior-Posterior	7.9	7.0	5.3	3.5	2.6
3.1	**8.0**	L19 D	Ankle, Anterior-Posterior	6.0	5.3	4.0	2.7	2.0
9.4	**24.0**	L20 D	Length of Foot	18.0	16.0	12.0	8.0	6.0
7.9	**20.0**	L21 D	Heel to 5th Toe	15.0	13.3	10.0	6.7	5.0
3.5	**9.0**	L22 T	Breadth Foot	6.8	6.0	4.5	3.0	2.3
2.4	**6.0**	L23 T	Width of Heel	4.5	4.0	3.0	2.0	1.5
		** These are also listed with Vertical Measurements.*						

EIGHTEEN-YEAR-OLD BOYS

The practical lists of body proportions are shown separately in the tables. The range of twenty-fifth to fiftieth and to seventy-fifth percentile is listed here for comparison:

Percentile	Twenty-fifth		Fiftieth		Seventy-fifth	
Boys:	Short		Medium		Tall	
Stature	172 cm,	67.8 in.	176.8 cm,	69.5 in.	181 cm,	71.3 in.
Weight	63 kg,	138 lb.	69 kg,	152 lb.	76 kg,	168 lb.[1]

By eighteen years of age, most boys have graduated from high school. Many are gainfully employed; others are pursuing career training in technical schools or colleges. Their physical development is near prime as they engage in vigorous competitive athletics or employment with physical demands. Social activities are also important as lasting friendships develop.

BODY MEASUREMENTS COMPARED TO HEIGHT

	Feature Measure in cm	Feature Compared to Stature	Approx. Ratio of Feature	Compared to Median Adult
Stature: Crown to sole	177			97 %
Head Height: Crown to chin	22	12%	1/8	96%
Sitting Height:	92.5	52%	1/2	96%
Trunk Measure: Sternal notch to symphysis pubis	53	30%	1/3	98%
Leg Length: Greater trochanter of the femur to sole	90	51%	1/2	92%
Thigh Length: Greater trochanter of the femur to tibia at knee	43	24%	1/4	92%
Tibia at Knee to Sole:	47	26%	1/4	92%
Arm Length: Acromion process at shoulder to dependent fingertip	80	45%	1/2+	98%
Hand Length:	19	11%	1/9	95%
Foot Length:	27	15%	1/7	97%

The increase in height for males of this age has changed little from age seventeen. There has been an average weight gain of over 6 kg, or about 14 pounds. The face continues to appear youthful, but some mature male features become apparent. There will be some incremental growth of about one inch between eighteen and twenty-five years. There will be an increase in weight as muscle mass and bone density increases. The face height will increase as a young man matures. During this time the bone growth of brows, maxilla, and mandible will result in more angular features characteristic of the mature adult male face. Proportion illustrations are omitted.

1. Nelson, W.E., p. 38.

Life size			EIGHTEEN-YEAR-OLD BOY	Fractions of life size				
				3/4	2/3	1/2	1/3	1/4
Inches	cm	Code		cm	cm	cm	cm	cm
			VERTICAL MEASUREMENTS					
69.7	**177.0**	1 V	Stature, Height, Vertex to Sole	132.8	118.0	88.5	59.0	44.3
34.3	**87.0**	2 V	Vertex to Symphysis Pubis	65.3	58.0	43.5	29.0	21.8
35.4	**90.0**	* 3 V	Symphysis Pubis to Sole	67.5	60.0	45.0	30.0	22.5
56.3	**143.0**	4 V	Sternal Notch to Sole	107.3	95.3	71.5	47.7	35.8
13.8	**35.0**	5 V	Vertex to Acromion Process	26.3	23.3	17.5	11.7	8.8
56.7	**144.0**	6 V	Acromion Process to Sole	108.0	96.0	72.0	48.0	36.0
39.0	**99.0**	* 7 V	Ant.Sup.Spine of Ilium to Sole	74.3	66.0	49.5	33.0	24.8
17.7	**45.0**	8 V	Acrom. proc. to Ant. Sup. Spine	33.8	30.0	22.5	15.0	11.3
19.3	**49.0**	9 V	Ant.Sup.Spine of Ilium to Knee	36.8	32.7	24.5	16.3	12.3
20.1	**51.0**	10 V	Knee at Patella to Sole	38.3	34.0	25.5	17.0	12.8
42.1	**107.0**	11 V	Umbilicus to Sole	80.3	71.3	53.5	35.7	26.8
33.1	**84.0**	* 12 V	Crotch to Sole	63.0	56.0	42.0	28.0	21.0
51.2	**130.0**	13 V	Kneeling Height	97.5	86.7	65.0	43.3	32.5
36.4	**92.5**	14 V	Sitting Height	69.4	61.7	46.3	30.8	23.1
36.6	**93.0**	15 V	Buttock, Straight Leg to Sole	69.8	62.0	46.5	31.0	23.3
23.4	**59.5**	16 V	Buttock to Flexed Knee	44.6	39.7	29.8	19.8	14.9
21.9	**55.5**	17 V	Flexed Knee to Sole	41.6	37.0	27.8	18.5	13.9
		**These are also listed with Lower Extremity Measurements.*						
			VERTICAL HEAD MEASUREMENTS					
13.0	**33.0**	H1 V	Vertex to Sternal Notch	24.8	22.0	16.5	11.0	8.3
8.9	**22.5**	H2 V	Vertex to Chin, Head Height	16.9	15.0	11.3	7.5	5.6
7.5	**19.0**	H3 V	Hairline to Chin, Face Height	14.3	12.7	9.5	6.3	4.8
4.5	**11.5**	H4 V	Vertex to Lateral Corner of Eye	8.6	7.7	5.8	3.8	2.9
5.1	**13.0**	H5 V	Vertex to Tragion of Ear	9.8	8.7	6.5	4.3	3.3
2.2	**5.5**	H6 V	Hairline to Brow, Glabella	4.1	3.7	2.8	1.8	1.4
5.3	**13.5**	H7 V	Brow, Glabella, to Chin, Menton	10.1	9.0	6.8	4.5	3.4
2.6	**6.7**	H8 V	Brow to Subnasal Point	5.0	4.5	3.4	2.2	1.7
2.0	**5.1**	H9 V	Sellion to Subnasal, Nose Length	3.8	3.4	2.6	1.7	1.3
4.7	**12.0**	H10 V	Sellion, Nose Saddle, to Chin	9.0	8.0	6.0	4.0	3.0
2.7	**6.8**	H11 V	Subnasal point to Chin	5.1	4.5	3.4	2.3	1.7
0.9	**2.2**	H12 V	Subnasal to Lip Aperature	1.7	1.5	1.1	0.7	0.6
0.7	**1.8**	H13 V	Lip Aperat. to Sublabial Crease	1.4	1.2	0.9	0.6	0.5
1.1	**2.8**	H14 V	Sublabial Crease to Chin	2.1	1.9	1.4	0.9	0.7
			TRANSVERSE HEAD MEASUREMENTS					
6.1	**15.5**	H15 T	Bi-Parietal, Head Breadth	11.6	10.3	7.8	5.2	3.9
5.5	**14.0**	H16 T	Bi-Zygomatic, Cheekbones	10.5	9.3	7.0	4.7	3.5
4.7	**12.0**	H17 T	Bi-Fronto Temporal, Forehead	9.0	8.0	6.0	4.0	3.0

Life size			EIGHTEEN-YEAR-OLD BOY	Fractions of life size				
				3/4	2/3	1/2	1/3	1/4
Inches	cm	Code		cm	cm	cm	cm	cm
			TRANSVERSE HEAD MEASUREMENTS					
4.2	10.6	H18 T	Inter Brow Ridge	8.0	7.1	5.3	3.5	2.7
5.1	13.0	H19 T	Inter Tragion of Ears	9.8	8.7	6.5	4.3	3.3
4.7	11.9	H20 T	Inter Angle of Mandible	8.9	7.9	6.0	4.0	3.0
3.9	9.9	H21 T	Inter Lateral Corners of Eyes	7.4	6.6	5.0	3.3	2.5
1.3	3.3	H22 T	Width of Eye	2.5	2.2	1.7	1.1	0.8
1.3	3.3	H23 T	Inter Inner Corner of eyes	2.5	2.2	1.7	1.1	0.8
1.4	3.6	H24 T	Breadth of Nose	2.7	2.4	1.8	1.2	0.9
2.0	5.0	H25 T	Breadth of Mouth	3.8	3.3	2.5	1.7	1.3
1.5	3.7	H26 T	Breadth of Chin	2.8	2.5	1.9	1.2	0.9
4.3	11.0	H27 T	Breadth of Neck	8.3	7.3	5.5	3.7	2.8
			ANTERIOR-POSTERIOR HEAD MEASUREMENTS					
7.9	20.0	H28 D	Brow to Occiput, Posterior Head	15.0	13.3	10.0	6.7	5.0
4.3	10.8	H29 D	Brow to Tragus of Ear	8.1	7.2	5.4	3.6	2.7
4.1	10.3	H30 D	Sellion to Tragus	7.7	6.9	5.2	3.4	2.6
3.6	9.2	H31 D	Tragus to Occiput	6.9	6.1	4.6	3.1	2.3
2.8	7.0	H32 V	Tragus to Angle Mandible	5.3	4.7	3.5	2.3	1.8
3.7	9.5	H33 V	Angle of Mandible to Chin	7.1	6.3	4.8	3.2	2.4
2.4	6.2	H34 V	Length of Ear, Vertical	4.7	4.1	3.1	2.1	1.6
1.4	3.6	H35 D	Width of Ear, Anterior-Posterior	2.7	2.4	1.8	1.2	0.9
4.5	11.5	H36 D	Anterior to Posterior of Neck	8.6	7.7	5.8	3.8	2.9
			VERTICAL BODY MEASUREMENTS					
4.7	12.0	B1 V	Chin to Sternal Notch, Neck Pit	9.0	8.0	6.0	4.0	3.0
3.1	8.0	B2 V	Sternal Notch to Axilla, Arm Pit	6.0	5.3	4.0	2.7	2.0
6.1	15.5	B3 V	Sternal Notch to Nipples	11.6	10.3	7.8	5.2	3.9
7.1	18.0	B4 V	Sternal Notch to Xyphoid Proc.	13.5	12.0	9.0	6.0	4.5
14.2	36.0	B5 V	Sternal Notch to Umbilicus	27.0	24.0	18.0	12.0	9.0
7.1	18.0	B6 V	Xyphoid Process to Umbilicus	13.5	12.0	9.0	6.0	4.5
14.6	37.0	B7 V	Sternal Notch to Crest Ilium	27.8	24.7	18.5	12.3	9.3
17.3	44.0	B8 V	Sternal Notch to Ant.S.Sp.Ilium	33.0	29.3	22.0	14.7	11.0
21.3	54.0	B9 V	Sternal Notch to Pubis	40.5	36.0	27.0	18.0	13.5
7.1	18.0	B10 V	Umbilicus to Pubis	13.5	12.0	9.0	6.0	4.5
			TRANSVERSE BODY MEASUREMENTS					
17.9	45.5	B11 T	Shoulder Breadth	34.1	30.3	22.8	15.2	11.4
7.7	19.5	B12 T	Clavicle to Acromion Process	14.6	13.0	9.8	6.5	4.9
15.7	40.0	B13 T	Biacromial Process Breadth	30.0	26.7	20.0	13.3	10.0
11.8	30.0	B14 T	Inter Anterior Axillary Fold	22.5	20.0	15.0	10.0	7.5

Life size			EIGHTEEN-YEAR-OLD BOY	Fractions of life size				
				3/4	2/3	1/2	1/3	1/4
Inches	cm	Code		cm	cm	cm	cm	cm
			TRANSVERSE BODY MEASUREMENTS					
11.0	**28.0**	B16 T	Breadth of Thorax	21.0	18.7	14.0	9.3	7.0
11.0	**28.0**	B17 T	Inter Costal Margin Breadth	21.0	18.7	14.0	9.3	7.0
11.4	**29.0**	B18 T	Inter Crest of Ilium Breadth	21.8	19.3	14.5	9.7	7.3
9.4	**24.0**	B19 T	Inter Anterior Superior Spine	18.0	16.0	12.0	8.0	6.0
12.8	**32.6**	B20 T	Inter Greater Trochanter Femur	24.5	21.7	16.3	10.9	8.2
			ANTERIOR-POSTERIOR BODY MEASUREMENTS					
8.1	**20.5**	B21 D	Manubrium to Dorsal Spine	15.4	13.7	10.3	6.8	5.1
8.3	**21.0**	B22 D	Xyphoid Process to Dorsum	15.8	14.0	10.5	7.0	5.3
7.3	**18.5**	B23 D	Umbilicus to Dorsum	13.9	12.3	9.3	6.2	4.6
8.3	**21.0**	B24 D	Pubis to Buttock	15.8	14.0	10.5	7.0	5.3
6.7	**17.0**	B25 D	Thigh to Gluteal Furrow	12.8	11.3	8.5	5.7	4.3
			VERTICAL BACK MEASUREMENTS					
9.4	**24.0**	B26 V	Vertex to 7th Cervical Spine	18.0	16.0	12.0	8.0	6.0
			7th Cervical Vertebra Spine to:					
7.9	**20.0**	B27 V	To Posterior Axillary Fold	15.0	13.3	10.0	6.7	5.0
8.3	**21.0**	B28 V	To Inferior Tip of Scapula	15.8	14.0	10.5	7.0	5.3
18.5	**47.0**	B29 V	To Posterior Superior Spine Ilium	35.3	31.3	23.5	15.7	11.8
22.4	**57.0**	B30 V	To Midline Upper Buttock Crease	42.8	38.0	28.5	19.0	14.3
27.2	**69.0**	B31 V	To Gluteal Furrow	51.8	46.0	34.5	23.0	17.3
40.6	**103.0**	B32 V	To Popliteal Crease of Knee	77.3	68.7	51.5	34.3	25.8
			TRANSVERSE BACK MEASUREMENTS					
15.0	**38.0**	B33 T	Inter Posterior Axillary Folds	28.5	25.3	19.0	12.7	9.5
6.7	**17.0**	B34 T	Inter Scapulae Medial Borders	12.8	11.3	8.5	5.7	4.3
3.5	**9.0**	B35 T	Inter Post. Sup. Spines Sacrum	6.8	6.0	4.5	3.0	2.3
			UPPER EXTREMITY MEASUREMENTS					
69.3	**176.0**	U1 V	Span of Arms	132.0	117.3	88.0	58.7	44.0
7.9	**20.0**	U2 V	Sternum to Acromion Process	15.0	13.3	10.0	6.7	5.0
31.5	**80.0**	U3 V	Acromion Process to Fingertip	60.0	53.3	40.0	26.7	20.0
87.8	**223.0**	U4 V	Vertical Reach to Fingertip	165.0	146.7	110.0	73.3	55.0
42.5	**108.0**	U5 T	Lateral Reach to Fingertip	81.0	72.0	54.0	36.0	27.0
32.3	**82.0**	U6 D	Forward Reach to Fingertip	61.5	54.7	41.0	27.3	20.5
14.6	**37.0**	U7 V	Shoulder to Elbow Flexed	27.8	24.7	18.5	12.3	9.3
13.2	**33.6**	U8 V	Acromion to Lat. Condyle Elbow	25.2	22.4	16.8	11.2	8.4
18.9	**48.0**	U9 D	Flexed Elbow to Fingertip	36.0	32.0	24.0	16.0	12.0
10.8	**27.5**	U10 D	Flexed Elbow Pivot to Wrist Pivot	20.6	18.3	13.8	9.2	6.9

Life size			EIGHTEEN-YEAR-OLD BOY	Fractions of life size				
				3/4	2/3	1/2	1/3	1/4
Inches	cm	Code		cm	cm	cm	cm	cm
			UPPER EXTREMITY MEASUREMENTS					
7.5	19.0	U11 V	Wrist to Fingertip	14.3	12.7	9.5	6.3	4.8
4.9	12.5	U12 V	Wrist to Thumb Tip	9.4	8.3	6.3	4.2	3.1
4.3	10.8	U13 V	Length of Palm	8.1	7.2	5.4	3.6	2.7
3.3	8.4	U14 V	Length 3rd Finger	6.3	5.6	4.2	2.8	2.1
5.2	13.2	U15 V	Thumb Crotch to 3rd Finger	9.9	8.8	6.6	4.4	3.3
2.6	6.7	U16 V	Thumb Crotch to Thumb Tip	5.0	4.5	3.4	2.2	1.7
3.5	9.0	U17 D	Anterior-Post. Arm at Biceps	6.8	6.0	4.5	3.0	2.3
3.5	8.8	U18 D	Anterior-Posterior at Forearm	6.6	5.9	4.4	2.9	2.2
1.7	4.3	U19 D	Depth of Wrist	3.2	2.9	2.2	1.4	1.1
2.2	5.5	U20 T	Breadth of Wrist	4.1	3.7	2.8	1.8	1.4
3.9	10.0	U21 T	Breadth of Hand	7.5	6.7	5.0	3.3	2.5
3.5	9.0	U22 T	Breadth of Palm	6.8	6.0	4.5	3.0	2.3
			LOWER EXTREMITY MEASUREMENTS					
41.7	106.0	L1 V	Crest of Ilium to Sole	79.5	70.7	53.0	35.3	26.5
39.0	99.0	* L2 V	Ant.Sup.Spine of Ilium to Sole	74.3	66.0	49.5	33.0	24.8
35.4	90.0	L3 V	Greater Trochanter Femur-Sole	67.5	60.0	45.0	30.0	22.5
35.4	90.0	* L4 V	Symphysis Pubis to Sole	67.5	60.0	45.0	30.0	22.5
33.5	85.0	* L5 V	Crotch to Sole	63.8	56.7	42.5	28.3	21.3
32.7	83.0	L5 V	Gluteal Furrow to Sole	62.3	55.3	41.5	27.7	20.8
19.7	50.0	L6 V	Posterior Knee Crease to Sole	37.5	33.3	25.0	16.7	12.5
18.5	47.0	L7 V	Tibia at Knee to Sole	35.3	31.3	23.5	15.7	11.8
2.8	7.0	L9 V	Med. Malleolus to Sole	5.3	4.7	3.5	2.3	1.8
2.2	5.5	L10 V	Lat. Malleolus to Sole	4.1	3.7	2.8	1.8	1.4
13.0	33.0	L11 T	Breadth of Hips, Sitting	24.8	22.0	16.5	11.0	8.3
6.5	16.6	L12 T	Breadth Mid Thigh	12.5	11.1	8.3	5.5	4.2
4.5	11.5	L13 T	Breadth of Knee	8.6	7.7	5.8	3.8	2.9
4.3	11.0	L14 T	Breadth of Calf	8.3	7.3	5.5	3.7	2.8
2.4	6.0	L15 T	Breadth Ankle	4.5	4.0	3.0	2.0	1.5
6.7	17.0	L16 D	Mid Thigh, Anterior-Posterior	12.8	11.3	8.5	5.7	4.3
5.9	15.0	L17 D	Knee, Anterior-Posterior	11.3	10.0	7.5	5.0	3.8
4.7	12.0	L18 D	Mid Calf, Anterior-Posterior	9.0	8.0	6.0	4.0	3.0
3.1	8.0	L19 D	Ankle, Anterior-Posterior	6.0	5.3	4.0	2.7	2.0
10.6	27.0	L20 D	Length of Foot	20.3	18.0	13.5	9.0	6.8
8.3	21.0	L21 D	Heel to 5th Toe	15.8	14.0	10.5	7.0	5.3
4.1	10.5	L22 T	Breadth Foot	7.9	7.0	5.3	3.5	2.6
2.4	6.0	L23 T	Width of Heel	4.5	4.0	3.0	2.0	1.5
		** These are also listed with Vertical Measurements.*						

Illustration of College Age Youth, by Avard Fairbanks

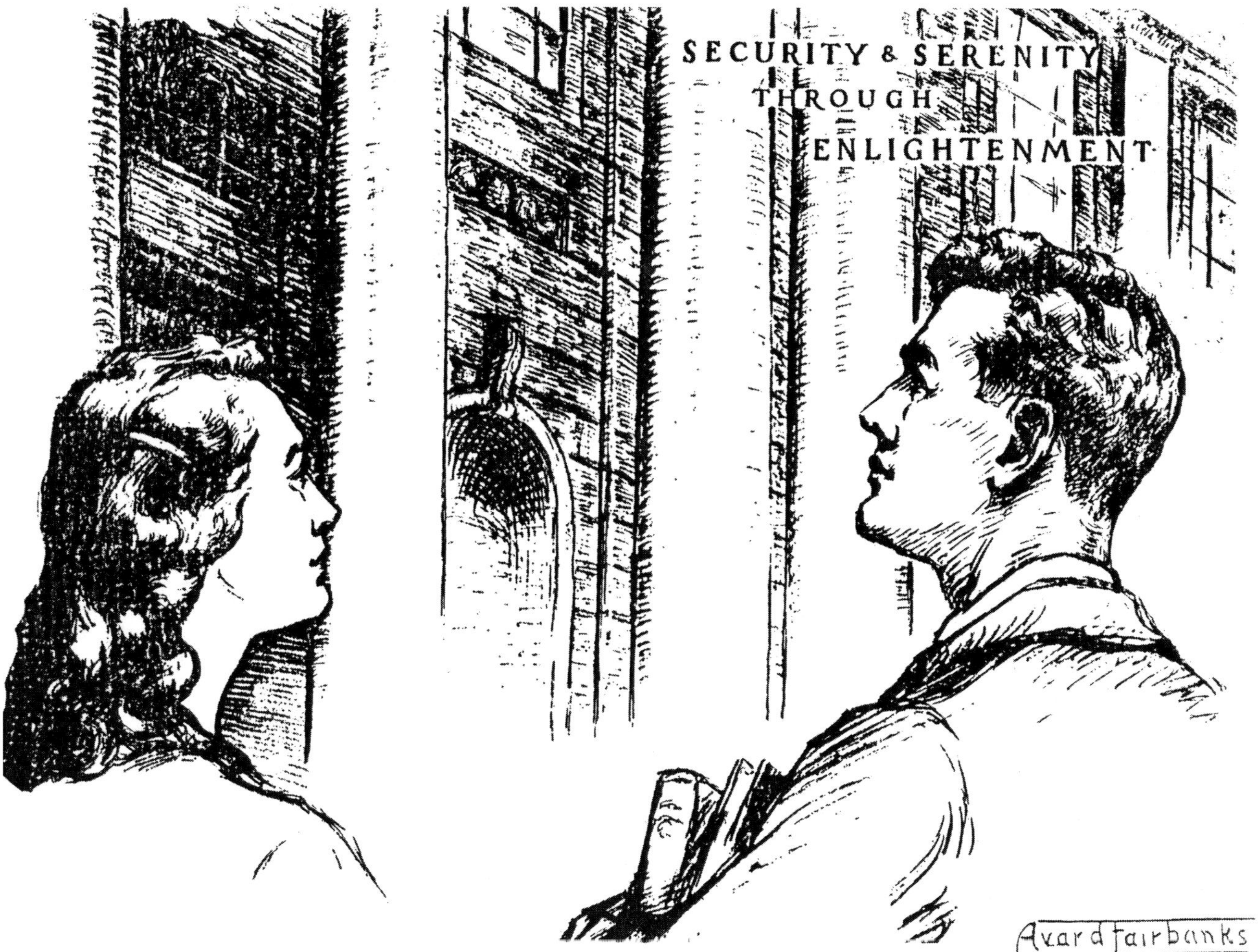

Appendix

SKELETAL TERMINOLOGY

A drawing of a human skeleton is included with a list of anatomic features for better comprehension. This list will attempt to correlate the common names of these bones with anatomic designations, starting from the neck and moving down. Not all of the bones will be listed. It is intended to be limited to those referred to in the discussion of proportions. Joints are designated by the bones they join, and the bone closest to the central body is usually named first.

The **clavicle** is known as the collar bone. It is attached to the sternum, or breast bone, near the midline and to the acromion process of the scapula, or shoulder blade, near the point of the shoulder.

The **sternum** is an anterior midline bone of the chest, called the breast bone. The upper ribs attach to it.

The **sternal notch** is an indentation of the upper end of the sternum between articulations of the sterno-clavicular joints.

The **acromioclavicular** joint is at the shoulder joining the clavicle, or collar bone, with the acromion process, a projection of the scapula, shoulder blade.

The **acromion process** is a forward projection of the scapula, or the shoulder blade, at the shoulder. It is part of the socket of the shoulder.

The **shoulder pivot** is on the lesser tuberosity, an anterior prominence at the head of the humerus, or arm bone. It can be felt at the anterior aspect of the humerus at the shoulder.

The **humerus** is the upper arm bone with joints at the shoulder and at the elbow with the forearm bones, the radius and ulna. The **medial epicondyle** is medial and the **lateral epicondyle** is lateral on the humerus near the elbow joint.

The **xiphoid** process is a small bone, of variable size, at the lower end of the sternum. It is firmly attached and serves as an attachment for abdominal muscles.

The **inferior costal margin**, or rib margin, is the lower border of the rib cage. The lateral border is the 10th and 11th ribs.

The **lateral epicondyle of the humerus** is a palpable prominence on the lower end of the humerus at the elbow that serves to locate the elbow pivot.

The **radius** and **ulna** are forearm bones.

The **spine of the ulna** is a prominence on the lateral aspect of the wrist

The **spine of the radius** is a palpable prominence at the wrist, proximal to the thumb. It is the location of wrist pivot. See details in the section on hands and feet.

The **pelvic girdle** is composed of the **sacrum,** the lower portion of the spine, at the posterior aspect that attaches to both **ilia** at the **sacroiliac joints**.

Each **ilium** is fused to the **pubis** and **ischium** bones early in development. Although these fused bones appear like one bone, the parts retain their individual names. The two pelvic bones join anteriorly at a firm, fibrous joint, the **symphysis pubis**. The upper border of the **symphysis** is a significant point of reference. Leonardo da Vinci called it the base of the genitals. The **crest of the ilium** is lateral, easily palpable, and is just below the belt line.

The **anterior superior spine of the ilium** is a prominence, anterior to, medial to, and a little below the level of the crest. It is easily palpable.

The **femur** is the thigh bone. The **greater trochanter** is easily palpated on the lateral aspect and is about on the level of the hip pivot. The distance between the trochanters is the widest measurement of the hips.

Human Skeleton, Anterior View

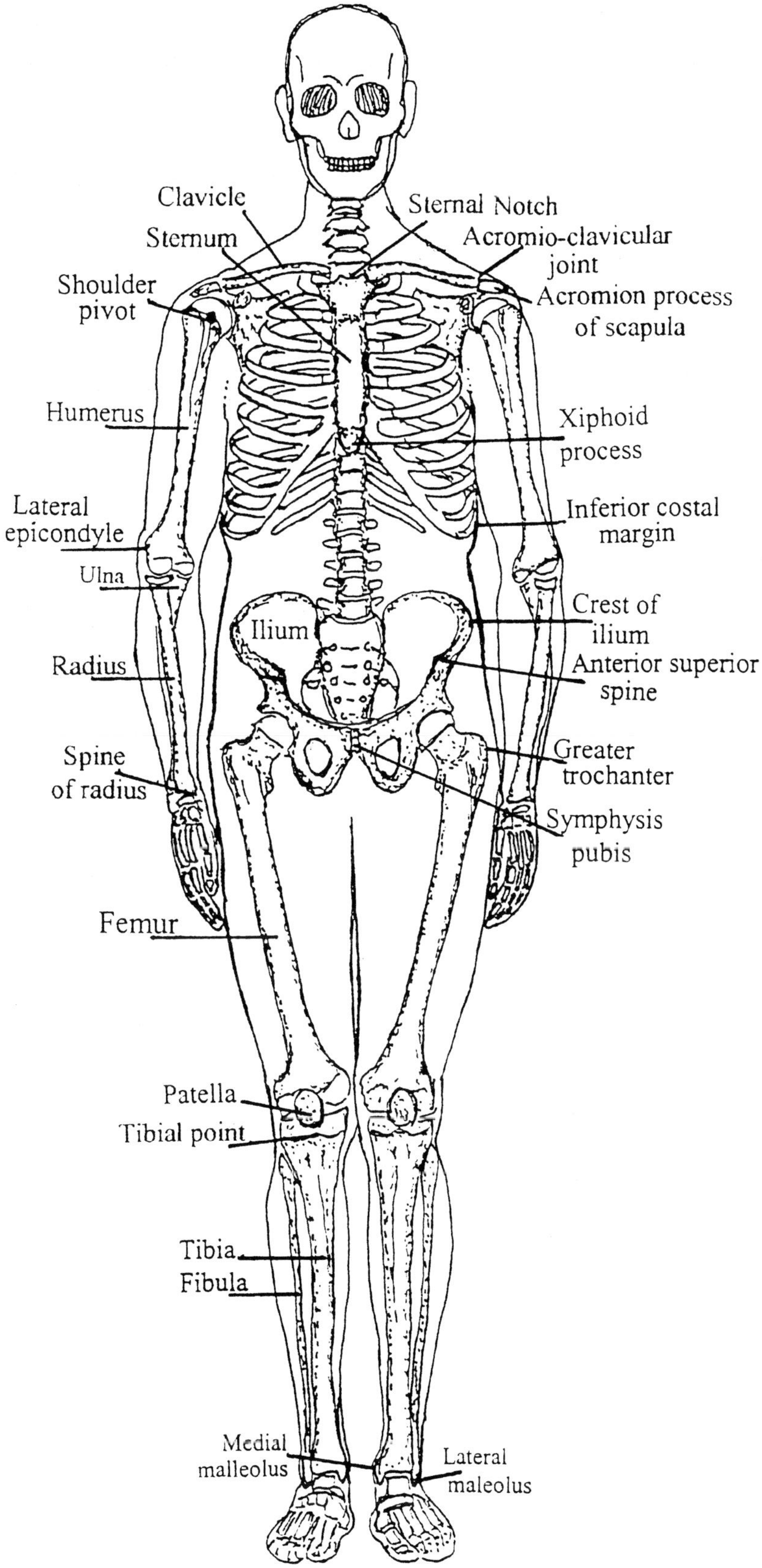

The Thorax and Shoulder Girdle, Viewed from Above

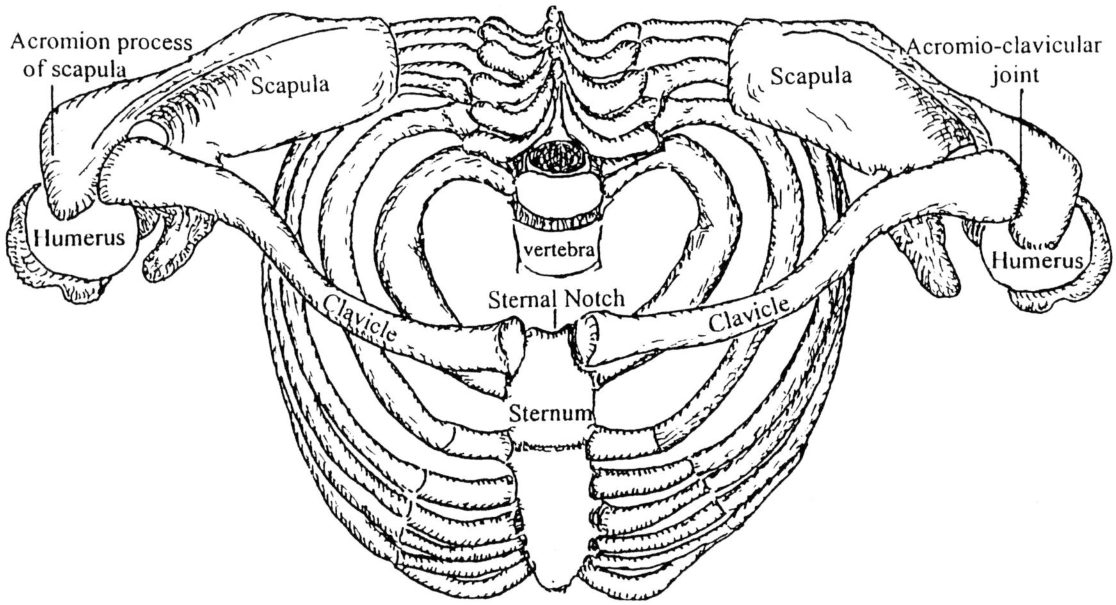

The **patella** is the knee cap. On a straight leg, the patella center is at the pivot.

The **tibial point** is the palpable ridge of the tibia at the knee joint. The measurement from this point is more accurate than measuring from the patella, since the patella moves on flexing the knee

The **tibia** and **fibula** are lower leg bones.

The **medial maleolus** is a prominence of the tibia at the ankle joint. The **lateral maleolus** is the lower tip of the fibula.

References

Backhouse, Kenneth M., Hutchings, Kenneth M. *Color Atlas of Surface Anatomy.* Baltimore: Williams and Wilkins 1986.

Huber, G. Carl. *Piersol's Human Anatomy.* Philadelphia: J.B. Lippincott Company, 1930.

Lockhart, R.D., Hamilton, G.F., Fyfe, F.D. *Anatomy of The Human Body.* Philadelphia: J.B. Lippincott Company, 1972.

Sbotta, Johanes, McMurric, J. Playfair. *Atlas of Human Anatomy.* New York: G.E. Strechert & Co., 1930.

Human Skeleton, Posterior View of Trunk

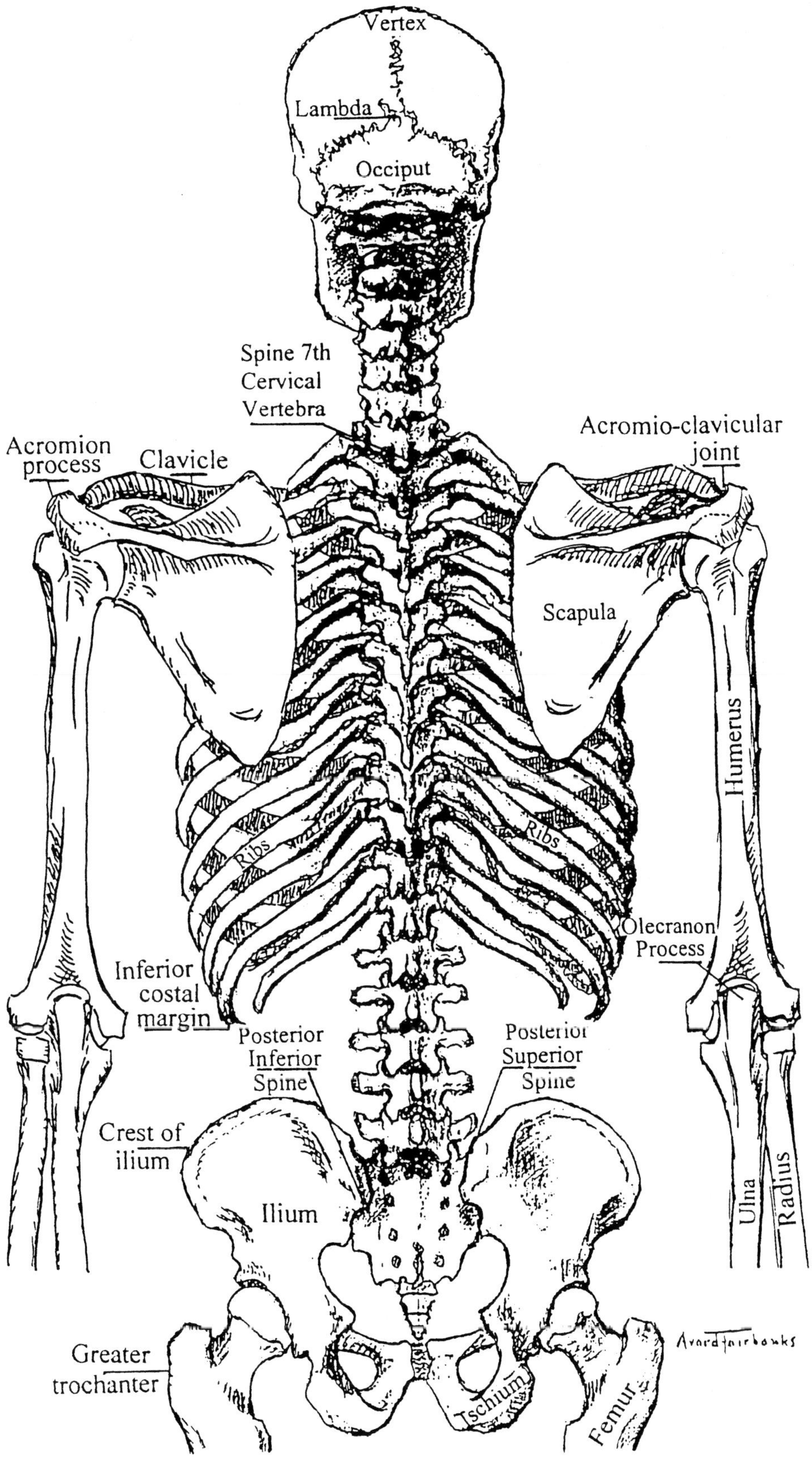

ANTHROPOMETRY

Points of Reference on the Head

Physical anthropologists are also interested in the measurements of the human body. Anthropometry is a subspecialty of anthropology, and it is directed to human measurements. Terms have been used to describe various points of reference to standardize information. When considering the skull, references are now made to the neurocranium, the portion containing the brain, and the viscerocranium, essentially the face. For simplicity, in this study, the skull will be construed to mean both the neurocranium and the viscerocranium. Precise points can often be determined by observing junctions of suture lines on some ancient skull. Sculptors deal with live human form and must try to use reference points that are covered with soft tissues. To facilitate understanding, these points will be described and illustrated in depicting a skull. It is interesting to note that Leonardo da Vinci chose many points that are now in use by anthropologists. Of prime interest to the serious artist is the human head. Beginning the list is the top of the head, the vertex.

Vertex is the uppermost portion of the skull vault in the midline, a little posterior to a vertical line to the ear.

Bregma is the junction of the midsaggital suture and the coronal sutures joining the frontal and parietal bones. It is anterior to the vertex and is often in a groove or depression. It is the soft spot or anterior fontanel, a quadrangle, of an infant skull until bone fusion occurs by growth.

Parietal Prominences are indefinite prominence of the parietal bones at the widest part of the cranium, posterior and above the ears. Biparietal measurement is between these two prominences.

Glabella is the midpoint of the frontal bones above the nasal root, essentially the midline of the brow.

Zygoma, or **zygomatic arch**, the cheekbone, is at the greatest breadth of the face. It is about the same breadth as the biparietal breadth of the upper part of the skull.

Nasion or **Selion** is the point at which the nasal bones join the frontal bones, commonly called the bridge of the nose. It is usually an indentation below the brow. The exception is in some ethnic groups with a high nasal bridge. In sculpture history, it is noted as the classic Greek nose.

Ala nasae, or wings of the nose, these semicircular structures are supported by cartilage and outline the nostrils.

The **subnasal point** is at the upper lip juncture and the nose. On a skull, it is a definite projection that supports the cartilage of the nasal septum in the midline.

Gonion is the angle of the jaw. Rather than a definite point, it is rounded, but it is a bony landmark that is easily palpable.

Menton is a designation for the bony border of the chin in the midline. Soft tissues on living subjects serve as a cushion in front and below the bone.

Porion is the ear canal. On living subjects, the tragus, the cartilaginous flap just anterior to the ear canal, is essentially the same point for anterior-posterior measurement.

Lambda of the posterior skull, so named for a Greek letter of three intersecting lines, is a slight indentation at the posterior skull, the location of three sutures that join. It is a remnant of the posterior or triangular fontanel of infancy.

Anthropometry Reference Points on the Head

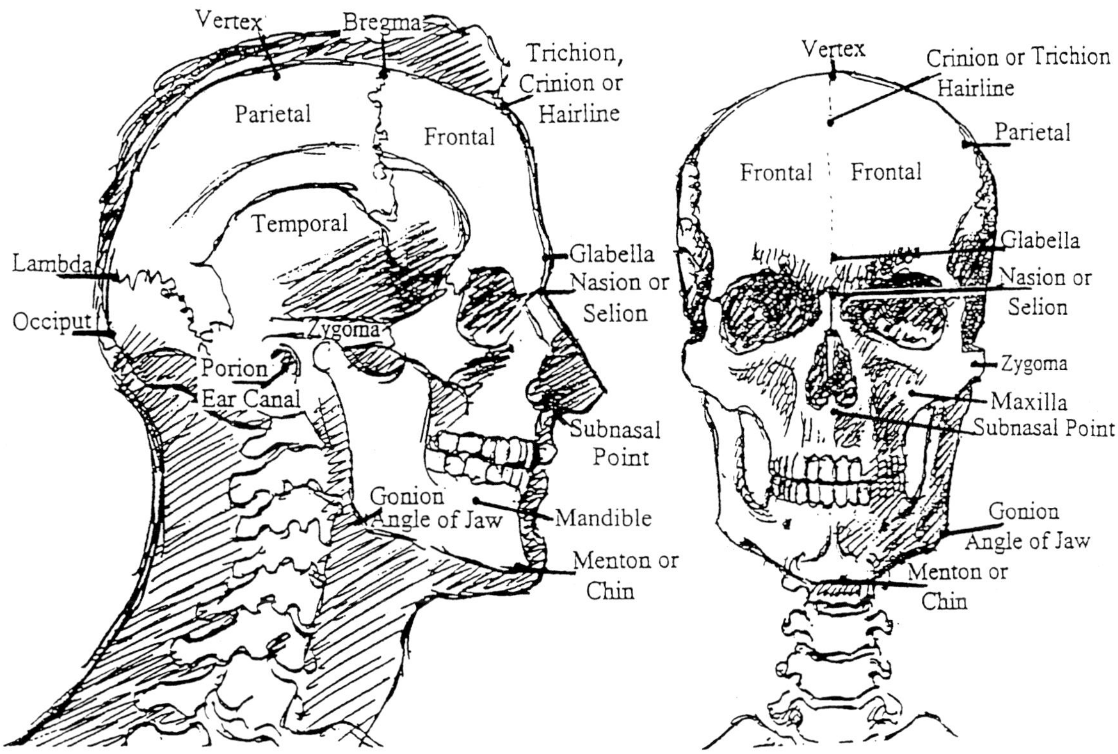

Occiput is the maximum posterior point of the skull.

There are other designated points for measurements of interest to anthropologists but of less significance to artists and therefore will not be listed. Many soft tissue features are supported by bony features and the localization of these need to be recognized. Other soft tissue features of living subjects have common names. Since they need the attention of serious students and amateurs, they are indicated in illustrations.

Reference

Hrdlicka, Arles. *Practical Anthropometry.* Philadelphia: Wistar Institute of Biology and Physiology, 1939. p 122-125

HAND FEATURES

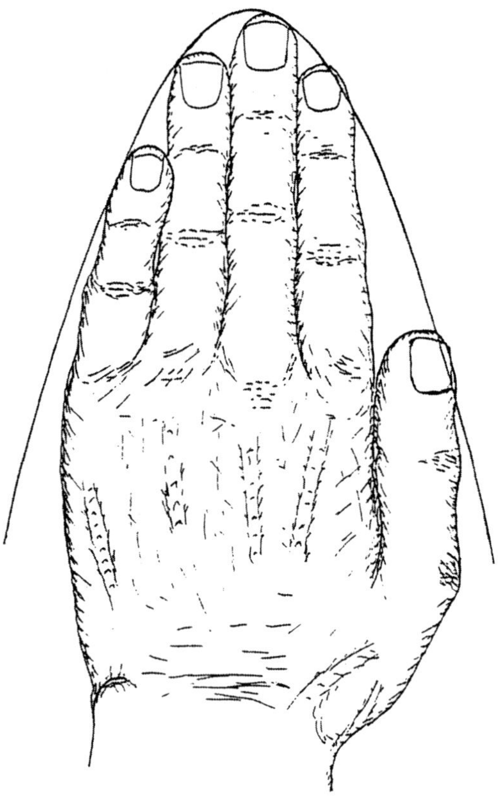

The outline of a hand conforms to a parabola.

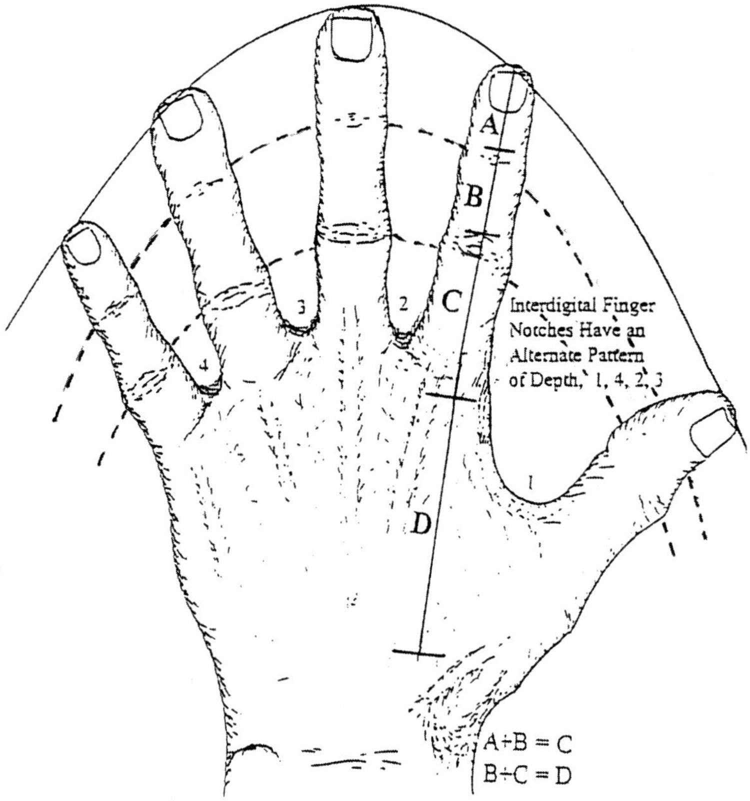

The outline of fingertips of the spread hand conforms to a parabola.

Since the hands are an important part of the expression of an artistic endeavor, many details, often overlooked, become important. Attention to special features may increase an intended message or interpretation. The overall fingertip outline of a hand, extended flat with the fingers together, conforms to a parabola. Lines connecting creases over the interphalangeal joints also resemble parabolic curves. When the fingers are spread, the lines connecting the fingertips and the interphalangeal joints still conform to parabolas, but to more broad parabolic curves. Even when fists are clenched, there is a broad parabolic curve evident bordering the knuckles.

On a dorsal, view of the hand in a gently relaxed position, all of the fingers are generally in the same plane, but the plane of the thumb's axis is at about a 45 to a 90 degree angle to the plane of the fingers. When all fingers of the hand are extended, the index finger is gently curved toward the middle finger. The thumb tip is near the middle of the index finger's proximal phalanx. The fingernails have an arch side to side and a slight curve from cuticle to fingernail tip. Fingernails are generally about half the length of the terminal phalanx.

Dynamic symmetry is present in the human hand, resulting in a very beautiful example of a summation series. The sum of the distal two phalanges of each finger is equal to the proximal phalanx that is one-half the length of a finger. The sum of the middle phalanx and the proximal phalanx equals the corresponding metacarpal in the hand. When the fist is clenched, the fingers form whorls resembling a snail shell. This configuration enables humans to grasp a tool or clench a fist.

THE FEET

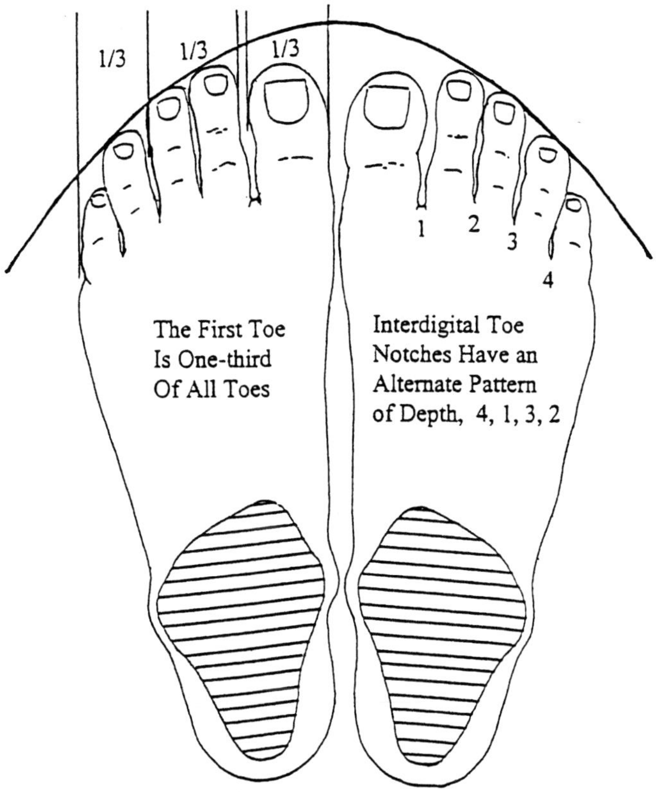

The outline of the feet conforms to a parabola.

When the two feet are together, the outline of the toes conforms to a parabola. An extended hand nearly equals three-fourths length of a foot. The width of a hand nearly equals the width of a foot. The breadth of a heel is about sixty-percent the foot breadth. The notches of toes have an alternation in a manner resembling the fingers. However, the deepest notch is the fourth to fifth toe and the next deepest is between the first and second toe. The third to fourth notch is nearly as deep as first to second and the shallowest notch is between second and third toes. As in the hand, the webs between the toes slope up, sole to dorsal, at about a 45-degree angle. The widths of each of the toes are nearly equal except for the first toe, which is twice the width of any other toe. When drawing or modeling, divide the toes in three parts, one for the big toe and one part for each other pair of toes. Other notches are then applied. Often, the second toe is longer than the big toe. The fifth toe sometimes curls under the fourth toe. The instep of the foot, the medial aspect of the sole, has a longitudinal arch. Above, on the dorsum of the foot, a transverse arching plane slopes from medial to lateral. The big toe has a terminal up-turn. The other four toes have first an up-turn and then a down turn.

PORTRAIT DETAILS OF AN EAR

In describing and discussing an ear, it is important to define the parts of an ear. The whole external ear is called a pinna. The outer border is called the helix. The groove inside the helix is the fossa of the helix. The ridge inside the helix is named the antihelix. The flap anterior to and partly covering the ear canal is the tragus. A notch above the tragus is named the tragion. Posterior and below the tragion is the antitragus that joins the antihelix. Between the tragus and the antitragus is a definite notch, the intertrageal notch. The cavity between the tragus and the antihelix is the concha, named for the resemblance of a sea shell. These structures are supported by cartilage, a firm flexible type of tissue. Below the tragus and antitragus is the ear lobule. Since it contains fibrous and fatty tissue, but no cartilage, it is soft and pliable.

The highest point of the ear, in teens and adults, is nearly the same level as the brow. The lobule and lower border is on the same level as the subnasal point. The helix originates about the middle half of the ear, just posterior to the upper border of the tragus. The helix, or outer border of the ear, begins directed anteriorly, curls up and out away from the temporal bone. It curves around the apex of the ear to become the outer border and it then descends nearly vertically, curving slightly medial as it approaches the lobule. Its course, similar to a whorl, resembles the coil of a snail shell, and it is a manifestation of dynamic symmetry.

The general outline is like a question mark, a character that was apparently so formed to resemble an ear. As the helix descends it curves inward to join the lobule. Anterior to the ear's posterior margin is another prominent ridge, the anti helix. That begins at the antitragus and is opposite to the external auditory canal. It progresses up and inward and divides like a "Y" into two ridges that approach the upper anterior concavity. Between the divisions of the anti helix is the antihelix fossa.

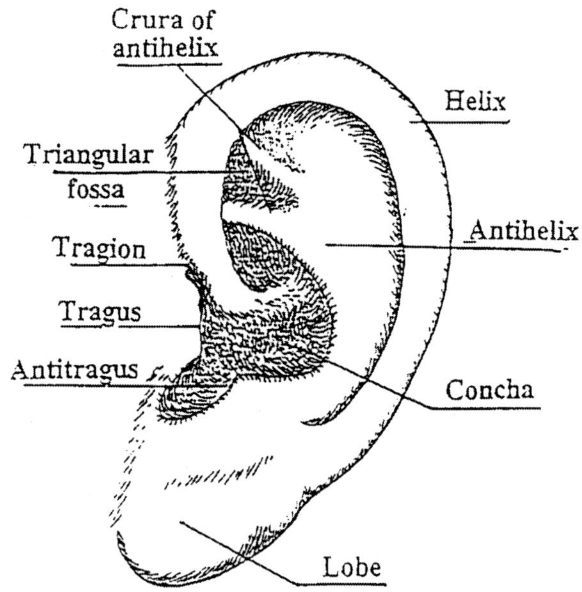

Terminology of Ear Features

The tragion is a small notch above the tragus between the tragus and the beginning of the helix. The tragion is approximately one-half the distance between the brow and the occiput. As a surface feature the tragion indicates the ear canal, it identifies the porion, the external auditory canal of the ear at the base of the cranium, and is a relatively fixed point during head growth. The position of the porion at the base of the skull is relatively unchanged during other facial growth and development.

In adults, the width of the pinna, the external ear, is about one-half its vertical height. Vertical growth is gradually from infancy until about age 13 when the ear resembles adult ear proportions. Cartilage in the nose and ears grows slowly in adult life. With advancing age it continues to grow vertically, gradually and becomes more flexible. Observation of these features is a common gauge of estimated age.

ANATOMY OF THE NOSE

ANTERIOR VIEW

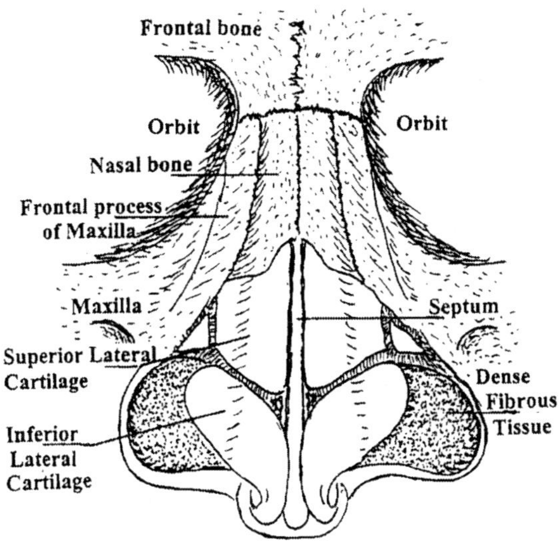

LATERAL VIEW

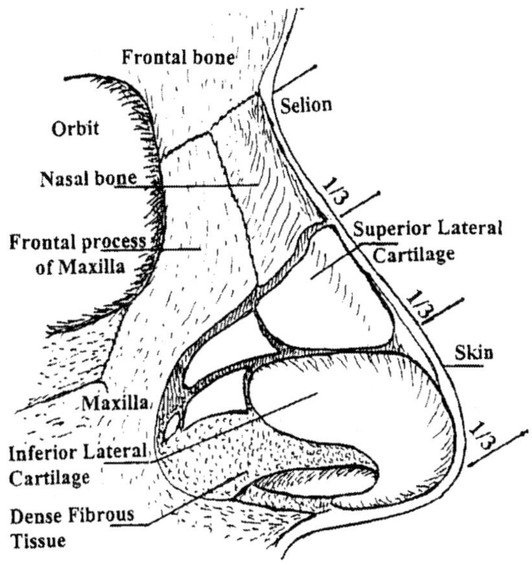

A better appreciation for the form of the nose involves a study of its development and anatomy. The purpose of the nose is to warm and moisturize air on the way to the lungs. The anatomy of the nose consists of some bone and some cartilage. The septum of the nose acts like a central longitudinal pillar that helps to support the lateral cartilages.

The supporting lateral structures are composed of three segments of joined, paired components. Uppermost, there is the lower border of the frontal bones of the skull that forms the medial aspect of the brow. It curves down and posterior to the selion, or saddle, of the nose. It joins the paired nasal bones there.

The upper third of the nose is composed of the nasal bones. These join the middle third of the nose with a dense fibrous tissue. The middle third consists of the paired superior lateral cartilages that slant laterally and are triangular in form. The bases of the triangles are along the leading edge or upper border of the nose. The apices of these triangles are down and lateral.

By palpating the flat leading border of the nose, these cartilage edges, separated by the septum, can be felt. The lower segment to the tip of the nose is supported by paired inferior lateral cartilages. These are attached to the superior lateral cartilages and the maxilla, wherever adjacent, by dense fibrous ligaments.

The nostrils in the lower third are supported by curved cartilages, somewhat elliptical, that form the bulbous tip and the flair of the wings or ala of the nose. The vertical central support is the septum that is composed of cartilage. At the subnasal point, there is a bony prominence of the maxilla just above the upper lip. It is essentially a pedestal for the cartilaginous septum of the nose in the midline.

GROWTH SUMMARY

At birth an infant's length, height is 51 cm, or 20 inches. When compared to a mature adult male of height 183 cm, or 72 inches, **the length of an infant is about 28% or about 1/4 an adult stature.** In the first few months an infant grows rapidly. The total stature growth of a person to adult is about four times its infant's stature. Leonardo da Vinci observed that a three-year-old child is one half adult's height. The following tables demonstrate the changes from birth of a male infant to fifteen-years of age. Female measurements are similar

Stature Comparisons

	Height or Length		Increments of Stature Growth	Percent of Adult Stature	Approximate Ratio of Adult
	Inches	cm			
Newborn Infant	20	51	1.00	28	1/4
3-Months-Old Infant	24	61	1.20	33	1/3
6-Months-Old Infant	27	69	1.37	38	2/5
1-Year-Old Infant	30	76	1.49	41	2/5
3-Year-Old Child	37	95	1.87	52	1/2
5-Year-Old Child	43	110	2.16	60	3/5
8-Year-Old Child	50	127	2.49	69	2/3
10-Year-Old Youth	54	138	2.70	75	3/4
12-Year-Old Youth	59	151	2.96	82	4/5
15-Year-Old Youth	66	169	3.31	92	9/10

Comparisons of Head Heights

	Inches	cm	Increments of Growth	Percent of Adult	Ratio of Adult
Newborn Infant	5	13	1.00	56	1/2+
6-Months-Old Infant	6.3	16	1.23	70	3/4–
1-Year-Old Infant	6.7	17	1.31	74	3/4
2-Year-Old Infant	6.9	17.5	1.35	76	3/4
5-Year-Old Child	7.3	18.5	1.42	80	4/5
8-Year-Old Child	7.7	19.5	1.5	89	9/10
10-Year-Old Youth	7.9	20.0	1.5	98	1
12-Year-Old Youth	8.3	21.0	1.6	100	1

The growth of the head, from the vertex to the chin, is more rapid than body growth in the first five years. **An infant's head height at 13 cm is about one half an adult's head height at 23 cm.**

Comparisons of Torso / Sitting Heights

	Comparisons of Torso				Sitting Heights			
	Inches	cm	Adult%	& Ratio	Inches	cm	Adult %	& Ratio
Newborn Infant	7.5	19	36	1/3+	13.8	35	38	2/5
1-Year-old Infant	10.6	27	51	1/2–	18.5	47	51	1/2
2-Year-old Infant	11.8	30	57	3/5–	20.5	52	56	1/2+
3-Year-old Infant	12.2	31	58	3/5	22.8	58	63	2/3
5-Year-old Child	13.8	35	66	2/3+	24	61	66	2/3
8-Year-Old Child	14.2	36	69	3/4–	27.2	69	72	3/4
10-Year-Old Child	16.5	42	79	4/5	28.7	73	79	3/4+
12-Year-Old Child	18	46	90	9/10	31	79	86	4/5+
15-Year-Old Child	20	51	95	1–	36.6	86	89	9/10+

Hand to Face and Foot to Head Comparisons

In Leonardo da Vinci's system of adult relative proportions, he demonstrated that the height of the face, chin to hairline, is approximately equal to the length of a hand, and the head height nearly equals the length of the foot. In children, these comparisons are not applicable. The hands, the feet, and the head are growing at different rates.

At birth and during early development a baby's hands are short and stubby. The breadth exceeds the metacarpal length. Digits and the segments or phalanges are also short. The lengths of the segments of each finger have a progression similar to a Fibonacci series, like an adult hand. The distal phalanx plus the middle phalanx equals the proximal phalanx; the middle phalanx plus the proximal phalanx equals the length of the metacarpal of the hand. The table shows a comparison of the hand length to face height.

Wrist to Fingertip Comparison with Chin to Hairline Height

Age	Hairline to Chin cm	Hand Length cm	Expressed in Percentage	Ratio
Birth	10	6.5	65	2/3
6 Months	12.2	8.0	65	2/3
1 Year	13.0	9.0	68	2/3+
2 Years	14.0	9.8	70	3/4–
3 Years	14.7	11.5	75	3/4
5 Years	15.0	12.0	78	3/4+
8 Years	15.8	13.8	85	4/5+
10 Years	16.4	15.0	91	9/10
13 Years	17.0	16.8	99	1
18 Years	19.0	19.0	100	1

The hand length is shorter than the face height until early teenage. The cranium grows more than the face during the first five years. After five years of age, dentition progresses and the height of the face height increases. These changes are subtle, but by early teenage, the hand length approximates face height. After age thirteen, in both sexes, the measurements are approximately equal. The table shows a comparison of the foot length with the height of the head.

Foot Length Compared to Head Height

Age	Vertex to Chin cm	Foot Length cm	Expressed in Percentage	Ratio
Birth	13	8	62	2/3
1 Year	16.5	11	75	3/4
2 Years	17.5	13	89	9/10
5 Years	18.5	16.8	93	9/10+
8 Years	19	19.5	103	1+
14 Years	21	25.5	120	6/5

At birth, the foot length is less than head height, but parents discover that growth is rapid in childhood. Throughout youth and early adulthood, the foot measurement continues to be slightly greater than head height.

Growth Comparisons of Head to Umbilicus and of Umbilicus to Sole

Age Years	Vertex To Sole cm	Vertex to Umbilicus cm	Umbilicus to Sole cm	Vrtx. to Umblicus ÷ Vrtx. to Sole Percent	Umb. to Sole ÷ Vrtx. to Sole Percent
Newborn	51	27	24	**53**	**47**
1 Year	75	36	39	48	52
2 Year	87	42	45	48	52
5 Year	110	49	61	45	55
8 Year	127	54	73	43	57
12 Year	150	60	90	40	60
18 Year	177	70	107	**40**	**60**

Note! During the growth of a child there is an interesting change in the percent of vertex to umbilicus measurement compared to the stature, and the umbilicus to sole measurement compared to the stature. That comparison begins in infancy as **53%** to **47%**, and is nearly equal. As legs grow faster than the torso it changes to a comparison of **40% to 60%** at eighteen-years of age. That is similar to the Golden Mean, or about two thirds.

Growth Comparisons of Head to Pubis and to Pubis to Sole

Age Years	Vertex to Sole cm	Vertex to Sym. Pub. cm	Sym. Pubis to Sole cm	Vrtx.-Pubis ÷ Pubis.-Sole Percent	Pubis-Sole ÷ Vrtx.-Sole Percent
Newborn	51	32	19	**63**	**37**
1 Year	75	45	30	60	40
2 Year	87	51.5	33.5	59	41
5 Year	110	60	50	55	45
8 Year	127	65	62	51	49
12 Year	150	74	76	49	51
18 Year	177	87	90	**49**	**51**

There is **an interesting reversal** of percentages; the vertex to the symphysis pubis compared to the stature, and the symphysis pubis to the sole compared to the stature. Those ratios begins at **63%** to **37%** in infancy, near the Golden Mean, and as legs grow, the ratios progress to become almost equal at **49%** to **51%**, when reaching eighteen-years of age. **Essentially, in infancy, the vertex to umbilicus nearly equals the umbilicus to sole; and in adulthood, the vertex to symphysis pubis nearly equals the symphysis pubis to sole.**

Comparison of Arm Length: Infants to Eighteen-Year-Old Youth

	Inches	cm	Increments of Growth	Percent of Adult	Ratio of Adult
New born Infant	8.25	21	1.00	25	**1/4**
6-Month-Old Infant	10.2	26	1.24	31	1/3
1-Year-Old Infant	11.4	29	1.38	35	1/3+
2-Year-Old Infant	14.2	36	1.71	43	2/5
3-Year-Old Infant	15.7	40	1.90	48	1/2
4-Year-Old Child	17.3	44	2.10	53	1/2+
5-Year-Old Child	19	48	2.29	58	3/5
8-Year–Old Child	21.7	55	2.6	72	3/4
10-Year-Old Youth	24.4	62	2.9	77	3/5
12-Year-Old Female	25.2	64	3.0	85	4/5+
Male	25.2	64	3.0	79	4/5
15-Year-Old Female	27.6	70	3.3	93	9/10+
Male	28.4	72	3.4	89	9/10
18-Year-Old Female	27.6	70	3.3	93	9/10+
Male	31.5	80	3.8	100	1.0

Measurements are from the acromion process to the fingertip. **An infant's arm length is about one-fourth an adult male's arm length.**

Comparisons of Leg Length: Infants to Eighteen-Year-Old Youth

	Inches	cm	Increments of Growth	Percent of Adult	Ratio of Adult
New born Infant	7.9	20	1.00	16	**1/6**
6-Month-Old Infant	10	25	1.25	20	1/5
1-Year-Old Infant	11.8	30	1.50	24	1/4
2-Year-Old Infant	15	38	1.90	30	1/3–
3-Year-Old Child	17	43	2.15	34	1/3
4-Year-Old Child	18.5	47	2.35	37	2/5–
5-Year-Old Child	20.5	52	2.60	41	2/5+
8-Year–Old Child	25.2	64	3.15	51	1/2
10-Year-Old Youth	28	71	3.6	75	3/4
12-Year-Old Female	29.5	75	3.8	87	9/10–
12-Year-Old Male	30.3	77	3.9	81	4/5
15-Year-Old Female	32	81.5	4.1	95	9/10+
15-Year-Old Male	34.6	88	4.4	93	9/10+
18-Year-Old Female	32.3	82	4.1	96	1.0–
18-Year-Old Male	35.4	90	4.3	95	1.0–

The leg length of an infant, measured from the greater trochanter to sole at 20 cm, **is about one-sixth of an adult male's leg length** at 126 cm. At fifteen years of age, a female youth has gained nearly all of the adult female leg measurements. After fifteen-years of age, a male youth will expect additional leg growth and some after eighteen-years of age.

BODY MEASUREMENTS OF MATURE ADULT MALE, Age Twenty-Five to Thirty, for Comparisons

	Feature Measure		Feature Compared to Stature	Approx. Ratio of Feature
	Inches	cm		
Stature: Crown to sole	72	183		
Head Height: Crown to chin	9	23	13%	1/8
Sitting Height: Crown to rump	38	96	54%	1/2
Trunk Measure: Sternal notch to symphysis pubis	21	53	29%	1/3–
Leg Length: Greater trochanter of femur to sole	37	98	52%	1/2
Thigh Length: Greater trochanter to tibia at knee	18	47	26%	1/4
Tibia at Knee to Sole:	19	51	28%	1/4
Arm Length: Acromion process at shoulder to dependant fingertip	32	81	44%	2/5+
Hand Length:	8	20	11%	1/9
Foot Length:	11	28	15%	1/6

BODY MEASUREMENTS OF MATURE ADULT FEMALE, Age Twenty-Five to Thirty, for Comparisons

	Feature Measure		Feature Compared to Stature	Approx. Ratio of Feature
	Inches	cm		
Stature: Crown to sole	66	168		
Head Height: Crown to chin	8	20.4	12%	1/8
Sitting Height: Crown to rump:	35	89	53%	1/2
Trunk Measure: Sternal notch to symphysis pubis	19	49	29%	1/3–
Leg Length: Greater trochanter of femur to sole	34	86	51%	1/2
Thigh Length: Greater trochanter to tibia at knee	17	43	26%	1/4
Tibia at Knee to Sole:	17	43	26%	1/4
Arm Length: Acromion process at shoulder to dependant fingertip	30	75	45%	2/5+
Hand Length:	7	18	11%	1/9
Foot Length:	9.6	24	14%	1/7[1]

1. Fairbanks, A.T., *Anatomical Design, A Doctorate Thesis.* Ann Arbor, MI, The University of Michigan, 1936. Plate 94 and 95.

CONVERSION SCALE OF INCHES TO CENTIMETERS

inch x 2.54 = cm cm x 0.9 = 1 inch

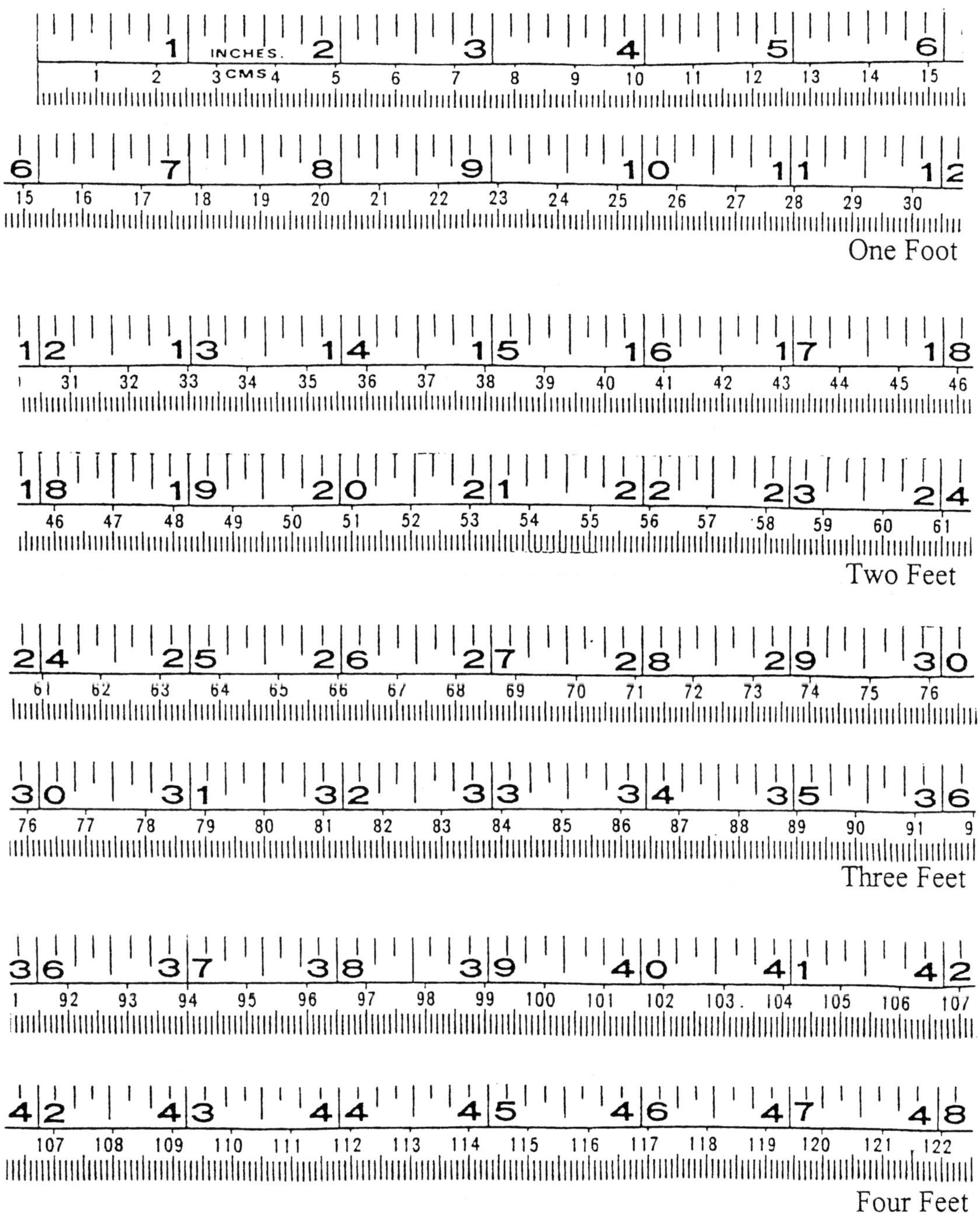

CONVERSION TABLES OF FRACTIONS AND PERCENT

The following list is for ease of comparison or conversion, when ratios are expressed in a fraction or in a percentage:

Ratio:	1/12	1/11	1/10	1/9	1/8	1/7	1/6	1/5	1/4	1/3	1/2
Percent:	8	9	10	11	12	14	16	20	25	33	50

Other common fractions may be useful when calculating percentages:

Ratio:	3/8	2/5	3/5	5/8	2/3	3/4	4/5	5/6	7/8
Percent:	38	40	60	63	67	75	80	83	88

CONVERSION SCALES TO DECIMAL INCHES

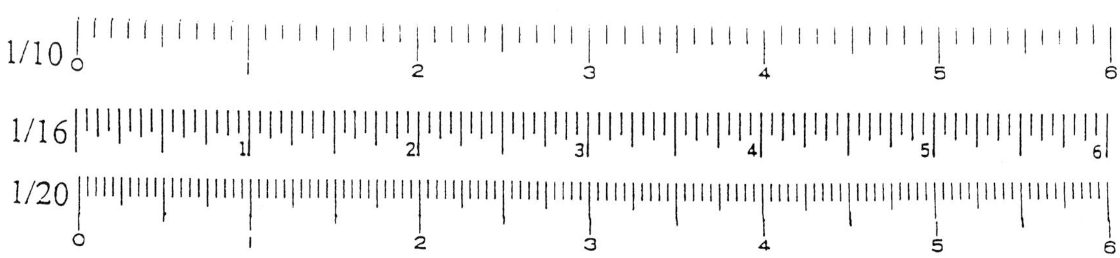

Conversions of Inch Fraction in 1/16ths to Inch Decimal

Inch fraction	1/16	1/8	3/16	1/4	5/16	3/8	7/16	1/2
Inch decimal	.06	.13	.19	.25	.31	.38	.44	.50

Inch fraction	9/16	5/8	11/16	3/4	13/16	7/8	15/16
Inch decimal	.56	.63	.69	.75	.81	.88	.94

GLOSSARY

Acromion process: the lateral point of the shoulder

Ala, Ala nasae: the lateral border of the nostrils, of the lower third of the nose, literally, wings of the nose.

Anterior aspect: ventral, the front of a standing human figure.

Anterior superior spine of ilium: a prominence of the ilium of the pelvic bone, anterior to the crest of the ilium.

Anthropometric, Anthropometry: measuring the human figure.

Axilla, Axillary: arm pit.

Axillary pivot: the center pivot for extension, or raising of the arm.

Braccio: an old Italian measure from the shoulder to the wrist.

Brachychepalic: a head configuration that is short, width to height, with a cephalic index of 80 to 86.

Brachymorphic: short, broad body build.

Carpals: bones in the wrist.

Caudal: also Inferior: toward the tail or coccyx.

Central: toward the central axis. The vertebral column which is generally vertical and generally but not entirely central. An axis is on a plane, but is not a planc, it is a theoretical straight line, but in body development it has developed gentle curves.

Cephalic: also Superior: toward the head.

Cephalic index: the ratio of the breadth to the vertical measurement of the head.

Cervical: refers to the neck.

Crest of the ilium: a prominent superior and lateral ridge of the pelvis bone that can be felt on the lateral aspect of the abdomen.

Cubit: an ancient unit of measure from the tip of the flexed elbow to the tip of the middle finger of the straightened hand.

Deep: in from the surface or skin.

Distal: a part of an extremity away or farther from the body.

Dolicocephlic: a type of head that is long and narrow with head width to height about index of 70 to 75.

Dolicomorphic: tall and slender body form.

Dorsal: the posterior aspect; the dorsum is the back of a standing human figure, the back of a hand and the top of a foot.

Frankfort Plane: A practical and theoretical plane for the base of the cranium, calculated from the ear canals to the lower borders of the eye sockets.

Frontal plane: any theoretical plane that intersects the erect human body from side to side, parallel to the central axis.

Glabella: The prominence of the brow in the midline of the forehead.

Gluteal cleft: the midline crease between each buttock.

Gluteal furrow: a prominent transverse crease below a buttock.

Gluteus maximus muscle: a prominent large muscle of the buttocks.

Greater trochanters: the widest prominences of the femurs, the hip bone.

Infra-: below, compare to supra, above.

Larynx: the voice box or Adam's apple, an anatomical feature of the neck composed primarily of the thyroid cartilage

Lateral: is away from the midline or the mid-sagittal plane.

Malleolus: a bony prominence at the ankle above and on each side of a heel, both medial and lateral.

Mastoid process: a bony prominence pointing down, just behind an ear.

Medial: is toward the midline or the mid-sagittal plane.

Mesocephlic: medium head, a head width to height index of 75 to 80.

Mesomorphic: medium body build.

Metacarpals: bones in the hand in the palmar area.

Metatarsals: bones in the arch of a foot.

Mid-saggital plane: a theoretical plane that intersects the human body from front to back through the central axis.

Nasion: the junction of the nasal bone and the frontal bone near the midline of the skull just below the glabella, or brow. It is very close to the sellion or saddle of the nose, also called bridge, or base of the nose.

Occiput: the low posterior prominence at back of the skull.

Olecranon process: the tip of the flexed elbow.

Plantar: the bottom of a foot.

Patella: the knee cap.

Pectoralis muscles: anterior chest muscles inserting on the upper part of a humerus.

Phalanx, Phalanges: segments or bones of digits, of fingers or toes.

Porion: a point on upper border of the opening of the ear canal on the side of a skull.

Posterior aspect: dorsal, it is the back of a standing human figure.

Pronasal point: the tip of the nose.

Proximal: a part of an extremity near or closer to the body.

Sellion: the soft tissue saddle of the nose, also considered the base or bridge of the nose. It is just inferior to the nasion, which is at the junction of the nasal bones and the frontal bone of the forehead. For practical purposes, they are nearly synonymous.

Septum of the nose: the vertical midline structure of the nose between the nostrils and between the nasal airways.

Sternal notch: the top of the sternum, or breast bone, between clavicles or collar bones.

Sternocleidomastoid muscle: a prominent muscle of the anterior neck originating at the sternal notch, directed laterally up, and inserting on the mastoid process behind an ear. These paired muscles assist in turning the head.

Subnasal point: the junction of the upper lip with the septum of the nose.

Superficial: is above, on, or toward the surface.

Supra-: above.

Symphysis pubis: a midline pelvic junction of both pubic bones of the pelvis.

Tarsals: bones in the ankle.

Tragus: a cartilaginous prominence or flap anterior to the ear canal.

Tragion: a soft tissue point just above the tragus along inferior arc of the helix of an ear. It is superficial to the upper external ear canal and indicates the porion of the skull.

Transverse plane: any theoretical horizontal plane that intersects the erect human body at right angles to the midsaggital plane (that is front to back).

Truncated angle: an angle whose apex of has been intersected, by a line or a plane.

Umbilicus: belly button.

Ventral aspect: anterior, it is the front of a standing human figure.

Xiphoid process: a midline bony prominence below and attached to the sternum or breast bone .

Zygomatic arch: a lateral cheek bone in front of an ear.

Symbols:

=	Equal to.
±	Approximately Equal to.
≠	Not Equal to
Φ	Phi, the ratio of 1 .62… in the Fibonacci series. The Golden Mean.

BIBLIOGRAPHY

Backhouse, Kenneth M., Hutchings, Kenneth M. *Color Atlas of Surface Anatomy.* Baltimore: Williams and Wilkins 1986.

Bakwin, Harry and Bakwin, Ruth Morris. "Body Build in Infants. V. Anthropometry in the Newborn." *Human Biology* 6 (1934): 612-624.

Breckenridge, Marion E., Murphy, Margaret N. *Growth and Development of the Young Child.* Philadelphia: W.B. Saunders Company, 1969.

Croney, John. *Anthropometry for Designers.* London: Bratsford, New York: Van Nostrand Reinhold, 1971.

Damon, Albert, Stoudt, Harry W., Mc Farland, Ross. *The Human Body in Equipment Design. Cambridge.* Boston: Harvard University Press, 1966.

Enlow, Donald H. *The Human Face.* New York, Evanston, London: Hoeber Medical Division, Harper and Row, 1968.

Enlow, Donald H., Ph.D. *Handbook of Facial Growth.* Philadelphia, London, Toronto: W.B. Saunders Company, 1975.

Fairbanks, Avard T. *Anatomical Design, A Doctoral Thesis.* Ann Arbor, Michigan. The University of Michigan, 1936.

Fairbanks, Jonathan L. "America's Measure of Mankind, Proportions and Harmonics." *Smithsonian Studies in American Art.* 2.1, 1988, 73-87.

Falkner, Frank. Tanner, J.M. *Human Growth, Vol II, Postnatal Growth.* New York: Plenum Press, 1978.

Goldstein, Marcus S. "Changes in Dimensions and Form of the Face and Head with Age." *American Journal of Physical Anthropology* xxii 1 (1936) 37-89.

Hrdlicka, Arles. *Old Americans, A Physiologic Profile.* New York: Arno Press and the New York Times, 1970.

Hrdlicka, Arles. *Practical Anthropometry* . Philadelphia: Wistar Institute of Biology and Physiology, 1939, 183.

Huber, G. Carl. *Piersol's Human Anatomy.* Philadelphia: J.B. Lippincott Company, 1930.

Krogman, Wilton Marion. "The Physical Growth of Children, Head, Face, Trunk and Limbs." *The Society for Research in Child Development*, Monographs. #136, Vol. 35 #3. Chicago: University of Chicago Press.

Krogman, Wilton Marion. "A Handbook of the Measurements and Interrelation of Height and Weight in the Growing Child." Evanston, Illinois: *Child Development Publications, Society for Research in Child Development.* 13 #3, Serial Number 48.

Krogman, Wilton Marion. "The Measurement of the Human Body." *Ciba Symposia*, (April-May, 1943) 1467-1477.

Lockhart, R.D., Hamilton, G.F., Fyfe, F.W. *Anatomy of the Human Body.* Philadelphia: J.B. Lippincott Company, 1972. (57)

Lowrey, George H. *Growth and Development of Children.* 8th ed., Chicago, London: Year Book Medical Publishers, Inc., 1986.

Martin, William Edgar. *Childrens' Body Measurements for Planning and Equipping Schools,* U.S. Department of Health Education and Welfare, 1953.

Moore, William M., M.D., Roche, Alex F., M.D. "Pediatric Anthropometry." *Ross Growth & Development Program* Columbus, Ohio: Ross Laboratories, 1983.

Moore, W.J., Lavelle, C.L.B., *Growth of the Facial Skeleton in Hominoidae.* London, New York, San Francisco: Academic Press, 1974.

Nelson, Waldo E., M.D., Vaughan, Victor C. III, M.D., Behrman, Richard E., M.D., et al, *Textbook of Pediatrics.* 11th ed.

Philadelphia, London, Toronto, Montreal, Sydney, Tokyo: W.B. Saunders Company, 1996 (38-41).

Oliver, George. *Practical Anthropology.* Springfield, IL: Charles C. Thomas, 1960.

Patterson, Carl N., M.D., Powell, Donald G., A.B. "Facial Analysis in Patient Evaluation for Physiologic and Cosmetic Surgery." *The Laryngoscope.* March 1974 (1004-1019).

Peck, Stephen R. *Atlas of Human Anatomy For The Artist.* New York: Oxford University Press, 1951.

Quetelet, Lambert Adolphe Jaques. *Anthropometrie, Ou Measure Des Differentes Facultes de l'Homme.* Brussels: 1870.

Sbotta, Johanes, McMurric, J. Playfair. *Atlas of Human Anatomy.* New York: G.E. Strechert & Co., 1930.

Schadow, Dr. Johan Gottfried. *The Sculptor and Art Student's Guide to the Proportions of the Human Form.* London: Chapman and Hall, Limited, 1883.

Schmidtke, Heinze. *Ergonomic Data For Equipment Design.* New York, London: Plenum Press, 1984.

Scott, James H., M.D. "Growth of the Human Face." *Proceedings of the Royal Society of Medicine* 47.91 (1953) 91-100 (68-183).

Simmons, Katherine. "Study of Child Growth and Development, Physiologic Growth and Development." *Monographs of the Society for Research in Child Development*, The Brush Foundation. National Research Foundation. IX No. 1, Series #37.

Shelton, E. Kost, M.D., F.A.C.P. and Skeels, M.D. "Growth and Development." *Ciba Clinical Symposia* September, 1951 (3-6).

Snyder, Richard G., Ph.D., Spencer, Martha L., M.D., Owings, Clyde, M.D. Ph.D., and Schneider, Ph.D. *Physical Characteristics of Children as Related to Death and Injury for Consumer Product Safety Design.* Washington D.C.: U.S. Government Printing Office, 1973 725-504/328.

Snyder, Richard G., Ph.D., et al. *Anthropometry of Infants, Children, and Youths to Age 18 For Product Safety Design.* Bethesda, Maryland: Consumer Product Safety Commission, 1977.

Strauss, Walter L. *The Human Figure by Albrecht Durer, The complete Dresden Sketchbook.* New York: Dover Publications, Inc. 1972.

Richter, Jean Paul. *The Notebooks of Leonardo da Vinci.* New York: The Dover Publications, Inc. 1970.

Watson, Ernest H., Lowry, George H. *Growth and Development of Children.* Chicago, Year Book Publishers, 1951.

Williams, P.L., Wendell-Smith, C.P. *Basic Human Embryology.* Philadelphia: J.B. Lippincott Company, 1969.

•

Tables for height, weight, and head circumference are listed in *Nelson's Textbook of Pediatrics,* see *Nelson.* The data is from: National Center for Health Statistics (NCHS), Health Resources Administration, Department of Health Education and Welfare (DHEW) 1976. United States Vital Health Statistics 1977 (Nov. 165): I-IV, 1979.

Anthrotech, Inc., 503 Xenia Avenue, Yellow Springs, OH 45387, is a company that offers comprehensive studies on human dimensions for design solutions. Although its services were not used in this book, it deserves a special mention as the leading commercial source of anthropometry data, and it recommends an informative text:

Woodson, Wesley E., Tillman, Barry, Tillman, Peggy. *Human Factors Design Handbook*, Second Edition. New York: McGraw Hill 1992.

About the Author

Eugene F. Fairbanks

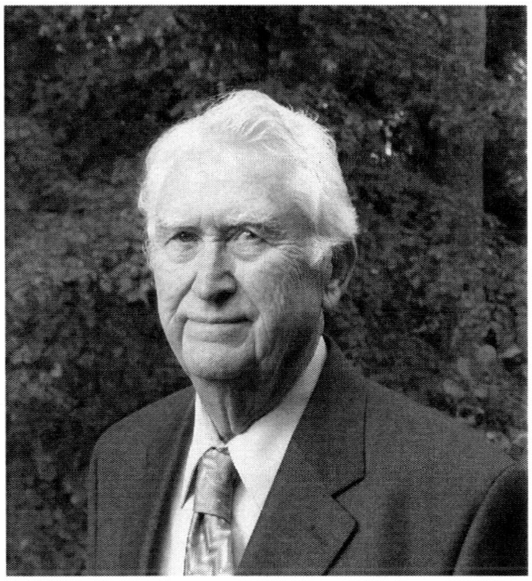

Eugene Fox Fairbanks, the second son of Maude Fox and Avard T. Fairbanks, was born 1921 in Eugene, Oregon. He was involved in the studio at an early age. When the family was in Florence Italy, he modeled his first sculpture under the direction of Dante Sodini, one of Italy's most famous sculptors at that time. When Sculptor Avard Fairbanks was appointed to the faculty of the University of Michigan in 1930, his sons often helped at the studio. They did some modeling, assisted the plaster caster, retouched casts, and were taught photography for sculpture. There was a progressive increased expectation in skill and responsibility. This continued from grade school through high school and into college. An older brother chose engineering as a vocation. Younger brothers studied engineering, art, and medicine.

Eugene chose to study Medicine. He married Florence Sundwall in 1944 in his last year of the University of Michigan Medical School, and graduated in 1945. He interned at Wayne County General Hospital near Detroit, Michigan. Subsequently, he served in the Army Medical Corps as a transport surgeon on a troop transport which sailed from the Seattle Port of Embarkation. Following military service he practiced medicine as a family physician in Pasco and Kennewick, Washington. After an Anesthesiology residency in Seattle, he moved to Bellingham, Washington, and practiced Anesthesiology and Family Practice. He was certified as a diplomat of the American Board of Anesthesiology and the American Board of Family Physicians. In time Florence and Eugene became the proud parents of five sons and five daughters. He practiced medicine in Bellingham, Washington, until retirement in 1995.

With Dr. Charles J. Flora he has co-authored an invertebrate marine biology species identification handbook, published in 1966 and 1977, titled *The Sound and The Sea.* His first book about Avard Fairbanks' sculpture, titled *A Sculptor's Testimony in Bronze and Stone,* featured sacred sculpture. He has collected photographs and made many enlargements of Fairbanks sculpture for projected publications. In 1995 he was asked to create a heroic bronze monument honoring fishermen of Bellingham who have been lost at sea. This was titled, *A Safe Return,* and was dedicated on Memorial Day 1999 in Zuanich Point Park in Bellingham, Washington. A second book, titled *A Sculpture Garden of Fantasy,* was published in 2001. A third book titled, *Abraham Lincoln Sculpture Created by Avard Fairbanks,* was published in 2003. A fourth book, *Human Proportions for Artists,* a studio guide for adult proportions, was published in 2005.

Please visit the website: fairbanksartbooks.com